STRUCTURE REPORTS

for 1981

Volume 48A

Structure Reports is prepared under the guidance of a Commission of the International Union of Crystallography. The members of the Commission sometime concerned with the preparation of this volume are listed below.

STRUCTURE REPORTS

for 1981

Volume 48A

METALS AND INORGANIC SECTIONS

General editor

G. Ferguson

Section editors

L. D. Calvert *and* J. Trotter

Springer Science+Business Media, B.V.

First published in 1984

ISSN 0166–6983
ISBN 978-94-017-3150-8   ISBN 978-94-017-3148-5 (eBook)
DOI 10.1007/978-94-017-3148-5

TABLE OF CONTENTS

[V]

INTRODUCTION

The present volume continues the aim of Structure Reports to present critical accounts of all crystallographic structure determinations. Details of the arrangement in the volumes, symbols used etc. are given in previous volumes (e.g. 41B or 42A, pages vi-viii).

University of Guelph,   G. FERGUSON
Guelph, Ontario, Canada

4 February, 1984

STRUCTURE REPORTS

SECTION I

METALS

Edited by

L. D. Calvert

(National Research Council of Canada)

with the assistance of

J. K. Byron

J. R. Rodgers

ARRANGEMENT

As in previous volumes the arrangement in the Metals section is approximately, but not strictly, alphabetical, and to find particular substances the subject index or formula index should be used.

ALKALI METAL - TRANSITION METAL - PNICTIDES
ABX_2 (A = Ba, Sr; B = Zn, Cd; X = Sb, Bi), $CaMnSb_2$

E. BRECHTEL, G. CORDIER and H. SCHÄFER, 1981. J. Less-Common Metals, 79, 131-138.

ABX_2, tetragonal, $SrZnBi_2$ type (1), I4/mmm, Z = 4. Mo radiation, diffractometer data. A in 4(e) 0,0,z; B in 4(d) 0,1/2,1/4; X(1) in 4(c) 0,1/2,0; X(2) in 4(e) 0,0,z. See Fig. 1.

ABX_2	a(Å)	c(Å)	c/a	D_x	R	refl.	z(A)	z(X(2))
$BaZnSb_2$	4.584	23.05	5.03	6.12	0.070	268	0.1202	0.3175
$BaCdSb_2$	4.558	24.16	5.30	6.56	0.115	218	0.1129	0.3255
$BaZnBi_2$	4.846	21.98	4.54	7.98	0.122	179	0.1283	0.3141
$SrCdBi_2$	4.635	22.88	4.94	8.35	0.069	260	0.1138	0.3364
$BaCdBi_2$	4.768	23.60	4.95	8.27	0.060	200	0.1158	0.3296

$CaMnSb_2$, orthorhombic, $SrZnSb_2$ type (2), Pnma, a = 22.11, b = 4.312, c = 4.344 Å, Z = 4, D_x = 5.43. Mo radiation, R = 0.069 for 624 reflexions, diffractometer data. All atoms in 4(c) x,1/4,z; Ca 0.1126, 0.7326; Mn 0.2502, 0.2324; Sb(1) 0.0008, 0.2254; Sb(1) 0.3280, 0.7343. See Fig. 1.

Interatomic distances (Å)

ABX_2	A-A	A-B	A-X	A-X	B-B	B-X	X-X
$BaZnSb_2$	-	3.77	3.60	3.54	3.24	2.77	3.24
$BaCdSb_2$	-	4.02	3.55	3.55	3.22	2.92	3.22
$BaZnBi_2$	-	3.61	3.72	3.65	3.43	2.80	3.43
$SrCdBi_2$	-	3.88	3.49	3.47	3.28	3.05	3.28
$BaCdBi_2$	-	3.96	3.63	3.61	3.37	3.04	3.37
$CaMnSb_2$	4.31	3.72	3.27	3.33	3.06	2.76	2.91
$CaMnSb_2$	4.34	3.74	3.31	3.34	-	2.78	3.22

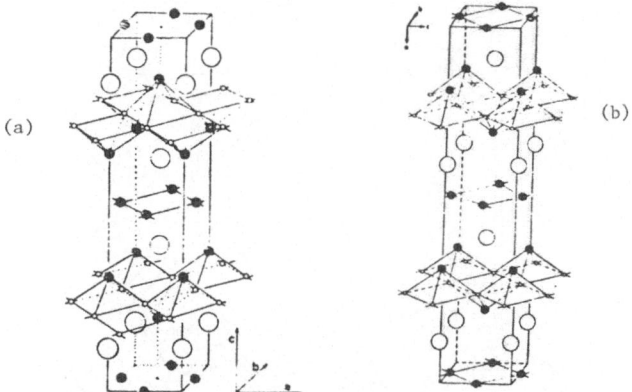

(a)   (b)

Fig. 1.   The $SrZnBi_2$ structure: a; the $SrZnSb_2$ structure: b; large open circles: alkali metal atoms (A); small open circles:transition metal atoms (B); solid circles: pnictides (X).

1. Structure Reports, 42A, 17; 44A, 10; 46A, 32.
2. Ibid., 45A, 20.

ALUMINUM ANTIMONY CALCIUM   ALUMINUM ARSENIC CALCIUM
$Al_2Ca_5Sb_6$ $Al_2As_4Ca_3$

G. CORDIER, E. CZECH, M. JAKOWSKI and H. SCHÄFER, 1981. Rev. Chim. Miner., 18, 9-18.

$Ca_5Al_2Sb_6$, orthorhombic, $Ca_5Ga_2As_6$ type (1), Pbam, a = 14.07, b = 12.09, c = 4.46 Å, D_m = 4.26, Z = 2. Mo radiation, R = 0.06$\overline{3}$ for 1214 reflexions, diffractometer data.

Atomic positions
Ca(1),(2) and Sb(2) in 4(h) x,y,1/2; Ca(3) in 2(d) 0,1/2,1/2; Al, Sb(1) and Sb(3) in 4(g) x,y,0.

	x	y			x	y
Ca(1)	0.4879	0.6754		Sb(1)	0.4057	0.8433
Ca(2)	0.2479	0.4074		Sb(2)	0.1796	0.6602
Al	0.2884	0.6693		Sb(3)	0.4006	0.4764

$Ca_3Al_2As_4$, monoclinic, C2/c, a = 12.86, b = 10.05, c = 6.57 Å, β = 90.7°, Z = 4, D_x = 3.71. Mo radiation, R = 0.094 for 1205 reflexions, diffractometer data.

Atomic positions
Ca(1) in 4(e) 0,y,1/4; remainder in 8(f) x,y,z.

	x	y	z			x	y	z
Ca(1)	0	0.0989	1/4		As(1)	0.8456	0.4122	0.0620
Ca(2)	0.1210	0.1282	0.4715		As(2)	0.3883	0.3294	0.0233
Al	0.2037	0.3758	0.0890					

Both structures are based on AlX_4 tetrahedra. In $Ca_5Al_2Sb_6$ (Fig. 1) $AlSb_4$ tetrahedra (Al-Sb = 2.67-2.82 Å) share corners to form chains parallel to c; these chains are close enough to form Sb-Sb bonds (2.86 Å); the Ca atoms are six-coordinated by Sb (3.16-3.35 Å) as nearest neighbours. In $Ca_3Al_2As_4$, Fig. 2, the $AlAs_4$ tetrahedra share corners to form a three dimensional array (Al-As = 2.41-2.49 Å with Ca atoms six-coordinated by As (Ca-As = 2.87-3.35 Å).

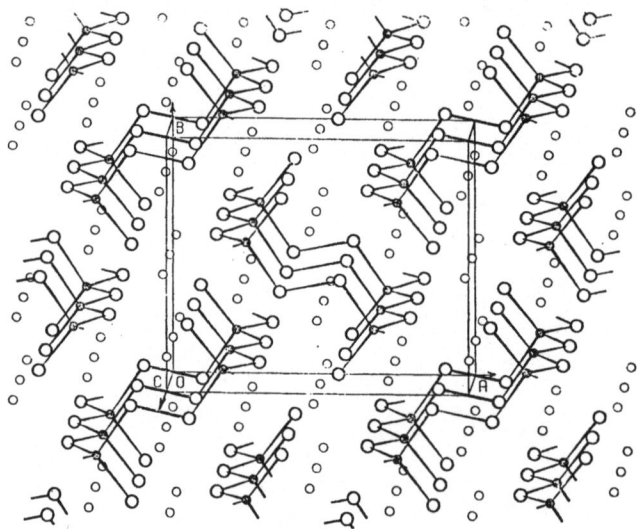

Fig. 1.   Perspective view of the $Ca_5Al_2Sb_6$ structure. Al atoms: shaded circles;
Sb atoms: large open circles; Ca atoms: small open circles.

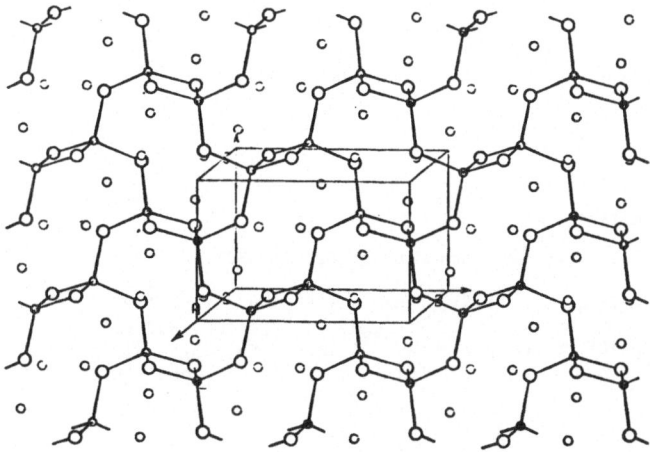

Fig. 2.   Perspective view of the $Ca_3Al_2As_4$ structure. Shaded circles: Al atoms;
           large open circles: As atoms; small open circles: Ca atoms.

<u>1</u>. Structure Reports, <u>42</u>A, 27.

ALUMINUM BARIUM SELENIUM   BARIUM GALLIUM SELENIUM
Al_2BaSe_4 $BaGa_2Se_4$

CALCIUM GALLIUM SELENIUM   CALCIUM INDIUM TELLURIUM
$CaGa_2Se_4$ $CaIn_2Te_4$

W. KLEE and H. SCHÄFER, 1981. Z. anorg. Chem., <u>479</u>, 125-133.

$BaAl_2Se_4$, tetragonal, P4/nnc, a = 11.35, c = 6.19 Å, c/a = 0.55, D_m = 4.18, Z = 4.
Mo radiation, R = 0.089 for 292 reflexions, diffractometer data. Ba(1) in 2(a)
1/4,1/4,1/4; 2Ba(2) in 4(c) 1/4,3/4,3/4; Al in 8(h) 0.9953,0.9953,1/4; Se in 16(k)
0.1605,0.9966,0.0079.

$BaGa_2Se_4$, orthorhombic, $SrAl_2Se_4$ type (<u>1</u>), Cccm, a = 6.48, b = 11.13, c = 11.24 Å, D_m
= 4.78, Z = 4. Cu radiation, R = 0.083 for 104 reflexions, photographic data. Ba in
4(b) 0,1/2,1/4; Ga in 8(ℓ) 0.0092,0.2429,0; Se(1) in 8(ℓ) 0.2303,0.4144,0; Se(2) in
8(k) 1/4,1/4,0.3375.

$CaGa_2Se_4$, orthorhombic, $BaIn_2Se_4$ [$PbGa_2Se_4$] type (<u>2</u>), Fddd, a = 21.06, b = 21.00, c =
12.66 Å, D_m = 4.68, Z = 32. Cu radiation, R = 0.098 for 263 reflexions, photographic
data. Ca(1) in 8(a) 1/8,1/8,1/8; Ca(2) in 8(b) 1/8,1/8,5/8; Ca(3) in 16(e) 0.8724,1/8,
1/8; Ga(1) in 32(h) 0.9998,0.8048,0.1665; Ga(2) in 32(h) 0.004,0.2382,0.3731; Se(1) in
32(h) 0.1630,0.2507,0.0181; Se(2) in 32(h) 0.1628,0.4993,0.2344; Se(3) in 32(h)
0.9993,0.3422,0.9959; Se(4) in 32(h) 0.0017,0.4162,0.2657.

$CaIn_2Te_4$, tetragonal, $SrIn_2Te_4$ type (<u>3</u>), I4/mcm, a = 8.42, c = 7.14 Å, c/a = 0.85, Z =
2, D_x = 5.12. Cu radiation, R = 0.088 for 72 reflexions, photographic data. 2Ca in
4(a) 0,0,1/4; In in 4(b) 0,1/2,1/4; Te in 8(h) 0.1813,0.6813,0.

     These AB_2X_4 structures (A = Ba,Ca; B = Al,Ga,In; X = Se,Te) are based on BX_4
tetrahedra and are derived from the TlSe (B37) structure (<u>4</u>). The larger A atoms
centre square antiprisms as described in <u>1-3</u>. Distances are summarised below.

	$BaAl_2Se_4$	$BaGa_2Se_4$	$CaGa_2Se_4$	$CaIn_2Te_4$	
B-4X	2.38,2.40	2.39-2.48	2.33-2.51	2.80	Å
A-8X	3.33-3.40	3.32,3.37	3.07-3.29	3.57	Å

1. Structure Reports, 44A, 5.
2. Ibid., 46A, 6, 25.
3. Ibid., 39A, 100.
4. Strukturbericht, 7, 6; Structure Reports, 42A, 98.

ALUMINUM BORON LITHIUM
$AlB_{14}Li$ $(Al_{0.956}B_{14}Li_{1.0})$

I. HIGASHI, 1981. J. Less-Common Metals, 82, 317-323.

Orthorhombic, $MgAlB_{14}$ type (1-3), Imam, a = 5.8469, b = 8.1429, c = 10.3542 Å, D_m = 2.46, Z = 4, for a composition of $Li_{0.96}Al_{1.06}B_{14}$. Mo radiation, R = 0.024 for 1871 reflexions, diffractometer data. Li in 4(e) 0,y,3/4; Al in 4(c) 1/4,1/4,1/4; B(1),(2) in 16(j) x,y,z; B(3),(4),(5) in 8(h) 0,y,z; [note setting].

Atomic positions

	x	y	z		x	y	z
Li*	0	0.35185	3/4	B(3)	0	0.16994	0.0856
Al	1/4	1/4	1/4	B(4)	0	-0.03188	0.16677
B(1)	0.24996	0.04368	0.07986	B(5)	0	0.37953	0.14953
B(2)	0.15940	0.83563	0.06236				

*occupancy = 95.6%

The structure is as described (1-3) with B-B = 1.72-2.08; Li-B = 2.42-2.79 and Al-B = 2.08-2.43 Å.

1. Structure Reports, 35A, 4, 39.
2. R. NASLAIN, A. GUETTE and P. HAGENMULLER, 1976. J. Less-Common Metals, 47, 1.
3. This volume, p. 34.

ALUMINUM BORON YTTERBIUM
AlB_6Yb_2

ALUMINUM BORON LUTETIUM
AlB_4Lu

S.I. MIKHALENKO, Ju.B. KUZ'MA, M.M. KORSUKOVA and V.N. GURIN, 1980. Izv. Akad. Nauk SSSR, Neorg. Mater., 16, 1941-1944 [Inorg. Mater., 16, 1325-1328].

Orthorhombic, Cu radiation, powder data, z = 4.

Phase	Space Group	a(Å)	b(Å)	c(Å)	R	type	ref.
Yb_2AlB_6	Pbam	9.127	11.46	3.584	0.085	Y_2ReB_6	1
$LuAlB_4$	Pbam	5.906	11.44	3.480	0.112	$YCrB_4$	2

Atomic positions

Yb_2AlB_6	site	x	y
Yb(1)	in 4(g)	0.822	0.086
Yb(2)	in 4(g)	0.444	0.133
Al	in 4(g)	0.148	0.192
B(1)	in 4(h)	0.050	0.060
B(2)	in 4(h)	0.250	0.075
B(3)	in 4(h)	0.300	0.240
B(4)	in 4(h)	0.140	0.310
B(5)	in 4(h)	0.480	0.290
B(6)	in 4(h)	0.110	0.470

$LuAlB_4$	site	x	y
Lu	in 4(g)	0.132	0.149
Al	in 4(g)	0.129	0.413
B(1)	in 4(h)	0.305	0.300
B(2)	in 4(h)	0.370	0.465
B(3)	in 4(h)	0.385	0.050
B(4)	in 4(h)	0.470	0.195

[Interatomic distances; mean (Å)]   Ln = Lu, Yb

A-B	Al-Al	Al-B	Al-Ln	B-B	B-Ln	Ln-Ln
AlB_4Lu	2.51	2.34	3.04	1.79	2.66	3.64
AlB_6Yb	-	2.38	2.99	1.82	2.66	3.51

1. Structure Reports, 38A, 53.
2. Ibid., 35A, 36.

ALUMINUM CHROMIUM SULPHUR
$AlCr_{5/3}S_4$

CHROMIUM GALLIUM SULPHUR
$Cr_2Ga_{2/3}S_4$

I. NAKATANI, 1980. J. Solid State Chem., 35, 50-58.

Cubic, Al_2MgO_4 (spinel) type (1), Fd3m, Z = 8. Cu radiation, powder diffractometer data. Al and Ga in 8(a) 0,0,0; Cr and Al in 16(d) 5/8,5/8,5/8; S in 32(e) x,x,x. Vacancies occur only in 8(a).

	a(Å)	R	x(S)	8(a)	Occupancies 16(d)
$AlCr_{5/3}S_4$	9.893	0.055	0.3825	16/3 Al	40/3 Cr + 8/3 Al
$Cr_2Ga_{2/3}S_4$	9.895	0.095	0.3825	16/3 Ga	16 Cr

1. Structure Reports, 16, 244; 33A, 272.

ALUMINUM COBALT NICKEL
Al_4CoNi_2

Lu XUE-SHAN and Li FANG-HUA, 1980. Acta Phys. Sinica, 29, 182-198.

Cubic, [Ga_4Ni_3 type (1),] Ia3d, a = 11.3962 Å, Z = 16, D_X = 5.06. Co radiation, photographic data. Al(1) in 16(a) 0,0,0; Al(2) in 48(f) 0.010,0,1/4; 2/3 Ni + 1/3 Co in 48(g) 1/8,0.369,0.119. [Distances are Al-Al = 2.62-3.08, Al-Co/Ni = 2.46 - 2.54, Co/Ni-Co/Ni = 2.78 Å].

1. Structure Reports, 34A, 137.

ALUMINUM COPPER RHENIUM
$AlCu_2Re_2$

A.P. PREVARSKIJ, Ju.B. KUZ'MA and M.S. ZRADA, 1979. Russ. Metallurgy (Metally), No. 4, 177-180.

Tetragonal, Mo_2FeB_2 type (1), P4/mbm, a = 5.716, c = 3.176 Å, c/a = 0.56, Z = 2, D_X = 16.85. Cr radiation, R = 0.087, photographic data. Al in 2(a) 0,0,0; Re in 4(h) 0.181,0.681,1/2; Cu in 4(g) 0.389,0.889,0; [the distances are Al-4Cu = 2.31, Al-8Re = 2.63; Re-6Cu = 2.31, 2.34, Re-4Al = 2.63, Re-7Re = 2.93-3.18; Cu-2Al = 2.31, Cu-6Re = 2.31, 2.34, Cu-1Cu = 1.79 Å; this Cu-Cu distance is short, suggesting that the x parameter should be modified].

1. E.I. GLADYŠEVSKIJ, 1966. Poroš. Metall., No. 4, 55.

ALUMINUM GERMANIUM TUNGSTEN
$Al_{1.3}Ge_{0.7}W$

V.V. MILJAN and Ju.B. KUZ'MA, 1980. Izv. Akad. Nauk SSSR, Met., No. 4, 231-232.

Hexagonal, $CrSi_2$ type (1), $P6_222$, a = 4.820, c = 6.711 Å*, c/a = 1.39, Z = 3, [D_x = 9.95]. Fe radiation, powder diffractometer data; 3.9 Al + 2.1 Ge in 6(j) 1/6,1/3,1/2; W in 3(d) 1/2,0,1/2; [W-9Al/Ge = 2.63-3.29, W-5W = 2.63, 2.78; Al/Ge-4W, Al/Ge-10 Al/Ge = 2.63, 2.78]. *[Also misprinted as 5.711 Å.]

1. Strukturbericht, 3, 35, 628; Structure Reports, 45A, 8.

ALUMINUM HAFNIUM ZINC
$Al_{1.5}HfZn_{0.5}$

ALUMINUM ZINC ZIRCONIUM
$Al_{1.5}Zn_{0.5}Zr$

A. DRASNER and Ž. BLAŽINA, 1981. Z. Naturf., 36B, 1547-1550.

Phase	Space Group	Type	a(Å)	Site	Occupancy
$Al_{1.5}Zn_{0.5}Zr$	Fd3m	Cu_2Mg (1)	7.473*	16(d)	4 Zn + 12 Al
$Al_{1.5}Zn_{0.5}Hf$	Fd3m	Cu_2Mg	7.414	16(d)	4 Zn + 12 Al
$Al_{0.5}Zn_{1.5}Zr$	Pm3m	$AuCu_3$ (2)	4.073	3(c)	2 Zn + 1/3 Zr + 2/3 Al

*Also given as 7.472.

1. Strukturbericht, 1, 490, 531.
2. Ibid., 1, 486, 505̄, 507, 508.

ALUMINUM MAGNESIUM YTTRIUM
Al_4MgY

O.S. ZAREČNJUK, M.E. DRIC, R.M. RYKHAL' and V.V. KINŽIBALO, 1980. Izv. Akad. Nauk SSSR, Met., No. 5, 242-244.

Hexagonal, $MgZn_2$ type (1), $P6_3/mmc$, a = 5.33, c = 8.57 Å, c/a = 1.61, Z = 2, [D_x = 3.48], composition given as $Mg_{16.1}Al_{67.2}Y_{16.7}$. Cr radiation with photographic data. 2 Y + 2 Mg in 4(f) 1/3,2/3,0.062*; Al(1) in 2(a) 0,0,0; Al(2) in 6(h) x,2x,1/4,x = 0.830; Al - 6 Mg/Y = 3.07, 3.12, Al - 6 Al = 2.61-2.72; Mg/Y - 4 Mg/Y = 3.26, Mg/Y - 12 Al = 3.07, 3.12 Å. *[z for Mg/Y misprinted as 0.620.]

1. Strukturbericht, 1, 183, 228, 564; 3, 311.

ALUMINUM MANGANESE
Al_6Mn

A. KONTIO and P. COPPENS, 1981. Acta Cryst., B37, 433-435.

Orthorhombic, Cmcm, a = 7.5551, b = 6.4994, c = 8.8724 Å, Z = 4, D_x = 3.31. Mo radiation, R = 0.0209 for 530 reflexions, three-dimensional counter data. See also 1, 2. Confirms the results of 3; corrects misprint in 3: Al(2)-Al(3) = 3.00 Å (not 2.77̄ Å).

Atomic positions

			x	y	z
Mn	in	4(c)	0	0.45686	1/4
Al(1)	in	8(e)	0.32602	0	0
Al(2)	in	8(f)	0	0.13917	0.10039
Al(3)	in	8(g)	0.31768	0.28622	1/4

1. Strukturbericht, 2, 684.
2. Structure Reports, 18, 23; 24, 25; 44A, 105.
3. Ibid., 17, 25.

ALUMINUM NICKEL PRASEODYMIUM
Al_5Ni_2Pr

Ja.P. JARMOLJUK, R.M. RYKHAL', L.G. AKSEL'RUD and O.S. ZAREČNJUK, 1981. Dop. Akad. Nauk Ukr. RSR, 43, No. 9, 86-90.

Tetragonal, Immm, a = 7.024, b = 9.562, c = 3.979 Å, D_m = 4.75, Z = 2. Fe radiation, R = 0.119 for 57 reflexions, diffractometer data. Pr in 2(a) 0,0,0; Ni in 4(h) 0,0.258,1/2; Al(1) in 8(n) 0.190,0.354,0; Al(2) in 2(b) 0,1/2,1/2.

The structure (Fig. 1) has Pr with CN = 20 and distances (Pr-Ni = 3.17, Pr-Al = 3.26-3.51, Pr-Pr = 3.98), Al with CN = 12 (Al-Ni = 2.31-2.57, Al-Al = 2.67-2.94, Al-Pr = 3.26-3.64), and Ni with CN = 9 and distances as above.

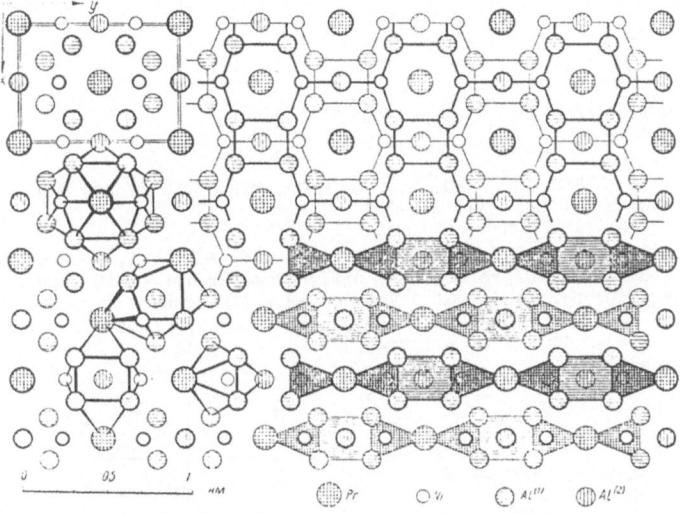

Fig. 1.   The Al_5Ni_2Pr structure projected onto 001.

ANTIMONY BARIUM
Ba_5Sb_4

E. BRECHTEL, G. CORDIER and H. SCHÄFER, 1981. Z. Naturf., 36B, 1341-1342.

Orthorhombic, Gd_5Si_4 type (1), Pnma, a = 9.012, b = 17.823, c = 9.041 Å, Z = 4, D_x = 5.37. Mo radiation, R = 0.089 for 1899 reflexions, diffractometer data.

Atomic positions
Ba(1), (2), Sb(1) in 8(d), Ba(3), Sb(2) and (3) in 4(c)

	x	y	z		x	y	z
Ba(1)	0.5895	0.1056	0.1625	Sb(1)	0.2555	0.4484	0.0072
Ba(2)	0.4294	0.3979	0.6694	Sb(2)	0.1452	1/4	0.6165
Ba(3)	0.2571	1/4	0.0020	Sb(3)	0.3715	1/4	0.3908

The structure contains Sb_2 pairs (Sb-Sb = 2.89 Å) which allow the formulation $Ba_5^{2+}Sb_2^{3-}Sb_2^{4-}$. It is as described in 1 with Ba-Ba = 4.00-4.29, Ba-Sb = 3.46-3.84 Å.

1. Structure Reports, 32A, 87; 38A, 91; 44A, 24.

ANTIMONY BARIUM SULPHUR
$Ba_8S_{17}Sb_6$

W. DÖRRSCHEIDT and H. SCHÄFER, 1981. Z. Naturf., 36B, 410-414.

Monoclinic, P2/c, a = 11.41, b = 13.73, c = 22.53 Å, β = 90.94°, D_m = 4.45, Z = 4. Mo radiation, R = 0.055 for 5290 reflexions, diffractometer data; 33 site-sets are given.

The structure contains isolated SbS_3 trigonal pyramids and Sb_3S_8 groups composed of a central SbS_4 trigonal bipyramid connected by the common axial S atoms to two further SbS_3 groups; Sb-S = 2.39-2.71 Å. The Ba atoms are six-to twelve-coordinated with Ba-S = 3.11-3.72 Å.

ANTIMONY CADMIUM SULPHUR
$Cd_2S_{11}Sb_6$

I.P. DEINEKO, Ju.K. EGOROV-TISMENKO, V.D. SPICYNA, M.A. SIMONOV and N.V. BELOV, 1980. Soviet Physics-Doklady, 25, 788-790.

Orthorhombic, $Pmn2_1$, a = 3.9177, b = 9.599, c = 12.497 Å, Z = 1, D_x = 4.62. Mo radiation, R = 0.064 for 1401 reflexions, diffractometer data. 10 site-sets given with 75% occupancy for S(2) and S(6). Both Cd and Sb(1) occupy distorted S_6 octahedra (Cd-S = 2.60-2.83, Sb(1)-S = 2.60-2.85 Å). The other two Sb atoms are surrounded by 7 S atoms (Sb-3S = 2.43-2.70, Sb-2S = 2.94-2.99, Sb-2S = 3.48-3.53 Å) as is common in sulpho-salts. See Fig. 1.

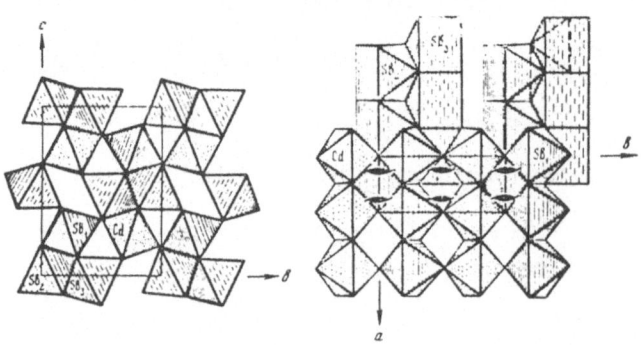

Fig. 1.   Projections of the $Cd_2Sb_6S_{11}$ structure. Left: onto 100; right: onto 001.

ANTIMONY CAESIUM SELENIUM
CsSbSe$_2$

A.S. KANIŠČEVA, Ju.N. MIKHAILOV, V.B. LAZAREV and N.A. MOŠČALKOVA, 1980. Soviet
Physics-Doklady, 25, 413-414.

Monoclinic, P2$_1$/a, a = 7.422, b = 10.161, c = 15.47 Å, β = 102.07°, D$_m$ = 4.92, Z = 8.
Mo radiation, R = 0.084 for 833 reflexions, diffractometer data.

Atomic positions.
All atoms in 4(e)

	x	y	z		x	y	z
Sb(1)	0.1373	0.1689	0.0443	Se(1)	0.4416	0.2574	0.1523
Sb(2)	0.1468	0.1643	0.5502	Se(2)	0.4697	0.2482	0.6685
Cs(1)	0.2057	0.5610	0.1638	Se(3)	0.3084	0.4229	0.4061
Cs(2)	0.2192	0.5705	0.6732	Se(4)	0.2892	0.4373	0.9146

The structure is similar to that of CsSbS$_2$ (1) with SbSe$_3$ trigonal pyramids which
share two corners to form chains; the Sb-Se distances are 2.47-3.12 Å; the
Cs atoms are seven- and eight-coordinated (Cs-Se = 3.57-4.01 Å).

1. Structure Reports, 46A, 10.

ANTIMONY CALCIUM SULPHUR
Ca$_2$S$_5$Sb$_2$

G. CORDIER and H. SCHÄFER, 1981. Rev. Chim. Minér., 18, 218-223.

Monoclinic, P2$_1$/c, a = 15.074, b = 5.694, c = 11.378 Å, β = 110.99°, Z = 4, D$_x$ = 3.52.
Mo radiation, R = 0.076 for 3247 reflexions, diffractometer data.

Atomic positions
All atoms in 4(e)

	x	y	z		x	y	z
Ca(1)	0.6677	0.2694	0.8327	S(2)	0.5001	0.4885	0.3474
Ca(2)	0.1060	0.7200	0.3078	S(3)	0.6951	0.2188	0.0957
Sb(1)	0.3856	0.1580	0.9478	S(4)	0.8794	0.1878	0.4363
Sb(2)	0.1248	0.2434	0.5732	S(5)	0.2980	0.2303	0.1588
S(1)	0.9024	0.7787	0.7264				

The Sb atoms form two types of groups, SbS$_3$ pyramids and Sb$_2$S$_4$ clusters made of
two SbS$_3$ pyramids sharing one edge (Fig. 1).

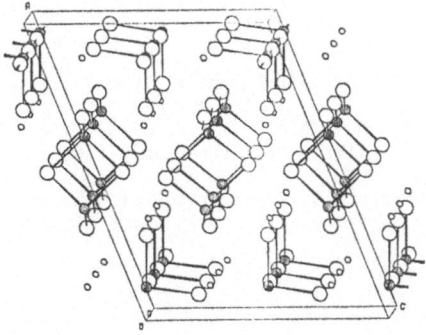

Fig. 1.   A perspective view of the Ca$_2$Sb$_2$S$_5$ structure.

ANTIMONY COPPER MERCURY SULPHUR (SCHWATZITE[mercurian tetrahedrite or mercurian
  tennantite])
$Cu_{10.4}Hg_{1.6}S_{12}Sb_4$

L.N. KAPLUNNIK, E.A. POBEDIMSKAJA and N.V. BELOV, 1980. Dokl. Akad. Nauk SSSR, 253,
105-107 [Soviet Physics-Doklady, 25, 506-507].

Cubic,$I\bar{4}3m$, a = 10.453 Å, Z = 2, D_x = 5.39. Mo radiation, R = 0.081 for 171
reflexions, diffractometer data. Cu in 12(e) 0.212,0,0; 8.8 Cu + 3.20 Hg in 12(d)
1/4,1/2,0; Sb in 8(c) 0.2666,0.2666,0.2666; S in 24(g) 0.1133,0.1133,0.361.

The structure is as described by 1 with Hg replacing the Cu in the 12(d) site
only. Distances are Cu-S = 2.29, Cu/Hg-S = 2.35, Cu-Cu = 3.13 and Sb-S = 2.48 Å.
Contrary to 2 there are no S atoms in 2(a). Rather there are vacancies in the metal
sites, not an excess of S atoms,i.e. the ratio M/S = 11/12 and not 12/13. [The data
in 2 appear to be more accurate; in addition the microprobe analysis quoted is
$Hg_{1.02}Cu_{9.79}Zn_{.66}Fe_{.45}Sb_{4.22}S_{12.96}$].

1. Strukturbericht, 1, 335; Structure Reports, 37A, 4.
2. Strukturbericht, 3, 413, 414; Structure Reports, 29, 15; 45A, 14.

ANTIMONY COPPER POTASSIUM SULPHUR THALLIUM (CHALCOTHALLITE)
$Cu_{6.35}(Tl,K)_2S_4Sb$

E. MAKOVICKY, Z. JOHAN and S. KARUP-MØLLER, 1980. Neues Jb. Miner. Abh., 138, 122-
146.

Tetragonal, I4/mmm, a = 3.827, c = 34.280 Å, c/a = 8.96, Z = 2, D_x = 6.67. Cu
radiation, R = 0.207 for 86 reflexions, photographic data. Sb in 2(a) 0,0,0; 3.57 Tl
+ 0.43 S in 4(e) 0,0,0.1483; Cu(1) in 8(g) 0,1/2,0.0570; Cu(2) in 4(d) 0,1/2,1/4;
0.70 Cu(3) in 2(b) 1/2,1/2,0; S(1) in 4(e) 1/2,1/2,0.093; S(2) in 4(e) 1/2,1/2,0.214.
See also 1.

The structure has two layers. In one, Cu(1) is tetrahedrally coordinated (Cu-2S =
2.28, Cu-2Sb = 2.74) and also Cu-4Cu = 2.71 and Cu-2Cu = 2.74 Å. Cu(3) is square
coordinated by Sb (Cu-4Sb = 2.71) and 8-coordinated by Cu at 2.74 Å; Sb is 8-coordin-
ated by Cu(1) plus 4 additional bonds to Cu(3). In the other layer,less well defined in
the structure determination, there is a mackinawite-like Cu-S layer. The Tl atoms
are 8-coordinated (Tl-S = 3.30, 3.50 Å) and lie between layers.

1. M. FLEISCHER, 1968. Amer. Min., 53, 1775; Idem, 1979. Ibid., 64, 658.

ANTIMONY EUROPIUM SULPHUR
$Eu_3S_9Sb_4$

P. LEMOINE, D. CARRÉ and M. GUITTARD, 1981. Acta Cryst., B37, 1281-1284.

Orthorhombic, Pnam, a = 16.495, b = 23.843, c = 4.031 Å, D_m = 5.16, Z = 4. Mo
radiation, R = 0.065 for 1526 reflexions, diffractometer data; 16 site-sets are given.

The Eu^{2+} atoms centre bicapped trigonal prisms of S atoms (Eu-S = 3.04-3.26);
Sb atoms centre the bases of square pyramids of S atoms (Sb-S = 2.41-3.10 Å).

ANTIMONY HAFNIUM   GERMANIUM TANTALUM
Hf_3Sb h-$GeTa_3$

J.-O. WILLERSTRÖM and S. RUNDQVIST, 1981. J. Solid State Chem., 39, 128-132.

Tetragonal, Fe3P type (1), I$\bar{4}$, Z = 8. Cu radiation, powder profile refinement by Rietveld method, Guinier photographs.

A_3X	a(Å)	c(Å)	c/a	D_X	R	refl.
Hf_3Sb	11.1899	5.6364	0.50	12.37	0.062	147
Ta_3Ge	10.3421	5.1532	0.50	14.83	0.055	121

A = Hf, Ta, X = Sb, Ge

Atomic positions
All atoms in 8(g)

Hf_3Sb	x	y	z	Ta_3Ge	x	y	z
A(1)	0.0830	0.1082	0.2081		0.0834	0.1002	0.2335
A(2)	0.3544	0.0238	0.9818		0.3562	0.0180	0.9891
A(3)	0.1880	0.2143	0.7641		0.1928	0.2128	0.7561
X	0.2902	0.0333	0.4805		0.2942	0.0261	0.4703

The structure is as previously described (1) with distances Hf-Sb = 2.82-3.09, Hf-Hf = 3.01-3.94 and Ta-Ge = 2.56-2.92, Ta-Ta = 2.70-3.66 Å

1. Structure Reports, 19, 237; 27, 97, 290.

ANTIMONY LEAD SULPHUR (HETEROMORPHITE)
$Pb_7S_{19}Sb_8$

A. EDENHARTER, 1980. Z. Kristallogr., 151, 193-202.

Monoclinic, C2/c, a = 13.628, b = 11.943, c = 21.285 Å, β = 90.55°, D_m = 5.73, Z = 4. Cu radiation, R = 0.105 for 2531 reflexions, diffractometer data. 18 site-sets given. See also 1 and 2.

The structure contains SbS_3 groups which share corners to form discrete chains $(Sb_6S_{13}$, Sb-S = 2.37-2.63). In addition one Sb is 5-coordinated (Sb-S = 2.60-3.04); the Pb atoms are 6-, 7- or 8-coordinated (Pb-S = 2.71-3.60 Å).

1. Strukturbericht, 6, 85.
2. Structure Reports, 43A, 99.

ANTIMONY LITHIUM SULPHUR
$Li_{3x}Sb_{6-x}S_9$ (x = 0.33)

J. OLIVIER-FOURCADE, L. IZGHOUTI and E. PHILIPPOT, 1981. Rev. Chim. Minér., 18, 207-217.

Triclinic, P$\bar{1}$, a = 4.085, b = 6.678, c = 14.700 Å, α = 96.84, β = 90.28, γ = 107.70°, Z = 1, D_X = 4.32. Mo radiation, R = 0.041 for 530 reflexions, diffractometer data.

Atomic positions
S(1) in 1(a) 0,0,0; Li in 1(b) 0,0,1/2; rest in 2(i) x,y,z

	x	y	z		x	y	z
Sb(1)	0.3534	0.7127	0.9250	S(3)	0.261	0.549	0.1317
Sb(2)	0.2672	0.5169	0.3834	S(4)	0.362	0.721	0.5355
Sb(3)*	0.9344	0.8736	0.1908	S(5)	0.437	0.863	0.7754
S(2)	0.109	0.214	0.6494				

*occupancy = 0.8333

Two of the Sb atoms centre distorted S_6 octahedra (Sb-S = 2.48-3.01 Å) while the third occupies the base of an S_5 square pyramid (Sb-S = 2.44-2.92 Å) with three other neighbours (Sb-S = 3.72-3.77 Å). These are linked by shared edges to form a framework with Li occupying octahedral holes (Li-S = 2.43-2.80 Å).

ANTIMONY POTASSIUM   LITHIUM PHOSPHORUS
KSb   LiP

W. HÖNLE and H.G. von SCHNERING, 1981.   Z. Kristallogr., 155, 307-314.

Monoclinic, LiAs type (1), $P2_1/c$, Z = 8. Mo radiation, single-crystal diffractometer data. Structures are as described in 1, 2.

	a(b)	b(Å)	c(Å)	β°	D_X	R	refl.	see also
KSb	7.156	6.917	13.355	115.17	3.571	0.028	1115	2
LiP	5.582	4.940	10.255	118.15	2.022	0.021	564	

Atomic positions (All atoms in 4(e))

LiP	x	y	z	KSb	x	y	z
Li(1)	0.2151	0.3876	0.3299	K(1)	0.2189	0.3990	0.3318
Li(2)	0.2257	0.6597	0.0293	K(2)	0.2395	0.6668	0.0317
P(1)	0.3165	0.8952	0.29198	Sb(1)	0.32222	0.89906	0.28645
P(2)	0.3050	0.1565	0.11252	Sb(2)	0.31909	0.16620	0.12312

1. Structure Reports, 23, 38.
2. Ibid., 26, 44.

ANTIMONY PRASEODYMIUM SULPHUR
$Pr_8S_{15}Sb_2$

G.G. GUSEINOV, F.K. MAMEDOV, I.R. AMIRASLANOV and K.S. MAMEDOV, 1981. Kristallografija, 26, 831-833 [Soviet Physics-Crystallography, 26, 470-471].

Tetragonal, $I4_1cd$, a = 15.626, c = 19.659 Å, c/a = 1.26, Z = 8, D_X = 5.12. Mo radiation, R = 0.037 for 1860 reflexions, diffractometer data. 13 site-sets given.

The structure contains chains of linked polyhedra sharing edges or vertices; they are centred by Pr atoms (Pr-7S = 2.71-3.07 or Pr-8S = 2.85-3.24 Å). These chains are cross-linked by SbS_5 trigonal bipyramids (Sb-S = 2.47-3.09 Å).

ANTIMONY SULPHUR THALLIUM
S_4SbTl_3 S_8Sb_5Tl

I. M. GOSTOJIĆ, W. NOWACKI and P. ENGEL, 1981. Z. Kristallogr., 157, 299-308.
II. P. ENGEL, 1980. Ibid., 151, 203-216.

I. S_4SbTl_3, triclinic, P1, a = 6.285, b = 6.364, c = 11.647 Å, α = 94.61, β = 98.51, γ = 103.93°, Z = 2, D_X = 6.45. Mo radiation, R = 0.07 for 2369 reflexions, diffractometer data. 16 site-sets are given.

The structure contains isolated SbS_4 tetrahedra (Sb-4S = 2.30-2.39) linked by TlS_6 (Tl-S = 3.02-3.68), TlS_7 (Tl-S = 3.11-3.64) and TlS_8 (Tl-S = 3.04-3.68 Å) polyhedra.

II. S_8Sb_5Tl, monoclinic, Pn, a = 8.098, b = 19.415, c = 9.059 Å, β = 91.96°, D_m = 5.07, Z = 4. Mo radiation, R = 0.07 for 4489 reflexions, diffractometer data. 28 site-sets are given; contrary to 1.

The structure contains infinite corrugated Sb_5S_8 double sheets (Sb-S = 2.40-2.86) with the Tl atoms 8- or 9-coordinated (Tl-S = 3.12-3.71).

1. Structure Reports, 41A, 132.

ARSENIC BARIUM   ARSENIC CALCIUM   ARSENIC EUROPIUM
As_3Ba As_3Ca As_3Eu

W. BAUHOFER, M. WITTMANN and H.G. von SCHNERING, 1981. J. Phys. Chem. Solids, 42, 687-695.

As_3Ba, As_3Eu, monoclinic, As_3Eu type (1), C2/m, Z = 4. Mo radiation, diffractometer data. See also 2.

As_3M	a(Å)	b(Å)	c(Å)	β(°)	D_X	R	refl.
As_3Ba	10.162	7.760	6.015	113.55	5.53	0.035	489
As_3Eu	9.471	7.598	5.778	112.53	6.51	0.063	463

Atomic positions
M, As(1) in 4(i), As(2) in 8(j)

As_3Ba	x	y	z	As_3Eu	x	y	z
Ba	0.1637	0	0.3370	Eu	0.1638	0	0.3327
As(1)	0.5042	0	0.7969	As(1)	0.4931	0	0.7830
As(2)	0.1386	0.2342	0.8165	As(2)	0.1382	0.2291	0.8022

As_3Ca, triclinic, CaP_3 type (3), P$\bar{1}$, a = 5.866, b = 5.838, c = 5.921 Å, Z = 2, D_X = 4.78. Mo radiation, R = 0.059 for 389 reflexions, diffractometer data. All atoms in 2(i) x,y,z; Ca 0.3153,0.3405,0.3757; As(1) 0.9896,0.2171,0.9851; As(2) 0.1449, 0.8130,0.5928; As(3) 0.4105,0.2199,0.9093. See also 4. The structures are as previously described (1-4) and bond distances are summarized below. The bonding is compared and discussed in detail.

MAs$_3$	BaAs$_3$	EuAs$_3$	CaAs$_3$
As-As (Å)	2.45-2.48	2.45-2.47	2.44-2.46
Av. (Å)	2.46	2.45	2.45
M-As (Å)	3.26-3.77	3.08-3.56	3.00-3.58
Av. (Å)	3.42	3.25	3.13
M-M (Å)	4.09-4.49	3.91-4.29	3.91-4.58
Av. (Å)	4.34	4.18	4.11
Av. angle ° As-As-As	105.2	102.7	101.3

<u>1</u>. Structure Reports, <u>37</u>A, 149; <u>42</u>A, 30.
<u>2</u>. Ibid., 42A, 135.
<u>3</u>. Ibid., <u>39</u>A, 42.
<u>4</u>. Ibid., <u>42</u>A, 25.

ARSENIC BARIUM GERMANIUM
As$_2$BaGe$_2$

BARIUM GERMANIUM PHOSPHORUS
BaGe$_2$P$_2$

B. EISENMANN and H. SCHÄFER, 1981. Z. Naturf., <u>36</u>B, 415-419.

Tetragonal, P4$_2$mc, Z = 4. Mo radiation, diffractometer data.

BaGe$_2$X$_2$	a(Å)	c(Å)	c/a	D$_m$	R	reflex.
BaGe$_2$As$_2$	7.786	8.664	1.11	5.37	0.069	414
BaGe$_2$P$_2$	7.618	8.500	1.12	4.60	0.075	428

Atomic positions (X = As or P)

		BaGe$_2$As$_2$			BaGe$_2$P$_2$		
		x	y	z	x	y	z
Ba(1)	in 2(a)	0	0	0	0	0	0
Ba(2)	in 2(c)	1/2	0	0.0241	1/2	0	0.0276
Ge(1)	in 4(e)	0.3351	1/2	0.8664	0.3318	1/2	0.8655
Ge(2)	in 4(e)	1/2	0.1632	0.6068	1/2	0.1656	0.6141
X	in 8(f)	0.2290	0.2453	0.2558	0.2315	0.2488	0.2564

The structure (Fig. 1) contains clusters formed of condensed 5- and 7-membered rings with Ge-Ge = 2.48-2.57, Ge-As = 2.42 or Ge-Ge = 2.46-2.56, Ge-P = 2.32,2.33 Å. The Ba atoms are 8-coordinated (Ba-As = 3.48-3.54 or Ba-P = 3.32-3.48 Å).

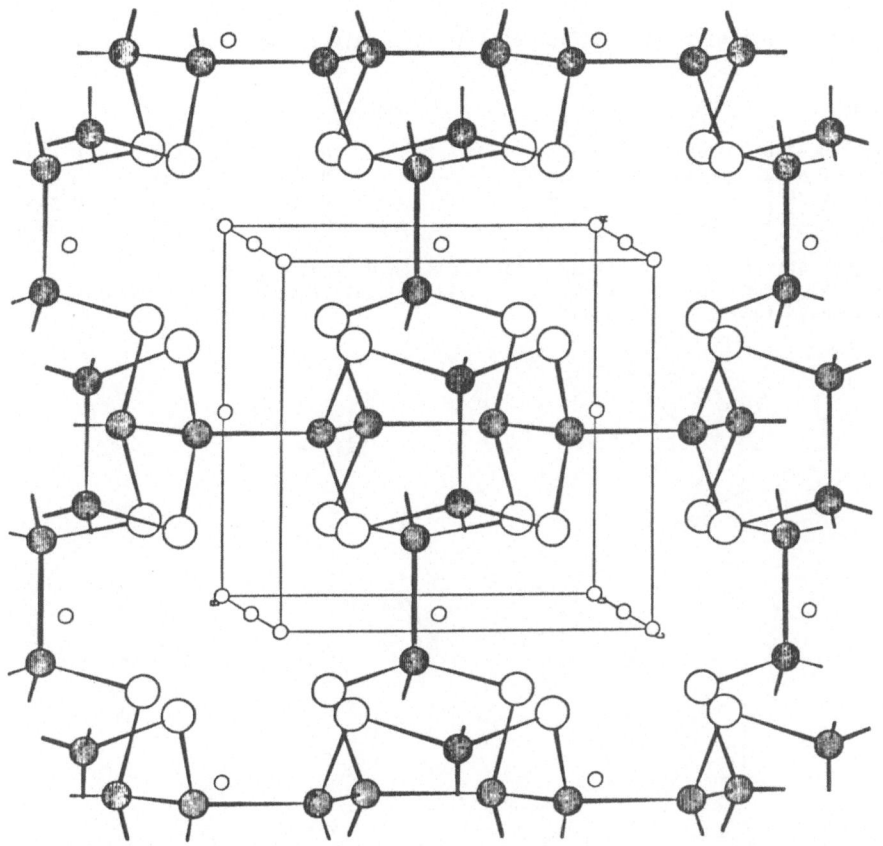

Fig. 1.   The BaGe$_2$As$_2$ structure; As: large open circles, Ge: large filled circles,
Ba: small open circles.

ARSENIC BARIUM SILICON   ARSENIC BARIUM GERMANIUM
As$_4$Ba$_4$Si   As$_4$Ba$_4$Ge

I.   B. EISENMANN, H. JORDAN and H. SCHÄFER, 1981. Z. anorg. Chem., <u>475</u>, 74-80.

II. B. EISENMANN, H. JORDAN and H. SCHÄFER, 1981. Angew. Chem., <u>93</u>, 211.

Cubic, P$\bar{4}$3n, Z = 8. Mo radiation, diffractometer data.

Ba$_4$XAs$_4$	a(Å)	D$_x$	R	refl.
Ba$_4$SiAs$_4$	13.307	4.82	0.094	574
Ba$_4$GeAs$_4$	13.419	5.04*	0.099	643
X = Si or Ge; *D$_m$				

Atomic positions

Ba(1), As(1) in 8(e) x,x,x; Ba(2), As(2) in 24(i) x,y,z; X(1) in 2(a) 0,0,0; X(2) in 6(d) 1/4,0,1/2.

Ba_4SiAs_4				Ba_4GeAs_4		
	x	y	z	x	y	z
Ba(1)	0.6446	0.6446	0.6446	0.6435	0.6435	0.6435
Ba(2)	0.4052	0.1451	0.3682	0.4037	0.1442	0.3688
As(1)	0.3962	0.3962	0.3962	0.3940	0.3940	0.3940
As(2)	0.1521	0.1120	0.3971	0.1497	0.1148	0.3954

These structures (Fig. 1) are characterised by isolated $GeAs_4$ or $SiAs_4$ tetrahedral groups (Ge-As = 2.46, 2.48; Si-As = 2.39, 2.41 Å); the Ba atoms are 6-coordinated (Ba-As = 3.34-3.45).

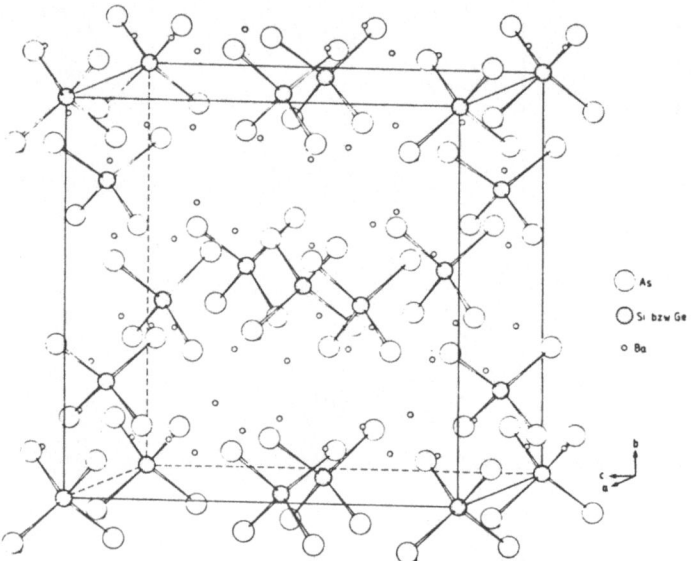

Fig. 1.   A perspective view of the cubic Ba_4XAs_4 structures.

ARSENIC COPPER SELENIUM
AsCuSe

H.J. WHITFIELD, 1981. J. Solid State Chem., <u>39</u>, 209-214.

Orthorhombic, Pbcn, a = 11.75, b = 6.79, c = 19.21 Å, D_m = 5.50, Z = 24. Mo radiation, R = 0.125 for 298 reflexions, photographic data.

Atomic positions

All atoms in 8(d)

	x	y	z		x	y	z
As(1)	0.0156	0.0108	0.1806	Se(3)	0.3354	0.3100	0.0144
As(2)	0.9832	0.3580	0.1459	Cu(1)	0.6685	0.0364	0.1965
As(3)	0.0176	0.3230	0.0170	Cu(2)	0.3343	0.3151	0.1402
Se(1)	0.3355	0.9773	0.1803	Cu(3)	0.6735	0.3644	0.0307
Se(2)	0.6635	0.3701	0.1545				

This structure is a 6-H stacking based on the B6 (ZnS) structure, contrary to 1, who gave the structure as lautite type; the Cu atoms (Fig. 1) are tetrahedrally bonded to 1 As + 3 Se (Cu-As = 2.34-2.41, Cu-Se = 2.37-2.42 Å); similarly As-(1 Cu + 1 Se + 2 As) and Se-(1 As + 3 Cu).

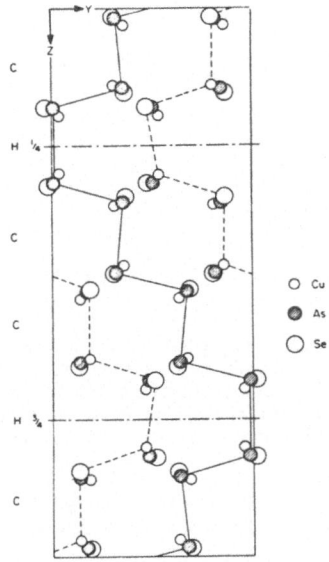

Fig. 1.   A projection of the CuAsSe structure onto (100) with the hcc stacking indicated.

1. J. KAMSU KOM, 1967. C.R. Acad. Sci., Paris, 265, 727, 918.

ARSENIC EUROPIUM SILVER
AgAsEu

C. TOMUSCHAT and H.-U. SCHUSTER, 1981. Z. Naturf., 36B, 1193-1194.

Hexagonal, Ni_2In type (1), $P6_3/mmc$, a = 4.516, c = 8.107 Å, c/a = 1.80, Z = 2, D_x = 7.76. Mo radiation, R = 0.06 for 76 reflexions, diffractometer data. Eu in 2(a) 0,0,0; Ag in 2(d) 1/3,2/3,3/4; As in 2(c) 1/3,2/3,1/4; distances are Ag-As = 2.61, Ag-Eu = 3.30, As-Eu = 3.30, Eu-Eu = 4.05-4.52 with Ag and As 9-coordinated and Eu 20-coordinated.

1. Structure Reports, 9, 91; 40A, 9.

ARSENIC GADOLINIUM SELENIUM   ARSENIC NEODYMIUM SELENIUM
AsGdSe   AsNdSe

R. SCHMELCZER, D. SCHWARZENBACH and F. HULLIGER, 1981. Z. Naturf., 36B, 463-469.

Monoclinic, $P2_1/n$, Z = 4. Mo radiation, diffractometer data.

	a(Å)	b(Å)	c(Å)	(γ°)	R	refl.	type	
GdAsSe*	3.9733	3.9250	17.432	90.00	0.032	787	CeAsS	(1)
NdAsSe	4.0358	4.0363	17.645	90.00	0.028	800	NdAsSe	

*This cell ignores certain satellite reflexions; the structure is thus an average one.

Atomic positions

All atoms in 4(e)

	GdAsSe				NdAsSe		
	x	y	z		x	y	z
Gd	0.2409	0.7506	0.36024	Nd	0.2454	0.7444	0.35971
As*	0.2258	0.2495	0.00137	As	0.2362	0.2638	-0.00003
Se	0.2420	0.7501	0.18699	Se	0.2480	0.7465	0.18591

*occupancy 0.966(7). All other atoms gave occupancies not significantly different from 1.0.

These structures belong to a sequence GdPS (2), CeAsS (1), ZrSiS (3), which are all derived from the tetragonal Cu_2Sb (4) structure. In GdPS there are infinite P chains of a cis-trans type; in CeAsS, which is less distorted, there are only trans-trans As chains (Fig. 1) while in NdAsSe, which is closest to ZrSiS, there are only As-As pairs (Fig. 1). Distances in (Å) are summarised below.

LnAsSe	Ln-Se	Ln-As	Ln-Ln	As-As
GdAsSe	2.91-3.02	3.09-3.16	3.93-3.97	2.66-2.94
NdAsSe	2.95-3.07	3.17-3.22	4.04	2.70-3.01

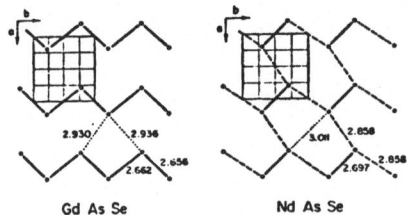

Gd As Se   Nd As Se

Fig. 1.   Left: the GdAsSe structure with the As chains shown against a grid marking the ideal (1/4,1/4,0) positions for the As atoms if they were in the ZrSiS structure. Right: the NdAsSe structure with the As-As pairs similarly marked.

1. Structure Reports, 39A, 8.
2. Ibid., 43A, 59.
3. Ibid., 19, 57; 29, 38.
4. Ibid., 40A, 20; 45A, 8; Strukturbericht, 3, 33.

ARSENIC GERMANIUM
AsGe

B.F. MENTZEN, R. HILLEL, A. MICHAELIDES, A. TRANQUARD and J. BOUIX, 1981. C.R. Acad. Sci. Paris, Ser. II, 293, 965-967.

Monoclinic, SiAs type (1), C2/m, a = 15.517, b = 3.775, c = 9.455 Å, β = 101.03°, D_m = 5.37, Z = 12. Ag radiation, R = 0.066 for 168 reflections, diffractometer data. All atoms in 4(i) x,0,z. The structure is as described (1) with Ge-As = 2.44-2.47 Å and Ge-Ge = 2.43, 2.46 Å.

Atomic positions

	x	y	z		x	y	z
Ge(1)	0.2548	0	0.2135	As(1)	0.1535	1/2	0.1870
Ge(2)	0.3718	0	0.0757	As(2)	0.4677	1/2	0.1672
Ge(3)	0.4309	1/2	0.4107	As(3)	0.3401	0	0.4584

1. Structure Reports, 30A, 22.

ARSENIC GERMANIUM SELENIUM
AsGeSe

F. HULLIGER and T. SIEGRIST, 1981. Mater. Res. Bull., 16, 1245-1251.

Orthorhombic, Pnna, a = 5.062, b = 10.117, c = 11.687 Å, Z = 8, D_x = 5.03. Mo
radiation, R = 0.089 for 376 reflexions, diffractometer data. Ge(1) in 4(c) 1/4,0,
0.7640; Ge(2) in 4(d) 0.2873,1/4,1/4; As in 8(e) 0.0781,0.4148,0.1282; Se in 8(e)
0.9288,0.8435,0.1091. See also 1.

The structure (Fig. 1) is a layer one derived from that of red HgI_2 with Ge
tetrahedrally coordinated (Ge-2 As = 2.44, 2.45, Ge-2 Se = 2.35, 2.38 Å). The
anions are in a distorted cubic close packed array with half of them (As) in contact
(As-As = 2.45 Å) forming anion-anion bonds and making the compound a semiconductor.

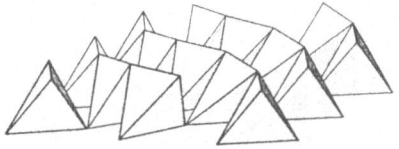

Fig. 1.   The array of corner-linked [$GeAs_2Se_2$] tetrahedra in one layer of the
GeAsSe structure.

1. Structure Reports, 42A, 135.

ARSENIC IRON POTASSIUM   ARSENIC POTASSIUM RHODIUM
As_2Fe_2K As_2KRh_2

S. RÓZSA and H.-U. SCHUSTER, 1981. Z. Naturf., 36B, 1668-1670.

Tetragonal, $ThCr_2Si_2$ type (1), I4/mmm, Z = 2. Mo radiation, diffractometer data. K
in 2(a) 0,0,0; X in 4(d) 0,$\bar{1}$/2,1/4; As in 4(e) 0,0,z. X = Fe or Rh

As_2KX_2	a(Å)	c(Å)	c/a	D_m	R	refl.	z(As)
As_2KFe_2	3.842	13.861	3.61	5.0	0.06	117	0.3525
As_2KRh_2	3.987	13.267	3.33	6.8	0.13	114	0.3538

Interatomic distances (Å)

As_2KX_2	K-X	K-As	X-X	X-As	As-As
As_2KFe_2	3.96	3.40	2.72	2.39	4.09
As_2KRh_2	3.87	3.42	2.82	2.42	3.87

1. Structure Reports, 28, 29; 43A, 99; 44A, 20.

ARSENIC NIOBIUM
As_3Nb_5

I. S. LAOHAVANICH, S. THANOMKUL and S. PRAMATUS, 1981. Acta Cryst., B$\underline{37}$, 227-228.

II. Idem, 1982. Ibid., B$\underline{38}$, 1398.

Orthorhombic, Hf_5As_3 type ($\underline{1}$), Pnma, a = 26.0701, b = 3.5198*, c = 11.7869 Å, Z = 8, D_x = 8.46. Mo radiation, R = 0.117 for 1199 reflexions, photographic data. 16 site-sets given. See also 2.
*[Misprinted as 5.5198. II]

     The structure is as described for Nb_5P_3 ($\underline{1}$) with slightly longer distances.

$\underline{1}$. Structure Reports, $\underline{37A}$, 118; $\underline{46A}$, 110.
$\underline{2}$. Ibid., $\underline{34A}$, 138.

ARSENIC SCANDIUM
As_3Sc_7

R. BERGER, B.I. NOLÄNG and L.-E. TERGENIUS, 1981. Acta Chem. Scand., A$\underline{35}$, 679-683.

Tetragonal, I4/mcm, a = 14.3751, c = 8.0281 Å, c/a = 0.56, Z = 8, D_x = 4.31. R = 0.078 for 1284 reflexions, diffractometer data.

Atomic positions

		x	y	z
Sc(1) in	16(1)	0.33176	0.83176	0.1956
Sc(2) in	16(k)	0.32822	0.01939	0
Sc(3) in	16(k)	0.11902	0.12547	0
Sc(4)* in	8(g)	0	1/2	0.2223
Sc(5) in	4(a)	0	0	1/4
As(1) in	16(j)	0.19499	0	1/4
As(2) in	8(h)	0.22006	0.72006	0

*50% occupancy

     The structure (Fig. 1) is similar to that of Ti_5Sb_3 and contains 14-, 15-, and 16-coordinated Sc (Sc-As = 2.66-3.15, Sc-Sc = 3.05-3.96) and one Sc which centres an Sc_8 cube (Sc-Sc = 3.20) capped with 4 As (Sc-As = 2.80); As atoms centre 2- or 3-capped trigonal prisms of Sc atoms (As-Sc = 2.66-3.15 Å)

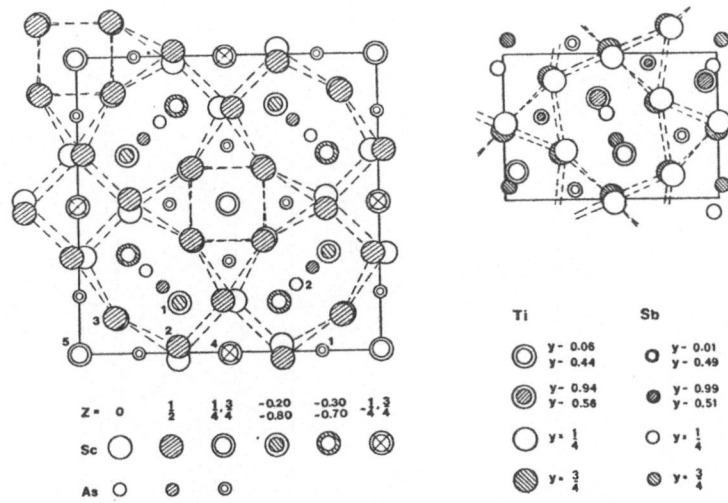

Fig. 1.   The Sc_7As_3 projected onto 001:(a); the Ti_5Sb_3 (ß-Yb_5Sb_3' type) structure
projected onto 010.

ARSENIC SODIUM SULPHUR
$AsNaS_2$

M. PALAZZI and S. JAULMES, 1977. Acta Cryst., B33, 908-910.

Monoclinic, $P2_1/b$, a = 5.859, b = 5.569, c = 11.291 Å, γ = 92.95°, D_m = 2.86, Z = 4.
Mo radiation, R = 0.033 for 764 reflexions, diffractometer data. All atoms in 4(e)
x,y,z; As 0.0394,0.2221,0.1416; S(1) 0.4078,0.2247,0.1242; S(2) -0.0232,0.6303,
0.1607; Na 0.4905,0.7313,0.1246.

    The structure contains AsS_3 trigonal pyramids (As-S = 2.17-2.33) linked to form
chains (Fig. 1a); the Na atoms (Fig. 1b) have 4 near S neighbours (Na-S = 2.81-
2.88) and two more distant neighbours (Na-S = 2.96, 3.06 Å).

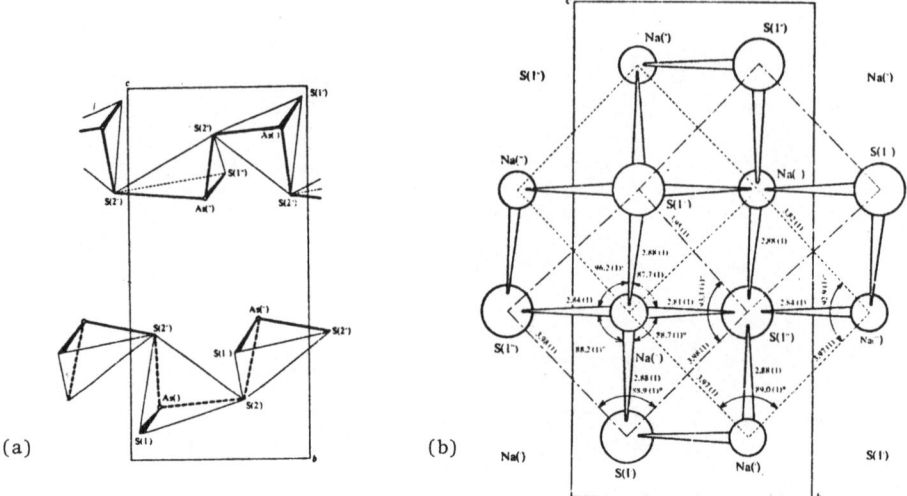

Fig. 1.   The NaAsS$_2$ structure.   a: a 010 projection showing the linkage of AsS$_3$
           pyramids; b: the 010 projection showing the Na-S arrangement.

ARSENIC SULPHUR THALLIUM (SYNTHETIC ELLISITE)
AsS$_3$Tl$_3$

M. GOSTOJIĆ, 1980. Z. Kristallogr., 151, 249-254.

Rhombohedral, R3m, a = 5.9967 Å, α = 105.88°, Z = 1, D$_x$ = 7.04; hexagonal axes a =
9.5707, c = 6.9888 Å. Mo radiation, R = 0.058 for 472 reflexions, diffractometer
data. Tl in 3(b) 0.9957,0.9957,0.6088; As in 1(a) 0.4458,0.4458,0.4458; S in 3(b)
0.431,0.431,0.067. See also 1.

     [This is essentially the structure of Tl$_3$AsSe$_3$ (2) and can be described in
similar terms]. The distances are Tl-3 S = 3.05, 3.1$\overline{0}$; Tl-2 S = 3.44; As-3 S = 2.25;
Tl-6 Tl = 3.70, 3.77 Å.

1. Structure Reports, 45A, 115.
2. Ibid., 40A, 24.

ARSENIC TANTALUM
As$_2$Ta

RONG GUO LING and C. BELIN, 1981. C.R. Acad. Sci. Paris, Ser. II, 292, 891-893.

Monoclinic, NbAs$_2$[Ge$_2$Os] type (1), C2/m, a = 9.331, b = 3.383, c = 7.752 Å, β =
119.71°, Z = 4, D$_x$ = 10.34. R = 0.039 for 247 reflexions, diffractometer data. All
atoms in 4(i) x,0,z; Ta 0.1575, 0.1959; As(1) 0.5946, 0.8918; As(2) 0.8611
0.4735. See also 2. The structure is as described previously (1) with distances
Ta-Ta = 3.00, 3.38, Ta-As = 2.65-2.72, As-As = 2.42-3.38 Å.

1. Structure Reports, 24, 144; 30A, 14; 39A, 8.
2. Ibid., 29, 18, 102.

BARIUM MERCURY SULPHUR
Ba_2HgS_3 $BaHgS_2$

I.   H.D. RAD and R. HOPPE, 1981. Z. anorg. Chem., **483**, 7-17.

II. Idem, 1981. Ibid., **483**, 18-25.

I. Ba_2HgS_3, orthorhombic, Pnma, a = 8.931, b = 4.357, c = 17.257 Å, D_m = 5.59, Z = 4.
Cu radiation, R = 0.065 for 976 reflexions, diffractometer data.

Atomic positions
All atoms in 4(c) x,1/4,z

	x	z		x	z
Ba(1)	0.4227	0.7850	S(1)	0.1180	0.4285
Ba(2)	0.2587	0.0400	S(2)	0.4858	0.5954
Hg	0.3747	0.3669	S(3)	0.3129	0.2227

II. $BaHgS_2$, orthorhombic, $Pmc2_1$, a = 4.215, b = 14.388, c = 7.338 Å, D_m = 5.92,
Z = 4. Mo radiation, R = 0.0884 for 1199 reflexions, diffractometer data.

Atomic positions
Ba(1), (2), S(3) in 2(a) 0,y,z; remainder in 2(b) 1/2,y,z

	y	z		x	y
Ba(1)	0.3913	0.8202	S(1)	0.9859	0.270
Ba(2)	0.1022	0.4977	S(2)	0.736	0.099
Hg(1)	0.3511	0.3347	S(3)	0.7794	0.610
Hg(2)	0.1402	0	S(4)	0.4695	0.112

[Interatomic distances (Å)]

	Ba-Ba	Ba-S	Ba-Hg	Hg-Hg	Hg-S
Ba_2HgS_3	4.36	3.11 - 3.38	3.71 - 3.88	-	2.53 - 2.59
$BaHgS_2$	4.21	3.17 - 3.32	-	3.90	2.31 - 3.44

BARIUM SELENIUM
$BaSe_3$, $BaSe_2$

F. HULLIGER and T. SIEGRIST, 1981. Z. Naturf., **36B**, 14-15.

$BaSe_3$, tetragonal, BaS_3 type (1), $P\bar{4}2_1m$, a = 7.2802, c = 4.2495 Å, c/a = 0.58, Z = 2,
D_x = 5.52. Cu radiation, R = $\overline{0}$.092 for 25 reflexions, powder diffractometer data.
Ba in 2(a) 0,0,0; Se(1) in 2(c) 0,1/2,0.175; Se(2) in 4(e) x,1/2+x,z, with x = 0.189,
z = 0.485.

$BaSe_2$, monoclinic, ThC_2 type (2), C2/c, a = 9.820, b = 4.929, c = 9.335 Å, β = 118.48°,
Z = 4, D_x = 4.94. Mo radiation, R = 0.13 for 364 reflexions, diffractometer data.
Ba in 4(e) 0,0.131,1/4; Se in 8(f) 0.156*,0.634,0.5248.
*[Misprinted as 0.1600 - private communication]

	Ba-Se (Å)	Se-Se (Å)
$BaSe_3$	3.36 - 3.45	2.35
$BaSe_2$	3.28 - 3.36	2.40 [2.33]

1. Structure Reports, **40A**, 35; **41A**, 28.
2. Ibid., **13**, 69; **33A**, 56.

BARIUM SILICON SULPHUR
Ba_3S_5Si

D. SCHMITZ, 1981. Acta Cryst., B$\underline{37}$, 518-525.

Orthorhombic, Pnma, a = 12.121, b = 9.527, c = 8.553 Å, Z = 4, D_X = 4.04. Mo
radiation, R = 0.044 for 820 reflexions, diffractometer data.

Atomic positions
Ba(1) and S(1) in 8(d), remainder in 4(c), x,1/4,z

	x	y	z		x	z
Ba(1)	0.6759	0.5193	0.4165	S(2)	0.5066	0.4770
Ba(2)	0.9754	1/4	0.3988	S(3)	0.7312	0.1785
Si	0.6003	1/4	0.6918	S(4)	0.7764	0.6431
S(1)	0.9388	0.9401	0.3127			

Ba$_3$S$_5$Si is a partly filled version (Fig. 1) of the Sb$_3$Yb$_5$ ($\underline{1}$) structure and
(NH$_4$)$_3$ZnCl$_5$ a partly filled version of the Bi$_3$Y$_5$ structure ($\underline{2}$); both Yb$_5$Sb$_3$ and
Y$_5$Bi$_3$ can be derived from the Rh$_5$Ge$_3$ structure ($\underline{3}$). The Si atoms occupy distorted
tetrahedral holes (Si - 4 S = 2.14-2.18 Å) in the open hexagonal channels already
occupied by [Ba$_2$S$_2$] rhombs. The Ba atoms have 8 near S neighbours (Ba-S = 3.08-3.51 Å)
and 1 or 3 near Si neighbours (Si-Ba = 3.60-4.11 Å); the Ba-Ba distances range from
4.40 to 5.27 Å.

$\underline{1}$. Structure Reports, $\underline{37}$A, 8; $\underline{40}$A, 9.
$\underline{2}$. Ibid., 42A, 50.
$\underline{3}$. Ibid., $\underline{19}$, 177.

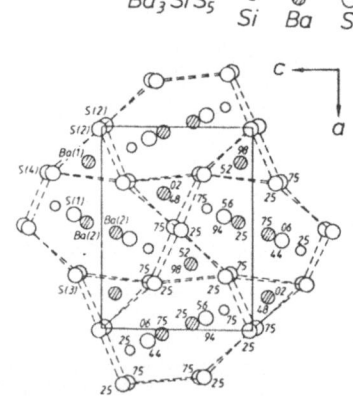

Fig. 1.   The (010) projection of the orthorhombic Ba$_3$SiS$_5$ structure.

BARIUM SILVER   CALCIUM COPPER   SILVER STRONTIUM
AgBa   CaCu- α&β(H.T. & L.T.)   AgSr

F. MERLO and M.L. FORNASINI, 1981. Acta Cryst., B$\underline{37}$, 500-503.

BaAg, orthorhombic, FeB type ($\underline{1}$). Mo radiation, photographic data.   α-CaCu,
orthorhombic, β-CaCu, monoclinic, Cu radiation, photographic data. SrAg, orthorhombic.
Mo radiation, diffractometer data.

A M[†]	a(Å)	b(Å)	c(Å)	Sp.Gr.	Z	D_x	R	refl.
BaAg	8.657	4.982	6.651	Pnma	4	5.68	0.095	85
α-CaCu	38.80	4.271	5.894	Pnma	20	3.52	0.061	171
β-CaCu[*]	19.47	4.271	5.880	P2₁/m	10	3.53	0.095	360
SrAg	16.558	4.788	6.385	Pnma	8	5.13	0.053	638

[*]β = 94.3°   [†]A = Ca, Sr, Ba; M = Cu, Ag

Atomic positions
BaAg atoms in 4(c) x,1/4,z

	x	z		x	z[*]
Ba	0.180	0.129	Ag	0.033	0.631

α-CaCu atoms in 4(c) x,1/4,z

	x	z		x	z
Ca(1)	0.4638	0.6321	Cu(1)	0.0931	0.1010
Ca(2)	0.3652	0.8735	Cu(2)	0.1948	0.3503
Ca(3)	0.2639	0.1259	Cu(3)	0.2938	0.6441
Ca(4)	0.1644	0.8726	Cu(4)	0.3934	0.3943
Ca(5)	0.0649	0.6362	Cu(5)	0.4940	0.1072

β-CaCu atoms in 2(e) x,1/4,z

	x	z		x	z
Ca(1)	0.1276	0.6550	Cu(1)	0.1882	0.1913
Ca(2)	0.3281	0.4626	Cu(2)	0.3882	0.9528
Ca(3)	0.5285	0.7576	Cu(3)	0.5882	0.2499
Ca(4)	0.7279	0.0515	Cu(4)	0.7876	0.5894
Ca(5)	0.9290	0.8565	Cu(5)	0.9871	0.3890

SrAg atoms in 4(c) x,1/4,z

	x	z[*]		x	z[*]
Sr(1)	0.2851	0.5127	Ag(1)	0.3574	0.0027
Sr(2)	0.0349	0.7467	Ag(2)	0.1078	0.2570

[*][Misprinted as y]

These structures are stacking variants of the FeB and CrB types such as those observed in the Dy-Gd-Ni system (1). The smaller M atoms (M = Cu, Ag) centre [A₆] trigonal prisms (A = Ca, Sr, Ba). These prisms have one face capped by one A atom and two faces capped by M atoms which form the zigzag chains characteristic of the parent FeB and CrB structures. These stacking variants are described in 1 and interatomic distances in Å are summarised below. The larger A atoms are 17-coordinated with two A neighbours further away than the others.

	α-CaCu	β-CaCu	SrAg	BaAg
A-A	3.78-4.27	3.78-4.27	4.12-4.79	4.33-4.98
A-M	2.95-3.34	3.03-3.30	3.34-3.47	3.49-3.57
M-M	2.50-2.52	2.50-2.53	2.92	3.09

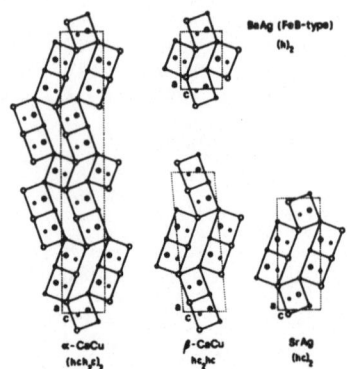

Fig. 1.   Projections of the BaAg, α-CaCu, β-CaCu and SrAg structures. The [A₆]
            prisms are lying with trigonal axes in the plane of the paper and are
            represented as rectangles. Open circles: A atoms; full circles: M
            atoms. Large circles are in the plane of the paper, small circles are
            above and below. See 1 for nomenclature.

1. Structure Reports, 46A, 75.

BARIUM SULPHUR VANADIUM
BaS_3V

M. GHEDIRA, J. CHENAVAS, F. SAYETAT, M. MAREZIO, O. MASSENET and J. MERCIER, 1981.
Acta Cryst., B37, 1491-1496.

Hexagonal, $P6_3/mmc$, a = 6.7283, c = 5.6263 Å, c/a = 0.84, Z = 2, D_X = 4.28. Ag
radiation, R = 0.011 for 216 reflexions, diffractometer data. Ba in 2(d) 1/3,2/3,
3/4; V in 2(a) 0,0,0; S in 6(h) 0.16531,0.33062,1/4. Confirms results of 1; there
is a transition to an orthorhombic form at about 250 K. Distances are Ba-$\overline{S}$ 3.36,
3.43; V-S = 2.39; V-V = 2.81 Å.

1. Structure Reports, 34A, 295; 45A, 34.

BISMUTH CADMIUM CALCIUM   BISMUTH CADMIUM STRONTIUM
$Bi_9Ca_9Cd_4$ $Bi_9Cd_4Sr_9$

CALCIUM STRONTIUM ZINC
$Ca_9Sb_9Zn_4$

E. BRECHTEL, G. CORDIER and H. SCHÄFER, 1981. Z. Naturf., 36B, 1099-1104.

Orthorhombic, $Ca_9Mn_4Bi_9$ type (1), Pbam, Z = 2. Mo radiation, diffractometer data.
12 site-sets for each compound given.

$A_9M_4X_9$	a(Å)	b(Å)	c(Å)	D_X	R	refl.
$Ca_9Cd_4Bi_9$	22.58	12.78	4.74	6.53	0.093	2167
$Sr_9Cd_4Bi_9$	23.47	13.27	4.89	6.80	0.134	1653
$Ca_9Zn_4Sb_9$	21.92	12.50	4.54	4.58	0.076	1774

	CN				Phase	Interatomic distances (Å)		
A	16,	15,	14,	13	$A_9M_4X_9$	A-M	A-X	A-A
M	10,	11,			$Ca_9Cd_4Bi_9$	3.3 - 4.2	3.2 - 4.0	4.0 - 4.4
X	10,	9,	8,	7	$Sr_9Cd_4Bi_9$	3.4 - 4.3	3.3 - 4.0	4.2 - 4.9
					$Ca_9Zn_4Sb_9$	3.2 - 4.0	3.1 - 3.7	3.8 - 4.5

1. Structure Reports, 45A, 35.

BISMUTH COPPER LEAD SULPHUR SELENIUM (NORDSTRÖMITE)
$Bi_7CuPb_3S_{10}Se_4$

W.G. MUMME, 1980. Canad. Miner., 18, 343-352.

Monoclinic, $P2_1/m$, a = 17.97, b = 4.11, c = 17.62 Å, β = 94.3°, Z = 2, D_X = 7.12.
Cu radiation, R = 0.117 for 1548 reflexions, single-crystal diffractometer data.
25 site-sets are given.

This unit cell was first given by 1 who, however, ascribed it to weibullite.
Subsequently the weibullite structure was determined by 2 and it was established that
the powder patterns (3) had been confused. The present crystal came from Falun,
Sweden (Royal Ontario Museum M12992) and the structure can be described in terms of
alternating slabs. In one the atoms have slightly distorted octahedral coordination
but in the other the coordination is irregular; the Cu atoms have threefold planar
coordination (Cu-S = 2.26, 2.39; Bi/Pb-S = 2.54 - 3.38 Å). The structure belongs to
the junoite series and lies on the tie-line between emplectite and galenobismutite.

1. M.A. PEACOCK and L.G. BERRY, 1940. Univ. Toronto Stud. Geol. Ser., 44, 48.
2. Structure Reports, 46A, 34. See references quoted.
3. L.G. BERRY and R.M. THOMSON, 1962. The Peacock Atlas. Geol. Soc. Amer. Mem. 85.

BISMUTH COPPER MANGANESE   BISMUTH MANGANESE NICKEL
$Bi_4Cu_4Mn_3$ at 80 K $Bi_4Mn_5Ni_2$ at 80 K

I.   J.C. SUITS, G.B. STREET, K. LEE and J.B. GOODENOUGH, 1974. Phys. Rev. B, 10,
     120-127.

II. A. SZYTUŁA, H. BIŃCZYCKA and J. TODOROVIĆ, 1981. Solid State Comm., 38, 41-43.

$Ni_2Mn_5Bi_4$
I. Cubic, Fm3m, a = 12.16 Å, Z = 8. R_I = 0.16, 52 reflections, powder diffractometer
data; 16 Ni + 16 Mn in 32(f) x,x,x, x = 0.104; Mn in 24(d) 0,1/4,1/4; Bi(1) in 24(e)
x,0,0, x = 0.285; Bi(2) in 8(c) 1/4,1/4,1/4.

II. Cubic, F$\bar{4}$3m, a = 12.161 Å, Z = 8, D_X = 9.11. R_I = 0.085, neutron powder data.
Ni in 16(e) x,x,x,x = -0.103; Mn(1) in 16(e), x = 0.075; Mn(2) in 24(g) x,1/4,1/4, x
= 0.036; Bi(1) in 24(f) x,0,0,x = 0.272; Bi(2) in 4(c) 1/4,1/4,1/4; Bi(3) in 4(d)
3/4,3/4,3/4, at 293 K. Supercedes the results of I.

II. $Cu_4Mn_3Bi_4$, cubic, Fm3m, a = 12.13 Å, Z = 8, D_X = 9.34, R_I = 0.059 for neutron
powder data. Cu in 32(f) x,x,x,x = 0.102; Mn in 24(d) 0,1/4,1/4; Bi(1) in 24(e)
x,0,0,x = 0.281; Bi(2) in 8(c) 1/4,1/4,1/4, at 80 K.

[Interatomic distances (Å)]

	Bi-Bi	Bi-Mn	Bi-Ni	Mn-Mn	Mn-Ni
$Ni_2Mn_5Bi_4$	3.92	2.60 - 3.69	2.71 - 3.0	2.58 - 3.68	2.22 - 3.04

	Bi-Bi	Bi-Cu	Bi-Mn	Cu-Cu	Cu-Mn
$Cu_4Mn_3Bi_4$	3.76	2.79 - 3.11	3.03 - 3.06	2.47	2.82

BISMUTH COPPER SULPHUR
Bi_3CuS_5, $Bi_5Cu_3S_9$

K. TOMEOKA, M. OHMASA and R. SADANAGA, 1980. Miner. J. Japan, 19, 57-70.

Bi$_3$CuS$_5$, monoclinic, C2/m, a = 13.217, b = 4.0327, c = 14.076 Å, β = 115.53°, Z = 2, D$_X$ = 4.17.   Mo radiation, R = 0.078 for 922 reflexions, diffractometer data.   12 site-sets given.

Bi$_5$Cu$_3$S$_9$, a re-examination of the data of 1, monoclinic, C2/m, a = 13.207, b = 3.993, c = 14.812 Å, β = 100.2°, Z = 2, D$_X$ = 6.58.   Cu radiation, R = 0.059 for 915 reflexions, 14 site-sets given.

     Bi$_3$CuS$_5$ is the Bi$_5$CuS$_8$ of 2 and Bi$_5$Cu$_3$S$_9$ is the Bi$_{6-x}$Cu$_{2+x}$S$_9$ (x = 1.21) of 1. The present study shows that in both structures different amounts of Cu and Bi are possible. The structures (Fig. 1) retain the S frameworks of 1 and 2 but with additional partly occupied Cu sites and reduced occupancy for Bi sites. Distances are not significantly changed but the short Cu-S distance (1.46 Å) quoted for Bi$_5$CuS$_8$ is no longer reported. There are satellite reflexions observed for Bi$_5$Cu$_3$S$_4$ and it is suggested that they are due to clustering of the disordered Cu atoms along b. The overall structure is as described in 1 and 2.

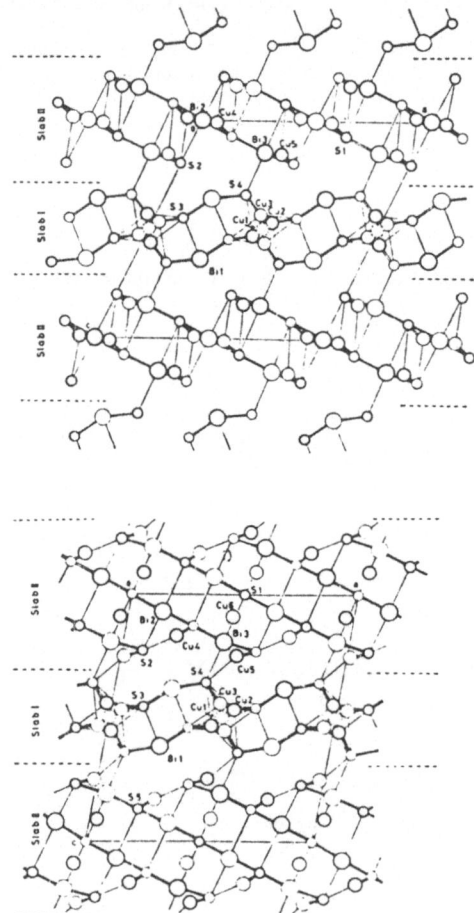

Fig. 1.   Top: the structure of Bi$_3$CuS$_5$ projected onto 010. The disordered Cu atoms occur in Slab I in channels parallel to b. Bottom: the structure of Bi$_5$Cu$_3$S$_4$ similarly outlined. Compare Figs. 2 and 3 in 1 and 2.

1.   Structure Reports, 39A, 31. IV.
2.   Ibid., 39A, 31. III.

BISMUTH INDIUM
BiIn

J.D. JORGENSEN and J.B. CLARK, 1980. Phys. Rev. B, $\underline{22}$, 6149-6154.

Tetragonal, PbO type ($\underline{1}$), P4/nmm, Z = 2. Neutron powder data. In in 2(a) 0,0,0; Bi
in 2(c) 0,1/2,z. See also $\underline{2}$; distances are summarised below.

kb	a(Å)	c(Å)	z(Bi)	Bi-Bi (Å)	Bi-In (Å)
0	5.0118	4.7790	0.3924	3.69	3.13
7.3	4.9979	4.7116	0.3972	3.66	3.12
13.9	4.9901	4.6527	0.4038	3.64	3.12
18.7	4.9841	4.6023	0.4024	3.64	3.10
25.6	4.9732	4.5524	0.4054	3.62	3.10

$\underline{1}$. Strukturbericht, $\underline{7}$, 71, 104.
$\underline{2}$. Structure Reports, $\underline{44A}$, 33.

BISMUTH SULPHUR
Bi_2S_3 (SYNTHETIC BISMUTHINITE)

A.S. KANISCEVA, Ju.N. MIKHAILOV and A.F. TRIPPEL', 1981. Izv. Akad. Nauk SSSR,
Neorg. Mater., $\underline{17}$, 1972-1975 [Inorg. Mater., $\underline{17}$, 1466-1468].

Orthorhombic,Sb_2S_3 type ($\underline{1}$, $\underline{4}$), Pmcn, a = 3.981, b = 11.147, c = 11.305 Å, Z = 4,
D_X = 6.80. Mo radiation, $\bar{R}$ = 0.050 for 561 reflexions, diffractometer data. See also
$\underline{2}$, confirms $\underline{3}$.

Atomic positions
All atoms in 4(c)

	x	y	z		x	y	z
Bi(1)	3/4	0.4655	0.1596	S(2)	3/4	0.0575	0.1230
Bi(2)	1/4	0.1748	0.0165	S(3)	3/4	0.1270	0.4508
S(1)	1/4	0.3063	0.2153				

$\underline{1}$. Strukturbericht, $\underline{1}$, 268; $\underline{3}$, 49, 349.
$\underline{2}$. Ibid., $\underline{3}$, 51; $\underline{6}$, $\overline{86}$.
$\underline{3}$. Structure Reports, $\underline{43A}$, 33; $\underline{44A}$, 108.
$\underline{4}$. Ibid., $\underline{38A}$, 20.

BORON CARBON EUROPIUM
$B_{6-x}C_xEu$ (x = 0,0.05,0.20)

J.M. TARASCON, J.L. SOUBEYROUX, J. ETOURNEAU, R. GEORGES, J.M.D. COEY and
O. MASSENET, 1981. Solid State Comm., $\underline{37}$, 133-137.

Cubic, CaB_6 type ($\underline{1}$), Pm3m, Z = 1. Neutron powder diffraction data. Eu in 1(a)
0,0,0; 6 B or 6-xB + xC in 6(f) x,1/2,1/2. See also $\underline{2}$. B - 5 B = 1.71, 1.75, B - 4
Eu = 3.08; Eu - 24 B = 3.08, Eu - 6 Eu = 4.19 Å.

$EuB_{6-x}C_x$	a(Å)	$[D_X]$	R	x	Magnetic structure
EuB_6	4.19	4.89	0.038	0.2043	ferromagnet
$EuB_{5.95}C_{0.05}$	4.17	4.97	0.038	0.2043	mixed
$EuB_{5.8}C_{0.2}$	4.16	5.01	0.032	0.2043	helimagnet

$\underline{1}$. Structure Reports, $\underline{17}$, 63; $\underline{18}$, 62.
$\underline{2}$. Ibid, $\underline{22}$, 64; $\underline{23}$, $\overline{61}$.

BORON COBALT URANIUM
B_4Co_4U $B_2Co_7U_3$

I.   I.P. VAL'OVKA and Ju.B. KUZ'MA, 1978.   Izv. Akad. Nauk SSSR, Neorg. Mater., 14,
      469-472 [Inorg. Mater., 14, 356-359].

II.   Ju.B. KUZ'MA and I.P. VAL'OVKA, 1980.   Ibid., 16, 1681-1683.

I. B_4Co_4U, tetragonal, $CeCo_4B_4$ type (1), $P4_2/nmc$, a = 5.097, c = 7.037 Å, c/a = 1.38,
Z = 2, D_x = 9.39. Cr radiation, photographic data. U in 2(b) 3/4,1/4,1/4; Co in
8(g) 1/4,-0.003,0.384; B in 8(g) 1/4,0.08,0.10. [Distances are U-Co = 2.87-3.01,
U-B = 2.89, 2.99; Co-Co = 2.52-2.60, Co-B = 2.04-2.17; B-B = 1.73 Å.]

II. $B_2Co_7U_3$, hexagonal, $Dy_3Ni_7B_2$ type, $P6_3/mmc$, a = 4.980, c = 13.947 Å, c/a = 2.80,
Z = 2, D_x = 12.73. Fe radiation, R = 0.209, diffractometer data. U(1) in 2(c) 1/3,
2/3,1/4; U(2) in 4(f) 1/3,2/3,0.033; Co(1) in 2(a) 0,0,0; Co(2) in 12(k) 0.841,
0.682,0.140; B(1) in 2(b) 0,0,1/4; B(2) in 2(d) 1/3,2/3,3/4. See also 2.

Interatomic distances (Å)

U(1) - 2 U 3.03		Co(2) - 5 U 2.84 - 2.92	
-12 Co 2.92		- 5 Co 2.39 - 2.60	
- 6 B 2.88		- 2 B 2.05 - 2.14	
U(2) - 4 U 3.02, 3.03		B(1) - 3 U 2.88	
-12 Co 2.84, 2.91		- 6 Co 2.05	
Co(1)- 6 U 2.91		B(2) - 3 U 2.88	
- 6 Co 2.39		- 6 Co 2.14	

1.   Structure Reports, 38A, 42; 46A, 37.
2.   Ibid., 45A, 115.

BORON ERBIUM   BORON TERBIUM
B_4Er B_4Tb

I.   G. WILL, W. SCHÄFER, F. PFEIFFER, F. ELF and J. ETOURNEAU, 1981.   J. Less-Common
      Metals, 82, 349-355.

Tetragonal, ThB_4 type (1), P4/mbm, Z = 4. Neutron single-crystal data. X in 4(g)
x,1/2 + x,0; B(1) in 4(e) 0,0,z; B(2) in 4(h) x,1/2 + x,1/2; B(3) in 8(j) x,y,1/2.
X = Tb or Er. See also 3.

XB_4	a(Å)	c(Å)	c/a	D_x	R	refl.
TbB_4	7.120	4.042	0.57	6.55	0.018	392
ErB_4	7.071	4.000	0.57	6.99	0.030	242

Atomic positions

TbB_4				ErB_4		
	x	y	z	x	y	z
X	0.3175	0.8175	0	0.3183	0.8183	0
B(1)	0	0	0.2017	0	0	0.2031
B(2)	0.0875	0.5875	1/2	0.0859	0.5859	1/2
B(3)	0.1758	0.0387	1/2	0.1767	0.0382	1/2
B_4Er						

II.   G. WILL, F. PFEIFFER, W. SCHÄFER, J. ETOURNEAU and R. GEORGES, 1980.   Rev. Chim.
      Minér., 17, 533-540.

Tetragonal, ThB_4 type (1), P4/mbm, a = 7.0705, c = 4.0000 Å, c/a = 0.57, Z = 4, D_x = 6.99. R = 0.037 for 43 reflexions, neutron single-crystal data. Er in 4(g) x,1/2 + x,0, x = 0.3183; B(1) in 4(e) 0,0,0.203; B(2) in 4(b) x,1/2 + x, 1/2, x = 0.086; B(3) in 8(j) 0.1767,0.0382,1/2. See also 2.

The structure (Fig. 1) contains B_6 octahedra, (B-B = 1.75 - 1.81) linked by pairs of B atoms (B-B = 1.71-1.76) to form 7-member rings. The voids between these rings are centred by RE atoms which are 18-coordinated (<Tb-B> = 2.83, <Er-B> = 2.81 Å). The octahedra are linked by B-B bonds (1.63 Å) parallel to c̲.

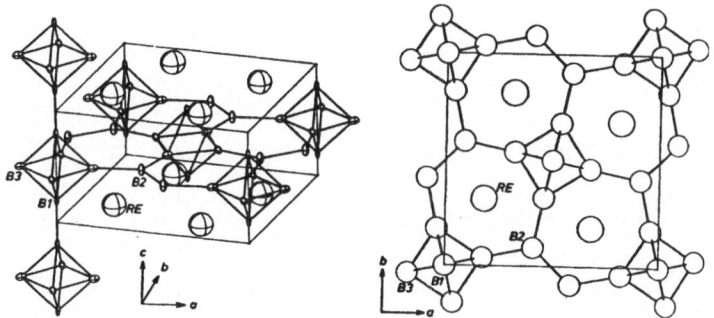

Fig. 1.   The ReB_4 structure. Left: A perspective view. Right: Projected onto 001.

1. Structure Reports, 13, 36; 17, 65; 18, 70; 30A, 31.
2. Ibid., 22, 183; 23, 62;, 42A, 56; 45A, 115.
3. This volume, p. 38.

BORON ERBIUM IRIDIUM
B_2ErIr_3

H.C. KU and G.P. MEISNER, 1981. J. Less-Common Metals, 78, 99-107.

Monoclinic, C2/m, a = 5.409, b = 9.379*, c = 3.101 Å, β = 91.2°, Z = 2, D_x = 16.16. Cu radiation, R = 0.10 for 22 reflexions, diffractometer data. Er in 2(a) 0,0,0; Ir(1) in 2(d) 0,1/2,1/2; Ir(2) in 4(f) 1/4,1/4,1/2; B in 4(g) 0,0.333,0. *[Also misprinted as 9.399 Å.]

Interatomic distances (Å)

Er -	2 Er	3.10	Ir(1) -	6 Ir	2.71, 3.10
	12 Ir	3.09 - 3.15		4 B	2.20
	6 B	3.12, 3.13		4 Er	3.09, 3.15
			Ir(2) -	6 Ir	2.70 - 3.10
B -	5 B	3.10 - 3.13		4 B	2.18, 2.22
	3 Er	3.12, 3.13		4 Er	3.11, 3.13
	6 Ir	2.18 - 2.22			

The structure is very similar to that of $CeCo_3B_2$ (1) and shows a magnetic ordering at 11.9 K.

1. Structure Reports, 34A, 38, 39.

BORON EUROPIUM NICKEL
B_6EuNi_{12}

Ju.B. KUZ'MA, G.V. ČERNJAK and N.F. ČABAN, 1981. Dop. Akad. Nauk Ukr. RSR, <u>43</u>, 80-83.

Rhombohedral, $SrNi_{12}B_6$ [B_6CeNi_{12}] type (<u>1</u>), R$\bar{3}$m, a = 9.551, c = 7.408 Å, c/a = 0.78, Z = 3, D_X = 7.84. Fe radiation, R = 0.13$\overline{4}$, diffractometer data. Eu in 3(a) 0,0,0; Ni(1) in 18(g) 0.368,0,1/2; Ni(2) in 18(h) 0.426,0.426,0.047; B in 18(h) 0.191,0.191, 0.042. The structure is as described with distances Ni-B = 1.99-2.10, Eu-B = 3.17-3.64, Eu-Ni = 3.21-3.27 and Ni-Ni = 2.34-2.66 Å

<u>1</u>. Structure Reports, <u>46A</u>, 21.

BORON GERMANIUM
$B_{\approx 90}Ge$

T. LUNDSTRÖM and L.-E. TERGENIUS, 1981. J. Less-Common Metals, <u>82</u>, 341-348.

Rhombohedral, R$\bar{3}$m, a = 10.9588, c = 23.8622 Å, c/a = 2.18, A = 312.11, D_X = 2.40. Mo radiation, R = 0.068 for 2263 reflexions, diffractometer data. 22 site-sets are given.

The B atom framework is that of β-rhombohedral boron (<u>1</u>) with Ge atoms partially occupying seven sites (20 to 0.4%). Five of these have been previously observed in solid solutions of boron (<u>2</u>) but two are new. Four of the sites are mutually exclusive and cannot be occupied simultaneously and one partially occupied boron site has some Ge content.

<u>1</u>. Structure Reports, <u>35A</u>, 34, 127, 128; <u>38A</u>, 37, 38; <u>43A</u>, 34.
<u>2</u>. Ibid., <u>35A</u>, 3, 34; <u>40A</u>, 41; <u>42A</u>, 55, 5$\overline{7}$; <u>44A</u>, 39.

BORON MAGNESIUM
$B_{14}Mg_2$

A. GUETTE, M. BARRET, R. NASLAIN, P. HAGENMULLER, L.-E. TERGENIUS and T. LUNDSTRÖM, 1981. J. Less-Common Metals, <u>82</u>, 325-334.

Orthorhombic, $MgAlB_{14}$ type (<u>1</u>, <u>2</u>), Imam, a = 5.970, b = 8.125, c = 10.480 Å, D_m = 2.59, Z = 4. Cu radiation, $\bar{R}$ = 0.115 for 116 reflexions, photographic powder data. B(1), (2) in 16(j) x,y,z; B(3), (4), (5) in 8(h) 0,y,z; Mg(1) in 4(e) 0,y,3/4; Mg(2) in 4(c)* 1/4,1/4,1/4. [Misprinted as 4(d) here and in <u>3</u>.]

Atomic positions

	x	y	z		x	y	z
B(1)	0.250	0.0407	0.0830	B(5)	0	0.382	0.142
B(2)	0.160	- 0.166	0.0671	Mg(1)*	0	0.3664	3/4
B(3)	0	0.181	0.081	Mg(2)	1/4	1/4	1/4
B(4)	0	- 0.028	0.1644				

*occupancy = 93%

The structure (Fig. 1) is characterised by B_{12} icosahedra (B-B = 1.77-1.95) joined by direct bonds (B-B = 1.73, 1.79) or bridges (B-B-B = 1.75, 1.73). The Mg atoms (Fig. 2) occupy interstices with CN = 12 (Mg-B = 2.16-2.44) and CN = 16 (Mg-B = 2.32-2.89 Å).

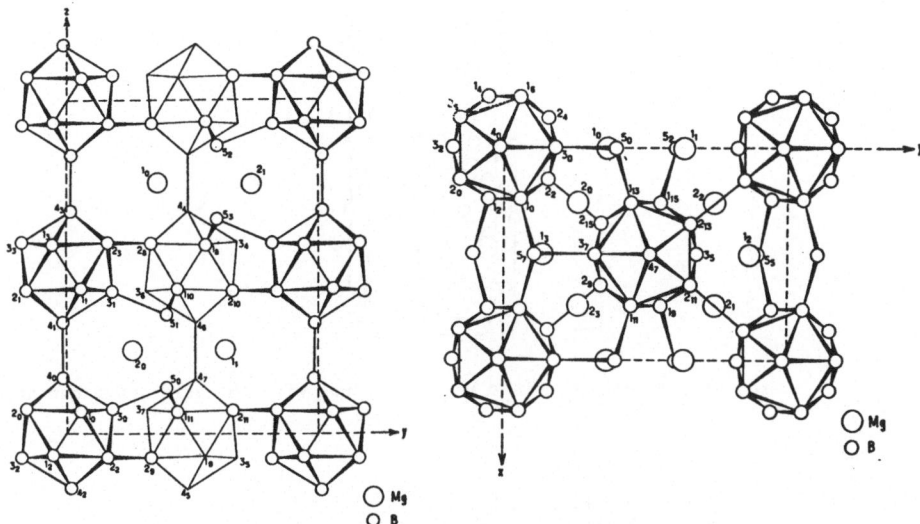

Fig. 1.   The B_{12} icosahedra in $B_{14}Mg_2$. Left: projected onto 100; - 1/4 < x < +
1/4. Right: projected onto 001; - 1/4 < z < + 1/4.

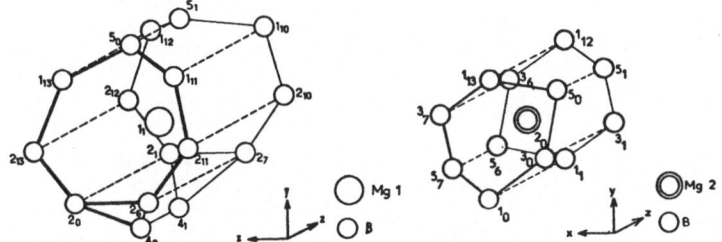

Fig. 2.   The environment of the 2 Mg atoms in $B_{14}Mg_2$.

1. Structure Reports, 35A, 4, 39.
2. This volume, p. 6.
3. R. NASLAIN, A. GUETTE and P. HAGENMULLER, 1976. J. Less-Common Metals, 47, 1.

BORON MANGANESE
BMn_2

L.-E. TERGENIUS, 1981. J. Less-Common Metals, 82, 335-340.

Orthorhombic, FeCrB type (1) Fddd, a = 14.5395, b = 7.2914, c = 4.2082 Å, Z = 16,
D_x = 7.19. Mo radiation, R = 0.029 for 676 reflexions, diffractometer data. Mn(1)
in 16(e) 0.08163,0,0; Mn(2) in 16(f) 0,0.33055,0; 15.712B in 16(e) 0.37550,0,0. This
is the "Mn₄B" of 2 with the B site nearly fully occupied as suggested by 3 for Cr_2B.
The original occupancy was assumed from the overall composition and neither the B
coordinates nor the occupancy were refined. Most Mn_2B crystals were twinned with
apparent space group Ab** as reported for Cr_2B by 4.

     The structure (Fig. 1) has B atoms centering an Archimedean antiprism (B - 8 Mn
= 2.187-2.199). The details are very similar to those given by 2 with distances Mn-
Mn = 2.38-2.73 and B-B = 2.10 Å.

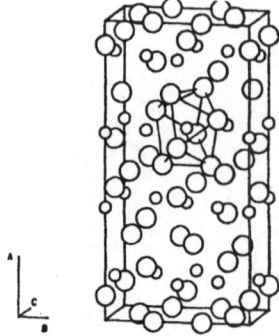

Fig. 1.   A view of the orthorhombic o-Mn₂B structure with the boron environment
           outlined.

1. Structure Reports, 29, 31.
2. Ibid., 13, 38; 23, 68.
3. Ibid., 22, 57. See Boron Chromium Iron.
4. Ibid., 13, 37. See δ-phase.

BORON NEODYMIUM NICKEL
$B_2Nd_3Ni_{13}$

Ju.B. KUZ'MA and N.S. BILONIŽKO, 1981. Dop. Akad. Nauk Ukr. RSR, 43, 87-90.

Hexagonal, P6/mmm, a = 5.005, c = 10.904 Å, c/a = 2.18, D_m = 8.41, Z = 1. Fe
radiation, R = 0.125 for 57 reflexions, diffractometer data. Nd(1) in 1(a) 0,0,0;
Nd(2) in 2(e) 0,0,0.328; Ni(1) in 3(g) 1/2,0,1/2; Ni(2) in 4(h) 1/3,2/3,0.323; Ni(3)
in 6(i) 1/2,0,0.134; B in 2(c) 1/3,2/3,0.

     The Nd atoms have CN = 20 (Nd-Nd = 3.58, 3.75, Nd-Ni = 2.89-3.28, Nd-B = 2.89),
the Ni atoms have CN = 12 (Ni-Ni = 2.41-2.89, Ni-Nd = 2.89-3.28, Ni-B = 2.05) and the
B atoms centre trigonal prisms which are tricapped (B-Ni = 2.05, B-Nd = 2.89 Å).

BORON NEODYMIUM OSMIUM
B_4NdOs_4

K. HIEBL, M.J. SIENKO and P. ROGL, 1981. J. Less-Common Metals, 82, 21-28.

Tetragonal, NdCo₄B₄ type (1), P4₂/n, a = 7.5591, c = 4.0026 Å, c/a = 0.53, Z = 2, D_x =
13.77. Mo radiation, R = 0.08 for 464 reflexions, diffractometer data. Nd in 2(b)
1/4,1/4,3/4; Os in 8(g) 0.6061,0.1418,0.1372; B in 8(g) 0.532,0.415,0.150. See also
2. The structure is as described with distances Nd-Os = 3.19, 3.21, Os-Os = 2.72-2.89,
Os-B = 2.13-2.22, Nd-B = 2.94, 3.05, B-B = 1.82 Å.

1. Structure Reports, 44A, 36; 45A, 45.
2. Ibid., 45A, 116.

BORON NICKEL THULIUM
$B_6Ni_{21}Tm_2$

N.F. CABAN, Ju.B. KUZ'MA and L.D. KOTOVSKAJA, 1980. Dop. Akad. Nauk Ukr. RSR, $\underline{42}$, 87-89.

Cubic, $W_2Fe_{21}C_6$ (τ-phase) type ($\underline{1}$), Fm3m, a = 10.633 Å, Z = 4, $[D_x = 9.04]$. Cr radiation, powder data. Tm in 8(c) 1/4,1/4,1/4; Ni(1) in 4(a) 0,0,0; Ni(2) in 32(f) 0.385,0.385,0.385; Ni(3) in 48(h) 0,0.165,0.165; B in 24(e) 0.275,0,0; Tm-16 Ni = 2.49,2.95; Ni(1)-12 Ni = 2.48; Ni(2)-3 B = 2.09, Ni(2)-1 Tm = 2.49, Ni(2)-9 Ni = 2.45, 2.69; Ni(3)-2 B = 2.11, Ni(3)-10 Ni = 2.48-2.69, Ni(3)-2 Tm = 2.95; B-8 Ni = 2.09, 2.11 Å.

$\underline{1}$. Strukturbericht, $\underline{3}$, 59; Structure Reports, $\underline{27}$, 122; $\underline{38A}$, 60.

BORON RHODIUM
B_4Rh_5

B.I. NOLÄNG, L.-E. TERGENIUS and I. WESTMAN, 1981. J. Less-Common Metals, $\underline{82}$, 303-308.

Hexagonal, $P6_3/mmc$, a = 3.3058, c = 20.394 Å, c/a = 6.17, Z = 2, $D_x = 8.01$. Mo radiation, R = 0.098 for 226 reflexions, diffractometer data. Rh(1) in 4(f) 1/3,2/3, 0.04670; Rh(2) in 4(e) 0,0,0.14737; Rh(3) in 2(c) 1/3,2/3,1/4; B(1) in 4(f) 2/3,1/3, 0.0872; B(2) in 4(f) 2/3,1/3,0.1957.

    This structure is based on a 10-layer stacking of Rh atoms, BABABCACAC, i.e. $(chhhc)_2$. The B atoms occupy the octahedral holes between the ch and hh layers; the holes between the cc layers are empty. This octahedral coordination for B is unusual (B-Rh = 2.08-2.27). The Rh-Rh distances (2.70-3.31) are normal. The B atoms form isolated B_4 strings (B-B = 2.21 Å) which could be considered to be distinct elements though the bonding is weak.

BORON RUTHENIUM TANTALUM
$B_6Ru_{19.9}Ta_{3.1}{}^{-\tau}$

W. STEURER, P. ROGL and H. NOWOTNY, 1979. Mh. Chem., $\underline{110}$, 791-798.

Cubic, $Cr_{23}C_6$ type ($\underline{1}$), Fm3m, a = 11.347 Å, Z = 4, $D_x = 12.53$. Cu radiation, R = 0.107 for 27 reflexions, diffractometer data. Ta in 4(a) 0,0,0; 1 Ta + 7 Ru(1) in 8(c) 1/4,1/4,1/4; 4 Ta + 28 Ru(2) in 32(f) 0.390,0.390,0.390; 4 Ta + 44 Ru(3) in 48(h) 0,0.171,0.171; B in 24(e) 0.275,0,0. Distances are Ta-Ta/Ru = 2.74; Ta/Ru-Ta/Ru = 2.50-2.87; Ta/Ru-B = 2.20, 2.27 Å

$\underline{1}$. Structure Reports, $\underline{38A}$, 60; $\underline{46A}$, 38.

BORON SILICON
$B_{\sim36}Si$

M. VLASSE and J.C. VIALA, 1981. J. Solid State Chem., $\underline{37}$, 181-188.

Rhombohedral, substituted β-B type ($\underline{1}$, $\underline{2}$), R3m, a = 11.01, c = 23.90 Å, c/a = 2.17, D_m = 2.37, Z = 3. Mo radiation, R = 0.054 for 815 reflexions, diffractometer data, 17 site-sets given. See also $\underline{3}$.

    The structure is based on that of β-B ($\underline{1}$) with two Si atoms occupying interstitial holes (A1, A2); a third silicon substitutes partially for B(1) - see $\underline{4}$ for numbering. A number of similar B-rich phases containing Fe, Cr, Cu, Zr, Sc, Mn, Al and Cu are known ($\underline{2}$). Distances are similar to those in β-B ($\underline{1}$).

1. Structure Reports, 35A, 34, 127, 128; 38A, 37; 43A, 34.
2. Ibid., 35A, 3, 34; 40A, 41; 42A, 55, 57; 44A, 39.
3. Ibid., 30A, 116.
4. Ibid., 38A, 38.

BORON TERBIUM
B_4Tb

F. ELF, W. SCHÄFER, G. WILL and J. ETOURNEAU, 1981. Solid State Comm., 40, 579-581.

Tetragonal, ThB_4 type (1), P4/mbm, a = 7.120, c = 4.031 Å, c/a = 0.57, Z = 4, D_x = 6.57. R = 0.044, neutron powder diffractometer data. Tb in 4(g) x,1/2 + x,0, x = 0.317; B(1) in 4(e) 0,0,z, z = 0.207; B(2) in 4(h) x,1/2 + x,1/2, x = 0.082; B(3) in 8(j) x,y,1/2, x = 0.176, y = 0.044. [The structure is as described with Tb-B = 2.74-3.11, Tb-Tb = 3.69 and B-B = 1.65-2.36 Å].

1. Structure Reports, 13, 36; 17, 65.

CADMIUM GERMANIUM. PHOSPHORUS
$CdGeP_2$

W. HÖNLE and H.G. von SCHNERING, 1981. Z. Kristallogr., 155, 319-320.

Tetragonal, $CuFeS_2$ (chalcopyrite) type (1), $I\bar{4}2d$, a = 5.738, c = 10.765 Å, c/a = 1.88, Z = 4, D_m = 4.626. Mo radiation, $\bar{R}$ = 0.032 for 192 reflexions, diffractometer data. P in 8(d) 0.2839,1/4,1/8; Ge in 4(b) 0,0,1/2; Cd in 4(a) 0,0,0. The distances are Cd - 4 P = 2.55, Ge - 4 P = 2.32 with angles P-Cd-P = 106, 116 and P-Ge-P = 110, 109°.

1. Structure Reports, 10, 122; 23, 141; 39A,51.

CAESIUM SULPHUR ZINC
$Cs_2S_4Zn_3$

W. BRONGER and U. HENDRIKS, 1980. Rev. Chim. Minér., 17, 555-560.

Orthorhombic, ($Cs_2Mn_3S_4$ type (1)), Ibam, a = 5.823, b = 11.283, c = 13.935 Å, D_m = 4.22, Z = 4. Mo and Cu radiation, R = 0.08 for 400 reflexions, diffractometer data.

Atomic positions

			x	y	z
Cs	in	8(j)	0.2325	0.1192	0
Zn(1)	in	4(b)	1/2	0	1/4
Zn(2)	in	8(g)	0	0.2238	1/4
S	in	16(k)	0.2183	0.3669	0.1557*

*[Misprinted as 0.1157 - private communication]

In $Cs_2Zn_3S_4$ the S atoms form edge-linked S_4 tetrahedra, three-quarters of which are occupied by the Zn atoms. In $CsMn_3S_4$ (1), there is a slightly different statistical distribution of the metal atoms; compare $TlFeS_2$ (2).

1. Structure Reports, 38A, 57.
2. Ibid., 45A, 91.

CALCIUM ZINC
CaZn   Ca$_3$Zn

M.L. FORNASINI, F. MERLO and K. SCHUBERT, 1981. J. Less-Common Metals, 79, 111-119.

CaZn, orthorhombic, CrB type (1), Cmcm, a = 4.202, b = 11.61, c = 4.442 Å, Z = 4,
D$_X$ = 3.23. Cu radiation, photographic data. Both atoms in 4(c) 0,y,1/4, y = 0.145,
0.435.

Ca$_3$Zn, orthorhombic, Re$_3$B type (2), Cmcm, a = 4.150, b = 13.258, c = 10.186 Å, Z = 4,
D$_X$ = 2.20. Mo radiation, R = 0.022 for 448 reflexions, diffractometer data. Ca(1)
in 4(c) 0,0.4323,1/4; Ca(2) in 8(f) 0,0.1414,0.0605; Zn in 4(c) 0,0.7524,1/4.

     Both structures are as previously described with distances Ca-Ca = 3.76-4.31 and
Ca-Zn = 3.16-3.46 Å.

1. Structure Reports, 12, 30; 29, 51.
2. Ibid., 24, 73; 28, 17.

CARBON HOLMIUM
CHo$_2$ at 296 K

M. ATOJI, 1981. J. Chem. Phys., 74, 1893-1897.

Rhombohedral, anti-CdCl$_2$ type (1), R$\bar{3}$m, a = 3.556, c = 17.70 Å, c/a = 4.98, Z = 3,
D$_X$ = 8.78; at 4 K a = 3.550, c = 17.67 Å. R = 0.018 for diffractometer data. C in
3(a) 0,0,0; Ho in 6(c) 0,0,0.2564. Confirms 2.

1. Strukturbericht, 1, 189; 2, 246, 249.
2. Structure Reports, 29, 106; 31A, 33.

CARBON SILICON
CSi-2H

H. SCHULZ and K.H. THIEMANN, 1979. Solid State Comm., 32, 783-785.

Hexagonal, wurtzite (ZnS) type (1), P6$_3$mc, a = 3.079, c = 5.053 Å, c/a = 1.64, Z = 2,
D$_X$ = 3.20. Mo radiation, R = 0.032 for 265 reflexions, diffractometer data. Si and C
in 2(b) 1/3,2/3,z, z = 0 and -0.3760* respectively; see also 2, 3; distances are
Si - 1 C = 1.930 [1.900], Si - 3 C = 1.875 [1.885] Å, C-Si-C = 108.5, 110.5°; the
absolute configuration was determined. [*Note that for ZnO z was positive.]

1. Strukturbericht, 1, 78, 128.
2. Ibid., 1, 80.
3. Structure Reports, 23, 280.

CERIUM GALLIUM
CeGa$_6$

Ju.M. GRIN', K.A. ČUNTONOV, Ja.P. JARMOLJUK, S.P. JACENKO and E.I. GLADYSEVSKIJ, 1981.
Dop. Akad. Nauk Ukr. RSR, 43, No. 5, 86-89.

Tetragonal, PuGa$_6$ type (1), P4/nbm, a = 6.052, c = 7.673 Å, c/a = 1.27, Z = 2, D$_X$ =
6.59. Fe radiation, R = 0.127 for 85 reflexions, diffractometer data. Ce in 2(c) 3/4,
1/4,0; Ga(1) in 4(g) 1/4,1/4,0.161; Ga(2) in 8(m) 0.047, 0.953, 0.355.

The structure has 20-coordinated Ce (Ce-Ga = 3.23-3.73, Ce-Ce = 4.28), 12-coordinated Ga (Ga-Ce = 3.23, 3.73, Ga-Ga = 2.64, 3.08) and 9-coordinated Ga (Ga-Ga = 2.64, 3.47, Ga-Ce = 3.27 Å).

<u>1</u>. Structure Reports, <u>30</u>A, 51.

CERIUM NICKEL TIN
$CeNi_5Sn$

R.V. SKOLOZDRA, V.M. MANDZYK and L.G. AKSEL'RUD, 1981. Kristallografija, <u>26</u>, 480-483 [Soviet Physics-Crystallography, <u>26</u>, 272-274.

Hexagonal, $P6_3/mmc$, a = 4.9049, c = 19.731 Å, c/a = 4.02, D_m = 8.78, Z = 4. Mo radiation, R = 0.117 for 205 reflexions, photographic data.

Atomic positions

	x	y	z		x	y	z
Ce(1) in 2(c)	1/3	2/3	1/4	Ni(3) in 2(d)	1/3	2/3	3/4
Ce(2) in 2(a)	0	0	0	Ni(4) in 2(b)	0	0	1/4
Ni(1) in 12(k)	0.831	0.662	0.1458	Sn in 4(f)	1/3	2/3	0.0873
Ni(2) in 4(f)	1/3	2/3	0.5425				

In this structure, which can be derived from that of $CaCu_5$, the Ce atoms have distances Ce-Ni = 2.83-3.32; Ce-Sn = 3.21, 3.31 (CN = 20, 18); Ni atoms with CN = 12, 13 have Ni-Ni = 2.42-2.83; Ni-Sn = 2.56, 2.97; Sn atoms have CN = 14 and distances as above.

CHROMIUM COPPER SULPHUR TIN
$Cr_xCu_xS_4Sn_{2-x}$

CHROMIUM COPPER SULPHUR TITANIUM
$Cr_xCu_xS_4Ti_{2-x}$

M. TREMBLET, P. COLOMBET, M. DANOT and J. ROUXEL, 1980. Rev. Chim. Minér. <u>17</u>, 183-191.

Cubic, Al_2MgO_4 (spinel) type (<u>1</u>), Fd3m, Z = 8. Photographic data. Cr/Sn and Cr/Ti in 16(d) 5/8,5/8,5/8; Cu in 8($\bar{a}$) 0,0,0; S in 32(e) x,x,x.

$Cr_xCu_xS_4M_{2-x}$*	a(Å)	R	x(S)	Cr/M occ.	Cu occ.
$CrCuS_4Ti$	9.943	0.035	0.383	0.5 Cr + 0.5 Ti	1
$CrCuS_4Sn$	10.170	0.064	0.380	0.5 Cr + 0.5 Sn	1
$Cr_{0.8}Cu_{0.8}S_4Ti_{1.2}$	9.916	0.043	0.383	0.4 Cr + 0.6 Ti	0.8
$Cr_{0.2}Cu_{0.2}S_4Sn_{1.8}$	10.289	0.040	0.379	0.1 Cr + 0.9 Sn	0.2
$Cr_{0.6}Cu_{0.6}S_4Sn_{1.4}$	10.236	0.051	0.381	0.3 Cr + 0.7 Sn	0.6
$CuCr_{1.2}S_4Ti_{0.8}$	9.918	0.071	0.385	0.6 Cr + 0.4 Ti	1
$CuCr_{1.2}S_4Sn_{0.8}$	10.125	0.032	0.381	0.6 Cr + 0.4 Ti	1

*M = Ti or Sn
<u>1</u>. Structure Reports, <u>11</u>, 497; <u>13</u>, 241; <u>15</u>, 207; <u>17</u>, 417.

CHROMIUM DEUTERIUM ZIRCONIUM
$Cr_2D_{3.5}Zr$

V.A. JARTYS', V.V. BURNAŠEVA, N.V. FADEEVA, S. SOLOV'EV and K.N. SEMENENKO, 1980. Dokl. Akad. Nauk SSSR, <u>255</u>, 582-586.

Cubic, modified HfV_2D_4 type (1), Fd3m, a = 7.697 Å, Z = 8, D_X = 5.89. Neutron powder data at 200°C, R = 0.08 for 18 reflexions; Zr in 8(a) 1/8,1/8,1/8, Cr in 16(d) 1/2,1/2,1/2; 28 D in 96(g) 0.9350,0.9350,0.1288.

The D atoms occupy tetrahedral sites (D - 2 Zr = 2.06, 2.07, D - 2 Cr = 1.73 [1.77] Å).

1. Structure Reports, 46A, 67.

CHROMIUM SELENIUM SODIUM
$Cr_{1.15}Na_{0.34}Se_2$

I.   D. TIGCHELAAR, R.J. HAANGE, G.A. WIEGERS and C.F. van BRUGGEN, 1981. Mater. Res. Bull., 16, 729-739.

II.   G.A. WIEGERS, 1980. Physica, 99B, 151-165.

Rhombohedral, R3m, a = 3.617, c = 38.99 Å, c/a = 10.78, D_m = 4.81, Z = 6. R = 0.082, diffractometer data. 1.91 Na in 6(c) 0,0,0.3360; 5.60 Cr(1) in 6(c) 0,0,0.4239; 1.28 Cr(2) in 3(b) 0,0,1/2; Se(1) in 6(c) 0,0,0.0564; Se(2) in 2(c) 0,0,0.2039.

The structure contains $CrSe_2$ sandwiches; the Se atoms are so stacked that half the interstices between the layers are trigonal-prismatic and half are octahedral; the octahedral holes are occupied by the extra Cr atoms (Cr-Se = 2.54 Å) the trigonal prismatic holes by the Na atoms (Na-Se = 2.95, 3.11 Å). The Cr atoms within sandwiches occupy octahedral holes (Cr-Se = 2.48, 2.58 Å).

COBALT GALLIUM GERMANIUM   GERMANIUM IRON
$Co_{13}Ga_2Ge_6$ $Fe_{13}Ge_8$, $Fe_{3.2}Ge_2$

B. MALAMAN, J. STEINMETZ and B. ROQUES, 1980. J. Less-Common Metals, 75, 155-176.

Hexagonal, $P6_3/mmc$, Z = 1. Mo radiation, diffractometer data.

η-$Co_{13}Ga_2Ge_6$ (approx. composition = $Co_{62}Ge_{28}Ga_{10}$), a = 7.853, c = 4.999 Å, c/a = 0.64, D_m = 8.20. R = 0.061, 96 reflexions. Co(1) in 2(a) 0,0,0; Co(2) in 6(g) 1/2,0,0; 5 Co(3) in 6(h) 0.8386,-0.8386,1/4; 1.5 Ge + 0.5 Ga in 2(d) 2/3,1/3,1/4; 4.5 Ge + 1.5 Ga in 6(h) 0.1912,-0.1912,1/4.

η-$Fe_{13}Ge_8$ (approx. composition = $Fe_{56.5}Ge_{43.5}$), a = 7.976, c = 4.993 Å, c/a = 0.63, D_m = 7.77. R = 0.077, 73 reflexions. Fe(1) in 2(a) 0,0,0; Fe(2) in 6(g) 1/2,0,0; 5 Fe(3) in 6(h) 0.8385,-0.8385,1/4; Ge(1) in 2(d) 2/3,1/3,1/4; Ge(2) in 6(h) 0.1922, -0.1922,1/4.

β-$Fe_{3.2}Ge_2$ (approx. composition = $Fe_{61.5}Ge_{38.5}$), a = 3.998, c = 5.010 Å, c/a = 1.25, D_m = 7.85. R = 0.083, 52 reflexions. Fe(1) in 2(a) 0,0,0; 1.2 Fe(2) in 6(h) 0.3377, -0.3377,1/4; 2 Ge(2) in 6(h) 0.3682,-0.3682,3/4.

These structures are based on a stacking of trigonal prisms and square prisms, similar to those found in CoGe, Fe_6Ge_5 and Co_2Ge. Distances are summarised below.

$A_{13}B_8$	A - B (Å)	A - A (Å)	B - B (Å)
$Co_{13}Ga_2Ge_6$	2.34 - 2.89	2.50 - 2.62	3.16 - 3.35
$Fe_{13}Ge_8$	2.37 - 2.93	2.50 - 2.65	3.17 - 3.38

COBALT GERMANIUM LITHIUM   COBALT GERMANIUM YTTRIUM
Co_6Ge_6Li Co_6Ge_6Y

COBALT GERMANIUM ZIRCONIUM   GERMANIUM IRON MAGNESIUM
Co_6Ge_6Zr Fe_6Ge_6Mg

GERMANIUM LITHIUM NICKEL   GERMANIUM NICKEL SCANDIUM
Ge_6LiNi_6 Ge_6Ni_6Sc

LITHIUM NICKEL SILICON
$LiNi_6Si_6$

W. BUCHHOLZ and H.-U. SCHUSTER, 1981. Z. anorg. Chem., <u>482</u>, 40-48.

Hexagonal, P6/mmm, Mo radiation, single-crystal diffractometer data.

AB_6X_6	$a(\overset{\circ}{A})$	$c(\overset{\circ}{A})$	D_m	Z	R	reflex.
$LiCo_6Ge_6$	5.048	7.729	7.67	1	0.108	259
YCo_6Ge_6	5.074	3.908	8.17	1/2	0.057	96
$ZrCo_6Ge_6$	5.061	7.808	8.45	1	0.060	166
$MgFe_6Ge_6$	5.067	8.045	7.30	1	0.070	253
$LiNi_6Ge_6$	8.734	7.797	7.58	3	0.087	223
$ScNi_6Ge_6$	10.152	7.813	7.72	4	0.057	147
$LiNi_6Si_6$	8.461	7.566	5.50	3	0.072	214

A = Li, Y, Zr, Mg, Sc; B = Co, Fe, Ni; X = Ge, Si

Atomic positions: P6/mmm
A in 1(a), B in 6(i), X(1) in 2(e), X(2) in 2(c), X(3) in 2(d)

$LiCo_6Ge_6$	x	y	z	$ZrCo_6Ge_6$ x	y	z	$MgFe_6Ge_6$ x	y	z
A	0	0	0	0	0	0	0	0	0
B	1/2	0	0.2498	1/2	0	0.2491	1/2	0	0.2496
X(1)	0	0	0.3343	0	0	0.3418	0	0	0.3408
X(2)	1/3	2/3	0	1/3	2/3	0	1/3	2/3	0
X(3)	1/3	2/3	1/2	1/3	2/3	1/2	1/3	2/3	1/2

YCo_6Ge_6

	x	y	z
0.45A in 1(a)	0	0	0
B in 3(g)	1/2	0	1/2
X(1) in 2(c)	1/3	2/3	0
1.0 X(2) in 2(e)	0	0	0.307

$LiNi_6Ge_6$	x	y	z	$LiNi_6Si_6$ x	y	z
B(1) in 12(o)	0.1608	0.3216	0.247	0.1623	0.3246	0.2485
B(2) in 6(i)	1/2	0	0.260	1/2	0	0.2546
X(1) in 6(j)	0.339	0	0	0.331	0	0
X(2) in 6(k)	0.323	0	1/2	0.326	0	1/2
X(3) in 4(h)	1/3	2/3	0.339	1/3	2/3	0.341
X(4) in 2(e)	0	0	0.163	0	0	0.155
A(1) in 2(c)	1/3	2/3	0	1/3	2/3	0
A(2) in 1(b)	0	0	1/2	0	0	1/2

$ScNi_6Ge_6$

	x	y	z		x	y	z
B(1) in 12(n)	0.238	0	0.258	X(4) in 2(d)	1/3	2/3	1/2
B(2) in 12(o)	0.247	0.494	0.250	X(5) in 2(e)	0	0	0.340
X(1) in 6(ℓ)	0.160	0.320	0	X(6) in 6(i)	1/2	0	0.158
X(2) in 6(m)	0.162	0.324	1/2	A(1) in 1(a)	0	0	0
X(3) in 2(c)	1/3	2/3	0	A(2) in 3(g)	1/2	0	1/2

These structures are based on that of CoSn (1) and like that of $LiFe_6Ge_6$ (2) contain B_6 trigonal prisms centred by Ge atoms; [A atoms have CN = 20, B atoms have CN = 12, X atoms have CN = 12, 13, 14, 15; distances are summarised below].

[Interatomic distances (Å)]

AB_6X_6

	A – B	A – X	B – B	B – X	X – X
$LiCo_6Ge_6$	3.18	2.58 - 2.91	2.52	2.42 - 2.61	2.56 - 3.18
$ZrCo_6Ge_6$	3.19	2.67 - 2.92	2.53	2.43 - 2.63	2.47 - 3.17
$MgFe_6Ge_6$	3.23	2.74 - 2.93	2.53	2.48 - 2.64	2.56 - 3.19
YCo_6Ge_6	3.20	2.71 - 2.93	2.54	2.44 - 2.65	2.40 - 3.17
$LiNi_6Ge_6$	3.13 - 3.24	2.63 - 2.89	2.43 - 2.57	2.43 - 2.71	2.51 - 3.22
$ScNi_6Ge_6$	3.15 - 3.26	2.66 - 2.97	2.42 - 2.63	2.37 - 2.77	2.47 - 3.24
$LiNi_6Si_6$	3.05 - 3.13	2.58 - 2.83	2.38 - 2.48	2.34 - 2.60	2.35 - 3.10

1. Strukturbericht, 6, 4.
2. Structure Reports, 43A, 62.

COBALT IRON THORIUM
$(Co,Fe)_{17}Th_2$ $(Co_{8.52}Fe_{8.04}Th_{2.22})$

W.J. JAMES and P.E. JOHNSON, 1980. Rare Earths Mod. Sci. Technol., 2, 333-338.

Rhombohedral, modified Th_2Zn_{17} type (1, 2), R3̄m, a = 8.521, c = 12.441 Å, c/a = 1.46, Z = 1, D_x = 9.33, R = 0.035 for 15 reflexions, neutron powder data. 0.66 Th(1) in 3(a) 0,0,0; 6 Th(2) in 6(c) 0,0,0.350; 4.80 Fe in 6(c) 0,0,0.103; 0.066 Co in 6(c) 0,0,0.103; 3.87 Fe + 5.13 Co(1) in 9(d) 1/2,0,1/2; 8.28 Fe + 9.72 Co(2) in 18(f) 0.286,0,0; 7.38 Fe + 10.62 Co(3) in 18(h) 0.166,-0.166,0.488.

There has been considerable discussion (1) on the detailed structures of rhombohedral phases Ln_2T_{17} (Ln = lanthanide or actinide, T = transition metal). The term hyperstoichiometric (2) has been used and it is possible that these phases are a group of closely related structures and not a single clearly defined one.

1. Structure Reports, 23, 47; 31A, 46; 38A, 76, 113.
2. Ibid., 32A, 50; 43A, 49; 46A, 58.

COBALT LANTHANUM PHOSPHORUS
Co_5LaP_3

V.N. DAVYDOV and Ju.B. KUZ'MA, 1981. Dop. Akad. Nauk Ukr. RSR, 43, 81-84.

Orthorhombic, Cmcm, a = 3.651, b = 11.573, c = 11.459 Å, Z = 4, D_x = 7.22. Mo radiation, R = 0.054 for 306 reflexions, diffractometer data. La in 4(c) 0,0.3384[*], 1/4; Co(1) in 4(b) 0,1/2,0; Co(2) in 8(f) 0,0.0543,0.1429; Co(3) in 8(f) 0,0.3023, 0.5670; P(1) in 8(f) 0,0.1176,0.5393; P(2) in 4(c) 0,0.6117,1/4.
[*][Misprinted as 0.3884.]

The structure (Fig. 1) is characterised by tri-capped trigonal prisms centred by P atoms and is similar to UNi_5Si_3 (1). The coordination numbers are La = 23, Co = 12, P = 9 with distances La-Co = 3.22-3.66, Co-Co = 2.41-2.62, La-P = 3.07-3.20, Co-P = 2.16-2.38 and La-La = 3.65 Å.

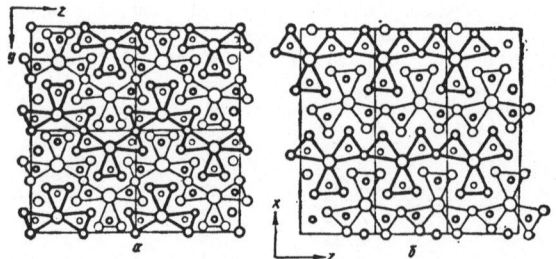

Fig. 1.   The LaCo$_5$P$_2$ structure (left) and the UNi$_5$Si$_3$ structure (right).

<u>1</u>. Structure Reports, <u>42</u>A, 112.

COBALT MANGANESE TIN
CoMnSn at 295 K

W. BAŻELA, A. SZYTUŁA and W. ZAJAC, 1981. Solid State Comm., <u>38</u>, 875-877.

Hexagonal, Ni$_2$In type (<u>1</u>), P6$_3$/mmc, a = 4.258, c = 5.360 Å, c/a = 1.26, Z = 2, D$_X$ =
9.17; at 80 K, a = 4.248, c = 5.344 Å. Fe radiation, powder diffractometer data.
1.508 Mn + 0.492 Co(1) in 2(a) 0,0,0; 0.492 Mn + 1.508 Co(2) in 2(d) 1/3,2/3,3/4; Sn
in 2(c) 1/3,2/3,1/4. See also <u>2</u>. [Distances are Sn-Co/Mn = 2.46, 2.80 Å.]

<u>1</u>. Structure Reports, <u>9</u>, 91.
<u>2</u>. Ibid., <u>17</u>, 171.

COBALT SILICON URANIUM
Co$_{51}$Si$_{33}$U$_{10}$

L.G. AKSEL'RUD, Ja.P. JARMOLJUK and E.I. GLADYŠEVSKIJ, 1980. Dop. Akad. Nauk Ukr.
RSR, <u>42</u>, 79-81.

Hexagonal, P6$_3$/m, a = 27.53, c = 3.678 Å, c/a = 0.13, Z = 2, D$_X$ = 8.68. Mo radiation,
R = 0.127 for 450 reflexions, diffractometer data. 32 site-sets are given.

    The structure (Fig. 1) belongs to a series based on [SiCo$_6$] trigonal prisms,
including Co$_5$Si$_3$U (<u>1</u>) and Co$_{30}$Si$_9$U$_6$ (<u>2</u>); distances are Co-Si = 2.14-2.54; Co-Co =
2.47-2.96; U-Co = 2.92-3.36; U-Si = 2.80-3.09; Si-Si = 2.46-2.66 Å.

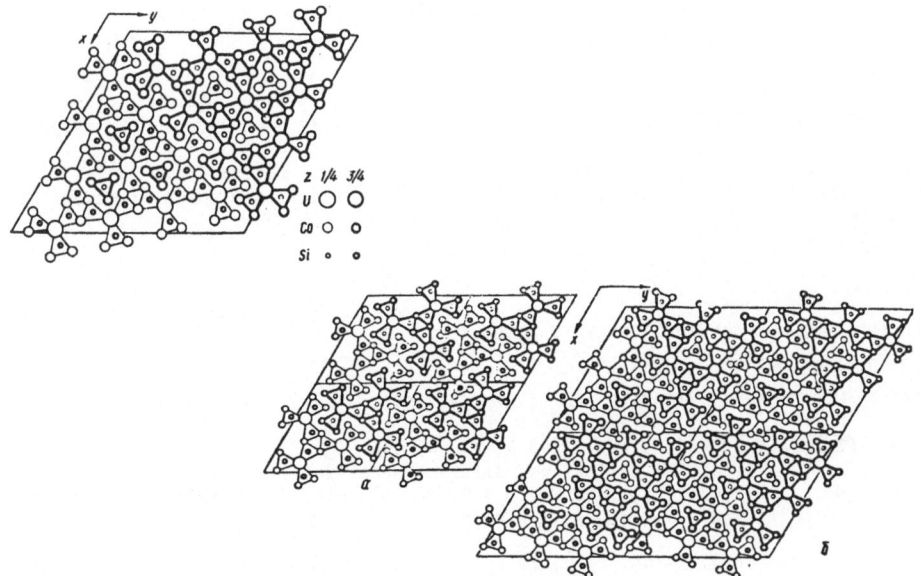

Fig. 1.   Projections of the $U_{10}Co_{51}Si_{33}$, UCo_5Si_3 and $U_6Co_{30}Si_{19}$ structures
           onto (010).

1. Structure Reports, 46A, 62.
2. Ibid., 45A, 58.

COPPER INDIUM
$Cu_{11}In_9$

T.P. RAJASEKHARAN and K. SCHUBERT, 1981. Z. Metallk., 72, 275-278.

Monoclinic, CuAl type (1), C2/m, a = 12.814, b = 4.3543, c = 7.353 Å, β = 54.49°, Z =
1, D_X = 8.61. Mo radiation, R = 0.087 for 436 reflexions, diffractometer data.
Cu(1) in 2(a) 0,0,0; Cu(2) in 4(i) 0.2553,0,0.7138; Cu(3) in 4(i) 0.1138,0,0.5477;
1 Cu + 1 In in 2(d) 0,1/2,1/2; In(1) in 4(i) 0.1494,0,0.1543; In(2) in 4(i) 0.3825,0,
0.2457. This is a disordered version of the type structure with distances Cu-In =
2.56-2.92, Cu-Cu = 2.56-2.75, In-In = 3.04-3.94 Å.

1. Structure Reports, 38A, 4.

COPPER LANTHANUM SULPHUR
$CuLaS_2$

M. JULIEN-POUZOL, S. JAULMES, A. MAZURIER and M. GUITTARD, 1981. Acta Cryst., B37,
1901-1903.

Monoclinic, $P2_1/b$, a = 6.646, b = 6.938, c = 7.325 Å, γ = 98.73°, D_m = 5.29, Z = 4.
Mo radiation, R = 0.063 for 999 reflexions, diffractometer data.

Atomic positions
All atoms in 4(e)

	x	y	z		x	y	z
La	0.6930	0.3024	0.0546	S(1)	0.4123	-0.0019	0.2280
Cu	0.0900	0.0695	0.1523	S(2)	0.9142	0.2204	0.3822

The structure is composed of two layers, one of $[CuS_4]$ tetrahedra (Cu-S = 2.34-2.54) and one of $[La_4S]$ tetrahedra (La-S = 2.89-3.08 Å; there are Cu-Cu distances of 2.64 Å.

COPPER IRON SULPHUR (SYNTHETIC NUKUNDAMITE)
$Cu_{3.39}Fe_{0.61}S_4$

A. SUGAKI, H. SHIMA, A. KITAKAZE and T. MIZOTA, 1981. Amer. Min., 66, 398-402.

Trigonal, P$\bar{3}$m1, a = 3.7830, c = 11.1950 Å, c/a = 2.96, D_m = 4.49, Z = 1. Mo radiation, R = 0.174 for 354 reflexions, diffractometer data. 1.695 Cu(1) + 0.305 Fe(1) in 2(d) 2/3,1/3,0.1547; 1.695 Cu(2) + 0.305 Fe(2) in 2(d) 2/3,1/3,0.5790; S(1) in 2(c) 0,0,0.0921; S(2) in 2(d) 2/3,1/3,0.3609.

The structure (Fig. 1) has S_2 pairs (S-S = 2.06) and Cu/Fe-S_4 tetrahedra (Cu/Fe-S = 2.29-2.44) with Cu/Fe-Cu/Fe = 2.81 Å.

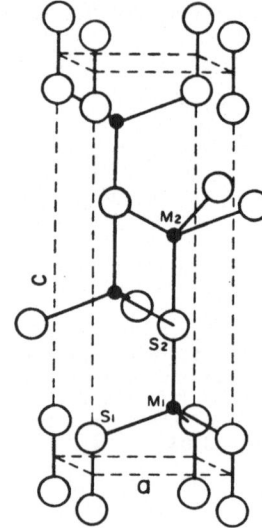

Fig. 1.   The nukundamite structure.

COPPER PHOSPHORUS TITANIUM
$Cu_{1-x}P_8Ti_{11+x}$ (x = 0.34)

W. CARRILLO-CABRERA and T. LUNDSTRÖM, 1981. Acta Chem. Scand., A35, 545-550.

Sample I, orthorhombic, Pbam, a = 18.596, b = 9.379, c = 3.4448 Å, Z = 2, D_x = 4.60. Mo radiation, R = 0.061 for 298 reflexions, diffractometer data. 11 site-sets are given. See also 1.

The structure is based on M_8 (M = Ti/Cu) cubes, centred mainly by Cu and M_6 trigonal prisms centred by P atoms. Distances are Ti-P = 2.42-2.70, Ti-Ti = 2.97-3.71, Cu-Ti = 2.74-2.79, Cu-Cu = 3.45, Cu-P = 2.49-2.76 Å.

1. Structure Reports, 45A, 119.

COPPER POTASSIUM TELLURIUM   COPPER SODIUM SULPHUR
Cu_3KTe_2 $Cu_4Na_2S_3$

G. SAVELSBERG and H. SCHÄFER, 1981. Mater. Res. Bull., 16, 1291-1297.

Monoclinic, C2/m, Z = 4. Mo radiation, diffractometer data.

	a(Å)	b(Å)	c(Å)	β(°)	D_m	R	refl.	type	
$Cu_4Na_2S_3$	15.63	3.86	10.33	107.6	4.45	0.089	317	$Ag_4K_2S_3$	(1)
Cu_3KTe_2	16.453	4.294	8.661	111.86	5.72*	0.052	926	Ag_3CsS_2	(2)

*D_x

Atomic positions
All atoms in 4(i) x,0,z

$Cu_4Na_2S_3$ Cu_3KTe_2

	x	z			x	z
Na(1)	0.8316	0.1542		K	0.1336	0.0446
Na(2)	0.6880	0.3474		Cu(1)	0.5843	0.6024
Cu(1)	0.5230	0.6365		Cu(2)	0.6959	0.4711
Cu(2)	0.9017	0.8938		Cu(3)	0.9304	0.5864
Cu(3)	0.0969	0.5141		Te(1)	0.7983	0.3061
Cu(4)	0.5085	0.8784		Te(2)	0.5213	0.2423
S(1)	0.3351	0.3833				
S(2)	0.9686	0.7182				
S(3)	0.6449	0.0432				

The structures are as described before (1, 2) with distances in $Na_2Cu_4S_4$: Na-S = 2.79-3.04, Na-Cu = 3.13-3.35, Na-Na = 3.42; Cu-S = 2.23-2.50, Cu-Cu = 2.58-3.04 Å. In KCu_3Te_2 the distances are K-Te = 3.54-3.65, K-Cu = 3.72, K-K = 4.17-4.70; Cu-Te = 2.57-2.90, Cu-Cu = 2.49-2.90 Å.

1. Structure Reports, 42A, 120.
2. Ibid., 44A, 43.

COPPER SCANDIUM SILICON
CuScSi

B.Ja. KOTUR, E.I. GLADYŠEVSKIJ and M. SIKIRICA, 1981. J. Less-Common Metals, 81, 71-78.

Orthorhombic phase, TiNiSi [Co_2Si] type (1), Pnma, a = 6.566, b = 3.976, c = 7.224 Å, Z = 4, D_x = 4.81. Mo radiation, R = 0.059 for 204 reflexions, diffractometer data.

Atomic positions
All atoms in 4(c) x,1/4,y

	x	z		x	z		x	z
Sc	0.5089	0.3046	Cu	0.1576	0.5665	Si	0.7702	0.6097

Hexagonal phase, Fe$_2$P type (2), P$\bar{6}$2m, a = 6.426, c = 3.922 Å, c/a = 0.61, Z = 3, D$_x$ = 4.85. Fe radiation, R = 0.1$\overline{4}$9, powder diffractometer data. Two other refinements for different atom positions also carried out, with less satisfactory results. Sc in 3(g) 0.574,0,1/2; Cu in 3(f) 0.241,0,0; Si(1) in 1(b) 0,0,1/2; Si(2) in 2(c) 1/3, 2/3,0.

Interatomic distances (Å)

Type	Cu - Cu	Cu - Sc	Cu - Si	Sc - Sc	Sc - Si
TiNiSi	3.03	2.85 - 3.10	2.41 - 2.56	3.38, 3.45	2.77, 2.84
Fe$_2$P	2.68	2.90 , 3.08	2.49, 2.50	3.32	2.74

1. Structure Reports, 30A, 75; 34A, 176.
2. Ibid., 23, 68; 39A, 76.

COPPER SELENIUM (SYNTHETIC KLOCKMANNITE)
α-CuSe

H. EFFENBERGER and F. PERTLIK, 1981. Neues Jb. Miner. Mh., 197-205.

Hexagonal, P6$_3$/mmc, a = 3.939, c = 17.25 Å, c/a = 4.38, D$_m$ = 6.03, Z = 6. Mo radiation, R = 0.043 for 121 reflexions, diffractometer data. 0.30 Cu(1) in 2(d) 2/3,1/3,1/4; 0.72 Cu(1) in 6(h) 0.751,0.502,1/4; 0.2 Cu(3) in 2(b) 0,0,1/4; 4 Cu(4) in 4(f) 1/3,2/3,0.1117; 0.50 Se(1) in 2(c) 1/3,2/3,1/4; 1.5 Se(2) in 6(h) 0.412, 0.824,1/4; 4 Se(2) in 4(e) 0,0,0.0685.

The unit cell agrees with the sub-cell reported by 1 but does not agree with that of 2. There is a supercell which was not solved; earlier, 3 reported no superstructure lines for freshly prepared samples but observed them in aged samples; most earlier workers observed superstructure lines (1, 2).

The structure (Fig. 1) has layers of ordered Cu$_2$Se$_2$ sheets alternating with disordered CuSe sheets. In the ordered sheets distances are Cu-Se = 2.39, 2.39, 2.45 Å and Se-Se 2.36; in the disordered sheets Cu-Se = 2.27-2.85 Å. The structure differs from those previously proposed (4).

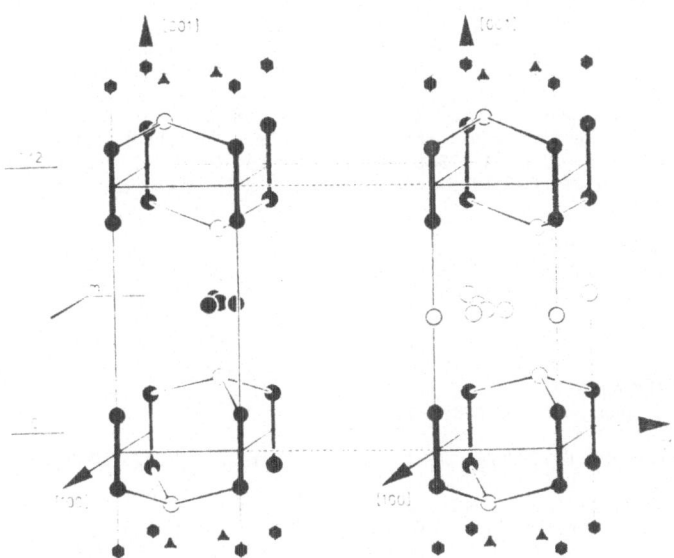

Fig. 1.   The klockmannite structure. a: A perspective view showing the disordered
           Se atoms (Se(1),(2)). b: the same view showing the disorder of the Cu
           atoms (Cu(1),(2),(3)). Cu atoms: open circles, Se atoms: solid circles.

<u>1</u>. Structure Reports, <u>11</u>, 255; <u>12</u>, 86; <u>18</u>, 380; <u>42</u>A, 75.
<u>2</u>. Ibid., <u>24</u>, 138; <u>27</u>, 183.
<u>3</u>. Ibid., <u>42</u>A, 75.
<u>4</u>. Ibid., <u>18</u>, 380; <u>27</u>, 183.

COPPER SILICON SULPHUR
Cu_8S_6Si

M. LEVALOIS and G. ALLAIS, 1981. Acta Cryst., B<u>37</u>, 1816-1819.

Orthorhombic, β'-Ag_8GeSe_6 type (<u>1</u>), $Pmn2_1$, a = 6.9928, b = 6.9000, c = 9.7723 Å, D_m
= 5.05, Z = 2. Mo radiation, R = 0.031 for 205 reflexions, diffractometer data.
11 site-sets are given. The structure is as described (<u>1</u>) with Si-S = 2.08-2.16 and
Cu-S = 2.24-2.94 Å.

<u>1</u>. Structure Reports, <u>46</u>A, 83.

DEUTERIUM IRIDIUM STRONTIUM
D_5IrSr_2

JIAN ZHUANG, J.M. HASTINGS, L.M. CORLISS, R. BAU, CHIAU-YU WEI and R.O. MOYER, 1981.
J. Solid State Chem., <u>40</u>, 352-360.

R.T. phase, cubic, Fm3m, a = 7.6464 Å, Z = 4, D_x = 5.60.   R = 0.0344, neutron powder data.   Sr in 8(c) 1/4,1/4,1/4; Ir in 4(a) 0,0,0; 20.88 D in 24(e) 0.2242,0,0. See also 1.   L.T. phase at 4.2 K, tetragonal, I4/mmm, a = 5.320, c = 7.796 Å, Z = 2, D_x = 5.68.   R = 0.0494, neutron powder data.   Sr in 4(d) 0,1/2,1/4; Ir in 2(a) 0,0,0; 7.68 D(1) in 8(h) 0.225,0.225,0; 2.8 D(2) in 4(e) 0,0,0.234.

The room temperature structure is as previously reported (1) and illustrated (2) with Ir-D = 1.714 Å; the tetragonal phase has the IrD_6 "octahedra" distorted with Ir-4 D = 1.693 and Ir-D = 1.82 Å.   Both structures can be considered to be built from IrD_5 square pyramids, 6-fold disordered in the R.T. phase and 2-fold disordered in the L.T. phase.

1.   Structure Reports, 37A, 83.
2.   Ibid., 46A, 72. Fig. 1.

DEUTERIUM MAGNESIUM NICKEL
$D_{3.9}Mg_2Ni$ (h.t. β' phase at 280°C and 22 bar D_2)

K. YVON, J. SCHEFER and F. STUCKI, 1981.   Inorg. Chem., 20, 2776-2778.

Cubic, defect D_6RuSr_2 type (1), Fm3m, a = 6.507 Å, Z = 4, D_x = 2.77.   R = 0.056 for 11 reflexions, diffractometer data.   Ni in 4(a) 0,0,0; Mg in 8(c) 1/4,1/4,1/4; ∿16 D in 24(e) 0.229,0,0.

The structure can be described as a defect D_6RuSr_2 type or a filled CaF_2 type (2, 3).   The maximum H(D) concentration in $M_2TH_{\sim6}$ phases appears to be governed by the valency of the T (transition metal) atoms.   The D atoms partially occupy sites which form octahedra around the Ni atoms, distances Ni - 6 D = 1.49 Å; Mg - 12 D = 2.305 Å.   The possibility of local $[NiD_4]^{4-}$ configurations averaging to an "octahedral" arrangement cannot be ruled out.

1.   Structure Reports, 37A, 83.
2.   Ibid., 46A, 72.
3.   Ibid., 43A, 69.

DEUTERIUM MANGANESE YTTRIUM   MANGANESE YTTRIUM
$D_{8.3}Mn_{23}Y_6, D_{18}Mn_{23}Y_6, D_{23}Mn_{23}Y_6$ $Mn_{23}Y_6$

I.   M. COMMANDRE, D. FRUCHART, A. ROUAULT, D. SAUVAGE, C.B. SHOEMAKER and D.P. SHOEMAKER, 1979.   J. de Phys.-Lett., 40, L639-L642.

II.   S.K. MALIK, T. TAKASHITA and W.E. WALLACE, 1977.   Solid State Comm., 23, 599-602.

I.   Cubic, Th_6Mn_{23} type (1), Fm3m, Z = 4.   Photographic data.   Mn(1) in 4(b) 1/2,1/2, 1/2; Mn(2) in 24(d) 0,1/4,1/4; Mn(3) in 32(f) x,x,x; Mn(4) in 32(f) x,x,x; Y in 24(e)* x,0,0; D(1) in 4(a) 0,0,0; D(2) in 32(f) x,x,x; D(3) in 96(j) y,0,z; D(4) in 96(k) x,x,z.   See also 2 and Fig. 1.
*[Misprinted as 32(f).]

	$Mn_{23}Y_6$	$D_{8.3}Mn_{23}Y_6$	$D_{18}Mn_{23}Y_6$	$D_{23}Mn_{23}Y_6$
a(Å)	12.5063	12.7853	12.799	12.840[†]
R	0.076	0.056	0.05	0.048
c°	250	307	R.T.	R.T.
x (Mn(3))	0.382	0.381	0.375	0.377
x (Mn(4))	0.181	0.181	0.184	0.186
x (Y)	0.206	0.228	0.232	0.235
x (D(2))	-	0.103	0.094	0.101
y (D(3))	-	-	0.173	0.167
z (D(3))	-	-	0.346	0.370
x (D(4))	-	-	-	0.153
z (D(4))	-	-	-	0.01

[†]$H_{25}Mn_{23}Y_6$ a = 12.842 (II)

Occupancies, in atoms per site

Atom Site		$D_{8.3}Mn_{23}Y_6$	$D_{18}Mn_{23}Y_6$	$D_{23}Mn_{23}Y_6$
D(1)	4(a)	2.5	4	4
D(2)	32(f)	30.8	32	32
D(3)	96(j)	-	36	38.4
D(4)	96(k)	-	-	17.28

Short Mn-Mn distances (Å)

	$Mn_{23}Y_6$	$D_{8.3}Mn_{23}Y_6$	$D_{18}Mn_{23}Y_6$	$D_{23}Mn_{23}Y_6$
Mn 1 - Mn 3	2.56	2.64	2.77	2.74
Mn 2 - Mn 3	2.76	2.82	2.77	2.80
Mn 2 - Mn 4	2.57	2.63	2.64	2.66
Mn 3 - Mn 3	2.95	3.04	3.20	3.16
Mn 3 - Mn 4	2.75	2.79	2.67	2.71
Mn 4 - Mn 4	2.44	2.50	2.39	2.32

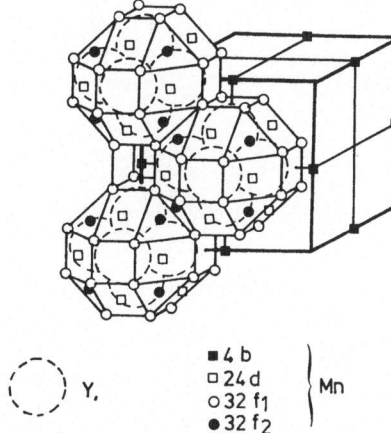

■ 4 b
□ 24 d   ⎫
○ 32 f₁ ⎬ Mn
● 32 f₂ ⎭

Y.

Fig. 1.   The face centred cubic arrangement of $Mn_{44}Y_6$ blocks in the $Mn_{23}Y_6$ structure.

1. Structure Reports, 16, 113.
2. Ibid., 30A, 154.

DEUTERIUM MANGANESE THORIUM
$D_{16.2}Mn_{23}Th_6$

K. HARDMAN, J.J. RHYNE, K. SMITH and W.E. WALLACE, 1980. J. Less-Common Metals, 74, 97-102.

Cubic, Fm3m, a = 12.922 Å, Z = 4, D_x = 8.22; the undeuterated cell has a = 12.518 Å.
R = 0.088, neutron powder data. Distances are Mn-Mn = 2.59- 2.95, Mn-Th = 3.13-3.70,
D-Mn = 1.71-2.56, D-Th = 2.36, 2.61 Å.

Atomic positions

				x	y	z
	Mn(1)	in	4(b)	1/2	1/2	1/2
	Mn(2)	in	24(d)	0	1/4	1/4
	Mn(3)	in	32(f)	0.179	0.179	0.179
	Mn(4)	in	32(f)	0.368	0.368	0.368
	Th	in	24(e)	0.214	0	0
0.4	D(1)	in	4(a)	0	0	0
	D(2)	in	32(f)	0.102	0.102	0.102
32.4	D(3)	in	48(i)	1/2	0.14	0.14

ERBIUM GALLIUM NICKEL
$Er_6Ga_{1.5}Ni_{2.5}$

Ju.M. GRIN', Ja.P. JARMOLJUK and E.I. GLADYŠEVSKIJ, 1980. Dop. Akad. Nauk Ukr. RSR,
42, 78-83.

Tetragonal, Y_3Rh_2 type (1), I4/mcm, a = 11.42, c = 24.47 Å, c/a = 2.14, Z = 14, [D_x =
9.14]. Fe radiation, R = 0.119 for 136 reflexions, powder diffractometer data. 11
site-sets given, with occupancies of Ga 0.1, Ni 0.9 in Ga/Ni(1)-(5).

In this structure the Er atoms are 14-, 15- and 17-coordinated with Er-Er =
3.23-3.91, Er-Ga = 2.94-3.33 and Er-Ni/Ga = 2.40-3.68; the Ga atoms are 10-coordinated;
the Ni/Ga atoms are 8-, 9- or 10-coordinated with Ni/Ga-Ni/Ga = 2.94, 3.18 Å. The
formula belongs to the series $R_{12m + 10n + 10p}$ $X_{6m + 8n + 6p}$ with m = 2, n = 4, p = 2
for $Er_6Ni_{1.5}Ga_{2.5}$ and m = 2, n = 4, p = 6 for $Pu_{31}Pt_{20}$ (2).

1. Structure Reports, 42A, 121.
2. Ibid., 43A, 92.

ERBIUM GALLIUM TITANIUM
$ErGa_4Ti_2$

Ju.M. GRIN', I.S. GAVRILENKO, V.Ja. MARKIV and Ja.P. JARMOLJUK, 1980. Dop. Akad.
Nauk Ukr. RSR, 42, 73-76.

Tetragonal, $YbMo_2Al_4$ type (1), I4/mmm, a = 6.706, c = 5.470 Å, c/a = 0.82, Z = 2,
[D_x = 7.32]. R = 0.10 for 51 reflexions, powder diffraction data. Er in 2(a) 0,0,0;
Ti in 4(d) 0,1/2,1/4; Ga in 8(h) 0.300,0.300,0. Er has 8 Ti at 3.62 and 12 Ga at
2.84, 3.33 Å; Ti has 4 Er at 3.62, 8 Ga at 2.78 and 2 Ti at 2.74 Å; Ga has 3 Er at
2.84, 3.33, 4 Ga at 2.69, 2.89 and 4 Ti at 2.78 Å.

1. Structure Reports, 42A, 9.

EUROPIUM GALLIUM SELENIUM
$EuGa_2Se_4$

R. RIMET, R. ROQUES, J.V. ZANCHETTA, J.P. DECLERQ and G. GERMAIN, 1981. Rev. Chim.
Minér., 18, 277-285.

Orthorhombic, $PbGa_2Se_4$ type (1), Fddd, a = 21.579, b = 21.336, c = 12.736 Å, Z = 32, D_X = 5.50. Mo radiation, R = 0.075 for 985 reflexions, diffractometer data. See also 2.

Atomic positions
Eu(1) in 16(e), Eu(2) in 8(a), Eu(3) in 8(b), remainder in 32(h)

	x	y	z		x	y	z
Eu(1)	0.8769	5/8	1/8	Se(1)	0.0014	0.5932	0.2508
Eu(2)	5/8	5/8	1/8	Se(2)	0.7510	0.5813	0.0148
Eu(3)	1/8	5/8	1/8	Se(3)	0.5850	0.7502	0.2486
Ga(1)	0.7506	0.4886	0.1267	Se(4)	0.9148	0.4998	0.0012
Ga(2)	0.7502	0.6956	0.4158				

The Eu atoms centre square antiprisms of Se atoms (Eu-Se = 3.18-3.24); Ga atoms centre Se_4 tetrahedra (Ga-Se = 2.37-2.43 Å). The Se atoms centre deformed Eu_2Ga_2 tetrahedra.

1. Structure Reports, 45A, 69.
2. Ibid., 37A, 156.

EUROPIUM NICKEL PHOSPHORUS
$EuNi_2P_2$

W. JEITSCHKO and B. JABERG, 1980. J. Solid State Chem., 35, 312-317.

Tetragonal, $ThCr_2Si_2$ type (1), I4/mmm, a = 3.938, c = 9.469 Å, c/a = 2.404, Z = 2, [D_X = 7.49]. Mo radiation, R = 0.049 for 118 reflexions, diffractometer data. Eu in 2(a) 0,0,0; Ni in 4(d) 0,1/2,1/4; P in 4(e) 0,0,0.3748.

Interatomic distances (Å)
Values for $EuCo_2P_2$ in brackets.

Eu - 8 P	3.03 (3.13)	Ni - 4 P	2.30 (2.23)	P - 4 Ni	2.30 (2.23)
- 8 Ni	3.08 (3.40)	- 4 Ni	2.79 (2.66)	- 4 Eu	3.03 (3.13)
		- 4 Eu	3.08 (3.40)	- 1 P	2.37 (3.27)

Contrary to $EuCo_2P_2$ (2), nominally isostructural, there are distinct P-P pairs in $EuNi_2P_2$ (Fig. 1) leading to the nominal formulation: $Eu^{2+}Ni^{+1}Ni^{+1}[P_2]^{4-}$. More detailed study indicates a valence for Eu between 2 and 3. Such an intermediate valence has been confirmed for related compounds by Mossbauer and X-ray photo-emission studies.

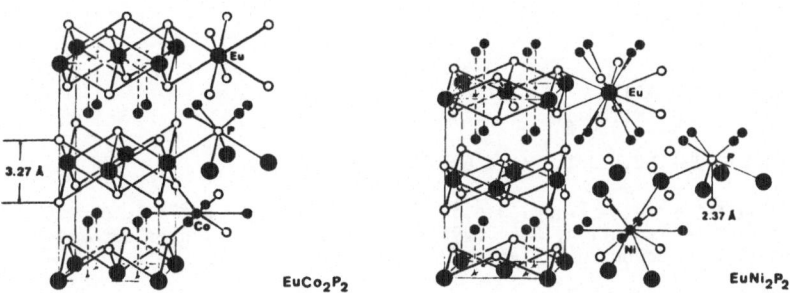

Fig. 1. The structures of $EuCo_2P_2$ (left) and $EuNi_2P_2$ (right) showing the differences in near-neighbour environments and the P-P pairs.

1. Structure Reports, 43A, 99.
2. Ibid., 44A, 50.

GADOLINIUM NICKEL SILICON
Gd_3NiSi_2

K. KLEPP and E. PARTHÉ, 1981. Acta Cryst., B37, 1500-1504.

Orthorhombic, Pnma, a = 11.398, b = 4.155, c = 11.310 Å, Z = 4, D_X = 7.27. Mo
radiation, R = 0.059 for 542 reflexions, diffractometer data.

Atomic positions
All atoms in 4(c) x,1/4,z

	x	z		x	z		x	z
Gd(1)	0.3814	0.4403	Gd(3)	0.2137	0.6976	Si(1)	0.473	0.685
Gd(2)	0.0576	0.3750	Ni	0.1285	0.1334	Si(2)	0.303	0.005

The structure (Fig. 1) can be considered as a filled version of the Hf_3P_2
structure (1) with the Gd atoms forming trigonal prisms centred by Ni and Si atoms,
which are 9-coordinated having 3 extra neighbours of Ni or Si. The average prism
distances are Ni-Gd = 2.92, Si-Gd = 3.01 while the Ni-Si distances are 2.45 and
2.46 Å.

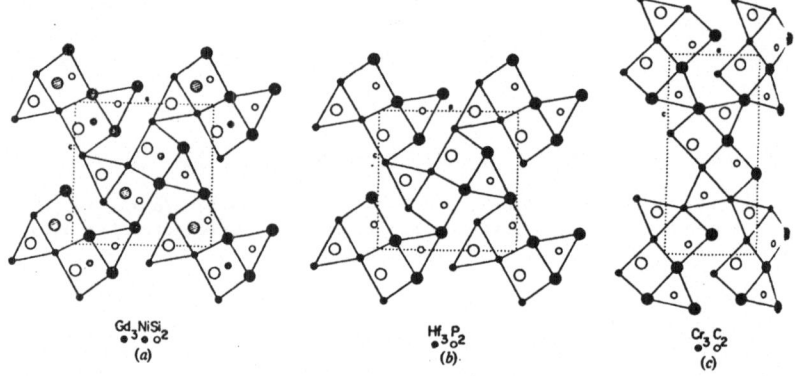

Fig. 1.   Projections of the crystal structures of (a) Gd_3NiSi_2, (b) Hf_3P_2 and
            (c) Cr_3C_2 along [010]. Atoms at y = 3/4 are indicated by large circles,
            those at y = 1/4 by small circles. The origin of the cell of Hf_3P_2 has
            been shifted by 1/2,0,0 to correspond with the origin of the unit cell
            chosen for Gd_3NiSi_2.

1. Structure Reports, 33A, 93.

GADOLINIUM NICKEL TERBIUM
$Gd_{0.4}NiTb_{0.6}$ $Gd_{0.2}NiTb_{0.8}$

K. KLEPP and E. PARTHÉ, 1981. Acta Cryst., B37, 495-499.

$Gd_{0.4}Tb_{0.6}Ni$, monoclinic, $P2_1/m$, a = 32.49, b = 4.226, c = 5.478 Å, β = 101.9°, Z =
18, D_X = 8.81. Mo radiation, R = 0.091 for 1144 reflexions, diffractometer data.
18 site-sets, with occupancies, given.

$Gd_{0.2}Tb_{0.8}Ni$, monoclinic, $P2_1/m$, a = 10.67, b = 4.195, c = 5.457 Å, β = 97.2°, Z = 6,
D_X = 8.93. Mo radiation, R = 0.071 for 244 reflexions, diffractometer data.

Atomic positions
All atoms in 2(e) x,1/4,z

	x	z		x	z
Gd/Tb*(1)	0.1198	0.8948	Ni(1)	0.025	0.380
Gd/Tb*(2)	0.4539	0.742	Ni(2)	0.357	0.211
Gd/Tb*(3)	0.7869	0.5833	Ni(3)	0.690	0.053
*20% Gd 80% Tb					

Both structures are stacking variants of FeB - CrB such as described in $\underline{1}$ for GdNi - DyNi. $Gd_{0.2}Tb_{0.8}Ni$ has the stacking sequence h_2c while $Gd_{0.4}Tb_{0.6}Ni$ has the sequence $h_2c_3h_2c_2$. The geometry is as described in $\underline{1}$ with distances Ni-Ni = 2.56-2.58, Gd/Tb-Gd/Tb = 3.57-3.78 and Gd/Tb-Ni = 2.86-2.$\overline{96}$ Å.

$\underline{1}$. Structure Reports, $\underline{46A}$, 75.

GALLIUM GERMANIUM NICKEL
$Ga_3Ge_6Ni_{13}$

G. NOVER and K. SCHUBERT, 1981. Z. Metallk., $\underline{72}$, 26-29.

Hexagonal, $P3_121$, a = 7.8487, c = 15.036 Å, c/a = 1.92, Z = 3, D_X = 8.74. Mo radiation, R = 0.082 for 1621 reflexions, diffractometer data. 13 site-sets are given. See also $\underline{1}$ and $\underline{2}$; $\underline{2}$ gave $GaGe_2Ni_4$ based on photographic data; present results are not significantly different for the atomic arrangement.

$\underline{1}$. Structure Reports, $\underline{39A}$, 115.
$\underline{2}$. Ibid., $\underline{42A}$, 82.

GALLIUM GERMANIUM SILVER SULPHUR
$AgGaGeS_4$

E.A. POBEDIMSKAJA, L.L. ALIMOVA, N.V. BELOV and V.V. BADIKOV, 1981. Dokl. Akad. Nauk SSSR, $\underline{257}$, 611-614 [Soviet Physics-Doklady, $\underline{26}$, 259-260].

Orthorhombic, Fdd2, a = 12.028, b = 22.918, c = 6.874 Å, Z = 12, D_X = 3.97. Mo radiation, R = 0.04 for 1700 reflexions, diffractometer data.

Atomic positions

	Site	x	y	z
*Ga/Ge(1)	8(a)	0	0	0
*Ga/Ge(2)	16(b)	0.12058	0.11454	0.2762
12 Ag	16(b)	0.08309	0.20925	0.7307
S(1)	16(b)	0.20150	0.17470	0.0564
S(2)	16(b)	0.15253	0.02253	0.1750
S(3)	16(b)	0.19037	0.11881	0.5730
*Occupancy 50:50				

The structure contains Ge/GaS_4 tetrahedra sharing edges (Ge/Ga - 4 S = 2.21-2.26) in an arrangement virtually identical to that of GeS_2 (C44 type) ($\underline{1}$). The Ag atoms occupy tetrahedral interstices (Ag - 4 S = 2.56-2.77 Å).

$\underline{1}$. Strukturbericht, 4, 11. [Compare the differing descriptions in Structure Reports $\underline{41A}$, 72; $\underline{42A}$, 81.]

GALLIUM HOLMIUM
δ-Ga$_3$Ho

M.M. CARNASCIALI, S. CIRAFICI and E. FRANCESCHI, 1981. J. Less-Common Metals, <u>81</u>, 115-119.

Hexagonal, β-PuAl$_3$ type (<u>1</u>), P6$_3$/mmc, a = 6..084, c = 14.061 Å, c/a = 2.31, Z = 6, D$_X$ = 8.26. Mo radiation, R = 0.084 for 119 reflexions, diffractometer data. Ho(1) in 2(b) 0,0,1/4; Ho(2) in 4(f) 1/3,2/3,0.0924; Ga(1) in 6(h) 0.5215,0.0430,1/4; Ga(2) in 12(k) 0.8325,0.6650,0.0809. The atoms are 12-coordinated with Ho-Ga = 2.96-3.05, Ga-Ga = 2.65-3.06 Å.

<u>1</u>. Structure Reports, <u>21</u>, 19.

GALLIUM LITHIUM
Ga$_4$Li$_5$

J. STÖHR and H. SCHÄFER, 1981. Z. anorg. Chem., <u>474</u>, 221-225.

Trigonal, P$\bar{3}$m1, a = 4.375, c = 8.257 Å, c/a = 1.89, Z = 1, D$_X$ = 3.80. Mo radiation, R = 0.078 for 202 reflexions, diffractometer data. Ga(1) in 2(c) 0,0,0.1671; Ga(2), Li(1) and (2) in 2(d) 1/3,2/3,z, z = 0.2765,0.612,0.945* respectively; Li(3) in 1(b) 0,0,1/2. *[Misprinted as 0.954.]

   The structure (Fig. 1) is based on b.c.c. packing with all atoms 14 = 6 + 8 coordinated; distances are Ga-Ga = 2.68, 2.76, Ga-Li = 2.69-3.13; Li-Li = 2.69-3.13 Å. LiGa (<u>1</u>) and Li$_3$Ga$_2$ (<u>2</u>) are also similar.

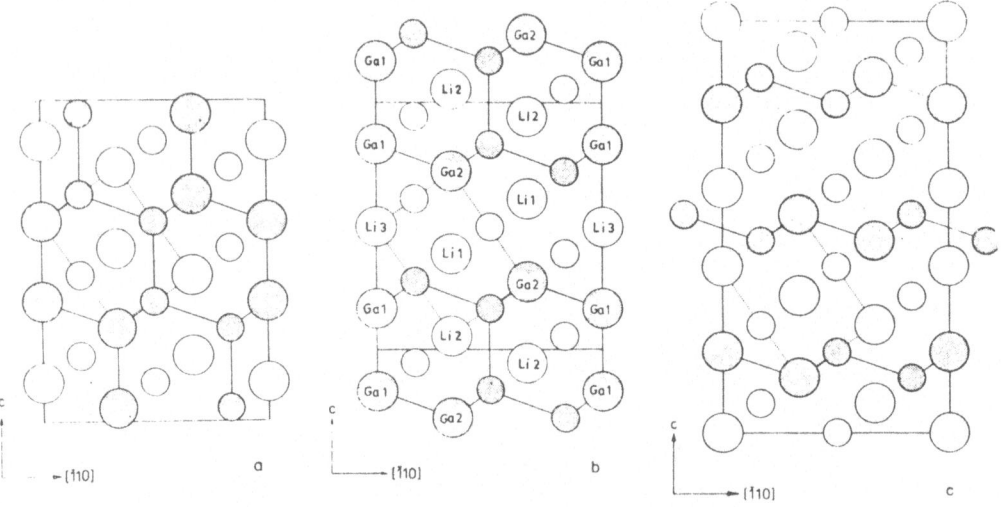

Fig. 1.   Projections of the LiGa (a), Li$_5$Ga$_4$ (b) and Li$_3$Ga$_2$ (c) structures.

<u>1</u>. Strukturbericht, <u>3</u>, 19, 267.
<u>2</u>. Structure Reports, <u>44A</u>, 62.

GALLIUM MANGANESE SULPHUR
α-Ga$_2$MnS$_4$

R. RIMET, R. BUDER, C. SCHLENKER, R. ROQUES and J.V. ZANCHETTA, 1981. Solid State
Comm., 37, 693-697.

Monoclinic, $MgGa_2S_4$ type (1), C2/c, a = 12.746, b = 22.609, c = 6.394 Å, β = 108.78°,
Z = 12. Mo radiation, R = 0.11 for 1413 reflexions, diffractometer data. 12 site-
sets are given. See also 2. The structure is based on a close packed S array with
Mn occupying octahedral sites (Mn-S = 2.52-2.65) and Ga in tetrahedral sites (Ga-S =
2.19-2.36 Å).

1. Structure Reports, 32A, 65.
2. Ibid., 37A, 157; 39A, 115.

GALLIUM NICKEL ZINC
Ga_3Ni_3Zn, $Ga_5Ni_8Zn_{36}$

N. SARAH, T. RAJASEKHARAN and K. SCHUBERT, 1981. Z. Metallk., 72, 732-735.

Ga_3Ni_3Zn, cubic, Ga_4Ni_3 type (1), Ia3d, a = 11.4302 Å, Z = 16, D_X = 8.01. Mo
radiation, R = 0.027 for 802 reflexions, diffractometer data. Ni in 48(g) 1/8,
0.3666,0.8834; Zn in 16(a) 0,0,0; Ga in 48(f) 0.0124,0,1/4.

$Ga_5Ni_8Zn_{36}$, cubic, P$\bar{4}$3m, a = 8.8483 Å, Z = 1, D_X = 7.60. Mo radiation, R = 0.056
for 761 reflexions, diffractometer data. Ni(1), (2) and Ga(1) in 4(e) x,x,x, x =
0.3290, 0.8254 and 0.6066 respectively; Zn(1) in 6(g) 1/2,1/2,0.8583; Zn(2) in 6(f)
0,0,0.323; Zn(3) and (4) in 12(i) x,x,z, x = 0.2959, 0.8137 respectively; z = 0.0517,
0.5438 respectively; Ga(2) in 1(a) 0,0,0. [Atoms are 10, 12, 13 or 15 coordinated
with Ga-Ga = 2.65, Ga-Ni = 2.68, Ga-Zn = 2.86, Ni-Zn = 2.49-2.71 and Zn-Zn = 2.51-
3.13 Å.]

1. Structure Reports, 34A, 137.

GALLIUM RUBIDIUM
Ga_7Rb Ga_3Rb

I.   C. BELIN, 1981. Acta Cryst., B37, 2060-2062.

II. R.G. LING and C. BELIN, 1981. Z. anorg. Chem., 480, 181-185.

I. Ga_7Rb, monoclinic, C2/m, a = 11.432, b = 6.603, c = 10.259 Å, β = 111.85°, Z = 4,
D_X = 5.29. Mo radiation, R = 0.058 for 584 reflexions, diffractometer data.

Atomic positions
Ga(1) and (5) in 8(j), remainder in 4(i)

	x	y	z		x	y	z
Ga(1)	0.1193	0.3019	0.5561	Ga(4)	0.4561	0	0.8679
Ga(2)	0.1823	0	0.4438	Ga(5)	-0.0073	-0.2095	0.2684
Ga(3)	0.2169	0	0.7321	Rb	0.1949	0	0.0842

The structure is composed of Ga icosahedra (Ga-Ga = 2.54-2.83 Å) linked in an
open array; the Rb atoms occupy channels (Rb-Ga = 3.71-3.84 Å) and form zig-zag
chains (Rb-Rb = 4.14 Å). [This structure has recently been revised (1).]

II. Ga_3Rb, tetragonal, I$\bar{4}$m2, a = 6.315, c = 15.000 Å, c/a = 2.38, Z = 6, D_X = 4.91.
Mo radiation, R = 0.049 for 159 reflexions, diffractometer data. Rb(1) in 2(a)
0,0,0; Rb(2) in 4(f) 0,1/2,0.3720; Ga(1) in 8(i) 0.2068,0,0.2229; Ga(2) in 8(i)
0.3067,0,0.3917; Ga(3) in 2(b) 0,0,1/2.

The structure (Fig. 1) contains Ga_8 dodecahedra (Ga-Ga = 2.44-2.81) linked to form a network (Ga-Ga = 2.53, 2.61 Å). The Rb atoms form an interpenetrating network (Rb-Rb = 3.66-3.70, Rb-Ga = 3.59-3.76 Å).

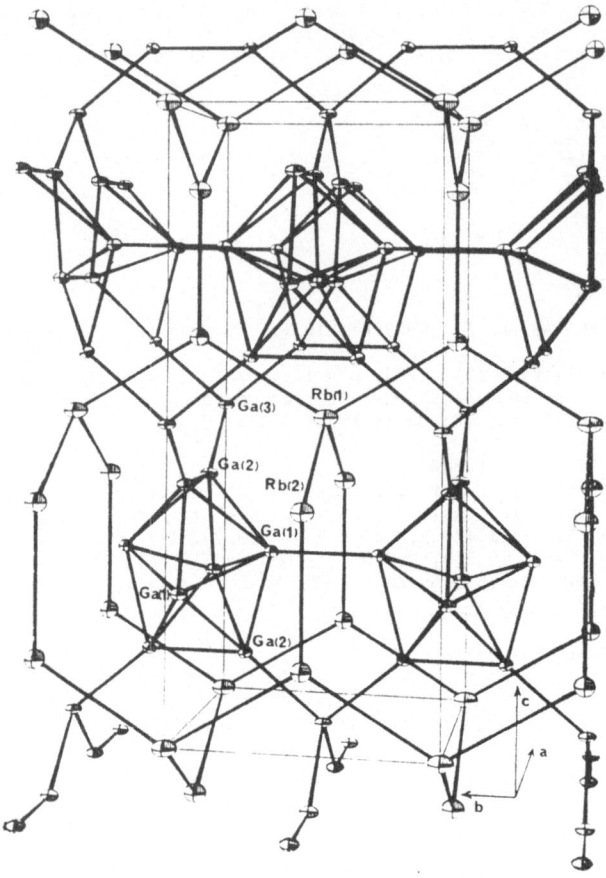

Fig. 1.   The $RbGa_3$ structure.

1.   R.E. MARSH and F.H. HERBSTEIN, 1983. Acta Cryst., B<u>39</u>, 280.

GALLIUM SCANDIUM
Ga_2Sc

I. N.N. BELJAVINA and V.Ja. MARKIV, 1980. Dop. Akad. Nauk Ukr. RSR, <u>42</u>, 87-89.

Orthorhombic, KHg_2 ($CeCu_2$) type (<u>1</u>), Imma, a = 4.140, b = 6.614, c = 7.914 Å, D_m = 5.60, Z = 4. R = 0.122 for 42 reflexions, photographic data. Sc in 4(e) 0,1/4, 0.557; Ga in 8(h) 0,0.043,0.159. [Note setting used. The distances are Sc - 4 Sc = 3.43, 3.69; Sc - 12 Ga = 2.95-3.44; Ga - 6 Sc = 2.95-3.44; Ga - 4 Ga = 2.52-2.74 Å.]

II. S.V. POPOVA, L.N. FOMIČEVA and V.G. PUTRO, 1980. Izv. Akad. Nauk SSSR, Neorg. Mater., <u>16</u>, 1563-1567 [Inorg. Mater., <u>16</u>, 1065-1068].

Orthorhombic, $ZrGa_2$ type (2), Cmmm, a = 13.31, b = 4.098, c = 4.242 Å, Z = 4, D_x = 5.29. Cu radiation, $\bar{R}$ = 0.19, photographic data. Sc in 4(g) 0.36,0,0; Ga(1) in 2(a) 0,0,0; Ga(2) in 2(c) 1/2,0,1/2; Ga(3) in 4(h) 0.20,0,1/2. Atoms are 12-coordinated with Sc-Sc = 3.57-3.73, Ga-Ga = 2.44-3.40, Ga-Sc = 2.77-3.00 Å.

1. Structure Reports, 19, 231; 26, 107; 39A, 87; 45A, 56.
2. Ibid., 27, 205.

GALLIUM SELENIUM SILVER
α-Ag_9GaSe_6

J.-P. DELOUME and R. FAURE, 1981. J. Solid State Chem., 36, 112-117.

Cubic, F$\bar{4}$3m, a = 11.126 Å, Z = 4, D_x = 7.30. Mo radiation, R = 0.063 for 103 reflexions, diffractometer data. Ga in 4(b) 1/2,1/2,1/2; Se(1) in 4(a) 0,0,0; Se(2) in 4(c) 1/4,1/4,1/4; Se(3) in 16(e) 0.623,0.623,0.623; 12 Ag(1) in 96(i) -0.160, 0.073,0.129; 12 Ag(2) in 48(h) 0.232,0.232,0.029; 12 Ag(3) in 48(h) -0.169,-0.169, 0.031.

As in β-Ag_9GaSe_6 (1) the structure is based on $GaSe_4$ tetrahedra (Ga - 4 Se = 2.37 Å); Ag atoms are statistically distributed, Ag(1) and (2) being close to the centres of Se_3 triangles (Ag-Se = 2.43-2.85 Å) whereas Ag(3) is tetrahedrally coordinated by Se (Ag-Se = 2.68-2.93 Å).

1. Structure Reports, 44A, 65.

GALLIUM THULIUM
Ga_2Tm

E.I. GLADYŠEVSKIJ, Ju.M. GRIN', S.P. JACENKO, Ja.P. JARMOLJUK and K.A. CUNTONOV, 1980. Dop. Akad. Nauk Ukr. RSR, 42, 81-84.

Orthorhombic, KHg_2 ($CeCu_2$) type (1), Imma, a = 6.887, b = 4.201, c = 8.078 Å, Z = 4, [D_x = 8.76]. Fe radiation, R = 0.129 for 48 reflexions, diffractometer data. Tm in 4(e) 0,1/4,0.695; Ga in 8(i) 0.197,1/4,0.089. Tm has 6 Tm at 3.55-4.20 and 12 Ga at 3.05-3.46 Å; Ga has 6 Tm at 3.05-3.46 and 4 Ga at 2.55-2.71 Å.

1. Structure Reports, 19, 229; 26, 107.

GERMANIUM HAFNIUM IRON
Fe_6Ge_6Hf

R.R. OLENIČ, L.G. AKSEL'RUD and Ja.P. JARMOLJUK, 1981. Dop. Akad. Nauk Ukr. RSR, 43, 84-88.

Hexagonal, $Mn_4Fe_3Ge_6$ type (1), P6/mmm, a = 5.065, c = 8.058 Å, c/a = 1.59, Z = 1, D_x = 8.80. Fe radiation, $\bar{R}$ = 0.118 for 52 reflexions, diffractometer data. Hf in 1(b) 0,0,1/2; Fe in 6(i) 1/2,0,0.2469; Ge(1) in 2(c) 1/3,2/3,0; Ge(2) in 2(d) 1/3, 2/3,1/2; Ge(3) in 2(e) 0,0,0.1562.

The structure is similar to $CaCu_5$ and $ThMn_{12}$; Hf atoms have CN = 20 (Hf - 12 Fe = 3.25, Hf - 8 Ge = 2.77, 2.92), Fe atoms have CN = 12 (Fe-Fe = 2.53, Fe-Ge = 2.46-2.63) while Ge atoms have CN = 12, 14, 15 with Ge-Ge = 2.51-3.18 Å.

1. Structure Reports, 42A, 87; 44A, 51.

GERMANIUM LITHIUM ZINC
GeLi$_2$Zn (L.T.β-phase)

H.-O. CULLMANN, H.-W. HINTERKEUSER and H.-U. SCHUSTER, 1981. Z. Naturf., **36B**, 917-921.

Hexagonal, Li$_2$ZnSi type (1), P$\bar{3}$m1, a = 4.326, c = 16.470 Å, c/a = 3.83, Z = 4, D$_X$ = 3.77. Mo radiation, R = 0.064 for 249 reflexions, diffractometer data. Zn(1) and (2) in 2(c) 0,0,z, z = 0.1355, 0.3855 respectively; Ge(1), (2), Li(1) - (4) in 2(d) 1/3,2/3,z, z = 0.3758, 0.8752, 0.0265, 0.2159, 0.5572, 0.7102 respectively. Interatomic distances are Zn-Ge = 2.50, Zn-Li = 2.67-2.95, Ge-Li = 2.49-2.99, Li-Li = 2.52, 2.78 Å.

1. Structure Reports, **37A**, 107.

GERMANIUM MANGANESE PALLADIUM
Ge$_8$Mn$_{5.33}$Pd$_{10.66}$

G. VENTURINI, B. MALAMAN, J. STEINMETZ and B. ROQUES, 1981. Mater. Res. Bull., **16**, 715-722.

Orthorhombic, Pnma, a = 6.910, b = 3.146, c = 16.504 Å, Z = 1, D$_X$ = 9.29. Mo radiation, R = 0.084 for 1027 reflexions, diffractometer data.

Atomic positions
All atoms in 4(c) x,1/4,z

	x	y	z		x	y	z
Pd(1)	0.4624	1/4	0.0741	Mn	0.0128	3/4	0.2902
Pd(2)	0.3310	1/4	0.3299	Ge(1)	0.2452	3/4	0.4486
Pd/Mn*	0.1040	3/4	0.0640	Ge(2)	0.1998	1/4	0.1830

*occupancy: Pd/Mn = 2/1

This structure (Fig. 1) is very similar to those of Fe$_2$P, Co$_2$P and Fe$_2$As. The Pd atoms occupy half the tetrahedral sites (Pd-Ge = 2.51-2.59) while Mn and Pd/Mn occupy half the square pyramidal sites (Mn-Ge = 2.69-2.71, Pd/Mn-Ge = 2.49-2.68 Å).

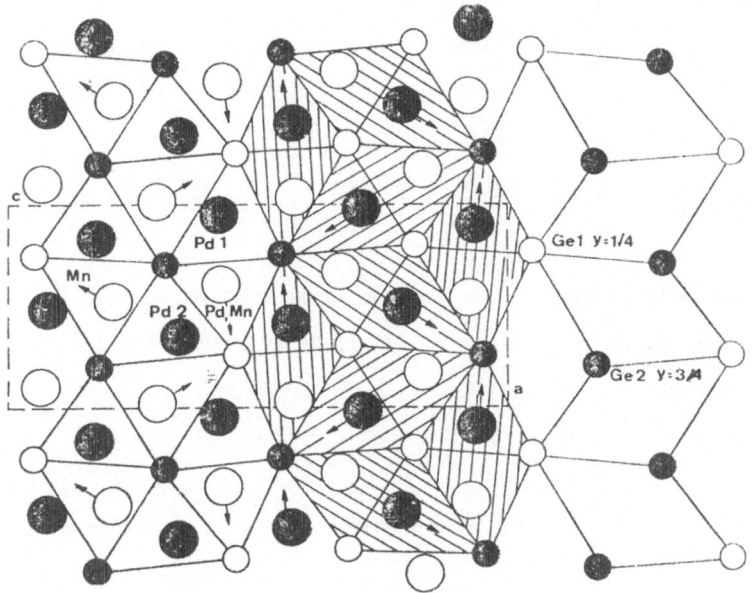

Fig. 1.   The structure of $Pd_{\sim10}Mn_{\sim5}Ge_8$ projected onto 010.   Arrows mark the directions of the axes of the pyramids.

GERMANIUM NICKEL TITANIUM
GeNiTi at 293 K

W. BAŻELA and A. SZYTUŁA, 1981. Phys. Stat. Solidi, A66, 45-52.

Orthorhombic, NiTiSi type (1), Pnma, a = 6.244, b = 3.747, c = 7.147 Å, Z = 4, D_x = 7.11. Fe radiation, R = 0.045, diffractometer data. All atoms in 4(c) x,1/4,z; Ni:x, z = 0.141, 0.560; Ti: 0.032, 0.187; Ge: 0.758, 0.623.   Other phases in the system $NiMn_{1-x}Ti_xGe$ were also examined for x = 0.05, 0.10, 0.15 and 0.25 at 80, 100, 110, 373 and 470 K. [Average distances are Ge-Ni = 2.39, Ge-Ti = 2.65, Ni-Ni = 2.71, Ni-Ti = 2.87, Ti-Ti = 3.27 Å with coordination number Ni = 12, Ti = 15, Ge = 9.]

1. Structure Reports, 19, 124; 30A, 75.

GERMANIUM NICKEL ZINC
$Ge_2Ni_7Zn_6$ $GeNi_2Zn_3$

N. SARAH, T. RAJASEKHARAN and K. SCHUBERT, 1981. Z. Metallk., 72, 652-656.

$Ge_2Ni_7Zn_6$, cubic, filled Ni_2Ti type (1), Fd3m, a = 11.4689 Å, Z = 8, D_x = 8.35. Mo radiation, R = 0.019 for 506 reflexions, diffractometer data. Ni(1) in 48(f) 0.2352,0,0; Ni(2) in 8(a) 0,0,0; Zn(1) in 32(e) x,x,x, x = 0.8582; Zn(2) in 16(c) 1/8,1/8,1/8; Ge in 16(d) 5/8,5/8,5/8; origin at $\bar{4}3m$.

$GeNi_2Zn_3$, cubic, Ni_2SiZn_3 type (2), Fd3m, a = 10.8807 Å, Z = 16, D_x = 7.96. Mo radiation, R = 0.031 for 817 reflexions, diffractometer data. Ni in 32(e) x,x,x, x = 0.8381; Zn in 48(f) 0.1848,0,0; Ge in 16(d) 5/8,5/8,5/8; origin at $\bar{4}3m$.

[Interatomic distance summary - average values (Å)]

	Ge-Ni	Ge-Zn	Ni-Ni	Ni-Zn	Zn-Zn
$Ge_2Ni_7Zn_6$	2.58	2.69	2.76	2.53	3.07
$GeNi_2Zn_3$	2.39	2.83	2.71	2.66	2.87

Atoms are 12-, 13- or 14-coordinated.

1. Structure Reports, <u>22</u>, 889; <u>23</u>, 195; <u>24</u>, 140; <u>27</u>, 296; <u>28</u>, 20.
2. Ibid., <u>33A</u>, 112.

GERMANIUM PLATINUM POTASSIUM   PALLADIUM POTASSIUM SILICON

PALLADIUM SILICON SODIUM   PLATINUM POTASSIUM SILICON

PALLADIUM SODIUM TIN

$A_xM_4X_4$ (A = Na, K; M = Pd, Pt; X = Si, Ge)
$NaPd_3Sn_2$
$NaPd_3Si_2$

W. THRONBERENS, H.-D. SINNEN and H.-U. SCHUSTER, 1980. J. Less-Common Metals, <u>76</u>, 99-108.

$A_xM_4X_4$, tetragonal, I4/mcm, Z = 2. Mo radiation, diffractometer data. A^* in 8(f) 0,0,z; M and X in 8(h) x,1/2+x,0.

	$Na_{1.25}Pd_4Si_4$	KPd_4Si_4	KPt_4Si_4	$K_{1.28}Pt_4Ge_4$
a(Å)	8.263	8.374	8.373	8.621
c(Å)	4.613	4.577	4.592	4.751
c/a	0.558	0.547	0.548	0.551
D_m	5.89	6.25	9.34	10.79
R	0.038	0.033	0.041	0.052
ref	219	270	253	164
z(A)	0.17[+]	0.155	0.2	0.14
x(M)	0.1793	0.1780	0.1750	0.1755
x(X)	0.3877	0.3882	0.3849	0.3821
occ(A)[*]	2.5	2	2	2.56

[+] [disordered and not refined; z(Na) set at 0.17]
[*] Number of A atoms in 8(f)

$NaPd_3Sn_2$, hexagonal, $CeCo_3B_2$ type (1), ordered $CaCu_5$ type (2), P6/mmm, a = 5.772, c = 4.229 Å, c/a = 0.732, D_m = 7.80, Z = 1. R = 0.058 for 238 reflexions, photographic data. Pd in 3(f)[*] 1/2,0,0; Sn in 2(d)[*] 1/3,2/3,1/2. The structure also contains Na in 0,0,z, so disordered that no z value was determined. [*Misprinted as 3(c) and 2(b); by an origin shift of 0,0,1/2 this description becomes that of the type structure, with atoms in 1(a), 2(c) and 3(g).]

$NaPd_3Si_2$, orthorhombic, I2mm, a = 5.685, b = 9.764, c = 7.047 Å, D_m = 6.68, Z = 4. R = 0.058 for 288 reflexions, diffractometer data. Na(1) in 2(a) 0.956,0,0; Na(2) in 2(b) 0.021,0,1/2; Pd(1) in 4(c)[*] 0.503,0,0.280; Pd(2) in 8(e) 0.246,0.250,0.299; Si(1) in 4(d) 0.070,0.311,0; Si(2) in 4(d) 0.439,0.156,0. [*Misprinted as (e).]

Interatomic distances (Å)

$A_xM_4X_4$	$Na_{1.25}Pd_4Si_4$	KPd_4Si_4	KPt_4Si_4	$K_{1.28}Pt_4Ge_4$
M - 5 X	2.44 - 2.47	2.42 - 2.49	2.40 - 2.49	2.48 - 2.58
M - 2 M	2.84	2.86	2.90	2.99
X - 5 X	2.63, 2.96	2.65, 2.96	2.72, 3.00	2.87, 3.13
X - 2 K [1]	-	3.46	3.48	3.51
M - 2 K [1]	-	3.16	3.22	3.25

NaPd$_3$Sn$_2$		NaPd$_3$Si$_2$	
Na	- 6 Pd	2.97	Na - 12 Pd	3.06 - 3.68
	- 6 Sn	3.61	- 6 Si	3.11 - 3.63
	- 2 Na	4.23	- 2 Na	3.54
Pd	- 4 Sn	2.69	Pd - 4 Si	2.40 - 2.55
	- 4 Pd	2.89	- 5 Pd	2.83 - 3.10
	- 2 Na	2.97	- 4 Na	3.06 - 3.68
Sn	- 6 Pd	2.69	Si - 1 Si	2.59
	- 3 Sn	3.33	- 6 Pd	2.40 - 2.55
	- 3 Na	3.61	- 3 Na	3.11 - 3.63

These structures can be thought of in terms of open hexagonal channels, made up of M and X atoms, and centred by large A atoms (Fig. 1) as the parent structure is described (2) or as [M$_6$] trigonal prisms centred by small X atoms as the CeCo$_3$B$_2$ structure is described (1).

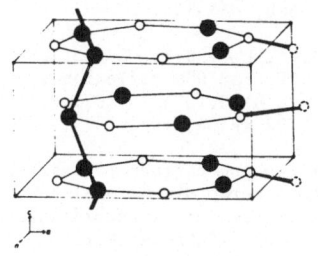 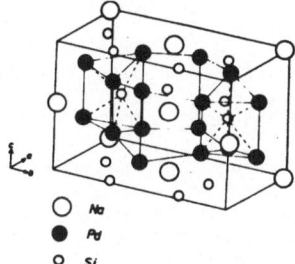

○ Na
● Pd
○ Si

Fig. 1.   Left: the open hexagonal channels in the AM$_4$X$_X$ phases (A = Na, K; M = Pd, Pt; X = Si, Ge; open circles = X, solid circles = M atoms). Right: the NaPd$_3$Si$_2$ structure; [Pd$_6$] trigonal prisms centred by Si atoms are outlined.

1. Structure Reports, 34, 38; 39A, 36.
2. Ibid., 11, 59.

GERMANIUM SELENIUM THALLIUM
Ge$_4$Se$_{10}$Tℓ$_4$

G. EULENBERGER, 1981. Z. Naturf., 36B, 521-523.

Monoclinic, Tℓ$_4$Ge$_4$S$_{10}$ type (1), C2/c, a = 15.602, b = 15.549, c = 9.052 Å, β = 107.10°, Z = 4, D$_X$ = 6.00. Mo radiation, R = 0.079 for 1345 reflexions, diffractometer data.

Atomic positions
S(5) and (6) in 4(e), remainder in 8(f)

	x	y	z		x	y	z
Tℓ(1)	0.17265	0.36946	0.90166	Se(2)	0.4553	0.2156	0.4350
Tℓ(2)	0.13860	0.06115	0.60060	Se(3)	0.2169	0.4559	0.5798
Ge(1)	0.1078	0.3749	0.4118	Se(4)	0.1206	0.1211	0.9468
Ge(2)	0.4319	0.2974	0.8896	Se(5)	0	0.1057	1/4
Se(1)	0.1841	0.2863	0.2728	Se(6)	0	0.4673	1/4

The structure is as previously described (1) with Ge-Se = 2.30-2.41 (mean = 2.36) and Tℓ - 9 Se = 3.12-4.05 (mean - 3.54) Å

1. Structure Reports, 42A, 92.

GERMANIUM SULPHUR THORIUM
GeSTh

K. STOCKS, G. EULENBERGER and H. HAHN, 1981. Z. anorg. Chem., $\underline{472}$, 139-148.

Tetragonal, UGeTe type ($\underline{1}$), I4/mmm, a = 3.9411, c = 17.1395 Å, c/a = 4.35, D_m = 8.32, Z = 4. Mo radiation, R = 0.078 for 172 reflexions, diffractometer data. Th in 4(e) 0,0,0.1394; Ge in 4(c) 0,1/2,0; S in 4(e) 0,0,0.3113; distances are Th-Th = 3.94, Th-Ge = 3.10, Th-S = 2.91, Ge-Ge = 2.79; S-S = 3.49 Å.

$\underline{1}$. Structure Reports, $\underline{34}$A, 92.

GOLD RUBIDIUM TIN
$Au_7Rb_4Sn_2$

H.-D. SINNEN and H.-U. SCHUSTER, 1981. Z. Naturf., $\underline{36}$B, 833-836.

Rhombohedral, R$\bar{3}$m, a = 6.801, c = 29.090 Å, c/a = 4.28, D_m = 8.34, Z = 3; rhombohedral axes a = 10.462 Å, α = 37.94°. Mo radiation, R = 0.092 for 698 reflexions, diffractometer data. Au(1) in 3(a) 0,0,0; Au(2) in 18(h) 0.1971,-0.1971,0.5896; Sn, Rb(1) and (2) in 6(c) 0,0,z, z = 0.4507, 0.6848, 0.8122 respectively.

The structure contains Au_7 clusters made up of two Au_4 tetrahedra sharing a corner (Au-Au = 2.76, 2.78 Å); these are linked into a three dimensional net by Sn_2 pairs (Sn-Sn = 2.87, Sn-Au = 2.60 Å). The Rb atoms are 14- and 16-coordinated (Rb-Au = 3.60-3.96, Rb-Sn 3.94, 4.01, Rb-Rb = 3.71 Å). The Au-Sn network resembles the Cu network in $MgCu_2$.

HAFNIUM IRON PHOSPHORUS
FeHfP

Ja.F. LOMNICKAJA and Ju.B. KUZ'MA, 1981. Ukrainskij Khim. Ž., $\underline{47}$, 103-104.

Orthorhombic, NiSiTi type ($\underline{1}$), Pnma, a = 6.227, b = 3.720, c = 7.138 Å, Z = 4, D_X = 10.65. R = 0.066, photographic data. All atoms in 4(c) x,1/4,z; Hf: 0.0199, 0.1830; Fe: 0.1468, 0.5615; P: 0.7782, 0.6160. [Distances are Fe-Fe = 2.75, Fe-Hf = 2.80-2.92, Fe-P = 2.30-2.44, Hf-Hf = 3.22-3.72, Hf-P = 2.66-2.67 Å.]

$\underline{1}$. Structure Reports, $\underline{30}$A, 75; $\underline{34}$A, 176.

HAFNIUM MOLYBDENUM NICKEL
$Hf_9Mo_4Ni(O)$-κ

A. HARSTA, 1981. Acta Chem. Scand., A$\underline{35}$, 43-47.

Hexagonal, κ(χ) type ($\underline{1}$, $\underline{2}$), P6_3/mmc, a = 8.6550, c = 8.4626 Å, Z = 2, D_X = 12.5. Mo radiation, R = 0.06 for 579 reflexions, diffractometer data. Hf(1) in 12(k) 0.20122, 0.40244,0.05670; Hf(2) in 6(h) 0.54240,0.08480,1/4; Mo(1) in 6(h) 0.88900,0.77800,1/4; Mo(2) in 2(a) 0,0,0; Ni in 2(c) 1/3,2/3,1/4; 1.68 O in 6(g) 1/2,0,0.

There is some variation in the descriptions of this phase, first reported as κ ($\underline{1}$) and later as χ ($\underline{2}$). The present description fits that of ($\underline{2}$) for χ-$W_{10}Co_3C_{\sim3}$ with Hf(1,2) = W(1,2), Mo(1) = Co, Mo(2) = W(3), Ni = C(2) and O = C(1). Distances are Hf-O = 2.21, 2.33, Hf-Ni = 2.57, 3.13, Hf-Mo = 2.89-3.09, Hf-Hf = 3.17-3.23; Mo-Mo = 2.69, 2.88. The descriptions differ in their distributions of atoms on the sites the structure occurs for carbides and borides as well as for the group Fe, Co and Ni.

1. Structure Reports, 16, 42; 37A, 39; 39A, 37.
2. Ibid., 18, 81; 43A, 43.

HYDROGEN MAGNESIUM NICKEL
H_4Mg_2Ni (L.T.)

I.   D. NORÉUS and P.-E. WERNER, 1981. Mater. Res. Bull., 16, 199-206.

II. J.-P. DARNAUDERY, M. PEZAT, B. DARRIET and P. HAGENMULLER, 1981. Mater. Res.
     Bull., 16, 1237-1244.

I. Monoclinic, Cm, a = 6.497, b = 6.414, c = 6.601 Å, β = 93.23°, Z = 4, D_X = 2.693.
Ni(1) in 2(a) 0,0,0; Ni(2) in 2(a) 0.427,0,0.524; Mg(1) in 4(b) 0.276,0.235,0.246;
Mg(2) 0.708,0.247,0.755. Reliable hydrogen positions could not be determined. See
also 1. The refinement was by the Rietveld method on high resolution Guinier data
recorded with $CuK\alpha_1$.

II. Monoclinic, a = 12.99, b = 6.390, c = 6.598 Å, β = 93.22°. Guinier data on a
sample prepared at 210°C and 30 bar. II concludes that I, who worked with a
multiphase sample, had not been able to assign the correct cell. II reported a
single phase sample.

1. Structure Reports, 46A, 72.

INDIUM LANTHANUM SULPHUR
$In_5La_4S_{13}$

G.G. GUSEINOV, F.K. MAMEDOV and K.S. MAMEDOV, 1979. Dokl. Akad. Nauk Azer. SSR, 35,
50-53.

Orthorhombic, $Nd_4In_5S_{13}$ type (1), Pbam, a = 21.393, b = 11.843, c = 4.061 Å, Z = 2,
$[D_X$ = 4.99]. Mo radiation, R = 0.082 for 464 reflexions, diffractometer data. 12
site-sets are given.

There are chains of $[InS_6]$ octahedra (In-S = 2.37-2.86, av. = 2.63 Å) and $[InS_4]$
tetrahedra (In-S = 2.45-2.55, av. = 2.50 Å) sharing edges and vertices; the La atoms
are 8-coordinated (La-S = 2.93-3.49, av. = 3.06 Å).

1. Structure Reports, 45A, 83.

INDIUM MOLYBDENUM SELENIUM
$In_2Mo_{15}Se_{19}$

M. POTEL, R. CHEVREL and M. SERGENT, 1981. Acta Cryst., B37, 1007-1010.

Rhombohedral, R3c, a = 20.159 Å, α = 27.808°, D_m = 6.59, Z = 2; hexagonal axes a =
9.688, c = 58.10 Å. Mo radiation, R = 0.055 for 942 reflexions, diffractometer data.
[This is essentially the same as reported by 1, on hexagonal axes.] The building
units are illustrated in Fig. 1.

Atomic positions (rhombohedral axes)

		x	y	z
Mo(1) in	12(f)	0.49988	0.68888	0.37056
Mo(2) in	12(f)	0.22250	0.04574	0.36373
Mo(3) in	6(e)	0.40973	0.09027	1/4
Se(1) in	12(f)	0.18982	0.80340	0.55668
Se(2) in	12(f)	0.53807	0.92007	0.18467
Se(3) in	6(e)	0.05981	0.44019	3/4
Se(4) in	4(c)	0.44543	0.44543	0.44543
Se(5) in	4(c)	0.17609	0.17609	0.17609
In in	4(c)	0.11425	0.11425	0.11425

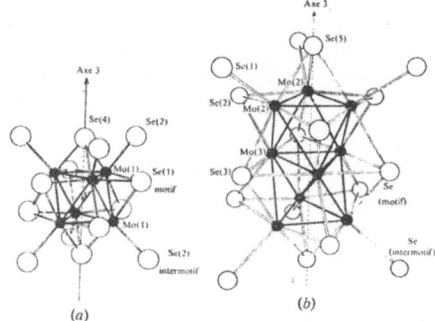

Fig. 1.   The $(Mo_6Se_8)Se_6$ and $(Mo_9Se_{11})Se_6$ units of the $In_2Mo_{15}Se_{19}$ structure.

1. Structure Reports, 46A, 86.

IRON MANGANESE YTTRIUM
$(Fe_{0.5}Mn_{0.5})_{23}Y_6$ at 295 K

K. HARDMAN, W.J. JAMES and W.B. YELON, 1980.   J. Phys. Chem. Solids, 41, 1105-1109.

Cubic, $Mn_{23}Th_6$ type (1), Fm3m, a = 12.188 Å, Z = 4, D_x = 6.28.   R = 0.050 on neutron
diffractometer data. $Y_6(Fe_{1-x}Mn_x)_{23}$ compounds with x = 0.2, 0.4, 0.75, 0.8 and 0.9,
were also refined between 295 and 573 K.   Y in 24(e) 0.205,0,0; 1.4 Fe + 2.6 Mn in
4(b) 1/2,1/2,1/2; 14.4 Fe + 9.6 Mn in 24(d) 0,1/4,1/4; 23.68 Fe + 8.32 Mn in 32(f)
x,x,x, x = 0.178; 8 Fe + 24 Mn in 32(f) x = 0.378.   The structure is as described (1)
with preferential ordering of Fe and Mn; [distances are Y - Fe/Mn = 2.98-3.60 and
Fe/Mn - Fe/Mn = 2.48-2.97 Å].

1. Structure Reports, 16, 113.

IRON MOLYBDENUM SULPHUR TIN
$Fe_{0.4}Mo_6S_8Sn$

J.D. JORGENSEN, D.G. HINKS and F.J. ROTELLA, 1981.   Ternary Superconductors, Ed.
Shenoy, Dunlap and Fradin, pp. 69-73.   .

Rhombohedral, R$\bar{3}$, a = 9.2256, c = 11.3418 Å, c/a = 1.23, D_m = 5.67, Z = 3.   R = 0.038
for 649 reflexions, diffractometer data.   Sn in 3(a) 0,0,0; 0.9 Fe in 6(f) 0.205,
0.045,0.112; Mo in 18(f) 0.0159,0.1739,0.4020; S(1) in 18(f) 0.3280,0.2907,0.4144;
S(2) in 6(c) 0,0,0.2395.

This is a "mixed" Chevrel phase with Sn occupying the origin position (Fig. 1) as in $PbMo_6S_8$ (1) and Fe partially occupying a six-fold position (Fig. 1) outside. The SnS_6 cube faces similar to that occupied by Mo. It is not certain whether the Fe position is occupied dynamically or statically; the Sn atom has very anisotropic "motion". Distances are Mo-Mo = 2.66, 2.70, S-S = 3.26-3.47, Fe-Fe = 2.62-3.07, Fe-S = 2.25-2.75 Å.

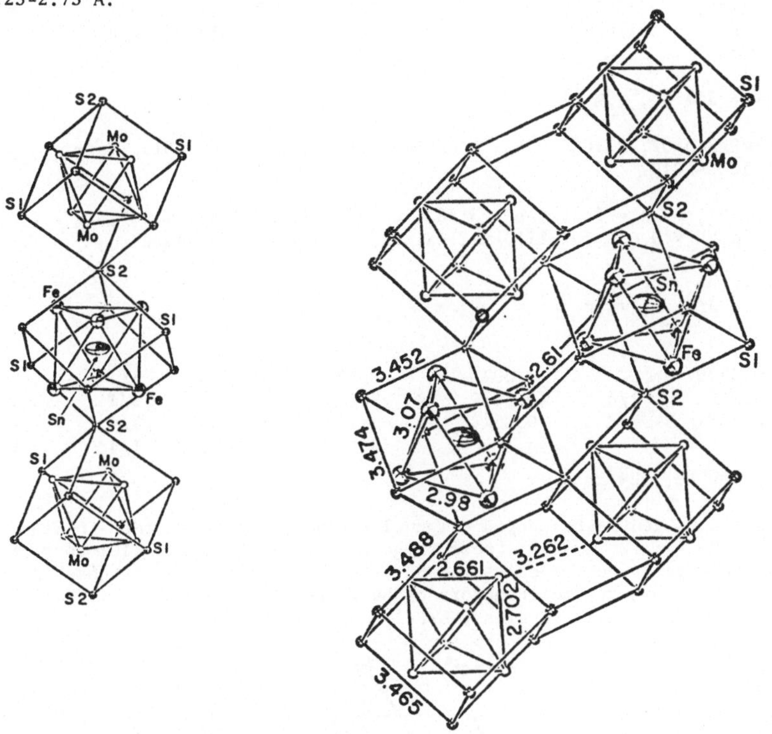

Fig. 1.　　Left:　The Mo_6S_8 and $SnFe_{.4}S_8$ cubes corner-linked along the c axis.
　　　　　　Right:　A perspective view with key distances marked.

1.　Structure Reports, 39A, 85.
2.　Ibid., 42A, 106; 46A, 102.

IRON NIOBIUM SELENIUM
$FeNb_3Se_{10}$

I.　R.J. CAVA, V.L. HIMES, A.D. MIGHELL and R.S. ROTH, 1981.　Phys. Rev., B, 24, 3634-3637.

II.　S.J. HILLENIUS, R.V. COLEMAN, R.M. FLEMING and R.J. CAVA, 1981.　Ibid., B, 23, 1567-1575.

I.　Monoclinic, $P2_1/m$, a = 9.235, b = 3.478, c = 10.271 Å, β = 114.18°, D_m = 6.15, Z = 1.　II gives a = 9.213, b = 3.4773, c = 10.299 Å, β = 114.52°.　Mo radiation, R = 0.041 for 689 reflexions, diffractometer data.

Atomic positions
All atoms in 2(e) ± x,1/4,z, from I

	x	y	z		x	y	z
Nb *	0.2759	3/4	0.8644	Se(3)	0.4890	1/4	0.8638•
Nb/Fe*	0.4493	1/4	0.5918	Se(4)	0.2448	3/4	0.5894
Se(1)	0.0407	1/4	0.7424	Se(5)	0.6618	3/4	0.6659
Se(2)	0.1607	1/4	0.9925				

* occupancy = 50% Nb, 50% Fe

III.   A. MEERSCHAUT, P. GRESSIER, L. GUEMAS and J. ROUXEL, 1981.   Mater. Res. Bull., 16, 1035-1040.

Monoclinic, $P2_1/m$, a = 9.213, b = 3.482, c = 10.292 Å, β = 114.46°, D_m = 6.12, Z = 1. Mo radiation, R = 0.034 for 1110 reflexions, diffractometer data.

Atomic positions
All atoms in 2(e) x,1/4,z

	x	y	z		x	y	z
Fe*/Nb	0.4480	3/4	0.0922	Se(2)	0.1608	3/4	0.4927
Nb	0.2750	1/4	0.3639	Se(3)	0.0388	3/4	0.2421
Se(1)	0.4887	3/4	0.3642	Se(4)	0.2430	1/4	0.0893
				Se(5)	0.6605	1/4	0.1654

* occupancy 50% Fe, 50% Nb

    The Nb atoms centre bi-capped trigonal prisms of Se atoms (Nb - 6 Se = 2.63-2.66, Nb - 2 Se = 2.72, 2.74 Å; these trigonal prisms stack parallel to b (Fig. 1) and the short edges give Se_2 pairs (Se-Se = 2.35 Å) comparable to those in $NbSe_3$ (1). The Fe/Nb atoms centre octahedra of Se atoms (Fe/Nb - Se = 2.42-2.67 Å), which share edges to form similar chains (Nb/Fe - Nb/Fe = 3.0 Å).

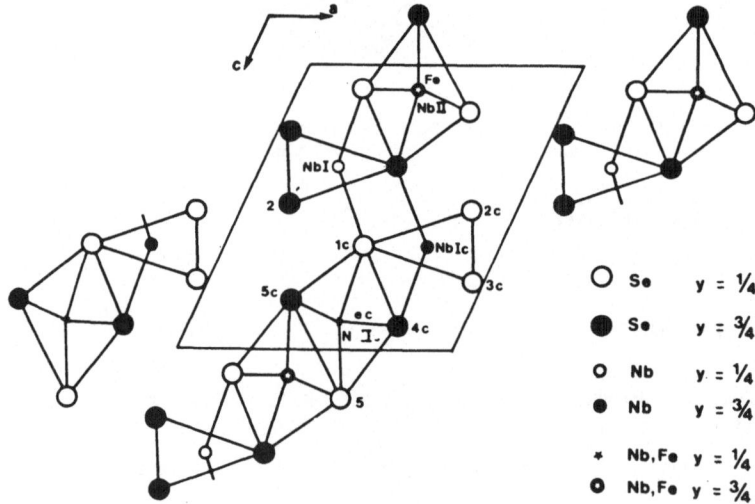

○	Se	y = 1/4
●	Se	y = 3/4
○	Nb	y = 1/4
●	Nb	y = 3/4
✳	Nb,Fe	y = 1/4
◉	Nb,Fe	y = 3/4

Fig. 1.   The $FeNb_3Se_{10}$ structure projected onto 010.

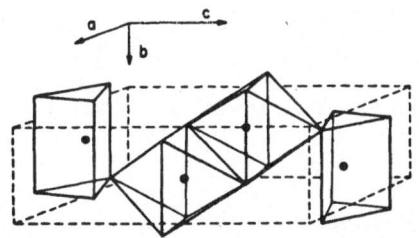

Fig. 2.   A perspective view with the coordination polyhedra outlined.

1. Structure Reports, 41A, 99; 45A, 98.

IRON PHOSPHORUS SELENIUM   MANGANESE PHOSPHORUS SELENIUM
PSe_3X (X = Fe, Mn)

A. WIEDENMANN, J. ROSSAT-MIGNOD, A. LOUISY, R. BREC and J. ROUXEL, 1981. Solid
State Comm., 40, 1067-1072.

Rhombohedral, R3̄, Z = 6, diffractometer data. Atomic positions in $MnPSe_3$ and $FePSe_3$:
X and P in 6(c) 0,0,z; Se in 18(f) x,y,z. Space group contrary to 1; see also 2.

$XPSe_3$	Temp.	a(Å)	c(Å)	z(X)	z(P)	x(Se)	y(Se)	z(Se)
$MnPSe_3$	100 K	6.394	20.019	0.1661	0.4443	0.3305	-0.0016	0.0818
$FePSe_3$	154 K	6.273	19.812	0.1665	0.4486	0.3282	0.0073	0.0801

     [The structure is basically as described by 1 with X - 6 Se : Fe-Se = 2.66,
2.70 and Mn-Se = 2.71, 2.74; P-Se = 2.21 and P-P = 2.04 Å.]

1. Structure Reports, 39A, 76.
2. Ibid., 30A, 136.

IRON PHOSPHORUS TANTALUM
$FePTa_4$

Ja.F. LOMNICKAJA and Ju.B. KUZ'MA, 1981. Izv. Akad. Nauk, SSSR, Neorg. Mater., 16,
1022-1025 [Inorg. Mater., 16, 705-707].

Tetragonal, Nb_4CoSi type (1), P4/mcc, a = 6.101, c = 5.006 Å, c/a = 0.821, Z = 2, D_x =
14.45. Cr radiation, R = 0.072, photographic data. Ta in 8(m) 0.162,0.662,0; Fe in
2(a) 0,0,1/4; P in 2(c) 1/2,1/2,1/4.

     The structure has Ta with CN = 15 (Ta - 11 Ta = 2.80-3.23; Ta - 2 Fe = 2.61; Ta -
2 P = 2.60), Fe with CN = 10 (Fe - 8 Ta = 2.61, Fe - 2 Fe = 2.50) and P with CN = 10
(P - 8 Ta = 2.60, P - 2 P = 2.50 Å).

1. Structure Reports, 30A, 44.

IRON POTASSIUM SULPHUR (BARTONITE)
$Fe_{20.27}K_{5.68}S_{26.93}$

I. H.T. EVANS and J.R. CLARK, 1981. Amer. Min., 66, 376-384.

II. G.K. CZAMANSKE, R.C. ERD, B.F. LEONARD and J.R. CLARK, 1981. Ibid., 66, 369-375.

I, II. Tetragonal, I4/mmm, a = 10.424, c = 20.626 Å, c/a = 1.98, Z = 2, D_m = 3.31.
Composition = $(K,Na)_{5.68}(Fe,Cu,Ni)_{20.27}S_{26}(S,Cl)_{0.93}$. Mo radiation, R = 0.068 for
504 reflexions, single-crystal diffractometer data. This is the $K_3Fe_{10}S_{14}$ of 1.

Atomic positions

			x	y	z	occ.
K(1)	in	8(i)	0.2996	0	0	0.920
K(2)	in	4(e)	0	0	0.1539	1.0
Fe(1)	in	32(o)	0.1299	0.3687	0.1841	0.841
Fe(2)	in	16(m)	0.3688	0.3688	0.0659	0.864
S(1)	in	16(m)	0.2307	0.2307	0.1159	1.0
S(2)	in	16(n)	0	0.2495	0.2510	1.0
S(3)	in	8(j)	0.2506	1/2	0	1.0
S(4)	in	8(g)	0	1/2	0.1237	1.0
S(5)	in	4(e)	0	0	0.3726	1.0
S(6)	in	2(a)	0	0	0	0.925

The structure (Fig. 1) is based on Fe_8S_{14} clusters formed from condensed FeS_4
tetrahedra; in addition there are distinctive SK_6 groups with an S atom centring a
K_6 cube (K-S = 3.15 Å). The average distances are Fe-S = 2.29, Fe-Fe = 2.72, K-S =
3.37 Å. The relationship with djerfisherite (2) and pentlandite (3) is discussed in
detail and compared with the discussion of 4.

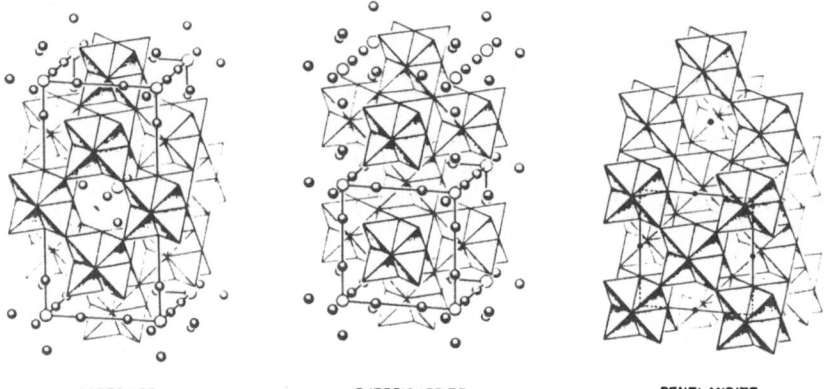

BARTONITE   DJERFISHERITE   PENTLANDITE

Fig. 1.   The structures of bartonite, djerfisherite and pentlandite showing Fe_8S_{14}
           clusters of condensed FeS_4 tetrahedra. Circle-shaded atoms: K; large
           circles: S; small filled circles: Fe, Ni, Ag (pentlandite only).

1. G.K. CZAMANSKE, R.C. ERD, M.N. SOKOLOVA, M.G. DOBROVOL'SKAYA and M.T. DMITRIEVA,
    1979. Amer. Min., 64, 776.
2. Structure Reports, 46A, 298.
3. Ibid., 20, 125; 42A, 116; 43A, 91.
4. Ibid., 42A, 116.

IRON SCANDIUM SILICON
$FeScSi_2$

Ja.P. JARMOLJUK, B.Ja. KOTUR and Ju.M. GRIN', 1980. Dop. Akad. Nauk Ukr. RSR, B6,
68-72.

Orthorhombic, Cmca, a = 5.115, b = 18.929, c = 14.298 Å, Z = 24, D_X = 4.52. Mo
radiation, R = 0.060 for 391 reflexions, diffractometer data. 12 site-sets given.
See also 1.

     Sc atoms are 16-and 17-coordinated (Sc-Sc = 3.23-3.67, Sc-Fe = 2.95-3.20, Sc-Si
= 2.70-2.90 Å), Fe atoms are 12-coordinated (Fe-Fe = 2.56, Fe-Si = 2.33-2.50 Å),
while Si atoms are 10-, 12- or 13-coordinated (Si-Si = 2.48-3.03 Å).

1. Structure Reports, 45A, 120.

IRON SELENIUM SODIUM
$FeNa_3Se_3$

P. MÜLLER and W. BRONGER, 1981. Z. Naturf., 36B, 646-648.

Monoclinic, Na_3FeS_3 type (1), $P2_1/n$, a = 7.452, b = 7.176, c = 13.081 Å, β = 90.5, Z
= 4. Mo radiation, R = 0.073 for 1488 reflexions, diffractometer data.

Atomic positions
All atoms in 4(e)

	x	y	z		x	y	z
Fe	0.5774	0.1313	0.0765	Na(1)	0.728	0.512	0.2199
Se(1)	0.9146	0.2952	0.4249	Na(2)	0.341	0.489	0.3866
Se(2)	0.0731	0.7292	0.2740	Na(3)	0.213	0.488	0.0962
Se(3)	0.8740	0.2496	0.0736				

     The structure is as previously described (1) with Fe-Se = 2.37-2.41, Fe-Fe =
2.97 and Na-Se = 2.87-3.56 Å.

1. Structure Reports, 45A, 89.

IRON SILICON URANIUM
$Fe_2Si_7U_3$

L.G. AKSEL'RUD, Ja.P. JARMOLJUK, I.V. ROŽDESTVENSKAJA and E.I. GLADYŠEVSKIJ, 1981.
Kristallografija, 26, 186-188 [Soviet Physics-Crystallography, 26, 103-104].

Orthorhombic, Cmmm, a = 4.020, b = 24.367, c = 4.028 Å, Z = 2, D_X = 8.60. Mo
radiation, R = 0.113 for 550 reflexions, diffractometer data. U(1) in 4(i) 0,0.31644,
0; U(2) in 2(d) 0,0,1/2; Fe in 4(j) 0,0.1283,1/2; Si(1) in 2(b) 1/2,0,0; Si(2) in 4(j)
0,0.4138,1/2; Si(3) in 4(i) 0,0.0873,0; Si(4) in 4(j) 0,0.2235,1/2.

     The structure can be built from elements of the Al_4Ba, Cu_3Au and AlB_2 structures.
The distances are U - 10 Si = 3.01-3.11, U - 4 Fe = 3.15, U - 6 U = 3.81-4.03, CN =
20; U - 12 Si = 2.85-2.93, U - 2 Fe = 3.13, U - 4 U = 4.02, 4.03, CN = 18; Fe - 5 Si =
2.25-2.32, Fe - 5 U = 3.13, 3.15, CN = 10; Si - 4 U = 2.85, Si - 8 Si = 2.91-2.93, CN
= 12; Si - 2 Fe = 2.26, Si - 6 Si = 2.85, 2.91, Si - 4 U = 2.91, 3.11, CN = 12; Si -
2 Fe = 2.25, Si - 6 Si = 2.85, 2.95, Si - 4 U = 2.93, 3.09, CN = 12; Si - 1 Fe = 2.32,
Si - 2 Si = 2.39, Si - 6 U = 3.01, 3.03 Å, CN = 9.

IRON SODIUM SULPHUR
$Fe_2Na_3S_4$

K. KLEPP and H. BOLLER, 1981. Mh. Chem., 112, 83-89.

Orthorhombic, Pnma, a = 6.6333, b = 10.675, c = 10.677 Å, Z = 4, D_X = 2.71. Mo
radiation, R = 0.028 for 640 reflexions, diffractometer data. Na(1) in 4(c) 0.0813,
1/4,0.7674; Na(2) in 8(d) 0.0426,0.4148,0.1428; Fe in 8(d) 0.0469,0.3786,0.4686; S(1)
in 8(d) 0.1708,0.5545,0.3731; S(2) and S(3) in 4(c) 0.3001,1/4,0.5396; S(3) 0.3731,
1/4,0.1639.

The structure contains FeS_4 tetrahedra (Fe-S = 2.28-2.34) linked by edges to
form chains; the Na atoms occupy octahedral interstices (Na-S = 2.78-3.08 Å).

IRON SULPHUR THALLIUM
$Fe_3S_4Tl_2$

M. ZABEL and K.-J. RANGE, 1980. Rev. Chim. Minér., 17, 561-568.

Orthorhombic, $Cs_2Mn_3S_4$ type (1), Ibam, a = 5.397, b = 10.587, c = 13.313 Å, Z = 4,
D_X = 6.15. Cu radiation, photographic data. Tl in 8(g) 0.243,0.123,0; S in 16(k)
0.264,0.128,0.347; Fe(1) in 8(g) 0,0.234,1/4; Fe(2) in 4(b) 1/2,0,1/4; distances are
Fe-Fe = 2.72, 2.81, Fe-S = 2.23-2.26; Tl-S = 3.33-3.41, Tl-Tl = 3.70 Å.

1. Structure Reports, 38A, 57.

IRON TITANIUM   MANGANESE TITANIUM
Fe_2Ti Mn_2Ti

M.J.M. de ALMEIDA, M.M.R. COSTA, L. ALTE da VEIGA, L.R. ANDRADE and A. MATOS BEJA,
1980. Portugal. Phys., 11, 219-227.

Hexagonal, $MgZn_2$ type (1), $P6_3/mmc$, Z = 4. Mo radiation, diffractometer data. A in
4(f) 1/3,2/3,z; B(1) in 2(a) 0,0,0; B(2) in 6(h) x,2x,1/4. A = Ti, B = Fe or Mn. See
also 2, 3.

AB_2	a(Å)	c(Å)	R	refl.	x(B(2))[*]	z(Å)	A-A, Å	A-B, Å	B-B, Å
$TiFe_2$	4.7870	7.8150	0.052	100	-0.1710	0.0640	2.91-4	2.80-1	2.33, 2.46
$TiMn_2$	4.8310	7.9390	0.051	86	-0.1701	0.0642	2.95-7	2.83-4	2.37, 2.47

* [Misprinted without minus sign]

1. Structure Reports, 46A, 96.
2. Ibid., 13, 91; 17, 114, 199; 18, 124; 22, 205.
3. Ibid., 17, 121, 122; 18, 124, 218; 22, 149, 205; 24, 180.

IRON ZINC
$Fe_{22}Zn_{78}$ (Γ_1)

A.S. KOSTER and J.C. SCHOONE, 1981. Acta Cryst., B37, 1905-1907.

Cubic, $F\bar{4}3m$, a = 17.963 Å, Z = 4, D_X = 3.62. Co radiation, R = 0.091 for 211
reflexions, diffractometer data. 14 site-sets given. See also 1.

This structure is closely related to the Γ phase (2); 8 of the 16 clusters in Γ
occur in the same sites in Γ_1, but for the remaining atoms the arrangement differs
although of the same kind. Distances are normal: Zn-Zn = 2.60-2.90, Fe-Zn = 2.49-
2.76, Fe-Fe = 2.38-2.60 Å.

1. Structure Reports, 40A, 108.
2. Ibid., 40A, 44.

LITHIUM and SODIUM with MANGANESE and GROUP V ELEMENTS
AMnX (A = Li, Na; X = P, As, Sb)

G. ACHENBACH and H.-U. SCHUSTER, 1981. Z. anorg. Chem., <u>475</u>, 9-17.

Tetragonal, P4/nmm, Z = 2. Cu radiation, diffractometer data.

AMnX	a($\mathring{A}$)	c($\mathring{A}$)	c/a	D_m	R	type	ref.
LiMnP	4.133	5.957	1.44	3.05	0.074	LiMnP	-
LiMnAs	4.263	6.179	1.45	4.02	0.093	LiMnP	-
NaMnP	4.086	6.884	1.68	3.13	0.036	Cu_2Sb	1,3
NaMnAs	4.206	7.077	1.68	4.09	0.080	Cu_2Sb	$\overline{1}$,2
NaMnSb	4.478	7.557	1.69	4.39	0.072	Cu_2Sb	$\overline{1}$,$\overline{3}$

Atomic positions [origin at centre]

LiMnP

	x	y	z
Li in 2(b)	0	0	1/2
Mn in 2(a)	0	0	0
P in 2(c)	0	1/2	0.763

NaMnAs

	x	y	z
Na in 2(c)	0	1/2	0.640
Mn in 2(a)	0	0	0
As in 2(c)	0	1/2	0.213

LiMnAs

Li in 2(b)	0	0	1/2
Mn in 2(a)	0	0	0
As in 2(c)	0	1/2	0.758

NaMnSb

Na in 2(c)	0	1/2	0.651
Mn in 2(a)	0	0	0
Sb in 2(c)	0	1/2	0.220

NaMnP

Na in 2(c)	0	1/2	0.637
Mn in 2(a)	0	0	0
P in 2(c)	0	1/2	0.205

Interatomic distances ($\mathring{A}$)

AMnX	A-A	A-X	A-X	A-Mn	Mn-X	Mn-Mn
LiMnP	2.92	2.59	-	2.98	2.50	2.92
LiMnAs	3.01	2.66	-	3.09	2.60	3.01
NaMnP	3.45	3.08	2.96	3.22	2.48	2.89
NaMnAs	3.58	3.14	3.02	3.30	2.59	2.97
NaMnSb	3.90	3.31	3.26	3.46	2.79	3.17

The structures (Fig. 1) differ only in the placement of the alkali metal atoms (Li or Na).

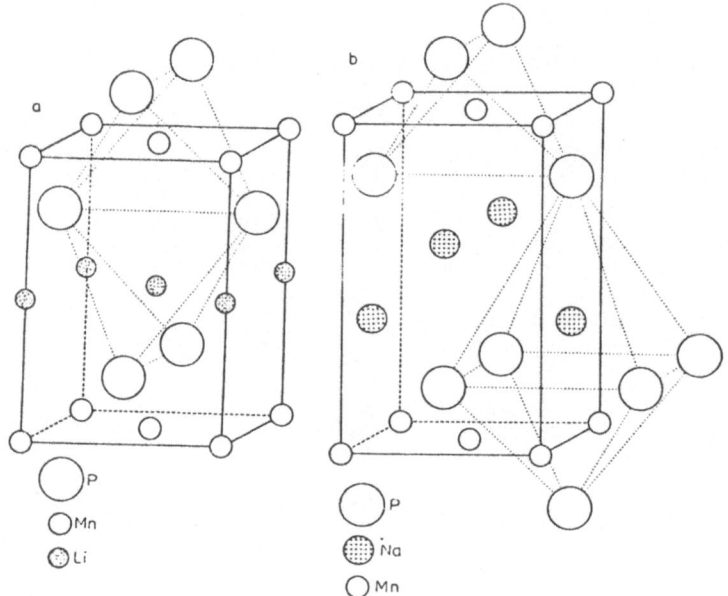

Fig. 1.   a: The LiMnP structure and b: the NaMnP structure (Cu_2Sb type).

1. Strukturbericht, 3, 33. Structure Reports, 29, 38; 40A, 20; 45A, 8.
2. Structure Reports, 44A, 106.
3. Ibid., 44A, 118.

LANTHANUM NICKEL PLATINUM
$LaNi_2Pt_3$

I.D. WEISMAN, L.H. BENNETT, A.J. McALISTER and R.E. WATSON, 1975. Phys. Rev. B, 11, 82-91.

Hexagonal, $CaCu_5$ type (1), P6/mmm, a = 5.22, c = 4.07 Å, [read from a small diagram] c/a = 0.78, Z = 1, D_x = 14.55. Diffractometer data. La in 1(a) 0,0,0; 1.88 Ni + 0.12 Pt in 2(c) 1/3,2/3,0; 2.88 Pt + 0.12 Ni in 3(g) 1/2,0,1/2.

1. Structure Reports, 11, 59; 20, 133.

LITHIUM SILICON
$Li_{12}Si_7$

H.G. von SCHNERING, R. NESPER, J. CURDA and K.-F. TEBBE, 1980. Angew. Chem. Int. Ed. Engl., 19, 1033-1034.

Orthorhombic, Pnma, a = 8.610, b = 19.738, c = 14.341 Å, Z = 8, D_x = 1.53. Mo radiation, R = 0.035 for 2190 reflexions, 22 site-sets given. This is the $Li_{13}Si_7$ of 1.

The structure contains planar Si_5 rings (Si-Si = 2.36-2.38) and star-shaped trigonal planar Si_4 groups (Si-Si = 2.37-2.39); Si-Li = 2.59-3.56; Li-Li = 2.40-3.76 Å.

1. A. AXEL, H. SCHÄFER and A. WEISS, 1965. Z. Naturforsch., B20, 1302.

LITHIUM SULPHUR YTTRIUM ZIRCONIUM
$Li_{0.45}S_2Y_{0.45}Zr_{0.55}$, $Li_{0.90}S_2Y_{0.90}Zr_{0.10}$

O. ABOU GHALOUN, P. CHEVALIER, L. TRICHET and J. ROUXEL, 1980. Rev. Chim. Minér., 17, 368-378.

$Li_{0.45}S_2Y_{0.45}Zr_{0.55}$, cubic, Al_2MgO_4 (spinel) type (1), Fd3m, a = 10.834 Å, Z = 16, D_X = 3.28. R = 0.07, diffractometer data. 7.2 Li in 8(a) 0,0,0; 7.2 Y + 8.8 Zr in 16(d) 5/8,5/8,5/8; S in 32(e) 0.382,0.382,0.382; Li-S = 2.48, Y/Zr-S = 2.63 Å.

$Li_{0.90}S_2Y_{0.90}Zr_{0.10}$, rhombohedral, R3̄m, a = 3.896, c = 18.55 Å, c/a = 4.76, Z = 3, D_X = 3.25. R = 0.132, diffractometer data. 2.7 Li in 3(a) 0,0,0; 2.7 Y + 0.3 Zr in 3(b) 0,0,1/2; S in 6(c) 0,0,0.249; Li-S = 2.74, Y/Zr-S = 2.72 Å.

1. Structure Reports, 11, 497; 13, 241; 15, 207; 17, 417.

LUTETIUM SULPHUR
Lu_3S_4

A.V. HARIHARAN, D.R. POWELL, R.A. JACOBSON and H.F. FRANZEN, 1981. J. Solid State Chem., 36, 148-150.

Orthorhombic, Fddd, a = 10.747, b = 22.813, c = 7.602 Å, Z = 12, D_X = 7.03. Mo radiation, R = 0.050 for 140 reflexions, diffractometer data. 4.32 Lu(1) in 8(a) 7/8,7/8,7/8; 6.64 Lu(2) in 8(b) 7/8,7/8,1/8; 13.28 Lu(3) in 16(f) 7/8,0.2077,7/8; 12 Lu(4) in 16(f) 7/8,0.0411,7/8; 16 S(1) in 16(e) 0.125,7/8,7/8; 32 S(2) in 32(h) 0.125,0.0417,7/8. [Origin shifted to 1̄]. See also 1.

The structure can be described as a sheared population wave defect ordering based on the NaCl structure of LuS. At a given height along a (x = 0, 1/4,1/2,3/4) the fractional occupancies in sequential (066) planes occur periodically in the order ... 0.54, 0.75, 0.83, 0.84, 0.83, 0.75, 0.54 ... . Further the occupation waves at x = 0 and x = 3/4 are in phase, as are those at x = 1/4 and x = 1/2 but the latter are exactly out of phase with the former. Distances are Lu - 6 S = 2.68-2.70, Lu - 5 Lu = 3.79, 3.80 Å.

1. Structure Reports, 44A, 118.

MANGANESE PHOSPHORUS
MnP_4-6

R. RÜHL and W. JEITSCHKO, 1981. Acta Cryst., B37, 39-44.

Triclinic, stacking variant of 8-MnP_4 (1), P1̄, a = 16.347, b = 5.847, c = 5.108 Å, α = 115.66, β = 95.15, γ = 89.21°, Z = 6, D_X = 4.07. Mo radiation, R = 0.060 for 1227 reflexions, diffractometer data. 15 site-sets given.

The near-neighbour environment is very like that in 8-MnP₄ (1), (Fig. 1(a)). As in related MP₄ structures (2), the structure can be considered as built from layers (Fig. 1(b)). The Mn atoms are displaced from the centres of the [P₆] octahedra, forming Mn-Mn bonds (2.92, 2.96 Å).

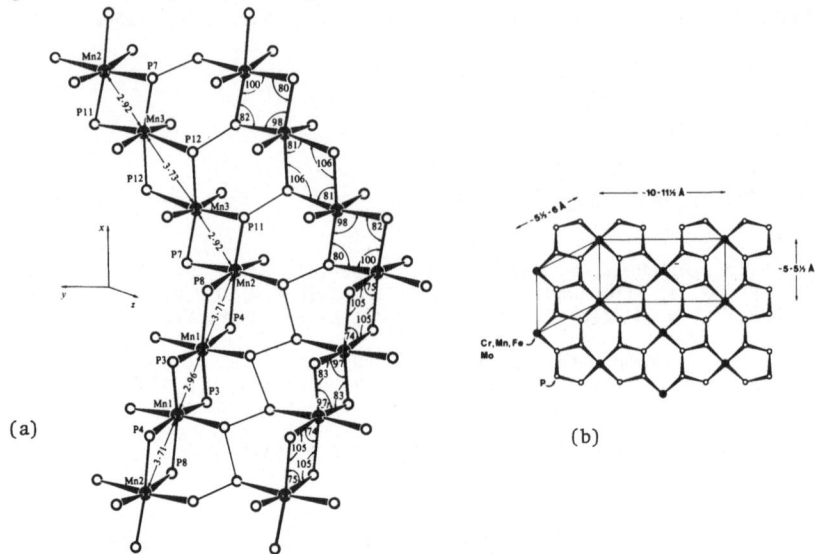

(a)   (b)

Fig. 1.   (a) The chains of [MnP₆] octahedra in 6-MnP₄. (b) The puckered net of
           hexagons and pentagons which occurs in several MP₄ compounds. The
           centred rectangular net corresponds to a and b axes in CrP₄, 8-MnP₄ and
           β-FeP₄. The primitive mesh on the left corresponds to the a and b axes
           in 2-MnP₄ and 6-MnP₄.

1.   Structure Reports, 41A, 95.
2.   Ibid., 38A, 74; 44A, 88; 46A, 98.

MANGANESE SCANDIUM SILICON
Mn₄Sc₄Si₇,   MnScSi

B.Ja. KOTUR, O.I. BODAK and O.Ja. KOTUR, 1980. Dop. Akad. Nauk Ukr. RSR, 42, 80-83.

Sc₄Mn₄Si₇, tetragonal, Zr₄Co₄Ge₇ type (1), I4/mmm, a = 13.06, c = 5.227 Å, c/a = 0.40, Z = 4, Dₓ = 4.44. R = 0.142. Contrary to 2.

ScMnSi, hexagonal, Fe₂P type (3), P6̄2m, a = 6.551, c = 3.861 Å, c/a = 0.59, Z = 3, Dₓ = 4.44. R = 0.105. Mo radiation, photographic data for both phases.

Atomic positions

Sc₄Mn₄Si₇		x	y	z
Sc(1) in	8(h)	0.137	0.137	0
Sc(2) in	8(j)	0.694	1/2	0
Mn in	16(k)	0.150	0.650	1/4
Si(1) in	8(j)	0.908	1/2	0
Si(2) in	8(i)*	0.215	1/2	1/2*
Si(3) in	8(h)	0.296	0.296	0
Si(4) in	4(e)	0	0	0.250

ScMnSi			x	y	z
Sc	in	3(g)	0.586	0	1/2
Mn	in	3(f)	0.234	0	0
Si(1)	in	2(c)	1/3	2/3	0
Si(2)	in	1(b)	0	0	1/2

*[Misprinted as 8(j), z = 0; by comparison with $Co_4Ge_7Zr_4$]

The structures are as described (1, 2) with distances: [$Sc_4Mn_4Si_7$: Sc-Sc = 3.25-3.58, Sc-Mn = 3.08-3.11, Sc-Si = 2.63-2.94; Mn-Mn = 2.61, Mn-Si = 2.42-2.50; Si-Si = 2.40-3.12; ScMnSi: Sc-Sc = 3.42, Sc-Mn = 3.00, 3.05; Sc-Si = 2.71, 2.76; Mn-Mn = 2.66, Mn-Si = 2.47, 2.57 Å.

1. Structure Reports, 34A, 67.
2. Ibid., 44A, 118.
3. Ibid., 23, 68; 33A, 13; 39A, 76.

MANGANESE SCANDIUM SULPHUR
MnS_4Sc_2 $Mn_{2.29}S_4Sc_{1.14}$

A. TOMAS, M. GUITTARD, E. BARTHÉLÉMY and J. FLAHAUT, 1981. Mater. Res. Bull., 16, 1213-1217.

$MnSc_2S_4$, cubic, normal spinel (Al_2MgO_4) type (1), Fd3m, a = 10.613 Å, Z = 8, D_X = 3.03. Mo radiation, R = 0.045 for 164 reflexions, diffractometer data. Mn in 8(a) 1/8,1/8, 1/8; Sc in 16(d) 1/2,1/2,1/2; S in 32(e) 0.2573,0.2573,0.2573. See also 2.

$Mn_{2.29}Sc_{1.14}S_4$, cubic, intermediate between rocksalt (MnS) (3) and spinel types, Fd3m, a = 10.523 Å, Z = 8, D_X = 3.48. Mo radiation, R = 0.055 for 128 reflexions, diffractometer data. 0.48 Mn in 8(a) 1/8,1/8,1/8; 9.76 Mn + 6.24 Sc(1) in 16(d) 1/2,1/2,1/2; 4.72 Mn + 2.88 Sc(2) in 16(c) 0,0,0; S in 32(e) 0.2548,0.2548,0.2548.

Interatomic distances (Å)

Phase	Tetrahedral		Octahedral	
$MnSc_2S_4$	Mn-S =	2.43	Sc-S =	2.58
$Mn_{2.29}Sc_{1.14}S_4$	Mn-S =	2.27	16(c) Mn/Sc-S =	2.68
		[2.37]	16(d) Mn/Sc-S =	2.58

1. Structure Reports, 11, 497; 13, 241; 15, 207; 33A, 272.
2. Ibid., 29, 123.
3. Ibid., 20, 151; 28, 48.

MAGNESIUM ZINC
$Mg_{51}Zn_{20}$ (Mg_7Zn_3)

I. HIGASHI, N. SHIOTANI, M. UDA, T. MIZOGUCHI and H. KATOH, 1981. J. Solid State Chem., 36, 225-233.

Orthorhombic, Immm, a = 14.083, b = 14.486, c = 14.025 Å, D_m = 3.0, Z = 2. Mo radiation, R = 0.048 for 1167 reflexions, diffractometer data; 18 site-sets given. This is the Mg_7Zn_3 of 1; two sites are occupied statistically.

The structure, (Fig. 1), is based on icosahedral coordination polyhedra; 6 of the 8 Zn atoms centre these icosahedra and distances are Zn-Zn = 2.71-3.07, Mg-Mg = 2.82-3.65, Zn-Mg = 2.60-3.20 Å.

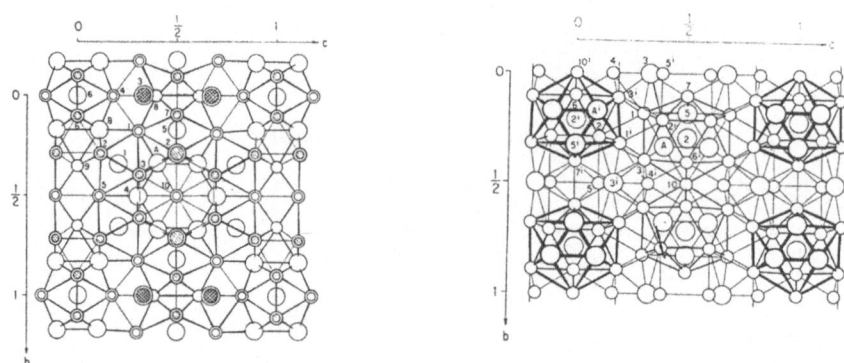

Fig. 1.   The $Mg_{51}Zn_{20}$ structure projected onto (100); the slab from $-0.25 < x < 0.25$ is on the left and the slab $0.05 < x < 0.45$ is on the right. The smaller circles represent Mg atoms and the larger ones Zn atoms.

<u>1</u>. Structure Reports, <u>21</u>, 149.

MOLYBDENUM NICKEL PHOSPHORUS
Mo_2Ni_3P $Mo_2Ni_6P_3$

I. S.V. ORIŠČIN, Ju.B. KUZ'MA and N.G. MARKIV, 1981. Dop. Akad. Nauk Ukr. RSR, <u>43</u>, No. 9, 80-83.

Mo_2Ni_3P, hexagonal, $MgZn_2$ type (<u>1</u>), $P6_3/mmc$, a = 4.688, c = 7.601 Å, c/a = 1.62, Z = 2, D_x = 9.15. Cr radiation, R = 0.178, diffractometer data. Mo in 4(f) 1/3,2/3, 0.063; 4 Ni + 2 P in 6(h) x,2x,1/4, x = 0.828; 1.5 Ni + 0.5 P in 2(a) 0,0,0. Distances are Mo-Mo = 2.84-2.87, Mo-Ni/P = 2.72-2.75, Ni/P-Ni/P = 2.36 Å.

II. S.V. ORIŠČIN and Ju.B. KUZ'MA, 1980. Kristallografija, <u>25</u>, 1066-1068 [Soviet Physics-Crystallography, <u>25</u>, 612-613].

$Mo_2Ni_6P_3$, orthorhombic, Pmmm*, a = 12.94, b = 3.568, c = 5.911 Å, D_m = 7.63, Z = 2. Cu radiation, R = 0.136 for 178 reflexions, photographic data. Mo, Ni(1) and (2), P(1) in 4(f) x,1/4,z; Ni(3) in 2(b) 1/4,3/4,z; Ni(4), P(2) in 2(a) 1/4,1/4,z.
*[Misprinted as Pmmm in English text.]

Atomic positions

	x	z		x	z
Mo	0.4187	0.384	Ni(4)	1/4	0.761
Ni(1)	0.4616	0.862	P(1)	0.611	0.320
Ni(2)	0.8493	0.946	P(2)	1/4	0.163
Ni(3)	1/4	0.431			

Interatomic distances (Å)

Mo	- 4 Mo	3.08, 3.57	Ni(1)	- 4 Mo	2.77 - 3.13	Ni(2)	- 2 Mo	2.79
	- 9 Ni	2.77 - 3.13		- 6 Ni	2.50 - 2.80		- 7 Ni	2.50 - 2.80
	- 4 P	2.52 - 2.54		- 2 P	2.29			

Ni(3)	- 4 Mo	2.83	Ni(4)	- 2 Mo	3.12	P(1)	- 3 Mo	2.52, 2.53
	- 4 Ni	2.57, 2.64		- 8 Ni	2.64, 2.80		- 6 Ni	2.27 - 2.58
	- 4 P	2.32, 2.39		- 5 P	2.37, 2.58			

P(2)	- 2 Mo	2.54
	- 7 Ni	2.29 - 2.39

The structure contains tricapped trigonal prisms (Fig. 1) centred by P atoms.

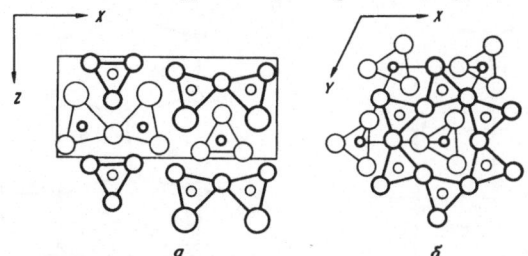

Fig. 1.   The orthorhombic $Mo_2Ni_6P_3$ structure (left) contains double chains of edge-shared prisms as compared to Fe_2P (right) where the prisms form six-fold channels.

1. Structure Reports, 46A, 95.

NICKEL NIOBIUM PHOSPHORUS
NbNiP$_2$ Nb$_6$Ni$_6$P$_9$

I. E.H.E. GHADRAOUI, R. GUÉRIN, M. POTEL and M. SERGENT, 1981. Mater Res. Bull., 16, 933-941.

NbNiP$_2$, orthorhombic, Pnma, a = 5.415, b = 3.348, c = 12.162 Å, D_m = 6.25, Z = 4. Mo radiation, R = 0.031 for 475 reflexions, diffractometer data. All atoms in 4(c) x,1/4,z; Ni in 0.02032,0.41459; Nb in 0.01510, 0.17038; P(1) in 0.13828,0.69096; P(2) in 0.19015,0.96408.

The P atoms centre trigonal prisms of Nb and Ni atoms (P-Ni = 2.28-2.37, P-Nb = 2.52-2.68) with zigzag P atom chains parallel to b (P-P = 2.79, P-P-P = 73.6°). The structure (Fig. 1) can be thought of as a mixture of the MnP and WC types. In addition -Ni-Nb-Ni- chains run parallel to a (Nb-Ni = 2.87-2.92 Å, Ni-Nb-Ni = 138°).

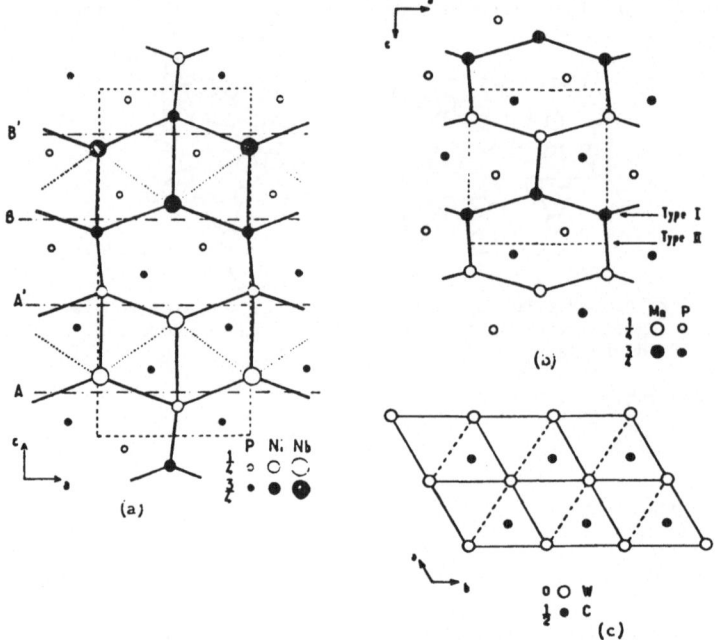

Fig. 1.   The structure of NiNbP$_2$ projected onto (010): a; the MnP structure
            projected onto (010): b; the structure of WC projected onto (001): c.

II.   R. GUERIN, M. POTEL and M. SERGENT, 1981. J. Less-Common Metals, $\underline{78}$, 177-187.

Nb$_6$Ni$_6$P$_9$, hexagonal, P6$_3$/m, a = 10.023, c = 3.408 Å, c/a = 0.34, D$_m$ = 6.60, Z = 1. Cu
radiation, R = 0.049 for 631 reflexions, diffractometer data.

Atomic positions

		x	y	z	occupancy
Nb	6(h)	0.3540	0.4882	1/4	1.0
Ni(1)	6(h)	0.2154	0.1157	1/4	1/2
Ni(2)	6(h)	0.2605	0.1371	1/4	1/2
P(1)	6(h)	0.0646	0.3688	1/4	1.0
P(2)	2(a)	0	0	1/4	1/2
P(3)	2(d)	2/3	1/3	1/4	1.0

    The structure is closely related to those of Cr$_{12}$P$_7$ ($\underline{1}$) and Fe$_{12}$Zr$_2$P$_7$ ($\underline{2}$). The
partly occupied sites give the arrangement of Fig. 2. The distances are Nb-$\overline{P}$ =
2.50-2.57, Ni-P = 2.25-2.53 Å.

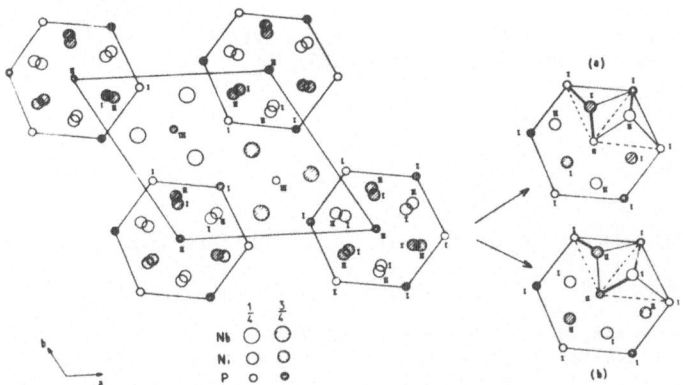

Fig. 2.   The $Nb_6Ni_6P_9$ structure showing the alternate positions of the Ni(1),
            Ni(2) and P(2) atoms.

1. Structure Reports, 45A, 54.
2. Ibid., 33A, 103.

NICKEL TITANIUM
NiTi (L.T.)

G.M. MICHAL and R. SINCLAIR, 1981. Acta Cryst., B37, 1803-1807.

Monoclinic, $P2_1/n$, a = 2.885, b = 4.622, c = 4.120 Å, γ = 96.8°, Z = 2, D_x = 6.48.
Cu radiation, R_I = 0.17 for 22 reflexions, powder diffractometer data from 1. Atoms
are in 2(e) with Ni at 0.0525,0.193,3/4* and Ti at 0.5274,0.279,1/4.
[*Incorrectly listed with z = 1/4.] See also 1 and references therein; distances
are Ti-Ni = 2.48-2.62, Ti-Ti = 2.89, 2.92 and $\overline{Ni}$-Ni = 2.72, 2.89 Å.

     There has been considerable discussion on the unit cell and structure of L.T.-
NiTi. The present unit cell agrees well with 1 and 2. The structure (Fig. 1) differs
from those of 2 and 3 and all previous workers (4). The structure is based on a
monoclinic distortion of a B2 to B19 type transformation.

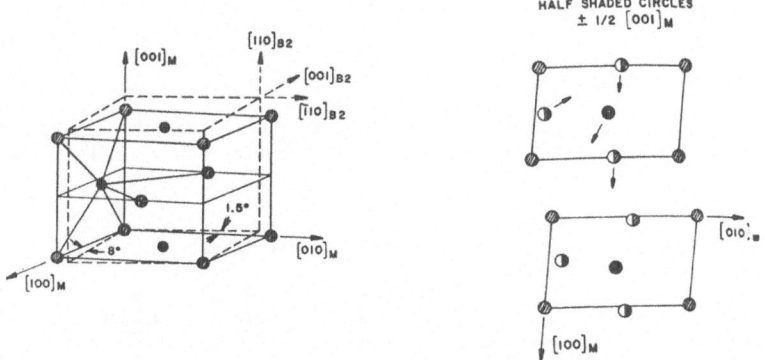

Fig. 1.   Left: The relation between the H.T.-NiTi cell (B2 type, CsCl) and the
            monoclinic L.T.-NiTi cell. Right: the shift involved in the
            transformation in terms of the monoclinic NiTi cell.

1. Structure Reports, 38A, 131.
2. Ibid., 37A, 161.
3. R.F. HEHEMANN and G. SANDROCK, 1971. Sripta Metall., 5, 801.
4. Structure Reports, 30A, 160, 181 refs. 285, 286.

NIOBIUM SULPHUR
3-R $Nb_{1.06}S_2$

I. D.R. POWELL and R.A. JACOBSON, 1981. J. Solid State Chem., 37, 140-143.

II. W.G. FISHER and M.J. SIENKO, 1980. Inorg. Chem., 19, 39-43.

I. Rhombohedral, R3m, a = 3.3285, c = 17.910 Å, c/a = 5.38, Z = 1, D_X = 4.71. Mo radiation, R = 0.026 for 79 reflexions, diffractometer data. All atoms in 3(a) 0,0,z; Nb(1) z = 0, 97% occupancy; S(1) z = 0.2463; S(2) z = 0.4201; Nb(2) z = 0.8171, 9% occupancy. II. Somewhat different lattice parameters are reported. See 1 for earlier work. Results for the basic structure (Nb(1), S(1) and S(2)) are as given in 2; Nb(2) centres a distorted octahedron of S atoms (Nb-S = 2.23, 2.58 Å).

1. Structure Reports, 18, 289, 387; 24, 191.
2. Ibid., 40A, 88.

NITROGEN VANADIUM
$N_{1.5}V_{16}$

B.V. KHAENKO, 1980. Dop. Akad. Nauk Ukr. RSR, 42, 85-89.

Tetragonal, $P4_2/mnm$, a = 8.801, c = 2.985 Å, c/a = 0.34, Z = 1, D_X = 6.00. R = 0.16, photographic data. V(1) in 8(i) 0.631,0.111,0; V(2) in 4(f) 0.151,0.151,0; V(3) in 4(f) 0.388,0.388,0; 1.5 N in 2(a) 0,0,0. This is the $NV_{~8}$ of 1.

    In this structure the N atoms are 6-coordinated (N-V = 1.88, 2.04 Å) while the V atoms are 8-, 9- or 10-coordinated (V-V = 2.58-2.79 Å).

1. Structure Reports, 42A, 144; 45A, 124.

PHOSPHORUS SILVER
Ag_3P_{11}

M.H. MÖLLER and W. JEITSCHKO, 1981. Inorg. Chem., 20, 828-833.

Monoclinic, Cm, a = 12.999, b = 7.555, c = 6.612 Å, β = 118.84°, Z = 2, D_X = 3.88. Mo radiation, R = 0.059 for 865 reflexions, diffractometer data. This is the AgP_3 of 1.

Atomic positions
Ag(1), (2), (3) and P(5), (6), (7) in 2(a) x,0,z, rest in 4(b) x,y,z.

	x	y	z		x	y	z
Ag(1)	0	0	0	P(3)	0.1676	0.2230	0.7282
Ag(2)	0.4108	0	0.1956	P(4)	0.4193	0.2267	0.7327
Ag(3)	0.7096	0	0.9773	P(5)	0.9185	0	0.2767
P(1)	0.4807	0.2529	0.4696	P(6)	0.5085	0	0.9490
P(2)	0.2363	0.1504	0.4926	P(7)	0.1888	0	0.9548

The structure (Fig. 1) can be regarded as an infinite polyanion of condensed 5-, 6- and 14-membered rings of P atoms (P-P = 2.17-2.27, mean = 2.21 Å). The Ag atoms are tetrahedrally coordinated by P atoms (Ag-P = 2.47-2.61, mean = 2.54 Å).

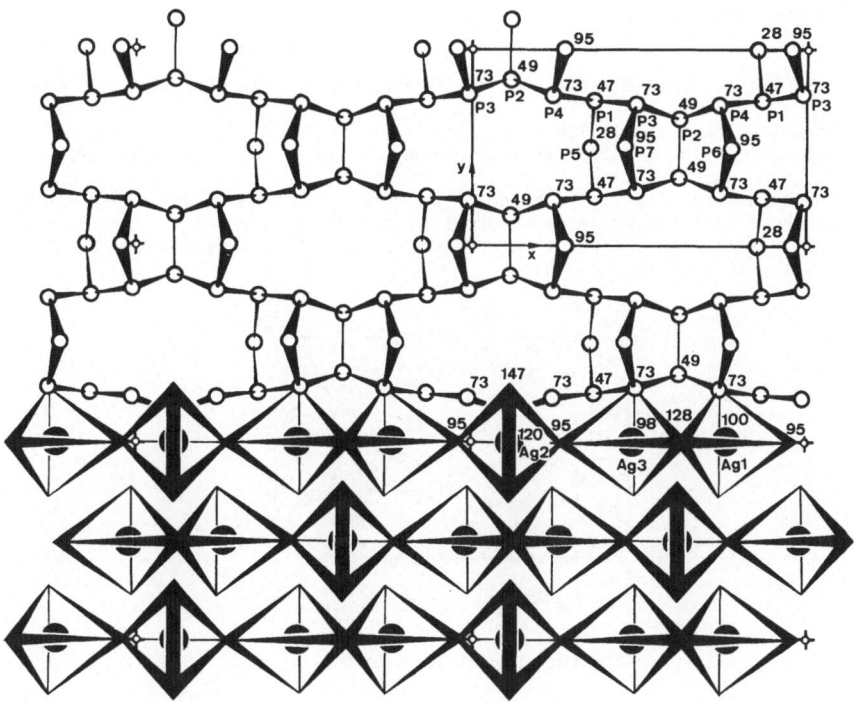

Fig. 1.   A projection of the Ag_3P_{11} structure. The top portion outlines the condensed P atom network and the lower half the chains of corner-shared AgP_4 tetrahedra.

1. W. HARALDSEN and W. BILTZ, 1931. Z. Elektrochem., 37, 502.

PHOSPHORUS SULPHUR THALLIUM
PS_4Tl_3

P. TOFFOLI, P. KHODADAD and N. RODIER, 1981. Bull. Soc. Chim. Fr., No. 11-12, I 429-I 432.

Orthorhombic, $K_3PS_4[K_3VS_4]$ type (1), Pnma, a = 8.733, b = 10.849, c = 8.959 Å, Z = 4, D_x = 6.04. Mo radiation, R = 0.057 for 724 reflexions, diffractometer data.

Atomic positions
Tl(1), S(1) in 8(d), remainder in 4(c)

	x	y	z		x	y	z
Tl(1)	0.5551	0.0488	0.1971	S(2)	0.489	1/4	0.938
Tl(2)	0.3781	1/4	0.6115	S(3)	0.298	1/4	0.255
S(1)	0.160	0.0964	0.965	P	0.276	1/4	0.027

The structure is as described with P - 4 S = 2.02-2.05, Tl - 5 S = 3.05-3.43, Tl - 7 S = 3.07-3.48 Å.

1. Structure Reports, 28, 153; 29, 82; 45A, 28.

PHOSPHORUS ZIRCONIUM
P_9Zr_{14}

L.-E. TERGENIUS, B.I. NOLÄNG and T. LUNDSTRÖM, 1981. Acta Chem. Scand., A35, 693-699.

Orthorhombic, Pnnm, a = 16.715, b = 27.572, c = 3.6742 Å, Z = 4, D_X = 6.10. Mo radiation, R = 0.108 for 1861 reflexions, photographic data. 24 site-sets are given.

The structure (Fig. 1) is based on Zr_6 trigonal prisms, centred by P atoms. Four of these have the prism axis normal to c with average P-Zr = 2.66-2.68 and five have axes parallel to c with average P-Zr = $\overline{2}$.72-2.81; the prisms are 1-, 2- or 3-capped with all Zr-P = $\overline{2}$.54-3.67 Å.

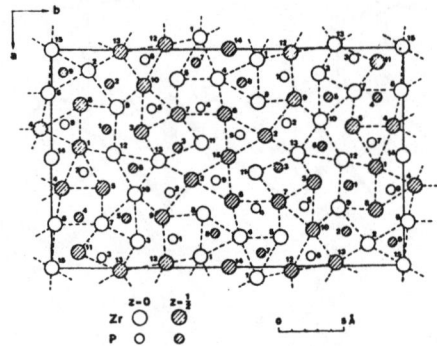

Fig. 1.   The $Zr_{14}P_9$ structure projected onto 001.

POTASSIUM SELENIUM SILVER
AgKSe

G. SAVELSBERG and H. SCHÄFER, 1981. J. Less-Common Metals, 80, P59-P69.

Tetragonal, PbFCl type (1), P4/nmm, a = 4.52, c = 7.59 Å, c/a = 1.68, Z = 2, D_X = 4.84. Cu radiation, R = 0.072 for 54 reflexions, photographic data. K and Se in 2(c) 1/4, 1/4,z, z = 0.6579 and 0.2156 respectively; Ag in 2(a) 3/4,1/4,0. The distances are K-Se = 3.32, 3.36, K-K = 3.99, 4.52, K-Ag = 3.44, Ag-Se = 2.79, Ag-Ag = 3.19 Å.

. This structure type, which has been variously known as the Cu_2Sb, Fe_2As, BiOCl, ZrSiS, ZrSiSe type, is discussed here in terms of the angles A_1, A_2 at the tetrahedral sites (2(a)) occupied by F in PbFCl (2). For a regular tetrahedron they are 109.5°; if the ratio A_1/A_2 is greater than $1.\overline{0}$ the tetrahedra will be compressed in the c-direction. Representative values are quoted in the table below; all known examples are given in the paper; the values of A_1/A_2 range from 1.15 for ClSmO to 0.60 for SUGe; for examples see 3. Others prefer to describe the structure in terms of the square pyramidal environments of the atoms in 2(c).

ABC	A_1,°	A_2,°	A_1/A_2	c/a
ClBiO	114.5	107.0	1.07	1.89
ClPbF	107.2	110.6	.97	1.80
SbCuCu	101.1	113.8	.89	1.52
FeAsFe	97.7	115.7	.84	1.65
CuAsCu	95.2	117.0	.81	1.57
SbThSb	81.6	125.0	.65	2.10
AsThAs	79.5	126.0	.63	2.15
SZrSi	77.8	127.3	.61	2.28

1. Strukturbericht, 2, 45; 3, 33.
2. Structure Reports, 40A, 20.
3. Ibid., 29, 38; 39A, 20; 41A, 26; 42A, 40; 43A, 16; 44A, 92; 45A, 8.

RARE-EARTH CARBIDES

C_3Ln_2 (Ln = Y, Er, Tm, Yb, Lu)

V.I. NOVOKŠONOV, 1980. Ž. Neorg. Khim., 25, 684-689 [Russ. J. Inorg. Chem., 25, 375-378].

Cubic, Pu_2C_3 type (1), $I\bar{4}3d$, Z = 8. Cu radiation, diffractometer data; Ln atoms in 16(c) x,x,x; C in $2\bar{4}$(d) x,0,1/4; samples prepared at 70Kbar and 1300°C.

Ln_2C_3	a(Å)	D_x	R	x(Ln)	y(c)	Ln-Ln, Å	Ln-C, Å
Y_2C_3	8.2335	5.09	0.049	0.0500	0.2821	3.40 - 3.57	2.51 - 2.86
Er_2C_3	8.1324	5.15	0.068	0.0499	0.2874	3.35 - 3.52	2.47 - 2.80
Tm_2C_3	8.0917	9.37	0.058	0.0499	0.2844	3.34 - 3.50	2.46 - 2.80
Yb_2C_3	8.0723	9.65	0.087	0.0499	0.2937	3.33 - 3.50	2.45 - 2.76
Lu_2C_3	8.0354	9.88	0.064	0.0499	0.2915	3.31 - 3.48	2.44 - 2.76

1. Structure Reports, 16, 48.

RARE-EARTH MANGANESE SILICIDES and GERMANIDE

$LnMn_2X_2$ (Ln = Pr, Nd, Y; X = Si, Ge) at 293 K

S. SIEK, A. SZYTUŁA and J. LECIEJEWICZ, 1981. Solid State Comm., 39, 863-866.

Tetragonal, $ThCr_2Si_2$ type (1), I4/mmm, Z = 2, powder neutron diffraction data. Refinements were also carried out at 1.8, 4.2, 80, 90 and 150 K. Ln in 2(a) 0,0,0; Mn in 4(d) 0,1/2,1/4; X in 4(e) 0,0,z.

$LnMn_2X_2$	a(Å)	c(Å)	c/a	D_x	R	X(z)
$PrMn_2Si_2$	4.030	10.559	2.67	5.94	0.083	0.375
$NdMn_2Si_2$	4.063	10.522	2.59	5.93	0.065	0.358
YMn_2Si_2	3.943	10.510	2.67	5.18	0.011	0.372
YMn_2Ge_2	3.984	10.850	2.72	6.63	0.045	0.386

[Interatomic distances (Å)]

$LnMn_2X_2$	Ln-Ln	Ln-Mn	Ln-X	Mn-Mn	Mn-X	X-X
$PrMn_2Si_2$	4.03	3.32	3.14	2.85	2.41	2.64
$NdMn_2Si_2$	4.06	3.32	3.24	2.87	2.33	2.99
YMn_2Si_2	3.94	3.28	3.10	2.79	2.35	2.69
YMn_2Ge_2	3.98	3.37	3.08	2.82	2.48	2.47

1. Structure Reports, 43A, 99; 46A, 17.

RHODIUM SCANDIUM SILICON   IRIDIUM SCANDIUM SILICON
$RhScSi_2$, Rh_3ScSi_7 Ir_3ScSi_7

I.   B. CHABOT, H.F. BRAUN, K. YVON and E. PARTHÉ, 1981. Acta Cryst., B37, 668-671.

II. B. CHABOT, N. ENGEL and E. PARTHÉ, 1981. Ibid., B37, 671-673.

I. $ScRhSi_2$, orthorhombic, ordered YZn_3 type (1), Pnma, a = 6.292, b = 4.025, c = 9.517 Å, Z = 4, D_X = 5.62. Mo radiation, R = $\overline{0}$.052 for 536 reflexions, diffractometer data. All atoms in 4(c) x,1/4,z, Sc 0.2385, 0.3181; Rh 0.5832, 0.9008; Si(1) 0.2318, 0.0282; Si(2) 0.9573, 0.8413.

II. $ScRh_3Si_7$, $ScIr_3Si_7$, rhombohedral, $R\overline{3}c$, Z = 6. Mo radiation, diffractometer data. Sc in 6(b) 0,0,0; Rh and Ir in 18(e) x,0,1/4, x = 0.3223, 0.3234 respectively, Si(1) in 36(f) x,y,z, x = 0.5380, 0.539, y = 0.6786, 0.680 and z = 0.0302, 0.029 respectively; Si(2) in 6(a) 0,0,1/4.

	a(Å)	c(Å)	c/a	D_X	R	refl.
$ScRh_3Si_7$	7.5056	19.691	2.62	5.706	0.037	272
$ScIr_3Si_7$	7.5010	19.909	2.65	8.403	0.072	250

The structure of $ScRhSi_2$ (Fig. 1(e)) is related to those of Re_3B, $MgAl_2Cu$, $CeZn_3$ and YZn_3 (Fig. 1(a-d); Si atoms centre tri-capped trigonal prisms (Si-Si = 2.48, 2.65, Si-Rh = 2.42-2.52, Si-Sc = 2.76-2.84 Å. If Fig. 1(e) is redrawn using only Si(2) centred prisms (Fig. 2) a similarity to TiNiSi is emphasised. The Sc atoms are 17-coordinated (Sc-Rh = 2.85-3.11; Sc-Sc 3.40 Å) while the Rh atoms are 12-coordinated.

In the $ScRh_3Si_7$ structure (Fig. 3) the Sc atoms are 18-coordinated (Sc - 6 Rh = 3.03, Sc - 6 Si = 3.08, Sc - 6 Si = 3.14 Å) and Rh atoms are 9-coordinated (Rh-Si = 2.42-2.48 Å). The Si atoms are 13- and 15-coordinated (Si-Si = 2.59-3.35 Å). The distances in $ScIr_3Si_7$ are very similar.

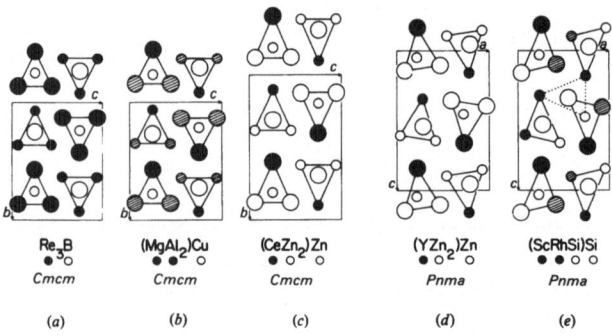

Re_3B	$(MgAl_2)Cu$	$(CeZn_2)Zn$	$(YZn_2)Zn$	$(ScRhSi)Si$
Cmcm	Cmcm	Cmcm	Pnma	Pnma
(a)	(b)	(c)	(d)	(e)

Fig. 1.   Structures related to $ScRhSi_2$.

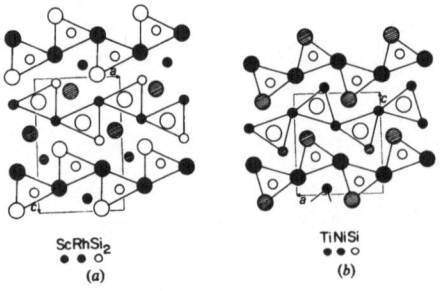

$ScRhSi_2$	TiNiSi
(a)	(b)

Fig. 2.   The $ScRhSi_2$ and TiNiSi structures.

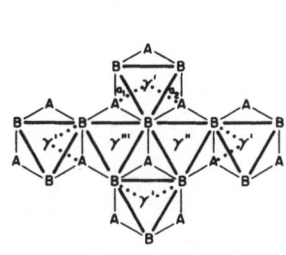

Fig. 3.   a: The ScRh$_3$Si$_7$ structure; Rh atoms are hexagonal close packed (A,B)
           and Sc atoms (γ) occupy one third of the octahedral interstices. The
           complete stacking is Aγ'Bγ''Aγ'''Bγ'Aγ''Bγ'''. b: Si atoms centre 4 of
           the 8 triangular faces of the empty Rh octahedra forming pairs of corner
           shared tetrahedra.

<u>1</u>. Structure Reports, <u>33A</u>, 100.

SCANDIUM SULPHUR
S$_3$Sc$_2$

I.   J.P. DISMUKES and J.G. WHITE, 1964. Inorg. Chem., <u>3</u>, 1220-1228.

II. Idem, 1963. Acta Cryst., <u>16</u>, A24.

Orthorhombic, Fddd, a = 10.41, b = 7.38, c = 22.05 Å, D$_m$ = 2.897, Z = 16. Cu
radiation, R = 0.078 for 244 reflexions, photographic data. Sc(1) and (2) in 16(g)
1/8,1/8,z, z = 0.0408 and 0.3761 respectively; S(1) in 16(f) 1/8,0.3697,1/8; S(2) in
32(h) 0.1231,0.3783, 0.4561. See also <u>1</u>; contrary to <u>2</u>.

     This structure is very closely related to the NaCl-type structure and contains
12 cation-deficient NaCl-type subcells with the voids ordered in such a way that each
sulphur-centered 'octahedron' has two empty positions. The distortion of the bond
angles is quite small, the average deviation from 90° being only 1-2°; the Sc-S
distances range from 2.57 to 2.61 Å.

<u>1</u>. Strukturbericht, <u>2</u>, 323.
<u>2</u>. Structure Reports, <u>26</u>, 245.

SELENIUM
Se

I.   Y. MIYAMOTO, 1979. Fukuoka Daigaku Rigaku Shuho, <u>9</u>, 1-13.

II. Idem, 1980. Jap. J. Applied Phys., <u>19</u>, 1813-1819.

Rhombohedral, rhombohedral-S type (<u>1</u>), R$\bar{3}$, A = 18. Photographic data for I,
diffractometer data for II. Se in 18(f) x,y,z.

	a(Å)	c(Å)	R	ref.	x	y	z
I	11.4	4.47	0.099		0.16018	0.20223	0.12038
II	11.362	4.429	0.099	578	0.16021	0.20227	0.12045

The structure is as described (1) with Se-Se = 2.356 Å, Se-Se-Se = 101.1° with dihedral angle = 76.2°.

1. Structure Reports, 26, 288; 29, 230.

SELENIUM SILVER THALLIUM
AgSeTl

K. KLEPP, 1980. Mh. Chem., 111, 1433-1436.

Orthorhombic, NiSiTi type (1), Pnma, a = 7.4756, b = 4.6375, c* = 8.390 Å, Z = 4, D_X = 8.93. Mo radiation, R = 0.045, diffractometer data. Atoms in 4(c) x,1/4,z; Tl 0.5005,0.6772; Ag 0.6706,0.0822; Se 0.3179,0.1048. *[Misprinted as 8.690 - Author.]

Interatomic distances (Å)

Tl - 6 Se	3.25 - 3.84	Ag - 4 Se	2.64 - 2.85	Se - 2 Se	3.99
- 6 Ag	3.29 - 3.63	- 6 Tl	3.29 - 3.63	- 4 Ag	2.64 - 2.85
- 4 Tl	3.77 - 3.93	- 2 Ag	3.71	- 6 Tl	3.25 - 3.84

1. Structure Reports, 30A, 75; 34, 176.

SELENIUM TELLURIUM TITANIUM   TELLURIUM TITANIUM
SeTeTi Te_2Ti

Y. ARNAUD and M. CHEVRETON, 1981. J. Solid State Chem., 39, 230-239.

Trigonal, CdI_2 type (1), P3̄m1, Z = 1. Mo radiation, diffractometer data. Ti in 1(a) 0,0,0; X in 2(d) 2/3,1̄/3,z; X = Te and Se/Te; occupancy for Se/Te = 50% Se, 50% Te. See also 2; Ti - 6 Te = 2.77 Å; Ti - 6 Se/Te = 2.67 Å.

TiX_2	a(Å)	c(Å)	c/a	D_X	R	refl.	z(X)
$TiTe_2$	3.777	6.498	1.720	6.26	0.050	195	0.2628
TiSeTe	3.651	6.317	1.730	5.79	0.039	247	0.26053

1. Strukturbericht, 1, 161, 189.
2. Ibid., 1, 163, 21̄3, 780; Structure Reports, 13, 228; 27, 366; 45A, 112.

SELENIUM THALLIUM
Se_3Tl_5

L.I. MAN, V.S. PARMON, R.M. IMAMOV and A.S. AVILOV, 1980. Kristallografija, 25, 1070-1072 [Soviet Physics-Crystallography, 25, 614-615].

Tetragonal, Cr_5B_3 [Tl_5Se_3] type, P4/n, a = 8.54, c = 12.38 Å, c/a = 1.45, Z = 4, D_X = 9.26. R = 0.15 for 108 reflexions, photographic data. Contrary to 1 who give the formula Tl_2Se and use P4/ncc.

Atomic positions
Origin at 1

		x	y	z
Tl(1) in 8(g)		0.407	0.905	0.091
Tl(2) in 8(g)		0.079	0.860	0.585
Tl(3) in 2(c)		1/4	1/4	0.235
Tl(4) in 2(c)		1/4	1/4	0.717
Se(1) in 8(g)		0.580	0.094	0.232
Se(2) in 2(c)		1/4	1/4	0.990
Se(3) in 2(c)		1/4	1/4	0.490

The structure has layers similar to those in Cr_5B_3; distances are Tl-Tl = 3.20, 3.23; Tl-Se = 2.80-3.38 Å.

1. Structure Reports, 22, 189.

SELENIUM TIN   SULPHUR TIN
β-SeSn at 835 K   β-SSn at 905 K

H.G. von SCHNERING and H. WIEDEMEIER, 1981. Z. Kristallogr., 156, 143-150.

Orthorhombic, TlI type (1), CmCm, Z = 4; high temperature Guinier-Simon data on powdered samples for the structure and high-temperature single-crystal photographs for the space group and Laue symmetry. See 2 for earlier discussions on the α-β change.

All atoms in 4(c) 0,y,1/4

SnX	a(Å)	b(Å)	c(Å)	R	refl.	y(Sn)	y(X)	t, °K
SnS	4.148	11.480	4.177	0.10	20	0.120	0.349	905
SnSe	4.310	11.705	4.318	0.06	15	0.120	0.356	825

X = S or Se

It was proposed previously that the λ-transition between α and β forms occurred along a reaction path GeS (B16 type)→TlI (B33 type). The present work confirms this from measurements between 705 and 905 K. The interatomic distances in Å are summarised below. A detailed discussion is given.

M-X	M-X	M-X	M-X	M-M	X-X	X-X	X-X	X-X
SnS	2.63	2.96	3.74	3.46	4.15	4.18	3.72	4.05
SnSe	2.76	3.06	3.77	3.54	4.31	4.32	3.93	4.00

1. Strukturbericht, 4, 6.
2. Structure Reports, 45A, 76, 110, 126.

SILICON SODIUM TELLURIUM
$Na_6Si_2Te_6$

B. EISENMANN, H. SCHWERER and H. SCHÄFER, 1981. Z. Naturf., 36b, 1538-1541.

Monoclinic, $K_6Sn_2Te_6$ type (1), $P2_1/c$, a = 8.786, b = 12.780, c = 8.864 Å, β = 119.71°, D_m = 3.70, Z = 2. Mo radiation, R = 0.064 for 2410 reflexions, diffractometer data.

Atomic positions
All atoms in 4(e) x,y,z

	x	y	z		x	y	z
Te(1)	0.7459	0.1887	0.8865	Na(1)	0.5053	0.3459	0.9872
Te(2)	0.8511	0.9744	0.6072	Na(2)	0.2750	0.9355	0.7944
Te(3)	0.6718	0.8735	0.9284	Na(3)	0.0396	0.2760	0.2307
Si	0.8549	0.0077	0.8867				

The structure has $Te_3Si-SiTe_3$ groups (Si-Si = 2.35, Si-Te = 2.50) and distances Te-Na = 3.07-3.53 Å.

1. Structure Reports, 45A, 75.

SILICON STRONTIUM ZINC
SiSrZn

SILICON SILVER STRONTIUM
$Ag_5Si_9Sr_8$

W. DÖRRSCHEIDT and H. SCHÄFER, 1981. J. Less-Common Metals, 78, 69-79.

SrZnSi, hexagonal, Ni_2In type (1), $P6_3/mmc$, a = 4.30, c = 9.02 Å, c/a = 2.10, D_m = 4.13, Z = 2. Mo radiation, R = 0.062 for 36 reflexions, photographic data. Sr in 2(a) 0,0,0; Zn in 2(c) 1/3,2/3,1/4; Si in 2(d) 1/3,2/3,3/4.

$Sr_8Ag_5Si_9$, orthorhombic, C222, a = 8.48, b = 14.69, c = 18.40 Å, Z = 4, D_X = 4.33. Mo radiation, R = 0.069 for 959 reflexions, diffractometer data; 21 site-sets are given.

SrZnSi has distances, Sr-Sr = 4.30, 4.51, Sr-Zn = 3.35, Sr-Si = 3.35 and Zn-Si = 2.48 Å. The $Sr_8Ag_5Si_9$ structure is a distorted variant of the AlB_2 type, with some disorder and vacancies on the sites; distances are Sr-Sr = 4.24-4.65; Ag-Sr = 3.33-3.39; Si-Sr = 3.18-3.53; Si-Si = 2.33; Ag/Si-Ag/Si = 2.32-2.57 Å.

1. Structure Reports, 9, 91.

STRONTIUM TIN
$SnSr_2$ SnSr

A. WIDERA and H. SCHÄFER, 1981. J. Less-Common Metals, 77, 29-36.

Sr_2Sn, orthorhombic, NiSiTi type (1), Pnma, a = 8.402, b = 5.378, c = 10.078 Å, Z = 4, D_X = 4.29. Mo radiation, R = 0.098 for 493 reflexions, diffractometer data. Sr(1), (2) and Sn in 4(c) x,1/4,z, x = 0.6561,0.5213,0.2512 respectively, z = 0.0723,0.6808, 0.1068 respectively. See also 2.

SrSn, orthorhombic, CrB type (3), Cmcm, a = 5.045, b = 12.040, c = 4.494 Å, Z = 4, D_X = 5.02. Mo radiation, R = 0.097 for 205 reflexions, diffractometer data. Sr and Sn in 4(c) 0,y,1/4, y = 0.1367, 0.4212 respectively. See also 4.

Structures are as previously described (2,4).

1. Structure Reports, 30A, 75.
2. Ibid., 44A, 3.
3. Ibid., 12, 30; 13, 48; 20, 88; 29, 51.
4. W. RIESER and E. PARTHÉ, 1967. Acta Cryst., 22, 919.

SULPHUR TANTALUM
S_3Ta

A. MEERSCHAUT, L. GUEMAS and J. ROUXEL, 1981. J. Solid State Chem., <u>36</u>, 118-123.

Monoclinic, $P2_1/m$, a = 9.515, b = 3.3412, c = 14.912 Å, β = 109.99°, Z = 6, D_x = 6.20.
Mo radiation, R = 0.025 for 1766 reflexions, diffractometer data. 12 site-sets, all
in 2(e), are given.

The structure is very similar to that of $NbSe_3$ (<u>1</u>) with Ta-S = 2.46-2.87 Å;
there are three short S-S distances (2.11, 2.07 and 2.84 Å.

<u>1</u>. Structure Reports, <u>41A</u>, 99; <u>45A</u>, 98.

SULPHUR THALLIUM TIN
$S_5Sn_2Tl_2$ S_3SnTl_4

I. G. EULENBERGER, 1981. Z. Naturf., <u>36B</u>, 687-690.

$S_5Sn_2Tl_2$, monoclinic, C2/c, a = 11.115, b = 7.723, c = 11.492 Å, β = 108.60°, Z = 4,
D_x = 5.73. Mo radiation, R = 0.049 for 1052 reflexions, diffractometer data.

Atomic positions
Tl, Sn, S(1) and S(2) in 8(f) x,y,z; S(3) in 4(e) 0,y,1/4

	x	y	z		x	y	z
Tl	0.22055	0.32650	0.31760	S(2)	0.0124	0.1391	0.1147
Sn	0.40613	0.34540	0.05926	S(3)	0	0.3199	1/4*
S(1)	0.3043	0.0517	0.9616				

*[Misprinted as 0]

The structure contains SnS_5 bipyramids (Sn-S = 2.41-2.62) sharing corners to
form a network; Tl atoms occupy interstices (Tl - 9 S = 3.07-3.90 Å).

II. S. DEL BUCCHIA, J.-C. JUMAS, E. PHILIPPOT and M. MAURIN, 1981. Rev. Chim.
Minér., <u>18</u>, 224-234.

S_3SnTl_4, tetragonal, P4/ncc, a = 8.305, c = 12.647 Å, c/a = 1.52, D_m = 7.85, Z = 4.
Mo radiation, R = 0.082 for 234 reflexions, diffractometer data. Tl in 16(g)
0.4288, 0.8687,0.1017; Sn and S(1) in 4(c) 1/4,1/4,z, z = 0.1866 and 0.436
respectively; S(2) in 8(f) 0.419,-0.419,1/4.

The Sn atoms are bonded to 6 S forming distorted octahedra (Sn-S = 3.15-3.19 Å);
the Tl atoms occupy the apices of TlS_3 pyramids (Tl-S = 2.87-3.04 Å). These pyramids
are linked to form 8-rings (Tl_4S_4) centred by an S atom, giving an overall unit of
composition Tl_4S_4 (Fig. 1). These rings are linked by the S atoms which are common
to two rings to form infinite sheets which are linked by a fourth Tl-S bond (3.47 Å).

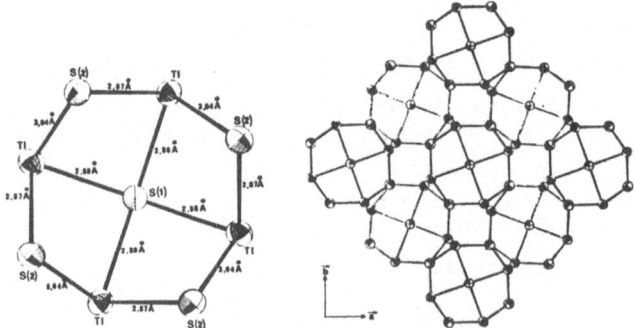

Fig. 1.   Left: the Tl_4S_4 rings in Tl_4SnS_3. Right: the sheets formed by the
             linked rings.

SULPHUR TITANIUM
$S_2Ti_{1.083}$ (1T-), S_3Ti_2

I. E. TRONC and R. MORET, 1981. Synthetic Metals, 4, 113-118.

$S_2Ti_{1.083}$, trigonal, $P\bar{3}m1$, a = 3.413, c = 5.717 Å, c/a = 1.68, Z = 1, D_x = 3.33.
Mo radiation, R = 0.016 for 304 reflexions, diffractometer data. Ti(1) in 1(a) 0,0,0;
0.083 Ti(2) in 1(b) 0,0,1/2; S in 2(d) 1/3,2/3,0.24981. There has been considerable
discussion on the nature of the charge carriers in $Ti_{1+x}S_2$ (1-3). The present
results, which are compared with earlier ones below, indicate a nearly ideal 1T-
structure and suggest that reinvestigation is desirable.

Sample	refl.	R	z(S)	Ti1-Ti2(Å)	Ti1-S(Å)	Ti2-S(Å)	S-S(Å)	Refer.
TiS	43	.047	.250	-	2.428	--	3.407	1
$Ti_{1.023}S$	861	.043	.24926	2.847	2.427	2.431	3.409	2
$Ti_{1.083}S$	304	.016	.24981	2.859	2.434	2.435	3.413	This work

II. M. ONODA and M. SAEKI, 1980. Chem. Lett. Japan, 665-666.

S_3Ti_2, rhombohedral, $R\bar{3}m$, a = 3.440, c = 17.100 Å, c/a = 4.97, Z = 2, [D_x = 3.64].
Cu radiation, R = 0.33, diffractometer data. S in 6(c) 0,0,0.25; Ti(1) in 3(a)
0,0,0; 1 Ti(2) in 3(b) 0,0,1/2. [S - 6 Ti = 2.44; Ti - 6 S, Ti - 6 Ti = 3.44 Å.
Compare "Zr_3Se_4", (4)]. See also 5.

1.   Structure Reports, 41A, 109. II, III.
2.   Ibid., 41A, 109. I.
3.   Ibid., 46A, 97.
4.   Ibid., 33A, 142.
5.   Ibid., 13, 228; 17, 447; 22, 196; 24, 233; 27, 355.

SULPHUR TIN
SSn

S. DEL BUCCHIA, J.-C. JUMAS and M. MAURIN, 1981. Acta Cryst., B37, 1903-1905.

Orthorhombic, GeS type (1), Pnma, a = 11.180, b = 3.982, c = 4.329 Å, Z = 4, D_x = 5.19.
Mo radiation, R = 0.041 for 418 reflexions, diffractometer data. Atoms in 4(c) Sn
with 0.11937,1/4,0.1194; S with 0.1493,3/4,0.5201. The structure is as previously
described (2) with Sn - 3 S = 2.62, 2.66 and Sn - 3 S = 3.29, 3.39 Å.

1. Strukturbericht, 2, 8.
2. Ibid., 3, 14, 253. Structure Reports, 45A, 76, 110, 126.

TITANIUM ZINC
TiZn$_{16}$

M. SAILLARD, G. DEVELEY, C. BECLE, J.M. MOREAU and D. PACCARD, 1981. Acta Cryst., B37, 224-226.

Orthorhombic, Cmcm, a = 7.720, b = 11.449, c = 11.755 Å, Z = 4, D_X = 6.99. Mo radiation, R = 0.11 for 407 reflexions, diffractometer data. This is the TiZn$_{15}$ phase of 1.

Atomic positions

		x	y	z				x	y	z
Ti	in 4(c)	0	0.953	1/4	Zn(4)	in	8(f)	0	0.1763	0.1399
Zn(1)	4(a)	0	0	0	Zn(5)	in	8(g)	0.166	0.3553	1/4
Zn(2)	4(c)	0	0.7114	1/4	Zn(6)	in	16(h)	0.3316	0.3135	0.0702
Zn(3)	8(f)	0	0.3807	0.0306	Zn(7)	in	16(h)	0.2942	0.0529	0.1352

The Ti atoms are 15-coordinated (Ti-Zn = 2.77-2.99) whereas Zn atoms are 11-, 12- or 13-coordinated (Zn-Zn = 2.52-3.18 Å). The structure can be described as in Fig. 1.

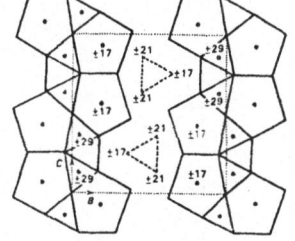

Fig. 1.   There are chains of linked pentagons and triangles in the x=0 plane of TiZn$_{16}$. These are centred as shown by atoms at x = ±.17 and ±.29. In addition, there are isolated trigonal prisms (x = ±.17, ±.21).

1. Structure Reports, 27, 372.

TABLE I

This Table lists substances for which structures have been assigned but not refined (usually on the basis of a powder pattern, and assumption of the atomic parameters of the type structure).

Phase	Structure type	Structure code	Reference
$Ag_4Bi_{20}Cu_{7.20}$ $Pb_{4.80}S_{40}$	$Ag_4Bi_{20}Cu_{7.2}$ $Pb_{4.8}S_{40}$	mC76	Canad. Miner., $\underline{18}$, 181
AgBiEu	$InNi_2$	hP6	Z. Naturforsch.,B36, 1193
$AgBiSe_2$	$AgBiSe_2$	hP48	Canad. Miner., $\underline{18}$, 353
AgCeSn	$CaIn_2$	hP6	J. Less-Common Metals, 80, 47
$Ag_2CoS_8Sn_3$	$Cu_2FeS_8Sn_3$	tI28	Rev. Chim. Minér., $\underline{18}$, $\overline{33}$
$AgCu_5PbS_4$	$AgCu_5PbS_4$	mC44	Bull. Soc. Fr. Minér. Crist., 104, 737
AgDySn	$CaIn_2$	hP6	J. Less-Common Metals, $\underline{80}$, 47
AgErSn	$CaIn_2$	hP6	Ibid., 80, 47
AgEuP	$InNi_2$	hP6	Z. Naturforsch., B36, 1193
AgEuSb	$InNi_2$	hP6	Ibid., B36, 1193
$Ag_2FeS_8Sn_3$	$Cu_2FeS_8Sn_3$	tI28	Rev. Chim. Minér., $\underline{18}$, 33
AgGdSn	$CaIn_2$	hP6	J. Less-Common Metals, $\underline{80}$, 47
Ag_8GeTe_6	Ag_8GeTe_6	cF60	J. Solid State Chem., $\underline{38}$, 259
AgHoSn	$CaIn_2$	hP6	J. Less-Common Metals, $\overline{80}$, 47
$AgIn_5S$	$A_{12}MgO_4$	cF56	J. Mater. Sci., $\underline{16}$, $171\overline{3}$
AgLaSn	$CaIn_2$	hP6	J. Less-Common Metals, 80, 47
$Ag_2MnS_8Sn_3$	$Cu_2FeS_8Sn_3$	tI28	Rev. Chim. Minér., $\underline{18}$, $\overline{33}$
$Ag_2NiS_8Sn_3$	$Cu_2FeS_8Sn_3$	tI28	Ibid., $\underline{18}$, 33
Ag_2PS_3	$Ag_4P_2S_6$	mP72	C. R. Acad. Sci. Paris, Ser. C, 291, 275
Ag_3S_3Sb	Ag_3AsS_3	hR14	J. Solid State Chem., $\underline{40}$, 203
$Ag_{0.67}S_2Ta$	$Ag_{0.67}S_2Ta$	hP7.34	Mater. Res. Bull., $\underline{15}$, $\overline{1703}$
$Ag_{0.33}S_2Ta$	$Ag_{0.33}S_2Ta$	hP20	Ibid., $\underline{15}$, 1703
$Ag_{0.42}S_2Ti$	$Ag_{0.42}S_2Ti$	hP3.42	Ibid., $\underline{15}$, 1703
$Ag_{0.20}S_2Ti$	$Ag_{0.2}S_2Ti$	hP6.40	Ibid., $\overline{15}$, 1703
AgSmSn	$CaIn_2$	hP6	J. Less-Common Metals, $\underline{80}$, 47
AgSnTb	$CaIn_2$	hP6	Ibid., $\underline{80}$, 47
AgSnYb	$CaIn_2$	hP6	Ibid., $\overline{80}$, 47
Al	Cu	cF4	J. Less-Common Metals, $\underline{44}$, 215
AlB_{12}	AlB_{12}	oP377	Ibid., $\underline{81}$, 135
$A_{14}C_3$	$A_{14}C_3$	hR7	J. Solid State Chem., $\underline{32}$, 213
$AlCCr_2$	$AlCCr_2$	hP8	Ibid., $\underline{32}$, 213
$AlCTi_2$	$AlCCr_2$	hP8	Ibid., $\overline{32}$, 213
$AlCTi_3$	CaO_3Ti	cP5	Ibid., $\overline{32}$, 213
$AlCV_2$	$AlCCr_2$	hP8	Ibid., $\overline{32}$, 213
Al_2Ce	Cu_2Mg	cF24	Dop. Akad. Nauk Ukr. RSR., A, 42, 84
Al_4CeCo	Al_4CoLa	oP12	Ibid., 42, 84
Al_8CeCo_2	Al_8CeFe_2	oP44	Ibid., $\overline{42}$, 84
$Al_2Ce_2Co_{15}$	Th_2Zn_{17}	hR19	Ibid., $\overline{42}$, 84
Al_5CeNi_2	Al_5Ni_2Pr	oI16	Ibid., $\overline{43}$, 86
$AlCr_2$	$MoSi_2$	tI6	J. Solid State Chem., $\underline{32}$, 213
$Al_3Ge_4Pd_{12}$	$S_{12}Th_7$	hP19	Z. Metallk., $\underline{72}$, 279
$Al_{0.70}Ge_{1.30}W$	$CrSi_2$	hP9	Izv. Akad. Nauk SSSR Metal., 231
$Al_{1.25}HfZn_{0.75}$	Cu_2Mg	cF24	Z. Naturforsch., B36, 1547
AlHfZn	Cu_2Mg	cF24	Ibid., $\underline{36}$, 1547
$Al_{0.75}HfZn_{1.25}$	Cu_2Mg	cF24	Ibid., $\overline{36}$, 1547
Al_5LaNi_2	Al_5Ni_2Pr	oI16	Dop. Akad. Nauk Ukr. RSR., A, $\underline{43}$, 86
Al_3Nb	Al_3Ti	tI16	Ibid., 44, 215
Al_3Nb	Al_3Ti	tI16	Ibid., $\overline{75}$, 227
Al_3Nb	Al_3Ti	tI16	Trans. Met. Soc. AIME, $\underline{236}$, 863

Phase	Structure type	Structure code	Reference
$AlNb_3$	Cr_3O	cP8	J. Less-Common Metals, 44, 215
$AlNb_3$	Cr_3O	cP8	Trans. Met. Soc. AIME, 236, 863
$AlNb_3$	Cr_3O	cP8	J. Less-Common Metals, 75, 227
$AlNb_2$	CrFe	tP30	Trans. Met. Soc. AIME, 236, 863
$AlNb_2$	CrFe	tP30	J. Mater Sci., 11, 760
$AlNb_2$	CrFe	tP30	J. Less-Common Metals, 44, 215
$AlNb_2$	CrFe	tP30	Ibid., 75, 227
$AlNi_4U$	Ni_5U	cF24	Fizika, 12, 200
$Al_2Pd_{12}Si_5$	$S_{12}Th_7$	hP19	Z. Metallk., 72, 279
Al_2S_3	In_2S_3	tI80	Z. Naturforsch., B36, 532
Al_2S_3	Al_2S_3	hP30	Ibid., B36, 532
Al_4Sr	Al_4Ba	tI10	Russian Metallurgy, 167
Al_2Sr_3	Ga_2Sr_3	cP50	Ibid., 167
Al_2Sr	$CeCu_2$	oI12	Ibid., 167
$AlTa_2$	CrFe	tP30	J. Mater. Sci., 11, 760
Al_3Ti	Al_3Ti	tI8	J. Solid State Chem., 32, 213
$AlTi_3$	Ni_3Sn	hP8	Ibid., 32, 213
$AlTi$	AuCu	tP4	Ibid., 32, 213
Al_8V_5	Cu_5Zn_8	cI52	Ibid., 32, 213
Al_3V	Al_3Ti	tI8	Ibid., 32, 213
$Al_{0.66}Zn_2Zr_{1.34}$	$AuCu_3$	cP4	Z. Naturforsch., B36, 1547
$Al_{1.60}Zn_{0.40}Zr$	Cu_2Mg	cF24	Ibid., B36, 1547
$Am_{1.20}Mo_6Se_8$	$Cu_2Mo_6S_8$	hR15.20	J. Solid State Chem., 39, 360
$AmMo_6Se_8$	Mo_6PbS_8	hR15	Ibid., 39, 360
$AsAuEu$	$InNi_2$	hP6	Z. Naturforsch., B36, 1193
$AsBk$	ClNa	cF8	J. Inorg. Nucl. Chem., 42, 995
$As_7Ce_2Ni_{12}$	$Fe_{12}P_7Zr_2$	hP21	J. Less-Common Metals, 79, 311
$AsCePd$	$CaIn_2$	hP6	Ibid., 78, 1
$AsCf$	ClNa	cF8	Inorg. Nucl. Chem. Lett., 16, 537
As_2Co_2K	Al_4Ba	tI10	Z. Naturforsch., B36, 1668
$As_7Dy_2Ni_{12}$	$Fe_{12}P_7Zr_2$	hP21	J. Less-Common Metals, 79, 311
$As_7Er_2Ni_{12}$	$Fe_{12}P_7Zr_2$	hP21	Ibid., 79, 311
$As_2Ga_5Pd_{12}$	$S_{12}Th_7$	hP19	Z. Metallk., 72, 279
$As_7Gd_2Ni_{12}$	$Fe_{12}P_7Zr_2$	hP21	J. Less-Common Metals, 79, 311
As_4GeSr_4	As_4Ba_4Si	cP72	Z. anorg. Chem., 475, 74
As_3Hf_5	As_3Hf_5	oP64	Acta Chem. Scand., 22, 2395
As_2Hf_3	Hf_3P_2	oP20	Ibid., 22, 2395
As_2Hf	Co_2Si	oP12	Ibid., 22, 2395
$AsHf_2$	PTa_2	oP36	Ibid., 22, 2395
$AsHf$	AsTi	hP8	Ibid., 22, 2395
$AsHf_9Mo_4$	Hf_9Mo_4Ni	hP28	Acta Chem. Scand. Ser., A, 35, 227
$As_7Ho_2Ni_{12}$	$Fe_{12}P_7Zr_2$	hP21	J. Less-Common Metals, 79, 311
$As_{0.50}In_{0.50}Ni_4U$	Ni_5U	cF24	Fizika, 12, 200
As_2K_2Pt	As_2K_2Pd	oC20	Z. Naturforsch., B36, 1666
$As_7Lu_2Ni_{12}$	$Fe_{12}P_7Zr_2$	hP21	J. Less-Common Metals, 79, 311
$As_7Nd_2Ni_{12}$	$Fe_{12}P_7Zr_2$	hP21	Ibid., 79, 311
$AsNdPd$	$CaIn_2$	hP6	Ibid., 78, 1
$As_7Ni_{12}Pr_2$	$Fe_{12}P_7Zr_2$	hP21	Ibid., 79, 311
$As_7Ni_{12}Sm_2$	$Fe_{12}P_7Zr_2$	hP21	Ibid., 79, 311

Phase	Structure type	Structure code	Reference
$As_7Ni_{12}Tb_2$	$Fe_{12}P_7Zr_2$	hP21	Ibid., $\underline{79}$, 311
$As_7Ni_{12}Tm_2$	$Fe_{12}P_7Zr_2$	hP21	Ibid., $\overline{79}$, 311
$As_7Ni_{12}Y_2$	$Fe_{12}P_7Zr_2$	hP21	Ibid., $\underline{79}$, 311
$As_7Ni_{12}Yb_2$	$Fe_{12}P_7Zr_2$	hP21	Ibid., $\overline{79}$, 311
AsP_3S_3	As_4S_3	oP28	Z. Naturforsch., B36, 1493
AsPdPr	$CaIn_2$	hP6	J. Less-Common Metals, 78, 1
As_4S_3	As_4S_3	oP28	Z. Naturforsch., B36, $\overline{1}493$
AsS	AsS	mP32	Canad. Miner., $\underline{18}$, $\overline{5}25$
$As_6S_{21}Sb_7Tl_3$	$As_6S_{21}Sb_7Tl_3$	aP259	Bull. Soc. Fr. $\overline{M}$iner. Crist., 104, 10
AsS_3Tl_3	AsS_3Tl_3	hR7	Z. $\overline{K}$ristallogr., $\underline{150}$, 169
As_4SiSr_4	As_4Ba_4Si	cP72	Z. anorg. Chem., $\overline{475}$, 74
AuEuP	$InNi_2$	hP6	Z. Naturforsch., $\overline{B}36$, 1193
AuEuSb	$InNi_2$	hP6	Ibid., B36, 1193
$AuSn_4$	$PtSn_4$	oC20	J. Less-Common Metals, $\underline{80}$, 53
$B_{6.10}Ce$	B_6Ca	cP7.10	Inorg. Mater. USSR, $\underline{16}$, $\overline{3}00$
B_6CeCo_{12}	$B_6Ni_{12}Sr$	hR19	Dop. Akad. Nauk Ukr. $\overline{R}$SR, A, 43, 80
B_2CeIr_3	B_2ErIr_3	mC12	J. Less-Common Metals, 78, 99
$B_2Ce_3Ni_{13}$	$B_2Nd_3Ni_{13}$	hP18	Dop. Akad. Nauk Ukr. RS$\overline{R}$, A, 43, 87
$B_6Co_{12}Dy$	$B_6Ni_{12}Sr$	hR19	I$\overline{b}$id., 43, 80
$B_6Co_{12}Er$	$B_6Ni_{12}Sr$	hR19	Ibid., $\overline{43}$, 80
$B_6Co_{12}Eu$	$B_6Ni_{12}Sr$	hR19	Ibid., $\overline{43}$, 80
$B_6Co_{12}Gd$	$B_6Ni_{12}Sr$	hR19	Ibid., $\overline{43}$, 80
$B_6Co_{12}Ho$	$B_6Ni_{12}Sr$	hR19	Ibid., $\overline{43}$, 80
$B_6Co_{12}La$	$B_6Ni_{12}Sr$	hR19	Ibid., $\overline{43}$, 80
$B_6Co_{12}Nd$	$B_6Ni_{12}Sr$	hR19	Ibid., $\overline{43}$, 80
$B_6Co_{12}Pr$	$B_6Ni_{12}Sr$	hR19	Ibid., $\overline{43}$, 80
$B_6Co_{12}Sm$	$B_6Ni_{12}Sr$	hR19	Ibid., $\overline{43}$, 80
$B_6Co_{22}Ta$	C_6Cr_{23}	cF116	Mh. Chem., 110, 791
$B_6Co_{12}Tb$	$B_6Ni_{12}Sr$	hR19	Dop. Akad. $\overline{N}$auk Ukr. RSR, A, 43, 80
$B_6Co_{12}Y$	$B_6Ni_{12}Sr$	hR19	I$\overline{b}$id., 43, 80
B_2DyIr_3	B_2ErIr_3	mC12	J. Less-Common Metals, $\underline{78}$, 99
B_6DyNi_{12}	$B_6Ni_{12}Sr$	hR19	Dop. Akad. Nauk Ukr. RS$\overline{R}$, A, 43, 80
B_2DyRh_3	B_2ErIr_3	mC12	J. Less-Common Metals, $\underline{78}$, 99
B_4DyRh_4	B_4CeCo_4	tP18	Solid State Comm., $\underline{35}$, $\overline{9}37$
B_4ErIr_4	B_4CeCo_4	tP18	Ibid., 32, 937
B_4ErNi	B_4CrY	oP24	Izv. Aka$\overline{d}$. Nauk SSSR, Neorg. Mater., $\underline{17}$, 1494
$B_2Er_3Ni_7$	$B_2Dy_3Ni_7$	hP24	Dop. Akad. $\overline{N}$auk Ukr. RSR, A, 42, 86
$B_6Er_2Ni_{21}$	C_6Cr_{23}	cF116	I$\overline{b}$id., 42, 87
B_2ErRh_3	B_2ErIr_3	mC12	J. Less-Common Metals, $\underline{78}$, 99
$B_6Gd_{0.91}$	B_6Ca	cP6.91	Inorg. Mater. USSR, $\underline{16}$, $\overline{1}060$
B_2GdIr_3	B_2ErIr_3	mC12	J. Less-Common Metal$\overline{s}$, $\underline{78}$, 99
$B_2Gd_3Ni_7$	$B_2Dy_3Ni_7$	hP24	Dop. Akad. Nauk Ukr. RS$\overline{R}$, A, 42, 86
$B_2Gd_3Ni_{13}$	$B_2Nd_3Ni_{13}$	hP18	I$\overline{b}$id., 43, 87
B_6GdNi_{12}	$B_6Ni_{12}Sr$	hR19	Ibid., $\overline{43}$, 80
B_2HoIr_3	B_2ErIr_3	mC12	J. Less-Common Metals, $\underline{78}$, 99
$B_2Ho_3Ni_7$	$B_2Dy_3Ni_7$	hP24	Dop. Akad. Nauk Ukr. RS$\overline{R}$, A, 42, 86
B_2HoRh_3	B_2ErIr_3	mC12	J. Less-Common Metals, $\underline{78}$, 99
B_4Ir_4La	B_4Co_4Nd	tP18	Ibid., 82, 21
$B_4Ir_4Lu_{0.80}Y_{0.20}$	B_4CeCo_4	tP18	Solid State Comm., 32, 937
B_2Ir_3Lu	B_2ErIr_3	mC12	J. Less-Common Metal$\overline{s}$, $\underline{78}$, 99
B_4Ir_4Nd	B_4Co_4Nd	tP18	Ibid., 82, 21
B_2Ir_3Nd	B_2ErIr_3	mC12	Ibid., $\overline{78}$, 99
B_4Ir_4Pr	B_4Co_4Nd	tP18	Ibid., $\overline{82}$, 21

Phase	Structure type	Structure code	Reference
B_2Ir_3Sc	B_2ErIr_3	mC12	Ibid., 78, 99
B_2Ir_3Sm	B_2ErIr_3	mC12	Ibid., $\overline{78}$, 99
B_2Ir_3Tb	B_2ErIr_3	mC12	Ibid., $\overline{78}$, 99
B_4Ir_4Tm	B_4CeCo_4	tP18	Solid State Comm., 32, 937
B_2Ir_3Tm	B_2ErIr_3	mC12	J. Less-Common Metals, 78, 99
B_2Ir_3Y	B_2ErIr_3	mC12	Ibid., 78, 99
B_2Ir_3Yb	B_2ErIr_3	mC12	Ibid., $\overline{78}$, 99
$B_6La_{0.97}$	B_6Ca	cP6.97	Inorg. Mater. USSR, 16, 1060
$B_2La_3Ni_{13}$	$B_2Nd_3Ni_{13}$	hP18	Dop. Akad. Nauk Ukr. RSR, A, 43, 87
B_4LaOs_4	B_4Co_4Nd	tP18	J. Less-Common Metals, 82, 21
B_4LuNi	B_4CrY	oP24	Izv. Akad. Nauk SSSR, Neorg. Mater., 17, 1494
$B_2Lu_3Ni_7$	$B_2Dy_3Ni_7$	hP24	Dop. Akad. Nauk Ukr. RSR, A, 42, 86
$B_6Lu_2Ni_{21}$	C_6Cr_{23}	cF116	Ibid., 42, 87
B_2LuRh_3	B_2ErIr_3	mC12	J. Less-Common Metals, 78, 99
$B_6Nb_{19.90}Ta_{3.10}$	C_6Cr_{23}	cF116	Mh. Chem., 110, 791
$B_2Ni_{13}Pr_3$	$B_2Nd_3Ni_{13}$	hP18	Dop. Akad. Nauk Ukr. RSR, A, 43, 87
$B_2Ni_{13}Sm_3$	$B_2Nd_3Ni_{13}$	hP18	Ibid., 43, 87
$B_2Ni_7Sm_3$	$B_2Dy_3Ni_7$	hP24	Ibid., $\overline{42}$, 86
$B_2Ni_7Tb_3$	$B_2Dy_3Ni_7$	hP24	Ibid., $\overline{42}$, 86
$B_6Ni_{12}Tb$	$B_6Ni_{12}Sr$	hR19	Ibid., $\overline{43}$, 80
$B_2Ni_7Tm_3$	$B_2Dy_3Ni_7$	hP24	Ibid., $\overline{42}$, 86
B_4NiTm	B_4CrY	oP24	Izv. Akad. Nauk SSSR, Neorg. Mater., 17, 1494
$B_2Ni_{13}Y_3$	$B_2Nd_3Ni_{13}$	hP18	Dop. Akad. Nauk Ukr. RSR, A, 43, 87
$B_2Ni_7Y_3$	$B_2Dy_3Ni_7$	hP24	Ibid., 42, 86
$B_6Ni_{12}Y$	$B_6Ni_{12}Sr$	hR19	Ibid., $\overline{43}$, 80
$B_6Ni_{21}Yb_2$	C_6Cr_{23}	cF116	Ibid., $\overline{42}$, 87
B_4Os_4Pr	B_4Co_4Nd	tP18	J. Less-Common Metals, 82, 21
B_2Rh_3Tb	B_2ErIr_3	mC12	Ibid., 78, 99
B_2Rh_3Tm	B_2ErIr_3	mC12	Ibid., $\overline{78}$, 99
B_2Rh_3Y	B_2ErIr_3	mC12	Ibid., $\overline{78}$, 99
B_2Rh_3Yb	B_2ErIr_3	mC12	Ibid., $\overline{78}$, 99
B_8Sm	B_8Sm	cP9	Inorg. Mater. USSR, 16, 44
B_6Sm	B_6Ca	cP7	Ibid., 16, 44
$B_{5.70}Sm$	B_6Ca	cP6.70	Ibid., $\overline{16}$, 300
BiCePd	AgAsMg	cF12	Gazz. Chim. Ital., 110, 357
BiCePt	AgAsMg	cF12	Ibid., 110, 357
BiCuEu	$InNi_2$	hP6	Z. Naturforsch., B36, 1193
$Bi_8Cu_2Pb_3S_{10}Se_6$	$Bi_8Cu_2Pb_3S_{10}Se_6$	mC29	Canad. Miner., 18, $\overline{353}$
BiDyPd	AgAsMg	cF12	Gazz. Chim. Ital., 110, 357
BiGd	ClNa	cF8	Ibid., 110, 357
BiGdPd	AgAsMg	cF12	Ibid., $\overline{110}$, 357
BiHoNi	AgAsMg	cF12	Ibid., $\overline{110}$, 357
BiHoPd	AgAsMg	cF12	Ibid., $\overline{110}$, 357
BiMnNa	LiMnSb	tP6	Z. anorg. Chem., 475, 9
BiNdPd	AgAsMg	cF12	Gazz. Chim. Ital., $\overline{110}$, 357
Bi_2PbS_4	Bi_2PbS_4	oP28	Dokl. Akad. Nauk Azerb. SSR, 36, 67
BiPdY	AgAsMg	cF12	Gazz. Chim. Ital., 110, 357
BiPdYb	AgAsMg	cF12	Ibid., 110, 357
BiPtYb	AgAsMg	cF12	Ibid., $\overline{110}$, 357
Bi_2S_3	S_3Sb_2	oP20	Russ. J. Inorg. Chem., 26, 1506
BkN	ClNa	cF8	J. Inorg. Nucl. Chem., 42, 995
BkN	ClNa	cF8	J. Less-Common Metals, 66, 201

Phase	Structure type	Structure code	Reference
BkP	ClNa	cF8	J. Inorg. Nucl. Chem., $\underline{42}$, 995
BkSb	ClNa	cF8	Ibid., $\underline{42}$, 995
C_2Ce	C_2Ca	tI6	J. Less-Common Metals, $\underline{81}$, 91
$C_2Ce_{0.52}Dy_{0.48}$	C_2Ca	tI6	Ibid., $\underline{81}$, 91
$C_2Ce_{0.67}U_{0.33}$	C_2Ca	cF12	J. Chem. Phys., $\underline{73}$, 5796
C_6Cr_{23}	C_6Cr_{23}	cF116	J. Solid State Chem., $\underline{32}$, 213
C_3Cr_7	C_3Cr_7	hP80	Ibid., $\underline{32}$, 213
C_2Cr_3	C_2Cr_3	oP20	Ibid., $\overline{32}$, 213
C_2Dy	C_2Ca	tI6	J. Less-Common Metals, $\underline{81}$, 91
$C_2Dy_{0.50}Er_{0.50}$	C_2Ca	tI6	Ibid., 81, 91
$C_2Dy_{0.49}Gd_{0.51}$	C_2Ca	tI6	Ibid., $\overline{81}$, 91
$C_2Dy_{0.50}H_{0.50}$	C_2Ca	tI6	Ibid., $\overline{81}$, 91
$C_2Dy_{0.46}Nd_{0.54}$	C_2Ca	tI6	Ibid., $\overline{81}$, 91
$C_2Dy_{0.50}Pr_{0.50}$	C_2Ca	tI6	Ibid., $\overline{81}$, 91
$C_2Dy_{0.50}Sm_{0.50}$	C_2Ca	tI6	Ibid., $\overline{81}$, 91
$C_2Dy_{0.50}Tb_{0.50}$	C_2Ca	tI6	Ibid., $\overline{81}$, 91
$C_2Dy_{0.44}Y_{0.56}$	C_2Ca	tI6	Ibid., $\overline{81}$, 91
C_2Er	C_2Ca	tI6	Ibid., $\overline{81}$, 91
C_2Eu	C_6Eu	hP14	Carbon, $\overline{18}$, 203
C_2Gd	C_2Ca	tI6	J. Less-Common Metals, $\underline{81}$, 91
C_2Ho	C_2Ca	tI6	Ibid., 81, 91
C_2La	C_2Ca	tI6	Ibid., $\overline{81}$, 91
C_2Tb	C_2Ca	tI6	Ibid., $\overline{81}$, 91
C_2Y	C_2Ca	tI6	Ibid., $\overline{81}$, 91
C_6Yb	C_6Eu	hP14	Carbon, $\overline{18}$, 203
Ca_5Pt_3	Si_3W_5	tI32	J. Less-Common Metals, $\underline{78}$, 49
Ca_5Pt_2	C_2Mn_5	mC28	Ibid., $\underline{78}$, 49
Ca_3Pt_2	Er_3Ni_2	hR15	Ibid., $\overline{78}$, 49
CdCeIn	$CaIn_2$	hP6	Ibid., $\overline{78}$, 1
CdDyIn	$CaIn_2$	hP6	Ibid., $\overline{78}$, 1
CdErIn	$CaIn_2$	hP6	Ibid., $\overline{78}$, 1
CdGdIn	$CaIn_2$	hP6	Ibid., $\overline{78}$, 1
CdHoIn	$CaIn_2$	hP6	Ibid., $\overline{78}$, 1
CdInLa	$CaIn_2$	hP6	Ibid., $\overline{78}$, 1
CdInNd	$CaIn_2$	hP6	Ibid., $\overline{78}$, 1
CdInPr	$CaIn_2$	hP6	Ibid., $\overline{78}$, 1
CdInSm	$CaIn_2$	hP6	Ibid., $\overline{78}$, 1
CdInYb	$CaIn_2$	hP6	Ibid., $\overline{78}$, 1
$CdLu_2S_4$	Al_2MgO_4	cF56	J. Solid State Chem., $\underline{37}$, 228
$CdLu_2Se_4$	Al_2MgO_4	cF56	Ibid., $\underline{37}$, 228
$CdPS_3$	$CrPSe_3$	mC20	Ann. Chim. Paris, $\underline{5}$, 499
$CdPSe_3$	$FePSe_3$	hR10	Ibid., $\underline{5}$, 499
CdS_4Sc_2	Al_2MgO_4	cF56	J. Solid State Chem., $\underline{37}$, 228
CdS_4Y_2	Al_2MgO_4	cF56	Ibid., $\underline{37}$, 228
$CdSc_2Se_4$	Al_2MgO_4	cF56	Ibid., $\overline{37}$, 228
$CdSe_4Y_2$	Al_2MgO_4	cF56	Ibid., $\overline{37}$, 228
Cd_5Th	$EuMg_5$	hP36	J. Less-Common Metals, $\underline{77}$, 215
$Cd_{23}Th_6$	$Mn_{23}Th_6$	cF116	Ibid., $\underline{77}$, 215
Cd_3Th	Ni_3Sn	hP8	Ibid., $\overline{77}$, 215
Cd_3Th	Ni_3Sn	hP8	Ibid., $\overline{77}$, 205
Cd_2Th	AlB_2	hP3	Ibid., $\overline{77}$, 215
$Cd_{11}Th$	$BaHg_{11}$	cP36	Ibid., $\overline{77}$, 215
$CeCo_5$	$CaCu_5$	hP6	Dop. Akad. Nauk Ukr. RSR., A, 42, 84
$CeCo_2$	Cu_2Mg	cF24	Ibid., 42, 84
$CeCrMg_{11}$	$CeMg_{12}$	tI338	Mater. Res. Bull., $\underline{15}$, 139
$CeFeMg_{11}$	$CeMg_{12}$	tI338	Ibid., $\underline{15}$, 139
$CeGa_6$	Ga_6Pu	tP14	J. Less-Common Metals, $\underline{81}$, 33
$CeMg_{11}$	$CeMg_{12}$	tI312	Mater. Res. Bull., $\underline{15}$, $\overline{139}$
$CeMg_{11}Mn$	$CeMg_{12}$	tI338	Ibid., $\underline{15}$, 139

Phase	Structure type	Structure code	Reference
$CeMg_{11}V$	$CeMg_{12}$	tI338	Ibid., 15, 139
$CeNi_2P_2$	Al_4Ba	tI10	J. Solid State Chem., 35, 312
$CePd_3$	$AuCu_3$	cP4	Mater. Res. Bull., 16, 1557
CePtSb	$CaIn_2$	hP6	J. Less-Common Metals, 78, 1
Ce_5Sb_3	Mn_5Si_3	hP16	Ibid., 79, 57
Ce_4Sb_3	P_4Th_3	cI28	Ibid., 79, 57
Ce_2Sb	La_2Sb	tI12	Ibid., 79, 57
$CeSb_2$	$LaSb_2$	oC24	Ibid., 79, 57
CeSb	ClNa	cF8	Ibid., 79, 57
CfSb	ClNa	cF8	Inorg. Nucl. Chem. Lett., 16, 537
CmN	ClNa	cF8	J. Less-Common Metals, 66, 201
Co_3Cu_2Ho	$CaCu_5$	hP6	Acta Phys. Pol., A55, 849
Co_3Cu_2Ho	Cu_5Tb	hP6	J. Magn. Magn. Mater., 15-, 1241
Co_4CuHo	Cu_5Tb	hP6	Ibid., 15-, 1241
Co_4CuHo	$CaCu_5$	hP6	Acta Phys. Pol., A55, 849
$Co_{4.50}Cu_{0.50}Ho$	Cu_5Tb	hP6	J. Magn. Magn. Mater., 15-, 1241
$Co_{4.80}Cu_{0.28}Ho_{0.92}$	Cu_5Tb	hP6	Ibid., 15-, 1241
$CoCuS_4Ti$	Al_2MgO_4	cF56	J. Solid State Chem., 31, 401
$CoCu_2S_8Sn_3$	$Cu_2FeS_8Sn_3$	tI28	Rev. Chim. Minér., 18, 33
$Co_{0.50}CuS_4Sn_{1.50}$	Al_2MgO_4	cF56	J. Solid State Chem., 31, 401
Co_6DyGe_6	Co_6Ge_6Y	hP13	Z. anorg. Chem., 482, 40
Co_6ErGe_6	Co_6Ge_6Y	hP13	Ibid., 482, 40
$Co_{4.38}Fe_{0.69}Ho_{0.92}$	Cu_5Tb	hP6	J. Magn. Magn. Mater., 15-, 1241
Co_5Gd	$CaCu_5$	hP6	J. Less-Common Metals, 78, 219
Co_6GdGe_6	Co_6Ge_6Y	hP13	Z. anorg. Chem., 482, 40
Co_6Ge_6Hf	Fe_3Ge_6Mn	hP13	Dop. Akad. Nauk Ukr. RSR., A, 43, 84
Co_6Ge_6Hf	Co_6Ge_6Li	hP13	Z. anorg. Chem., 482, 40
Co_6Ge_6Ho	Co_6Ge_6Y	hP13	Ibid., 482, 40
Co_6Ge_6Lu	Co_6Ge_6Y	hP13	Ibid., 482, 40
Co_6Ge_6Mg	Co_6Ge_6Li	hP13	Ibid., 482, 40
CoGeMn	Co_2Si	oP12	Solid State Comm., 39, 1081
Co_6Ge_6Sc	Co_7Ge_7Li	hP13	Z. anorg. Chem., 482, 40
Co_6Ge_6Tb	Co_6Ge_6Y	hP13	Ibid., 482, 40
Co_6Ge_6Ti	$Fe_3Ge_6Mn_4$	hP13	Dop. Akad. Nauk Ukr. RSR., A, 43, 84
Co_6Ge_6Tm	Co_6Ge_6Y	hP13	Z. anorg. Chem., 482, 40
Co_6Ge_6U	Co_6Ge_6Li	hP13	Ibid., 482, 40
Co_6Ge_6Yb	Co_6Ge_6Y	hP13	Ibid., 482, 40
Co_6Ge_6Zr	$Fe_3Ge_6Mn_4$	hP13	Dop. Akad. Nauk Ukr. RSR., A, 43, 84
$CoHf_4P$	$CoNb_4Si$	tP12	Inorg. Mater. USSR, 16, 705
CoHfP	Co_2Si	oP12	Ukr. Khim. Zh. Russ. Ed., 47, 103
$Co_{5.08}Ho_{0.92}$	Cu_5Tb	hP6	J. Magn. Magn. Mater., 15-, 1241
$Co_{5.08}Ho_{0.92}$	Cu_5Tb	hP6	Acta Phys. Pol., A55, 849
$Co_{4.62}Ho_{0.92}Mn_{0.46}$	Cu_5Tb	hP6	J. Magn. Magn. Mater., 15-, 1241
Co_4HoNi	Cu_5Tb	hP6	Ibid., 15-, 1241
Co_4HoNi	$CaCu_5$	hP6	Acta Phys. Pol., A55, 849
Co_3HoNi_2	Cu_5Tb	hP6	J. Magn. Magn. Mater., 15-, 1241
Co_3HoNi_2	$CaCu_5$	hP6	Acta Phys. Pol., A55, 849
Co_2HoSi_2	Al_4Ba	tI10	Ukr. Fiz. Zh. Russ. Ed., 25, 1683

Phase	Structure type	Structure code	Reference
$CoIn_3$	$CoGa_3$	tP16	J. Less-Common Metals, 79, 1
$Co_{0.40}Ni_{1.60}U$	$MgNi_2$	hP24	Phys. Stat. Solidi, A57, K17
$CoPTa_4$	$CoNb_4Si$	tP12	Inorg. Mater. USSR, 16, 705
$Co_{12}P_7Ti_2$	$Fe_{12}P_7Zr_2$	hP21	Ukr. Fiz. Zh. Russ. Ed., 47, 142
$CoPTi$	Co_2Si	oP12	Ibid., 47, 142
$CoPZr_4$	$CoNb_4Si$	tP12	Inorg. Mater. USSR, 16, 705
Co_2Sc	Cu_2Mg	cF24	Ibid., 17, 704
$CoSc_2$	Al_2Cu	tI12	Dop. Akad. Nauk Ukr. RSR., A, 39, 664
$Co_{0.50}Sc_2Si_{0.50}$	$NiTi_2$	cF96	Ibid., 39, 666
Co_3Sc_2Si	Cu_3Mg_2Si	hP12	Ibid., 39, 666
$CoScSi$	Co_2Si	oP12	Ibid., 39, 666
$Co_{16}Sc_6Si_7$	$Mn_{23}Th_6$	cF116	Ibid., 39, 664
$Co_2Sc_3Si_5$	$Co_2Sc_3Si_5$	tP40	Ibid., 39, 666
Co_2ScSi_2	Al_4Ba	tI10	Ibid., 39, 666
Co_3Sc_2Si	$Co_3Si_5U_2$	oI40	Ibid., 39, 666
Co_7ScSi_4	Co_7ScSi_4	hP24	Ibid., 39, 664
$Co_2Sc_{0.40}Ti_{0.20}Y_{0.40}$	Cu_2Mg	cF24	Inorg. Mater. USSR, 17, 704
$Co_2Si_7U_3$	$Fe_2Si_7U_3$	oC24	Sov. Phys. Crystallogr., 26, 103
Co_5Sm	$CaCu_5$	hP6	Russian Metallurgy, 188
$CoTi_2$	$NiTi_2$.	cF96	Mh. Chem., 111, 535
Co_2U	Cu_2Mg	cF24	Phys. Stat. Solidi, A, 57, K17
$Cr_{0.50}Cu_{0.50}PS_3$	$Cr_{.50}Cu_{.50}PS_3$	mC40	C. R. Acad. Sci. Paris, Ser. C, 291, 263
$CrCuS_4Sn$	Al_2MgO_4	cF56	J. Solid State Chem., 31, 401
$CrCuS_4Ti$	Al_2MgO_4	cF56	Ibid., 31, 401
$CrCuS_4Zr$	Al_2MgO_4	cF56	Ibid., 31, 401
$CrFeInS_4$	Al_2MgO_4	cF56	Ibid., 38, 40
$CrFeRhS_4$	Al_2MgO_4	cF56	Ibid., 35, 77
Cr_2FeS_4	Al_2MgO_4	cF56	Mater. Res. Bull., 16, 65
Cr_2FeS_4	Al_2MgO_4	cF56	J. Solid State Chem., 35, 77
Cr_2FeS_4	Al_2MgO_4	cF56	Ibid., 38, 40
$Cr_{1.87}Ga_{0.80}S_4$	Al_2MgO_4	cF53.34	Izv. Akad. Nauk SSSR, Neorg. Mater., 16, 926
$Cr_2Ga_{0.67}S_4$	Al_2MgO_4	cF53.36	Ibid., 16, 926
$CrGdS_3$	$CrLaS_3$	mP50	J. Solid State Chem., 38, 165
$CrLaS_3$	$CrLaS_3$	mP50	Ibid., 38, 165
$CrNdS_3$	$CrLaS_3$	mP50	Ibid., 38, 165
$CrPSe_3$	$CrPSe_3$	mC20	Ann. Chim. Paris, 5, 499
Cr_3S_4	Cr_3S_4	mC14	J. Phys. Soc. Japan, 50, 413
Cr_2S_4Zn	Al_2MgO_4	cF56	Mater. Res. Bull., 16, 65
Cr_2Se_4Zn	Al_2MgO_4	cF56	Inorg. Mater. USSR, 16, 817
$CrVZr$	Cu_2Mg	cF24	J. Less-Common Metals, 78, 725
$CuDyTl$	$CaIn_2$	hP6	Ibid., 80, 47
$CuErTl$	$CaIn_2$	hP6	Ibid., 80, 47
$CuEuP$	$InNi_2$	hP6	Z. Naturforsch., B36, 1193
$CuEuSb$	$InNi_2$	hP6	Ibid., B36, 1193
$Cu_2FeS_8Sn_3$	$Cu_2FeS_8Sn_3$	tI28	Rev. Chim. Miner., 18, 33
$CuFe_2S_3$	FeS	hP24	Mater. Res. Bull., 15, 907
$CuGaS_2$	$CuFeS_2$	tI16	Inorg. Mater. USSR, 17, 378
$CuGdTl$	$CaIn_2$	hP6	J. Less-Common Metals, 80, 47
Cu_8GeS_6	Cu_8GeS_6	cP60	Neues Jb. Miner. Mh., 442
$CuHoTl$	$CaIn_2$	hP6	J. Less-Common Metals, 80, 47
Cu_9In_4	Cu_9In_4	aP39	C.R. Acad. Sci. Paris, Ser. B, 266, 1397
$CuInS_2$	$CuFeS_2$	tI16	Inorg. Mater. USSR, 17, 378
$Cu_{13}La$	$NaZn_{13}$	cF112	J. Less-Common Metals, 79, 323

Phase	Structure type	Structure code	Reference
$Cu_2MnS_8Sn_3$	$Cu_2FeS_8Sn_3$	tI28	Rev. Chim. Minér., $\underline{18}$, 33
$Cu_2Mo_3S_4$	$Cu_2Mo_3S_4$	hR18	J. Less-Common Metals, $\underline{72}$, 193
$CuMo_6S_8$	$CuMo_6S_8$	hR15	Ibid., $\underline{72}$, 193
$Cu_2NiS_8Sn_3$	$Cu_2FeS_8Sn_3$	tI28	Rev. Chim. Minér., $\underline{18}$, 33
Cu_2NiSn	$AlCu_2Mn$	cF16	J. Therm. Anal., $\underline{17}$, 489
$Cu_4P_{10}Sn$	$Cu_4P_{10}Sn$	cF60	Phys. Stat. Solidi, A, $\underline{3}$, 75
$Cu_{39}S_{28}$	$Cu_{39}S_{28}$	hP1206	Canad. Miner., $\underline{18}$, 511
Cu_9S_8	Cu_9S_8	hP51	Ibid., $\underline{18}$, 511
$Cu_{1.60}S$	$Cu_{1.60}S$	cF10.40	Ibid., $\underline{18}$, 519
$Cu_{4.33}S_2SbTl$	$Cu_{4.335}S_2SbTl$	oP33.34	Neues Jahrb. Mineral. Abh., $\underline{138}$, 122
CuS_4SnTi	Al_2MgO_4	cF56	J. Solid State Chem., $\underline{31}$, 401
CuS_4SnV	Al_2MgO_4	cF56	Ibid., $\underline{31}$, 401
CuS_4TiV	Al_2MgO_4	cF56	Ibid., $\underline{31}$, 401
CuS_4TiZr	Al_2MgO_4	cF56	Ibid., $\underline{31}$, 401
CuS_4VZr	Al_2MgO_4	cF56	Ibid., $\underline{31}$, 401
Cu_6Sm	$CeCu_6$	oP28	Phys. Met. Metall., $\underline{49}$, 185
Cu_5Sm	$CaCu_5$	hP6	Ibid., $\underline{49}$, 185
$Cu_{2.40}Sn_2$	$Cu_{2.4}Sn_2$	hP4.40	Neues Jb. Miner. Mh., 117
$CuTbTl$	$CaIn_2$	hP6	J. Less-Common Metals, $\underline{80}$, 47
$Cu_{10}Zr_7$	$Ni_{10}Zr_7$	oC68	Ibid., $\underline{78}$, 45
Dy_2EuSe_4	P_4Th_3	cI28	Z. Neorg. Khim., $\underline{26}$, 2258
$DyFe_2Si_2$	Al_4Ba	tI10	Ukr. Fiz. Zh. Russ. Ed., $\underline{25}$, 1683
$DyGa_6$	Ga_6Pu	tP14	J. Less-Common Metals, $\underline{81}$, 33
$DyGa_3$	$HfNi_3$	hP40	Ibid., $\underline{77}$, 269
$DyGa_3$	Ga_3Pu	hR16	Ibid., $\underline{77}$, 269
$DyGa_3$	$AuCu_3$	cP4	Ibid., $\underline{77}$, 269
$DyGa_4Ti_2$	Al_4Mo_2Yb	tI14	Dop. Akad. Nauk Ukr. RSR., A, $\underline{42}$, 73
Dy_3In_5	Pd_5Pu_3	oC32	J. Less-Common Metals, $\underline{81}$, 45
$Dy_5Ir_4Si_{10}$	$Co_5Sc_5Si_{10}$	tP38	Ternary Superconductors, Proc. Int. Conf., 1980, 239
$DyNi_2P_2$	Al_4Ba	tI10	J. Solid State Chem., $\underline{35}$, 312
$DyPd_3$	$AuCu_3$	cP4	Mater. Res. Bull., $\underline{16}$, 1557
Dy_2Se_3	S_3Sb_2	oP29	Russ. J. Inorg. Chem., $\underline{25}$, 1130
$DySe_2$	$ErSe_2$	oC144	Ibid., $\underline{25}$, 1130
$DySe_{1.35}$	P_4Th_3	cI28.20	Ibid., $\underline{26}$, 626
$DySe_{1.35}$	P_4Th_3	cI28.20	Ibid., $\underline{25}$, 1130
$DySe_2Tl$	$FeNaO_2$	hR4	Mater. Res. Bull., $\underline{16}$, 1219
$DySe_4U$	$DySe_4U$	oP12	Russ. J. Inorg. Chem., $\underline{26}$, 621
$Dy_2Se_7U_2$	$Se_7Tb_2U_2$	mC22	Ibid., $\underline{26}$, 626
$Dy_4Se_{16}U_5$	$Dy_4S_{16}U_5$	mC100	Z. Neorg. Khim., $\underline{26}$, 1961
Dy_3Tl_5	Pd_5Pu_3	oC32	J. Less-Common Metals, $\underline{81}$, 45
Er_6Fe_{23}	$Mn_{23}Th_6$	cF116	Solid State Comm., $\underline{40}$, 117
$ErGa_6$	Ga_6Pu	tP14	J. Less-Common Metals, $\underline{81}$, 33
$ErGa_3$	$AuCu_3$	cP4	Ibid., $\underline{77}$, 269
Er_3In_5	Pd_5Pu_3	oC32	Ibid., $\underline{81}$, 45
$Er_5Ir_4Si_{10}$	$Co_4Sc_5Si_{10}$	tP38	Ternary Superconductors, Proc. Int. Conf., 1980, 239
$ErNi_2P_2$	Al_4Ba	tI10	J. Solid State Chem., $\underline{35}$, 312
Er_2PbS_4	Bi_2PbS_4	oP28	Dokl. Akad. Nauk Azerb. SSR, $\underline{36}$, 67
Er_2PbSe_4	Bi_2PbS_4	oP28	Ibid., $\underline{36}$, 67
$ErPd_3$	$AuCu_3$	cP4	Mater. Res. Bull., $\underline{16}$, 1557
Er_3Tl_5	Pd_5Pu_3	oC32	J. Less-Common Metals, $\underline{81}$, 45
$EuPd_3$	$AuCu_3$	cP4	Mater. Res. Bull., $\underline{16}$, 1557
Eu_5Pt_4	Pu_5Rh_4	oP36	J. Less-Common Metals, $\underline{80}$, 71
Eu_5Pt_2	C_2Mn_5	mC28	Ibid., $\underline{80}$, 71

Phase	Structure type	Structure code	Reference
Eu_3Pt_2	Er_3Ni_2	hR5	Ibid., 80, 71
Eu_2Pt_7	Ce_2Ni_7	hP36	Ibid., $\overline{80}$, 71
$EuPt_2$	Cu_2Mg	cF24	Ibid., $\overline{80}$, 71
$Eu_3S_9Sb_4$	$Eu_3Sb_4Se_9$	oP64	Inorg. $\overline{M}$ater. USSR, 17, 1469
$Eu_3Sb_4Se_9$	$Eu_3Sb_4Se_9$	oP64	Ibid., 17, 1469
$Fe_{1.70}Ga_{0.30}Sc$	Cu_2Mg	cF24	Ibid., $\overline{17}$, 704
$Fe_{1.95}Ge$	$InNi_2$	hP5.90	Phys. Rev., B, 18, 4860
Fe_6Ge_6Nb	$Fe_3Ge_6Mn_4$	hP13	Dop. Akad. Nau$\overline{k}$ Ukr. RSR., A, 43, 84
Fe_6Ge_6Sc	$Fe_3Ge_6Mn_4$	hP13	I$\overline{b}$id., 43, 84
Fe_6Ge_6Ti	$Fe_3Ge_6Mn_4$	hP13	Ibid., $\overline{43}$, 84
Fe_6Ge_6Zr	$Fe_3Ge_6Mn_4$	hP13	Ibid., $\overline{43}$, 84
$FeHf_4P$	$CoNb_4Si$	tP12	Inorg. $\overline{M}$ater. USSR, 16, 705
$Fe_{23}Ho_6$	$Mn_{23}Th_6$	cF116	Solid State Comm., $4\overline{0}$, 117
Fe_2HoSi_2	Al_4Ba	tI10	Ukr. Fiz. Zh. Russ. $\overline{E}$d., 25, 1683
$FeIn_2S_4$	Al_2MgO_4	cF56	J. Solid State Chem., 38, 40
Fe_2KS_3	Fe_2KS_3	oC24	Amer. Min., 64, 776
$Fe_{23}Lu_6$	$Mn_{23}Th_6$	cF116	Solid State $\overline{C}$omm., 37, 635
$Fe_{13}Ni_3Ti_{32}$	$NiTi_2$	cF96	Mh. Chem., 111, 535$\overline{}$
$FePS_3$	$CrPSe_3$	mC20	Ann. Chim. $\overline{P}$aris, 5, 499
$FePSe_3$	$FePSe_3$	hR10	Ibid., 5, 499
Fe_2S_4V	Cr_3S_4	mC14	Physica B+C, 105, 223
FeS_4V_2	Cr_3S_4	mC14	Ibid., 105, 2$\overline{23}$
Fe_2Sc	$MgNi_2$	hP24	Inorg. $\overline{M}$ater. USSR, 17, 704
$Fe_2Sc_{0.50}Y_{0.50}$	Cu_2Mg	cF24	Ibid., 17, 704
Fe_2Si_2Y	Al_4Ba	tI10	Ukr. Fiz. Zh. Russ. Ed., 25, 1683
$FeSi_2Zr$	$FeScSi_2$	oC96	Dop. Akad. Nauk. Ukr. RSR, B, 6, 68
Fe_2Zr	Cu_2Mg	cF24	J$\overline{.}$ Less-Common Metals, 81, 293
Fe_2Zr	Cu_2Mg	cF24	Ibid., 79, 243
$FeZr_3$	$FeZr_3$	oC16	Ibid., $\overline{81}$, 293
$FeZr_3$	BRe_3	oC16	Ibid., $\overline{79}$, 243
$FeZr_2$	Al_2Cu	tI12	Ibid., $\overline{81}$, 293
$FeZr_2$	$NiTi_2$	cF96	Ibid., $\overline{79}$, 243
Ga_6Gd	Ga_6Pu	tP14	Ibid., $\overline{81}$, 33
Ga_2Gd	AlB_2	hP3	Ibid., $\overline{77}$, 197
Ga_4HfV_2	Al_4Mo_2Yb	tI14	Dop. Ak$\overline{a}$d. Nauk Ukr. RSR., A, 42, 73
Ga_6Ho	Ga_6Pu	tP14	J. Less-Common Metals, 81, 33
Ga_3Ho	Al_3Pu	hP24	Ibid., 77, 269
Ga_3Ho	Al_3Ho	hR20	Ibid., $\overline{77}$, 269
Ga_3Ho	Ni_3Ti	hP16	Ibid., $\overline{77}$, 269
Ga_3Ho	$AuCu_3$	cP4	Ibid., $\overline{77}$, 269
Ga_4HoTi_2	Al_4Mo_2Yb	tI14	Dop. Ak$\overline{a}$d. Nauk Ukr. RSR., A, 42, 73
$Ga_{0.50}In_{1.50}Se_3$	In_2Se_3	hP30	J. Appl. Cryst., 13, 24
Ga_6La	Ga_6Pu	tP14	Dop. Akad. Nauk Uk$\overline{r}$. RSR., A, 43, 86
Ga_6La	Ga_6Pu	tP14	J. $\overline{}$Less-Common Metals, 81, 33
$GaLi$	$NaTl$	cF16	Russian Metallurgy, 192$\overline{}$
$GaLiMg_2$	$AlCu_2Mn$	cF16	Ibid., 192
Ga_2Lu	$CeCu_2$	oI12	Dop. Akad. Nauk Ukr. RSR., A, 42, 81
Ga_4LuTi_2	Al_4Mo_2Yb	tI14	I$\overline{b}$id., 42, 73
$Ga_3Mn_{17}Nb_{19}$	Fe_7W_6	hR13	J. Less-$\overline{C}$ommon Metals, 75, 261
$Ga_{7.50}Mn_3Nb_{19.50}$	$CrFe$	tP30	Ibid., 75, 261
$Ga_{0.17}Mo_{0.66}W_{0.17}$	Cr_3O	cP8	Ibid., $\overline{77}$, 137
Ga_3Nb	Si_3W_5	tI32	Ibid., $\overline{75}$, 261

Phase	Structure type	Structure code	Reference
GaNb$_3$	Cr$_3$O	cP8	Ibid., 75, 261
Ga$_6$Nd	Ga$_6$Pu	tP14	Dop. Akad. Nauk Ukr. RSR., A, 43, 86
Ga$_6$Nd	Ga$_6$Pu	tP14	J. Less-Common Metals, 81, 33
Ga$_3$Pd$_{12}$Si$_4$	S$_{12}$Th$_7$	hP19	Z. Metallk., 72, 279
Ga$_6$Pr	Ga$_6$Pu	tP14	J. Less-Common Metals, 81, 33
Ga$_6$Pr	Ga$_6$Pu	tP14	Dop. Akad. Nauk Ukr. RSR., A, 43, 86
Ga$_2$S$_5$Sn$_2$	Ga$_2$S$_5$Sn$_2$	oP36	C. R. Acad. Sci. Paris, Ser. C, 293, 275
Ga$_3$Sc$_5$	Mn$_5$Si$_3$	hP16	Inorg. Mater. USSR, 16, 1065
Ga$_3$Sc	AuCu$_3$	cP4	Ibid., 16, 1065
Ga$_2$Sc	CeCu$_2$	oI12	Dop. Akad. Nauk Ukr. RSR., A, 42, 81
Ga$_4$ScTi$_2$	Al$_4$Mo$_2$Yb	tI14	Ibid., 42, 73
Ga$_4$ScV$_2$	Al$_4$Mo$_2$Yb	tI14	Ibid., 42, 73
Ga$_{3.20}$Se$_{4.80}$	SZn	cF8	J. Solid State Chem., 40, 312
Ga$_2$Se$_3$	Ga$_2$Se$_3$	mC20	Ibid., 40, 312
GaSe	GaS	hP8	Russ. J. Inorg. Chem., 26, 1056
Ga$_6$Sm	Ga$_6$Pu	tP14	J. Less-Common Metals, 81, 33
Ga$_6$Tb	Ga$_6$Pu	tP14	Ibid., 81, 33
Ga$_3$Tb	AuCu$_3$	cP4	Ibid., 77, 269
Ga$_3$Tb	Ni$_3$Sn	hP8	Ibid., 77, 269
Ga$_4$TbTi$_2$	Al$_4$Mo$_2$Yb	tI14	Dop. Akad. Nauk Ukr. RSR., A, 42, 73
Ga$_4$Ti$_2$Tm	Al$_4$Mo$_2$Yb	tI14	Ibid., 42, 73
Ga$_4$Ti$_2$Zr	Al$_4$Mo$_2$Yb	tI14	Ibid., 42, 73
Ga$_6$Tm	Ga$_6$Pu	tP14	J. Less-Common Metals, 81, 33
Ga$_3$Tm	AuCu$_3$	cP4	Ibid., 77, 269
Ga$_4$V$_2$Zr	Al$_4$Mo$_2$Yb	tI14	Dop. Akad. Nauk Ukr. RSR., A, 42, 73
Ga$_5$W$_2$	Hg$_5$Mn$_2$	tP14	J. Less-Common Metals, 77, 137
Ga$_6$Yb	Ga$_6$Pu	tP14	Ibid., 81, 33
Gd$_3$In$_5$	Pd$_5$Pu$_3$	oC32	Ibid., 81, 45
GdNi$_5$	CaCu$_5$	hP6	Ibid., 78, 219
GdNi$_2$P$_2$	Al$_4$Ba	tI10	J. Solid State Chem., 35, 312
Gd$_{0.30}$NiTb$_{0.70}$	Dy$_{0.45}$Gd$_{0.55}$Ni	mP16	Acta Crystallogr., Sec. B, 37, 495
GdPd$_3$	AuCu$_3$	cP4	Mater. Res. Bull., 16, 1557
Gd$_3$S$_4$	P$_4$Th$_3$	cI28	Ibid., 16, 975
GdS$_3$V	CrLaS$_3$	mP50	J. Solid State Chem., 38, 165
Gd$_3$Tl$_5$	Pd$_5$Pu$_3$	oC32	J. Less-Common Metals, 81, 45
GeHf$_9$Mo$_4$	Hf$_9$Mo$_4$Ni	hP28	Acta Chem. Scand. Ser. A, 35, 227
GeLiMg$_2$	AlCu$_2$Mn	cF16	Russian Metallurgy, 192
Ge$_6$MgNi$_6$	Ge$_6$Ni$_6$Y	hP13	Z. anorg. Chem., 482, 40
GeMnNi	Co$_2$Si	oP12	Solid State Comm., 39, 1081
Ge$_2$Mo	MoSi$_2$	tI6	J. Less-Common Metals, 78, 235
Ge$_5$Ni$_3$Zn	PdSn$_2$	tI27	Z. Metallk., 72, 652
Ge$_{1.50}$Ni$_2$Zn$_{1.50}$	Al$_3$Ni$_2$	hP5	Ibid., 72, 652
Ge$_3$Rh$_2$S$_3$	Co$_2$Ge$_3$S$_3$	hR32	J. Solid State Chem., 40, 64
Ge$_3$Rh$_2$Se$_3$	Co$_2$Ge$_3$S$_3$	hR32	Ibid., 40, 64
GeS$_3$Yb	EuGeS$_3$	aP30	Inorg. Mater. USSR, 17, 1160
GeS$_4$Yb$_2$	GeS$_4$Sr$_2$	mP14	Ibid., 17, 1160
GeSe$_2$	GeS$_2$	mP48	Phys. Stat. Solidi B, 108, 153
GeSeTh	La$_2$Sb	tI12	Z. anorg. Chem., 472, 139
GeSe$_3$Yb	EuGeS$_3$	aP30	Inorg. Mater. USSR, 17, 1160
GeSe$_4$Yb$_2$	GeS$_4$Sr$_2$	mP14	Ibid., 17, 1160

Phase	Structure type	Structure code	Reference
GeTeTh	La_2Sb	tI12	Z. anorg. Chem., 472, 139
Ge_2W	$MoSi_2$	tI6	J. Less-Common Metals, 78, 235
H_2Mg	O_2Ti	tP6	Mater. Res. Bull., 15, 1779
H_2Mg	O_2Pb	oP12	Ibid., 15, 1779
H_4Mg_2Ni	D_6RuSr_2	cF28	J. Phys. Chem. Solids, 42, 611
Hf_9Mo_4P	Hf_9Mo_4Ni	hP28	Acta Chem. Scand. Ser. A, 35, 227
Hf_9Mo_4S	Hf_9Mo_4Ni	hP28	Ibid., 35, 227
Hf_9Mo_4Se	Hf_9Mo_4Ni	hP28	Ibid., 35, 277
Hf_9Mo_4Si	Hf_9Mo_4Ni	hP28	Ibid., 35, 227
Hf_4NiP	$CoNb_4Si$	tP12	Inorg. Mater. USSR, 16, 705
HfNiP	Co_2Si	oP12	Ukr. Khim. Zh. Russ. Ed., 47, 103
Hf_9PW_4	Hf_9Mo_4Ni	hP28	Acta Chem. Scand. Ser. A, 35, 227
$HfSe_2$	CdI_2	hP3	Inorg. Chem., 20, 3655
$HfTe_2$	CdI_2	hP3	Ibid., 20, 3655
HfV_2	Cu_2Mg	cF24	Phys. Lett., A, 38, 1
HfV_2	HfV_2	oI12	Ibid., A, 38, 1
$HfZn_2$	Cu_2Mg	cF24	Z. Naturforsch., B36, 1547
Ho_3In_5	Pd_5Pu_3	oC32	J. Less-Common Metals, 81, 45
$Ho_5Ir_4Si_{10}$	$Co_4Sc_5Si_{10}$	tP38	Ternary Superconductors, Proc. Int. Conf., 1980, 239
$HoNi_2P_2$	Al_4Ba	tI10	J. Solid State Chem., 35, 312
Ho_2PbS_4	Bi_2PbS_4	oP28	Dokl. Akad. Nauk Azerb. SSR, 36, 67
$HoPd_3$	$AuCu_3$	cP4	Mater. Res. Bull., 16, 1557
Ho_3Tl_5	Pd_5Pu_3	oC32	J. Less-Common Metals, 81, 45
In_3Ir	$CoGa_3$	tP16	Ibid., 79, 1
In_2Ir	$CuMg_2$	oF48	Ibid., 79, 1
In_5La_3	Pd_5Pu_3	oC32	Ibid., 81, 45
$In_{3.30}Mo_{15}Se_{19}$	$In_{3.3}Mo_{15}Se_{19}$	hP74.60	Ternary Superconductors, Proc. Int. Conf., 1980, 7
$In_2Mo_{15}Se_{19}$	$In_2Mo_{15}Se_{19}$	hR72	Ibid., 7
In_5Nd_3	Pd_5Pu_3	oC32	J. Less-Common Metals, 81, 45
$In_{0.50}Ni_4Sb_{0.50}U$	Ni_5U	cF24	Fizika, 12, 200
In_2PbS_4 (β)	In_2PbS_4	oP56	An. Quim., 75, 787
In_2PbS_4 (α)	In_2PbS_4	oP28	Ibid., 75, 787
InSe	InSe	hR4	Fukuoka Daigaku Rigaku Shuho, 9, 89
In_5Sm_3	Pd_5Pu_3	oC32	J. Less-Common Metals, 81, 45
In_5Tb_3	Pd_5Pu_3	oC32	Ibid., 81, 45
In_5Y_3	Pd_5Pu_3	oC32	Ibid., 81, 45
$Ir_4Lu_5Si_{10}$	$Co_4Sc_5Si_{10}$	tP38	Ternary Superconductors, Proc. Inst. Conf., 1980, 239
$Ir_4Si_{10}Tm_5$	$Co_4Sc_5Si_{10}$	tP38	Ibid., 239
$Ir_4Si_{10}Y_5$	$Co_4Sc_5Si_{10}$	tP38	Ibid., 239
KP_2Rh_2	Al_4Ba	tI10	Z. Naturforsch., B36, 1668
$LaNi_5$	$CaCu_5$	hP6	Phys. Rev., B, 11, 82
$LaNi_2P_2$	Al_4Ba	tI10	J. Solid State Chem., 35, 312
$LaPd_3$	$AuCu_3$	cP4	Mater. Res. Bull., 16, 1557
$LaPt_5$	$CaCu_5$	hP6	Phys. Rev., B, 11, 82
LaPtSb	$CaIn_2$	hP6	J. Less-Common Metals, 78, 1
LaS_2	LaS_2	oP24	J. Solid State Chem., 37, 44
LaS_2	$LaSe_2$	mP12	Ibid., 37, 44
LaS_3V	$CrLaS_3$	mP50	Ibid., 38, 165
La_3Tl_5	Pd_5Pu_3	oC32	J. Less-Common Metals, 81, 45
LiMnSb	LiMnSb	tP6	Z. anorg. Chem., 475, 9
$Li_{0.95}S_2Y_{0.95}Zr_{0.05}$	ClNa	cF7.90	Rev. Chim. Minér., 17, 368

Phase	Structure type	Structure code	Reference
Lu_2PbS_4	Bi_2PbS_4	oP28	Dokl. Akad. Nauk Azerb. SSR, 36, 67
Lu_2PbSe_4	Bi_2PbS_4	oP28	Ibid., 36, 37
$LuPd_3$	$AuCu_3$	cP4	Mater. Res. Bull., 16, 1557
Mg_2Ni	Mg_2Ni	hP18	Ibid., 16, 199
$MgNi_6Si_6$	Ni_6Si_6Y	hP13	Z. anorg. Chem., 482, 40
Mn_2Nb	$MgZn_2$	hP12	J. Less-Common Metals, 75, 261
$MnPS_3$	$CrPSe_3$	mC20	Ann. Chim. Paris, 5, 499
$MnPSe_3$	$FePSe_3$	hR10	Ibid., 5, 499
$MnSb_2Sr$	Sb_2SrZn	oP16	J. Less-Common Metals, 79, 131
$Mn_{0.25}ScSi_{1.75}$	Si_2Zr	oC12	Dop. Akad. Nauk Ukr. RSR., A, 42, 80
$MnSe$	$ClNa$	cF8	Indian J. Pure Appl. Phys., 18, 950
$MnTe$	$AsNi$	hP4	Ibid., 18, 950
$Mn_{23}Th_6$	$Mn_{23}Th_6$	cF116	Solid State Comm., 23, 599
$Mn_{23}Y_6$	$Mn_{23}Th_6$	cF116	Ibid., 23, 599
$Mo_6Np_{1.20}Se_8$	$Cu_2Mo_6S_8$	hR15.20	Ibid., 38, 437
$Mo_6Np_{1.20}Se_8$	$Cu_2Mo_6S_8$	hR15.20	J. Solid State Chem., 39, 360
Mo_6NpSe_8	Mo_6PbS_8	hR15	Solid State Comm., 38, 437
Mo_6NpSe_8	Mo_6PbS_8	hR15	J. Solid State Chem., 39, 360
$Mo_6Pu_{1.20}Se_8$	$Cu_2Mo_6S_8$	hR15.20	Ibid., 39, 360
Mo_6PuSe_8	Mo_6PbS_8	hR15	Ibid., 39, 360
Mo_3S_4	Mo_3Se_4	hR14	J. Less-Common Metals, 72, 193
$Mo_9S_{11}Tl_2$	$Mo_9S_{11}Tl_2$	hR44	Ternary Superconductors, Proc. Int. Conf., 1980, 7
Mo_3Se_3Tl	Mo_3Se_3Tl	hP7	Ibid., 7
Nb	W	cI2	J. Less-Common Metals, 75, 227
Nb	W	cI2	Trans. Met. Soc. AIME, 236, 863
Nb	W	cI2	J. Less-Common Metals, 44, 215
$Nb_5S_3Se_3$	Te_4Ti_5	tI18	Phys. Stat. Solidi, A, 65, K179
Nb_5Se_4	Te_4Ti_5	tI18	Ibid., A, 65, K179
Nb_5Si_3	Si_3W_5	tI32	Russian Metallurgy, 184
Nb_5Si_3	B_3Cr_5	tI32	Ibid., 184
Nb_5Si_3	B_3Cr_5	tI32	Z. Metallk., 72, 720
Nb_3Si	PTi_3	tP32	Russian Metallurgy, 184
$NbSi_2$	$CrSi_2$	hP9	Ibid., 184
$NbSi_2$	$CrSi_2$	hP9	Z. Metallk., 72, 720
$NdNi_2P_2$	Al_4Ba	tI10	J. Solid State Chem., 35, 312
$NdPd_3$	$AuCu_3$	cP4	Mater. Res. Bull., 16, 1557
$NdPtSb$	$CaIn_2$	hP6	J. Less-Common Metals, 78, 1
NdS_3V	$CrLaS_3$	mP50	J. Solid State Chem., 38, 165
Ni_2P_2Pr	Al_4Ba	tI10	Ibid., 35, 312
$NiPS_3$	$CrPSe_3$	mC20	Ann. Chim. Paris, 5, 499
$NiPSe_3$	$CrPSe_3$	mC20	Ibid., 5, 499
Ni_2P_2Sm	Al_4Ba	tI10	J. Solid State Chem., 35, 312
NiP_2Ta	$NbNiP_2$	oP16	Mater. Res. Bull., 16, 933
$NiPTa_4$	$CoNb_4Si$	tP12	Inorg. Mater. USSR, 16, 705
Ni_2P_2Tb	Al_4Ba	tI10	J. Solid State Chem., 35, 312
Ni_2P_2Tm	Al_4Ba	tI10	Ibid., 35, 312
Ni_2P_2Y	Al_4Ba	TI10	Ibid., 35, 312
Ni_2P_2Yb	Al_4Ba	tI10	Ibid., 35, 312
$NiPZr_4$	$CoNb_4Si$	tP12	Inorg. Mater. USSR, 16, 705
Ni_2Sc	Cu_2Mg	cF24	Ibid., 17, 704
Ni_4SnU	Ni_5U	cF24	Fizika, 12, 200

Phase	Structure type	Structure code	Reference
NiTb	NiTb	mP24	Acta Crystallogr., Sect. B, 37, 495
$NiTi_2$	$NiTi_2$	cF96	Mh. Chem., 111, 535
Ni_2U	$MgZn_2$	hP12	Phys. Stat. Solidi, A, 57, K17
OsS_2	FeS_2	cP12	Inorg. Chem., 20, 501
PS_3Zn	$CrPSe_3$	mC20	Ann. Chim. Paris, 5, 499
P_3Sc_7	B_3Ru_7	hP20	Acta Chem. Scand. Ser. A, 35, 635
PSc_3	CFe_3	oP16	Ibid., A, 35, 635
PbS_4Tm_2	Bi_2PbS_4	oP28	Dokl. Akad. Nauk Azerb. SSR, 36, 67
PbS_4Yb_2	Bi_2PbS_4	oP28	Ibid., 36, 67
$PbSe_4Tm_2$	Bi_2PbS_4	oP28	Ibid., 36, 67
$PbSe_4Yb_2$	Bi_2PbS_4	oP28	Ibid., 36, 67
Pb_4Sr_5	Gd_5Si_4	oP36	J. Less-Common Metals, 81, 155
Pd_3Pr	$AuCu_3$	cP4	Mater. Res. Bull., 16, 1557
PdS_2	PdS_2	oP12	Inorg. Chem., 20, 501
Pd_3Sm	$AuCu_3$	cP4	Mater. Res. Bull., 16, 1557
Pd_3Tb	$AuCu_3$	cP4	Ibid., 16, 1557
Pd_3Tm	$AuCu_3$	cP4	Ibid., 16, 1557
Pd_3Yb	$AuCu_3$	cP4	Ibid., 16, 1557
Pr_5Tl_3	Si_3W_5	tI32	J. Less-Common Metals, 79, 47
Pr_3Tl_5	Pd_5Pu_3	oC32	Ibid., 79, 47
Pr_3Tl	$AuCu_3$	cP4	Ibid., 79, 47
$PrTl_3$	$AuCu_3$	cP4	Ibid., 79, 47
$PrTl$	$ClCs$	cP2	Ibid., 79, 47
PtS_2	CdI_2	hP3	Inorg. Chem., 20, 501
Pt_4Sr_5	Pu_5Rh_4	oP36	J. Less-Common Metals, 78, 49
Pt_3Sr_7	Pt_3Sr_7	oP40	Ibid., 78, 49
Pt_2Sr_3	Er_3Ni_2	hR15	Ibid., 78, 49
$Rb_2S_4Zn_3$	$Cs_2Mn_3S_4$	oI36	Rev. Chim. Minér., 17, 555
RuS_2	FeS_2	cP12	Inorg. Chem., 20, 501
$S_9Sb_4Sm_3$	$Eu_3Sb_4Se_9$	oP64	Inorg. Mater. USSR, 17, 1469
S_4Sb_2Sn	S_4Sb_2Sn	mP84	Chem. d. Erde, 35, S179
$S_5Sb_2Sn_2$	$Pb_2S_5Sb_2$	oP36	Ibid., 35, S179
$S_6Sb_2Sn_3$	$S_6Sb_2Sn_3$	oP132	Ibid., 35, S179
S_4SbSn_2	BiS_4Sn_2	cP63	Ibid., 35, S179
S_8Sb_5Tl	S_8Sb_5Tl	mP56	Z. Kristallogr., 150, 169
S_2SbTl	S_2SbTl	aP16	J. Solid State Chem., 40, 203
$S_2Sc_{1.37}$	$S_2Sc_{1.37}$	hR3.37	Inorg. Chem., 3, 1220
SSc	ClNa	cF8	Ibid., 3, 1220
SSiTh	La_2Sb	tI12	Z. anorg. Chem., 472, 139
S_4Sm_3	P_4Th_3	cI28	Mater. Res. Bull., 16, 975
S_4Sm_3	P_4Th_3	cI28	Inorg. Mater. USSR, 16, 806
$S_4Sm_{2.67}$	P_4Th_3	cI26.68	Ibid., 16, 806
SSm	ClNa	cF8	Ibid., 16, 806
$S_3Sn_2Tl_2$	$S_3Sn_2Tl_2$	mC28	Rev. Chim. Minér., 18, 224
$S_7Sn_2Yb_3$	$Eu_3S_7Sn_2$	oP24	Inorg. Mater. USSR, 17, 1160
$S_{12}Sn_3Yb_5$	$Eu_5S_{12}Sn_3$	oP40	Ibid., 17, 1160
$S_9Sn_2Yb_4$	$Eu_4S_9Sn_2$	oP30	Ibid., 17, 1160
S_4SnYb_2	Eu_2S_4Sn	oP56	Ibid., 17, 1160
S_2Ta	NbS_2	hP6	Mater. Res. Bull., 15, 1703
S_2Ti	CdI_2	hP3	Ibid., 15, 1703
S_4V_3	Cr_3S_4	mC14	J. Phys. Soc. Japan, 50, 413
$Sb_4Se_9Sm_3$	$Eu_3Sb_4Se_9$	oP64	Inorg. Mater. USSR, 17, 1469
$SbSe_2Tl$	$SbSe_2Tl$	oC32	Mater. Res. Bull., 15, 1105
Sb_3Tb_5	Mn_5Si_3	hP16	J. Less-Common Metals, 77, 81
Sb_3Tb_4	P_4Th_3	cI28	Ibid., 77, 81
SbTb	ClNa	cF8	Ibid., 77, 81
$Sb_{9.33}Te_{16.33}$ $Yb_{2.34}$	P_4Th_3	CI28	Inorg. Mater. USSR, 17, 692

Phase	Structure type	Structure code	Reference
Sb_2Te_4Yb	P_4Th_3	cI28	Ibid., 17, 692
SeSiTh	La_2Sb	tI12	Z. anorg. Chem., 472, 139
$Se_7Tb_2U_2$	$Se_7Tb_2U_2$	mC22	Russ. J. Inorg. Chem., 26, 622
$Se_{1.90}U$	S_2U	tI29	Ibid., 25, 1130
$Se_{1.90}U$	S_2U	tI29	Ibid., 26, 626
SiTeTh	La_2Sb	tI12	Z. anorg. Chem., 472, 139
Sm_3Tl_5	Pd_5Pu_3	oC32	J. Less-Common Metals, 81, 45
Tb_3Tl_5	Pd_5Pu_3	oC32	Ibid., 81, 45
Ti	Mg	hP2	J. Solid State Chem., 32, 213
Tl_5Y_3	Pd_5Pu_3	oC32	J. Less-Common Metals, 81, 45
Zn_2Zr	Cu_2Mg	cF24	Z. Naturforsch., B36, 1547

STRUCTURE REPORTS

SECTION II

INORGANIC COMPOUNDS

Edited by

J. Trotter

(University of British Columbia)

ARRANGEMENT

To find particular inorganic compounds the subject index or formula index should be used. The general arrangement is: elements, boron hydrides, carbonyls, phosphorus-nitrogen and sulphur-nitrogen compounds, halides, cyanides, oxides, double oxides, hydroxides, sulphides, borates, carbonates, nitrates, phosphates, arsenates, sulphates, perchlorates, iodates, silicates, silicate minerals, electron-diffraction studies. Only complete structure analyses are described; incomplete structural data are given in a Table, and compounds which have been described only in preliminary communications are tabulated.

CARBON
C

N.N. MATJUŠENKO, V.E. STREL'NICKIJ and V.A. GUSEV, 1981. Kristallografija, 26, 484-487 [Soviet Physics - Crystallography, 26, 274-276].

Cubic, Im3, a = 4.28 Å, Z = 16. Electron diffraction and microscopy. C in 16(f): x,x,x, x = 1/6. The material is obtained by condensation of carbon plasma. The structure contains diamond-like layers, minimum C-C = 1.23 Å.

NITROGEN (49 kbar, 299K)
N_2

D.T. CROMER, R.L. MILLS, D. SCHIFERL and L.A. SCHWALBE, 1981. Acta Cryst., B37, 8-11.

Cubic, Pm3n, a = 6.164 Å, Z = 8. Mo radiation, R = 0.041 for 19 reflexions. 4 N in 16(i): (0.042,0.042,0.042); 12 N in 48(ℓ): (0.239,0.531,0.080).

The structure (Fig. 1) is the same as that of β-F_2 and γ-O_2 at 50K and atmospheric pressure (1). It contains two types of disordered N_2 molecules.

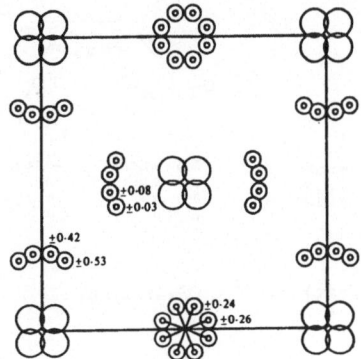

Fig. 1.   Structure of N_2 at 49 kbar.

1. Structure Reports, 29, 229; 39A, 125.

OXYGEN (5.5 GPa, 299K)
β-O_2

D. SCHIFERL, D.T. CROMER and R.L. MILLS, 1981. Acta Cryst., B37, 1329-1332.

Rhombohedral, R3m, a = 2.8467, c = 10.2249 Å, Z = 3 molecules. Mo radiation, R_W = 0.050 for 39 reflexions. O in 6(c): z = 0.0577.

The structure is the same as that determined at normal pressure at 28K (1). 0-0 = 1.202(2) Å (corrected for thermal motion).

1. Structure Reports, 27, 394.

CYCLODODECASULPHUR   CYCLODODECASULPHUR - CARBON DISULPHIDE
S_{12} $S_{12} \cdot CS_2$

J. STEIDEL and R. STEUDEL, 1981. Z. anorg. Chem., **476**, 171-178.

S_{12}, orthorhombic, Pnnm, a = 4.725, b = 9.104, c = 14.532 Å, Z = 2. Mo radiation, R = 0.035 for 581 reflexions.

$S_{12} \cdot CS_2$, rhombohedral, R$\bar{3}$m, a = 10.668, c = 11.551 Å, Z = 3. Mo radiation, R = 0.033 for 422 reflexions.

Both structures contain cyclic S_{12} molecules, which occupy sites of C_{2h} and D_{3d} symmetry in the two structures (Fig. 1); the CS_2 molecule interacts only weakly with the S_{12} units. Mean S-S = 2.052, 2.054(1) Å, S-S-S = 106.6, 106.2, S-S-S-S = 88.0, 87.2°.

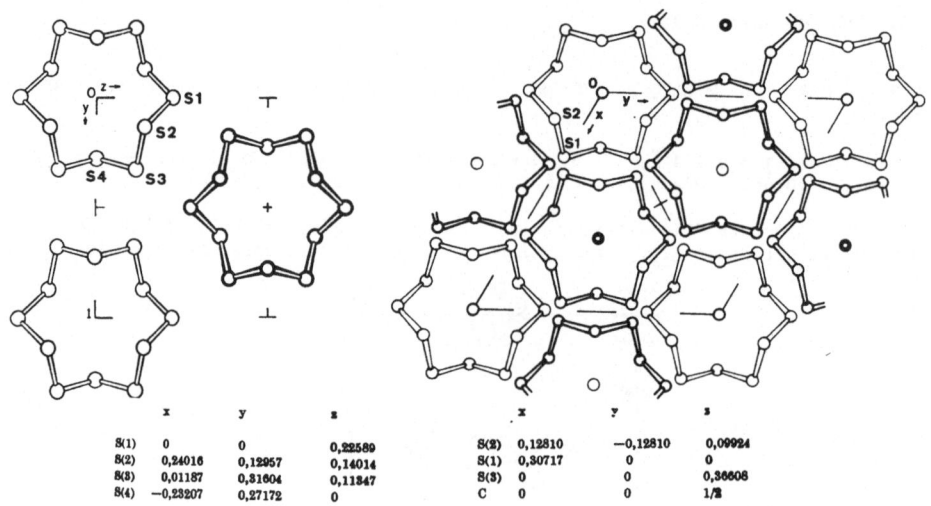

	x	y	z		x	y	z
S(1)	0	0	0,22589	S(2)	0,12810	−0,12810	0,09924
S(2)	0,24016	0,12957	0,14014	S(1)	0,30717	0	0
S(3)	0,01187	0,31604	0,11847	S(3)	0	0	0,36608
S(4)	−0,23207	0,27172	0	C	0	0	1/2

Fig. 1.   Structures of S_{12} (left) and $S_{12} \cdot CS_2$ (right).

IODINE (High-pressure)
I

K. TAKEMURA, S. MINOMURA, O. SHIMOMURA and Y. FUJII, 1980. Phys. Rev. Lett., **45**, 1881-1884.

Orthorhombic, Immm, a = 3.031, b = 5.252, c = 2.904 Å, at 30 GPa, Z = 2. Mo radiation, powder data. I in 2(a). Pressure-induced molecular dissociation appears to occur at 21 GPa.

LITHIUM SILICON OXYNITRIDE
α-LiSiON

Y. LAURENT, J. GUYADER and G. ROULT, 1981. Acta Cryst., B**37**, 911-913.

Orthorhombic, $Pca2_1$, a = 5.1991, b = 6.3889, c = 4.7390 Å, D_m = 2.72, Z = 4. Neutron time-of-flight powder data.

Structure as previously described (1). Si-O = 1.63, Si-N = 1.73-1.78, Li-O = 1.92-2.07, Li-N = 2.25 Å.

1. Structure Reports, 46A, 137.

TETRADECABORANE(20)

$B_{14}H_{20}$

J.C. HUFFMAN, D.C. MOODY and R. SCHAEFFER, 1981. Inorg. Chem., 20, 741-745.

Orthorhombic, $P2_12_12_1$, a = 13.119, b = 9.976, c = 8.963 Å (at -164°C), Z = 4. Mo radiation, R = 0.047 for 913 reflexions.

The molecule (Fig. 1) has close to C_{2v} symmetry, and consists of two cis-fused B_8H_{12}-type fragments; B-B = 1.72-1.93, B-H = 1.05-1.31 Å.

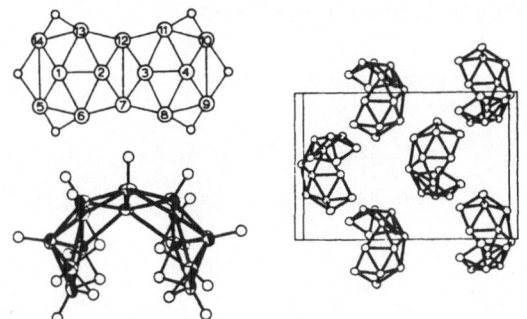

Fig. 1.   Structure of tetradecaborane(20).

DIAMMINEDECABORANE

$B_{10}H_{12}(NH_3)_2$

DIAMMINEDECABORANE-HYDRAZINE

$B_{10}H_{12}(NH_3)_2 \cdot N_2H_4$

V.I. PONOMAREV and L.O. ATOVMJAN, 1981. Kristallografija, 26, 505-511 [Soviet Physics - Crystallography, 26, 287-291].

$B_{10}H_{12}(NH_3)_2$, orthorhombic, Pnma, a = 18.096, 17.945, b = 7.373, 7.335, c = 7.223, 7.191 Å, at 298, 123K, Z = 4. Cu, Mo radiations, R = 0.048, 0.042 for 913, 843 reflexions.

$B_{10}H_{12}(NH_3)_2 \cdot N_2H_4$, monoclinic, B2/b, a = 11.596, b = 9.772, c = 10.135 Å, γ = 81.85°, at 298K, Z = 4. R = 0.044 for 1023 reflexions.

The structures (1) contain $B_{10}H_{12}(NH_3)_2$ molecules, linked by van der Waals forces in the parent compound, and by weak hydrogen bonding via the hydrazine molecules in the adduct (Fig. 1).

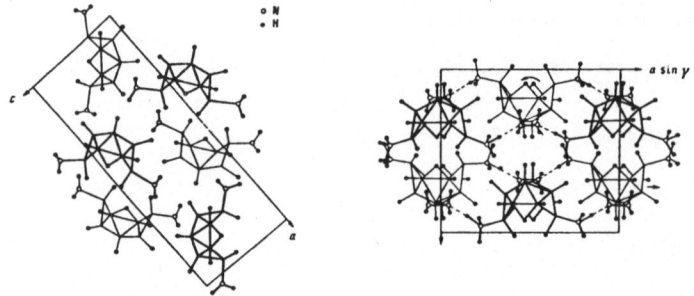

Fig. 1.   Structures of $B_{10}H_{12}(NH_3)_2$ (left) and $B_{10}H_{12}(NH_3)_2 \cdot N_2H_4$ (right).

<u>1</u>.   Structure Reports, <u>44</u>A, 126, 331.

POTASSIUM 1,7-DIFLUORODECAHYDRODODECABORATE MONOHYDRATE
$K_2B_{12}H_{10}F_2 \cdot H_2O$

K.A. SOLNCEV, N.A. ŽUKOVA, L.A. BUTMAN and N.T. KUZNECOV, 1980.   Izv. Akad. Nauk SSSR, Neorg. Mater., <u>16</u>, 1882-1884.

Orthorhombic, Pccn, a = 7.992, b = 11.424, c = 13.407 Å, D_m = 1.48, Z = 4.

The structure contains $1,7\text{-}B_{12}H_{10}F_2^{2-}$ anions linked by zigzag chains made up of -F-K-O-K-F- rings.

HEXACARBONYLDIIRON-TRIBORANE
$(CO)_6Fe_2B_3H_7$

K.J. HALLER, E.L. ANDERSEN and T.P. FEHLNER, 1981.   Inorg. Chem., <u>20</u>, 309-313.

Monoclinic, $P2_1/c$, a = 9.006, b = 10.878, c = 12.479 Å, β = 99.54°, at -20°C, Z = 4.   Mo radiation, R = 0.055 for 3157 reflexions.   Preliminary report in <u>1</u>.

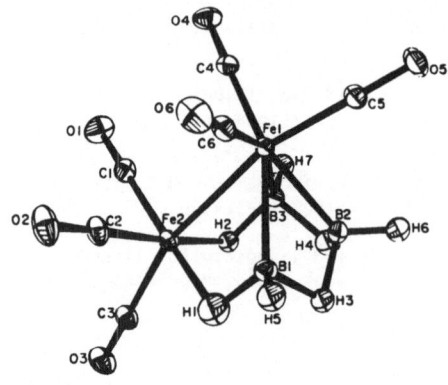

Fig. 1.   The $(CO)_6Fe_2B_3H_7$ molecule.

The molecule (Fig. 1) has approximate C_s symmetry, with six- and seven-coordinate Fe atoms; it may be considered as derived from B_5H_9 by replacing one apical and one basal BH group by $Fe(CO)_3$ groups. Fe-Fe = 2.559(2), Fe(1)-B = 2.070, 2.054, 2.065(4) Å.

1. Structure Reports, 45A, 391.

CHLOROBIS(PENTACARBONYLCHROMIUM)ARSINIDENE
$ClAs[Cr(CO)_5]_2$

J. von SEYERL, B. SIGWARTH, H.-G. SCHMID, G. MOHR, A. FRANK, M. MARSILI and G. HUTTNER, 1981. Chem. Ber., 114, 1392-1406.

Monoclinic, $P2_1/c$, a = 6.991, b = 11.430, c = 21.361 Å, β = 107.85°, Z = 4. Mo radiation, R = 0.052 for 1365 reflexions, at 193K.

Arsenic has trigonal planar coordination, and the Cr atoms have octahedral coordinations. As-Cl = 2.230(3), As-Cr = 2.319, 2.329(2) Å, Cl-As-Cr = 109.8, Cr-As-Cr = 138.8°. Three related organic compounds are also described [see 1].

1. Structure Reports, 48B.

μ-DIBROMOGERMANDIYL-BIS(PENTACARBONYLTUNGSTEN)(W-W)
$Br_2Ge[W(CO)_5]_2$

C. BURSCHKA, K. STROPPEL and P. JUTZI, 1981. Acta Cryst., B37, 1397-1399.

Monoclinic, $P2_1/c$, a = 9.475, b = 13.101, c = 17.537 Å, β = 120.28°, D_m = 3.14, Z = 4. Mo radiation, R = 0.030 for 1753 reflexions.

The structure (Fig. 1) contains a WGe_2 ring, W-W = 3.370(1), W-Ge = 2.573, 2.580(2) Å, Ge-W-W = 49.2, W-Ge-W = 81.7°; Ge-Br = 2.304, 2.321(2), W-C = 1.99-2.12, C-O = 1.08-1.16 Å.

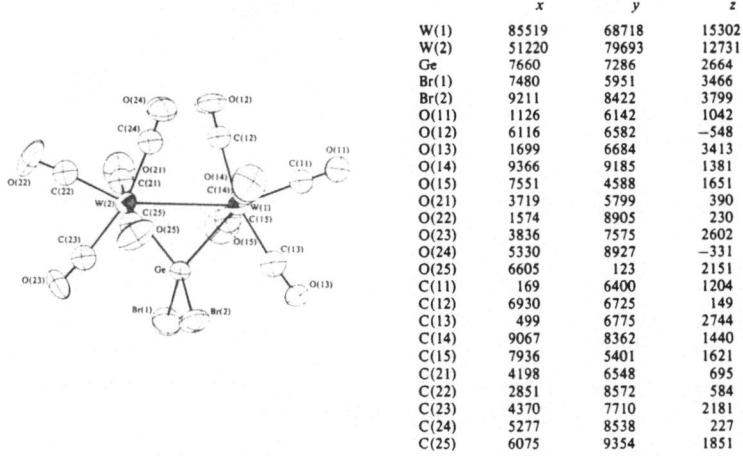

	x	y	z
W(1)	85519	68718	15302
W(2)	51220	79693	12731
Ge	7660	7286	2664
Br(1)	7480	5951	3466
Br(2)	9211	8422	3799
O(11)	1126	6142	1042
O(12)	6116	6582	-548
O(13)	1699	6684	3413
O(14)	9366	9185	1381
O(15)	7551	4588	1651
O(21)	3719	5799	390
O(22)	1574	8905	230
O(23)	3836	7575	2602
O(24)	5330	8927	-331
O(25)	6605	123	2151
C(11)	169	6400	1204
C(12)	6930	6725	149
C(13)	499	6775	2744
C(14)	9067	8362	1440
C(15)	7936	5401	1621
C(21)	4198	6548	695
C(22)	2851	8572	584
C(23)	4370	7710	2181
C(24)	5277	8538	227
C(25)	6075	9354	1851

Fig. 1.   The $Br_2Ge[W(CO)_5]_2$ molecule, and atomic positional parameters (× 10^5 for W; × 10^4 for others).

DIMANGANESE DECACARBONYL
$Mn_2(CO)_{10}$

DIRHENIUM DECACARBONYL
$Re_2(CO)_{10}$

M.R. CHURCHILL, K.N. AMOH and H.J. WASSERMAN, 1981. Inorg. Chem., <u>20</u>, 1609-1611.

Monoclinic, I2/a, a = 14.135, 14.658, b = 7.100, 7.119, c = 14.628, 14.815 Å, β = 105.17, 105.80°, Z = 4. Mo radiation, R = 0.023, 0.022 for 1107, 1207 reflexions.

Structures as previously determined (<u>1</u>). Mn-Mn = 2.904(1), Re-Re = 3.041(11), Mn-C = 1.81 (axial), 1.86 (equatorial), Re-C = 1.93 (axial), 1.99 Å (equatorial).

<u>1</u>. Structure Reports, <u>21</u>, 256; <u>28</u>, 67.

μ_4-CARBIDO-μ_3-CARBONYL-DODECACARBONYLTETRAIRON
$Fe_4(CO)_{13}C$

J.S. BRADLEY, G.B. ANSELL, M.E. LEONOWICZ and E.W. HILL, 1981. J. Amer. Chem. Soc., <u>103</u>, 4968-4970.

Monoclinic, P2$_1$/n, a = 9.337, b = 21.440, c = 9.519 Å, β = 98.73°, Z = 4. Mo radiation, R = 0.045 for 2566 reflexions.

The molecule (Fig. 1) contains an Fe$_4$ butterfly with a μ^4-carbon atom forming an Fe$_4$C cluster; each Fe has three terminal CO ligands, and there is one bridging CO. Fe(1)-Fe(4) (bridged by CO) = 2.545, other Fe-Fe = 2.637-2.647(1) Å.

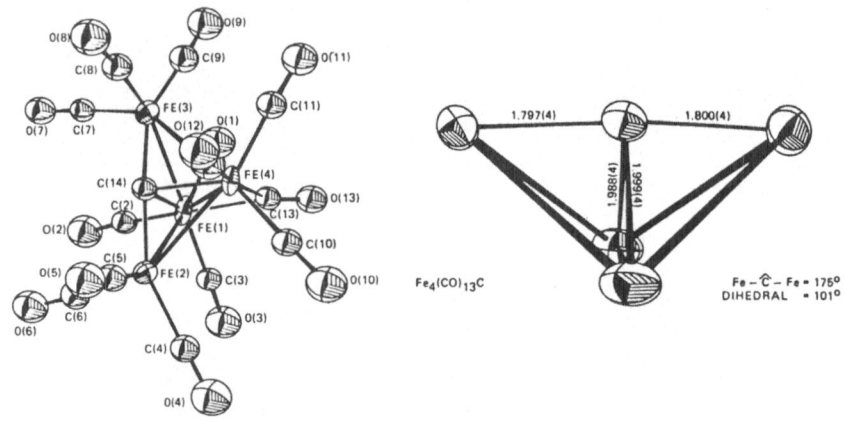

Fig. 1.   Structure of Fe$_4$(CO)$_{13}$C.

IRON CARBONYL HYDRIDE NITRIDES
$HFe_4N(CO)_{12}$ $HFe_5N(CO)_{14}$

M. TACHIKAWA, J. STEIN, E.L. MUETTERTIES, R.G. TELLER, M.A. BENO, E. GEBERT and J.M. WILLIAMS, 1980. J. Amer. Chem. Soc., <u>102</u>, 6648-6649.

$HFe_4N(CO)_{12}$, triclinic, P$\bar{1}$, a = 7.491, b = 9.214, c = 13.974 Å, α = 88.09, β = 86.98, γ = 73.76°, Z = 2. Mo radiation, R = 0.071 for 1415 reflexions.

$HFe_5N(CO)_{14}$, monoclinic, $P2_1/n$, a = 8.473, b = 15.056, c = 16.048 Å, β = 95.42°, Z = 4. Mo radiation, R = 0.031 for 3967 reflexions.

The Fe_4 compound (which has a disordered crystal structure) contains a butter-fly shaped iron cluster, and the Fe_5 compound is square pyramidal (Fig. 1).

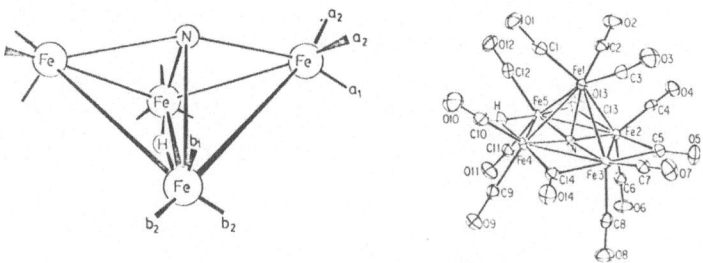

Fig. 1.   The $HFe_4N(CO)_{12}$ and $HFe_5N(CO)_{14}$ molecules.

DODECACARBONYLTETRA-μ-HYDRIDO-tetrahedro-TETRAOSMIUM
$Os_4(CO)_{12}H_4$

B.F.G. JOHNSON, J. LEWIS, P.R. RAITHBY and C. ZUCCARO, 1981. Acta Cryst., B37, 1728-1731.

Triclinic, P$\bar{1}$, a = 9.811, b = 9.893, c = 10.240 Å, α = 85.56, β = 82.71, γ = 88.71°, Z = 2. Mo radiation, R = 0.043 for 1815 reflexions.

Isostructural with the Ru analogue (1), the molecule (Fig. 1) containing a distorted tetrahedron of Os atoms with four long bonds (probably H-bridged) and two short bonds: Os(1)-Os(2) = Os(3)-Os(4) = 2.817, other Os-Os = 2.964(2), Os-C = 1.82-1.93(3), C-O = 1.10-1.17(3) Å. There is a small amount of disorder (19:1).

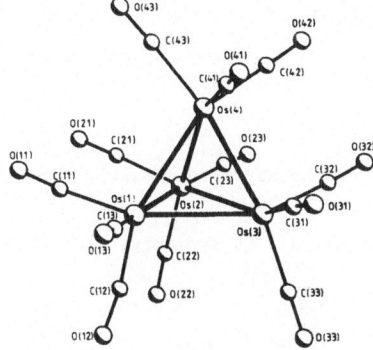

Fig. 1. $Os_4(CO)_{12}H_4$.

1. Structure Reports, 44A, 130.

DI-μ_3-SELENIDO-TRIS(TRICARBONYLOSMIUM)(2Os-Os)

$Os_3(CO)_9Se_2$

B.F.G. JOHNSON, J. LEWIS, P.G. LODGE and P.R. RAITHBY, 1981. Acta Cryst., B37, 1731-1733.

Triclinic, P$\overline{1}$, a = 6.804, b = 9.620, c = 13.527 Å, α = 94.206, β = 95.570, γ = 110.469°, Z = 2. Mo radiation, R = 0.042 for 4106 reflexions.

The molecular structure is similar to that of $Fe_3(CO)_9S_2$ (1), containing an isosceles triangle of Os atoms (Os-Os = 2.836, 2.847, 3.791(1) Å), with the long Os...Os edge bridged on both sides by Se atoms to give a trigonal bipyramidal cluster core (Fig. 1). Os-Se = 2.493-2.551(2), Os-C = 1.88-1.92(1), C-O = 1.11-1.13(1), Se...Se = 3.254(1) Å.

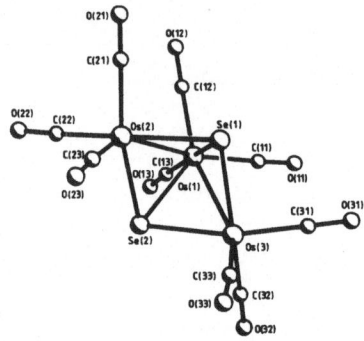

Fig. 1. $Os_3(CO)_9Se_2$.

1. Structure Reports, 30B, 384.

BIS[μ-CARBONYL-BIS(TRICARBONYLCOBALTIO)(Co-Co)]GERMANIUM(4Co-Ge)

$Ge[Co_2(CO)_7]_2$

R.F. GERLACH, K.M. MACKAY, B.K. NICHOLSON and W.T. ROBINSON, 1981. J. Chem. Soc., Dalton, 80-84.

Triclinic, P$\overline{1}$, a = 10.396, b = 16.495, c = 12.879 Å, α = 90.33, β = 97.68, γ = 95.03°, D_m = 2.11, Z = 4 (2 molecules per asymmetric unit). Mo radiation, R = 0.056 for 2852 reflexions.

The molecule (Fig. 1) has approximate C_2 symmetry, with Ge bridging two Co-Co bonds asymmetrically.

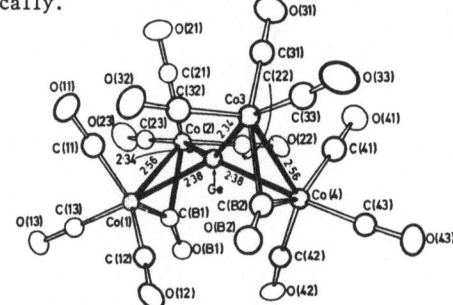

Fig. 1.   The $Ge[Co_2(CO)_7]_2$ molecule.

CAESIUM STRONTIUM AZIDES
$CsSr(N_3)_3$ $Cs_2Sr(N_3)_4$

I.   H. KRISCHNER, H.E. MAIER and O. BAUMGARTNER, 1981. Z. Kristallogr., 155, 201-206.
II. Idem, 1981. Ibid., 155, 211-216.

$CsSr(N_3)_3$, monoclinic, C2/c, a = 9.174, b = 10.173, c = 9.008 Å, β = 120.15°, Z = 4. Mo radiation, R = 0.042 for 626 reflexions.

$Cs_2Sr(N_3)_4$, orthorhombic, Cmcm, a = 12.642, b = 14.545, c = 11.833 Å, Z = 8. Mo radiation, R = 0.063 for 677 reflexions.

Atomic positions

		$CsSr(N_3)_3$						$Cs_2Sr(N_3)_4$		
		x	y	z				x	y	z
Cs	4e	0.	0,2033	0,25	Cs(1)	8g		0,2444	0,1209	0,25
					Cs(2)	8f		0	0,3817	0,0825
Sr	4e	0,	0,7606	0,25	Sr	8f		0	0,8667	0,0776
N1	4b	0,	0,50	0,50	N1			0,215	0	0
					N2	8e		0,123	0	0
N2		0,058	0,605	0,048	N3			0,310	0	0
					N4			0,282	0,381	0,25
N3		0,696	0,513	0,103	N5	8g		0,370	0,410	0,25
	8f				N6			0,198	0,352	0,25
N4		0,716	0,615	0,051	N7	8d		0,25	0,25	0
					N8	16h		0,337	0,261	0,038
N5		0,176	0,913	0,154	N9	4c		0	0,167	0,25
					N10	8f		0	0,163	0,154
					N11			0	0,649	0,25
					N12	4c		0	0,726	0,25
					N13			0	0,566	0,25

Both structures (Fig. 1) contain linear, symmetric azide anions, N-N = 1.18(1), 1.17(2) Å, linked by Cs and Sr cations which are coordinated to 8 azide groups, mean Cs-N = 3.31, 3.22, 3.58, mean Sr-N = 2.70 Å.

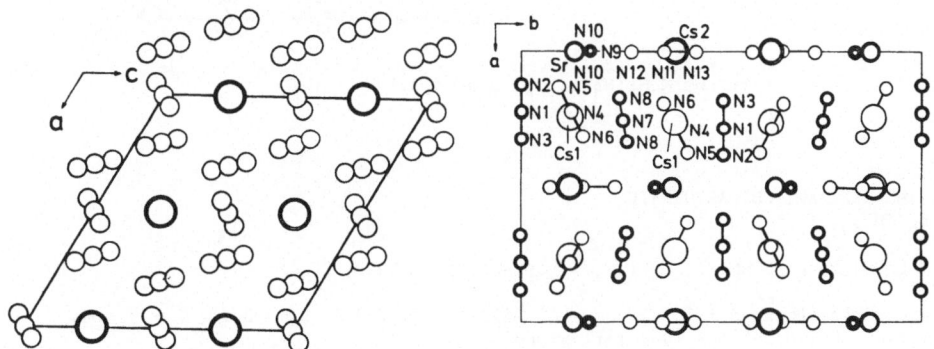

Fig. 1.   Structures of $CsSr(N_3)_3$ (left) and $Cs_2Sr(N_3)_4$ (right).

POTASSIUM HEXAAZIDONICKELATE DIHYDRATE
RUBIDIUM HEXAAZIDONICKELATE DIHYDRATE
CAESIUM HEXAAZIDONICKELATE DIHYDRATE
$M_4Ni(N_3)_6 \cdot 2H_2O$ (M = K, Rb, Cs)

CAESIUM TETRAAZIDONICKELATE MONOHYDRATE
$Cs_2Ni(N_3)_4 \cdot H_2O$

H.E. MAIER, H. KRISCHNER and H. PAULUS, 1981. Z. Kristallogr., <u>157</u>, 277-289.

Hexaazides, triclinic, $P\bar{1}$, a = 6.535, 6.716, 6.995, b = 10.819, 10.984, 11.162, c = 11.454, 11.808, 12.489 Å, α = 99.78, 99.81, 100.18, β = 89.31, 89.41, 89.31, γ = 94.55, 94.91, 95.12°, Z = 2. Mo radiation, R = 0.032, 0.053, 0.041 for 2803, 2992, 1920 reflexions.

Tetraazide, orthorhombic, Pbca, a = 17.336, b = 11.249, c = 11.291 Å, Z = 8. Mo radiation, R = 0.039 for 1925 reflexions.

    All four structures contain Ni coordinated octahedrally to six azide groups; in the hexaazides the octahedra are isolated, and in the tetraazide the octahedra are connected by azide bridges (Fig. 1). Alkali-metal cations have irregular tri-capped trigonal prismatic 9-coordination to N and H_2O. Ni-N = 2.08-2.13 Å.

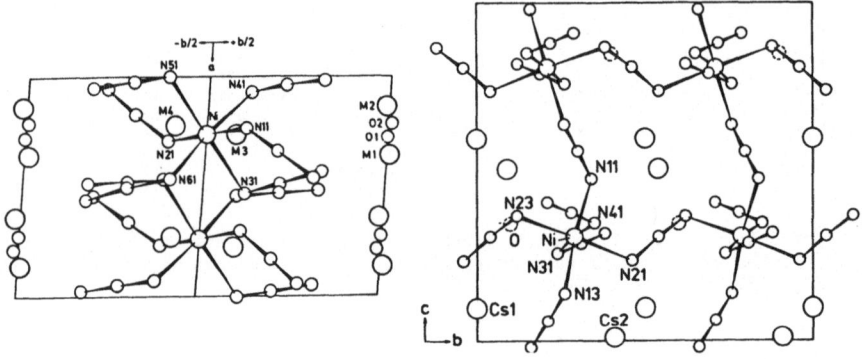

Fig. 1.   Structures of $M_4Ni(N_3)_6 \cdot 2H_2O$ and $Cs_2Ni(NH_3)_4 \cdot H_2O$.

CAESIUM HEPTAAMIDODILANTHANATE
$CsLa_2(NH_2)_7$

H. JACOBS and D. SCHMIDT, 1981. J. Less-Common Metals, <u>78</u>, 51-59.

Orthorhombic, Pbca, a = 10.414, b = 15.161, c = 13.709 Å, D_m = 3.20, Z = 8. Mo radiation, R = 0.059 for 1458 reflexions.

    The structure (Fig. 1) shows a similarity to that of the K salt (<u>1</u>). It contains alternating sheets of La and Cs ions parallel to (010); La ions are coordinated to 8 and Cs to 9 NH_2 groups.

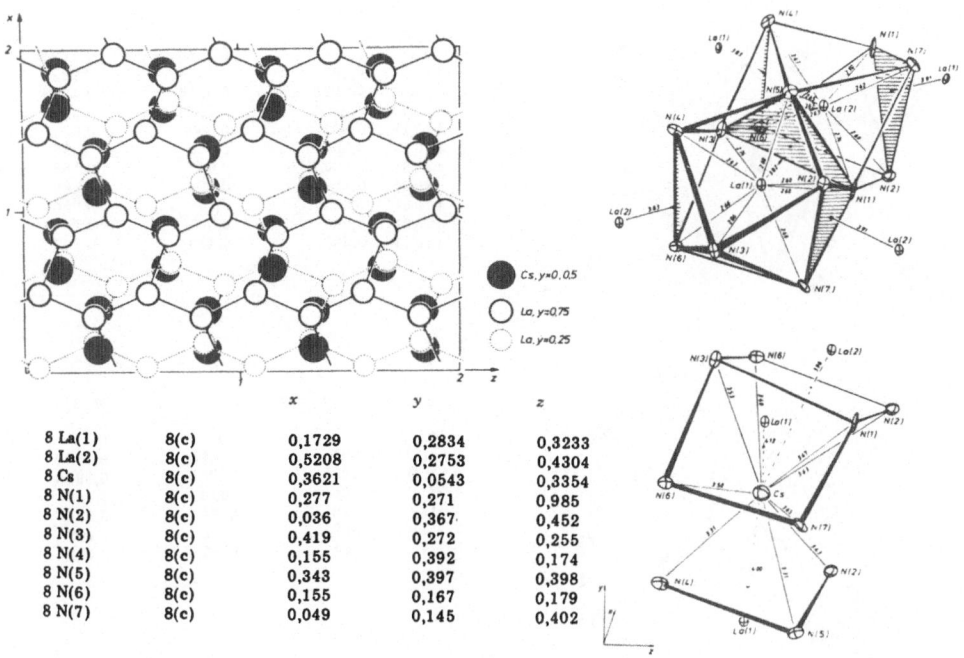

		x	y	z
8 La(1)	8(c)	0,1729	0,2834	0,3233
8 La(2)	8(c)	0,5208	0,2753	0,4304
8 Cs	8(c)	0,3621	0,0543	0,3354
8 N(1)	8(c)	0,277	0,271	0,985
8 N(2)	8(c)	0,036	0,367	0,452
8 N(3)	8(c)	0,419	0,272	0,255
8 N(4)	8(c)	0,155	0,392	0,174
8 N(5)	8(c)	0,343	0,397	0,398
8 N(6)	8(c)	0,155	0,167	0,179
8 N(7)	8(c)	0,049	0,145	0,402

Fig. 1.   Structure of $CsLa_2(NH_2)_7$.

<u>1</u>. Structure Reports, <u>40</u>A, 127.

CAESIUM LANTHANUM AMIDES
$Cs_3La(NH_2)_6 \cdot NH_3$ $Cs_4La(NH_2)_7 \cdot NH_3$

H. JACOBS, D. SCHMIDT, D. SCHMITZ, J. FLEISCHHAUER and W. SCHLEKER, 1981. J. Less-Common Metals, <u>81</u>, 121-133.

$Cs_3La(NH_2)_6 \cdot NH_3$, monoclinic, $P2_1/n$, a = 7.155, b = 12.358, c = 15.595 Å, β = 94.78°, D_m = 3.09, Z = 4. Mo radiation, R = 0.057 for 1504 reflexions.

$Cs_4La(NH_2)_7 \cdot NH_3$, monoclinic, $P2_1/c$, a = 7.139, b = 9.783, c = 11.867 Å, β = 98.06°, D_m = 3.27, Z = 2. Mo radiation, R = 0.088 for 802 reflexions.

Both structures (Fig. 1) contain $La(NH_2)_6$ octahedra, Cs ions, and ammonia molecules; in the $(NH_2)_7$ compound $N(4) = (NH_2 + NH_3)/2$, and may be a $N_2H_5^-$ ion. La-N = 2.49-2.62 Å.

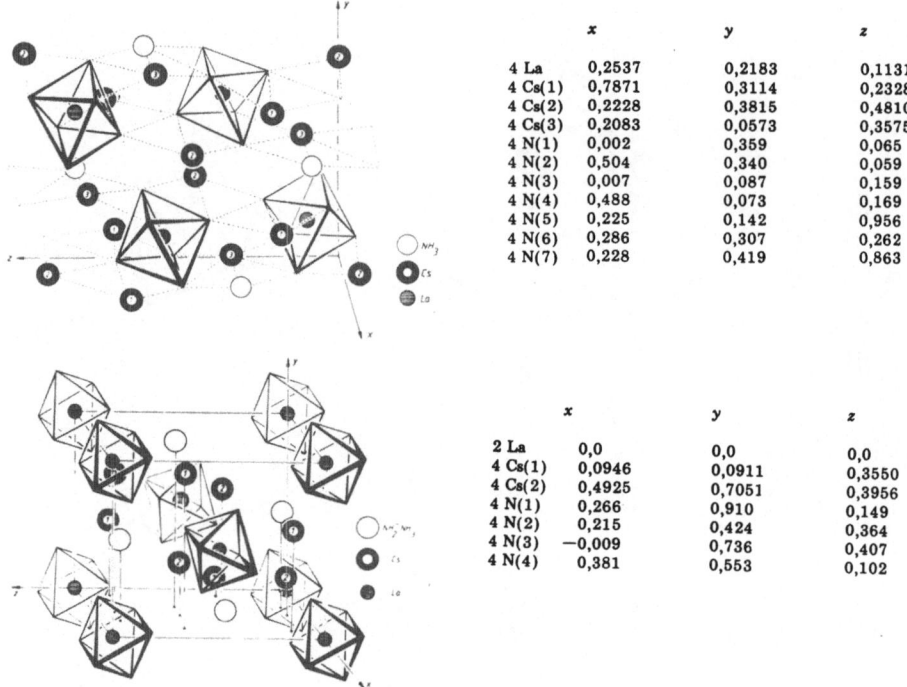

	x	y	z
4 La	0,2537	0,2183	0,1131
4 Cs(1)	0,7871	0,3114	0,2328
4 Cs(2)	0,2228	0,3815	0,4810
4 Cs(3)	0,2083	0,0573	0,3575
4 N(1)	0,002	0,359	0,065
4 N(2)	0,504	0,340	0,059
4 N(3)	0,007	0,087	0,159
4 N(4)	0,488	0,073	0,169
4 N(5)	0,225	0,142	0,956
4 N(6)	0,286	0,307	0,262
4 N(7)	0,228	0,419	0,863

	x	y	z
2 La	0,0	0,0	0,0
4 Cs(1)	0,0946	0,0911	0,3550
4 Cs(2)	0,4925	0,7051	0,3956
4 N(1)	0,266	0,910	0,149
4 N(2)	0,215	0,424	0,364
4 N(3)	−0,009	0,736	0,407
4 N(4)	0,381	0,553	0,102

Fig. 1.   Structure of $Cs_3La(NH_2)_6.NH_3$ (top) and $Cs_4La(NH_2)_7.NH_3$ (bottom).

CAESIUM CALCIUM AMIDE
$CsCa(NH_2)_3$

CAESIUM EUROPIUM(II) AMIDE
$CsEu(NH_2)_3$

H. JACOBS and J. KOCKELKORN, 1981.   J. Less-Common Metals, <u>81</u>, 143-154.

$CsCa(NH_2)_3$, orthorhombic, Pnma, a = 6.516, b = 7.078, c = 12.069 Å, D_m = 2.58, Z = 4. Mo radiation, R = 0.055 for 152 reflexions.

$CsEu(NH_2)_3$, monoclinic, C2/c, a = 7.201, b = 12.510, c = 6.981 Å, β = 106.37°, D_m = 3.68, Z = 4. Mo radiation, R = 0.083 for 235 reflexions.

Atomic positions

$CsCa(NH_2)_3$		x	y	z
Cs	in 4(c)	0.1806	1/4	0.1078
Ca	4(c)	0.095	1/4	0.7608
N(1)	8(d)	0.143	0.974	0.336
N(2)	4(c)	0.331	1/4	0.598

$CsEu(NH_2)_3$		x	y	z
Cs	in 4(e)	0	0.3014	1/4
Eu	4(a)	0	0	0
N(1)	4(e)	0	0.839	1/4
N(2)	8(f)	0.742	0.922	0.672

The materials are distorted perovskites, with structures related to that of BaNiO$_3$ (<u>1</u>). Ca and Eu have octahedral coordinations, Cs has 12-coordination in the Ca compound and 10-coordination in the Eu compound. Ca-N = 2.42-2.50, Eu-N = 2.67-2.69, Cs-N = 3.37-4.92 Å.

<u>1</u>. Structure Reports, <u>37</u>A, 263; <u>42</u>A, 287.

THIONITROSYL HEXAFLUOROANTIMONATE(V)
NS$^+$.SbF$_6^-$

THIONITROSYL UNDECAFLUORODIANTIMONATE(V)
NS$^+$.Sb$_2$F$_{11}^-$

W. CLEGG, O. GLEMSER, K. HARMS, G. HARTMANN, R. MEWS, M. NOLTEMEYER and G.M. SHELDRICK, 1981. Acta Cryst., B<u>37</u>, 548-552.

NS$^+$.SbF$_6^-$
Orthorhombic, Pbca, a = 13.999, b = 8.363, c = 10.468 Å, at 293K, Z = 8. Mo radiation, R = 0.032 for 2025 reflexions.

NS$^+$.Sb$_2$F$_{11}^-$
Monoclinic, P2$_1$/c, a = 8.374, 8.191, b = 11.792, 11.760, c = 10.108, 9.935 Å, β = 91.89, 92.88°, at 293, 121.5K, Z = 4. Mo radiation, R = 0.036, 0.037 for 2873, 1691 reflexions.

Both structures are ionic, containing NS$^+$ ions, N-S = 1.42(1) Å; S has 5 F neighbours at 2.6-2.8 Å. The SbF$_6^-$ ion is octahedral, Sb-F = 1.88 Å, and Sb$_2$F$_{11}^-$ contains two octahedra sharing a corner, Sb-F = 2.03 (bridging), 1.86 Å (terminal), Sb-F-Sb = 151°.

trans-BIS(HEXAFLUOROARSENATO)TETRAKIS(THIAZYL TRIFLUORIDE)MANGANESE(II)
Mn(NSF$_3$)$_4$(AsF$_6$)$_2$

B. BUSS, W. CLEGG, G. HARTMANN, P.G. JONES, R. MEWS, M. NOLTEMEYER and G.M. SHELDRICK, 1981. J. Chem. Soc., Dalton, 61-63.

Monoclinic, P2$_1$/n, a = 7.496, b = 10.378, c = 13.979 Å, β = 94.33°, Z = 2. Mo radiation, R = 0.048 for 1226 reflexions.

Mn is coordinated octahedrally to four N and 2 F (Fig. 1), Mn-N = 2.187, Mn-F = 2.193(4) Å; N-S = 1.365, S-F = 1.506, As-F(1) = 1.740, As-F(3) = 1.701, other As-F = 1.654-1.683(6) Å, Mn-N-S = 161.6, Mn-F-As = 150.6°.

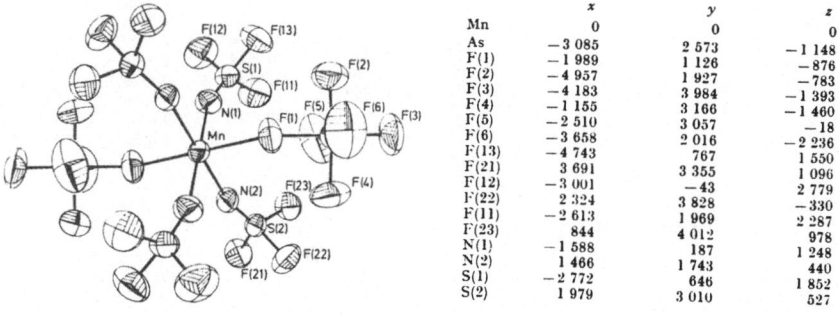

	x	y	z
Mn	0	0	0
As	−3 085	2 573	−1 148
F(1)	−1 989	1 126	−876
F(2)	−4 957	1 927	−783
F(3)	−4 183	3 984	−1 393
F(4)	−1 155	3 166	−1 460
F(5)	−2 510	3 057	−18
F(6)	−3 658	2 016	−2 236
F(13)	−4 743	767	1 550
F(21)	3 691	3 355	1 096
F(12)	−3 001	−43	2 779
F(22)	2 324	3 828	−330
F(11)	−2 613	1 969	2 287
F(23)	844	4 012	978
N(1)	−1 588	187	1 248
N(2)	1 466	1 743	440
S(1)	−2 772	646	1 852
S(2)	1 979	3 010	527

Fig. 1.   Structuré of Mn(NSF$_3$)$_4$(AsF$_6$)$_2$, and atomic positional parameters (x 10^4).

TRISILVER(I) SULPHAMIDE - AMMONIA - WATER
$Ag_3(HN_2O_2S) \cdot NH_3 \cdot H_2O$

C. KRATKY, E. NACHBAUR and A. POPITSCH, 1981. Acta Cryst., B37, 654-656.

Orthorhombic, Pcam, a = 11.813, b = 9.665, c = 6.578 Å, at 108K, Z = 4. Mo radiation, R = 0.06 for 1076 reflexions.

The structure (Fig. 1) contains alternating layers of Ag^+ ions, and NH_3, H_2O, and sulphamide ions (H atoms not located, and N and O distinguished by chemical considerations).

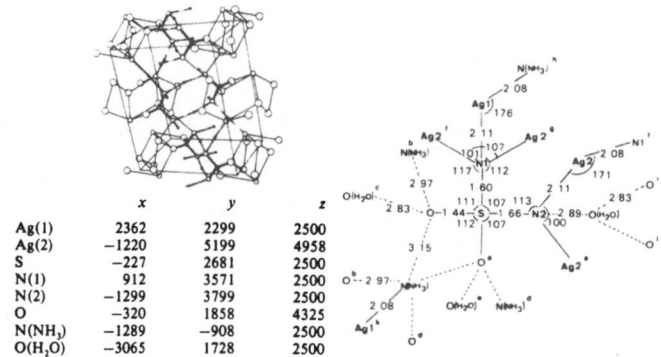

	x	y	z
Ag(1)	2362	2299	2500
Ag(2)	−1220	5199	4958
S	−227	2681	2500
N(1)	912	3571	2500
N(2)	−1299	3799	2500
O	−320	1858	4325
N(NH₃)	−1289	−908	2500
O(H₂O)	−3065	1728	2500

Fig. 1.   Structure of $Ag_3(HN_2O_2S) \cdot NH_3 \cdot H_2O$, and atomic positional parameters (x 10^4).

COBALT NITROSYL SULPHUR NITRIDE
$Co(NO)_2(S_3N)$

M. HERBERHOLD, L. HAUMAIER and U. SCHUBERT, 1981. Inorg. Chim. Acta, 49, 21-24.

Monoclinic, $P2_1/c$, a = 6.164, b = 15.32, c = 7.349 Å, β = 103.47°, Z = 4. Mo radiation, R = 0.038 for 868 reflexions.

The structure (Fig. 1) contains an almost-planar five-membered CoS_3N ring, Co having distorted tetrahedral coordination. Co-S = 2.25, 2.26, Co-N = 1.65, 1.66, S-S = 2.00, S-N = 1.55, 1.63 Å, S-Co-S = 91.8°.

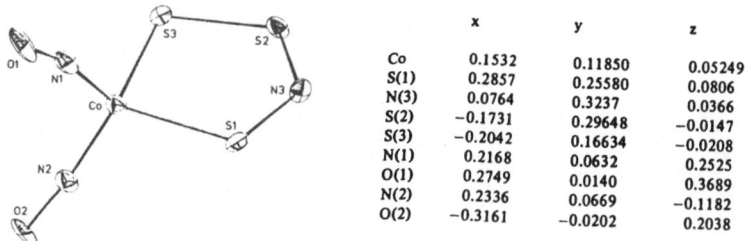

	x	y	z
Co	0.1532	0.11850	0.05249
S(1)	0.2857	0.25580	0.0806
N(3)	0.0764	0.3237	0.0366
S(2)	−0.1731	0.29648	−0.0147
S(3)	−0.2042	0.16634	−0.0208
N(1)	0.2168	0.0632	0.2525
O(1)	0.2749	0.0140	0.3689
N(2)	0.2336	0.0669	−0.1182
O(2)	−0.3161	−0.0202	0.2038

Fig. 1.   Structure of $Co(NO)_2(S_3N)$.

TETRASULPHUR DINITRIDE
S_4N_2

R.W.H. SMALL, A.J. BANISTER and Z.V. HAUPTMAN, 1981. J. Chem. Soc., Dalton, 2188-2191.

Tetragonal, $P4_2nm$, a = 11.25, c = 3.836 Å, Z = 4. Mo radiation, R = 0.057 for 424 reflexions, at 278K.

Atomic positions

			x	y	z
S(1)	in	4(c)	0.1115	0.1115	0.1000
S(2)		4(c)	0.3200	0.3200	0.1525
S(3)		8(d)	0.3530	0.1505	0.3278
N		8(d)	0.2425	0.0707	0.1616

The structure (Fig. 1) contains non-planar molecules (C_S symmetry), with short intermolecular S...N contacts, 3.02 Å. S-S = 2.055(3), S-N = 1.56, 1.66(1) Å. Slow decomposition results in polymerization to fibrous $(SN)_x$.

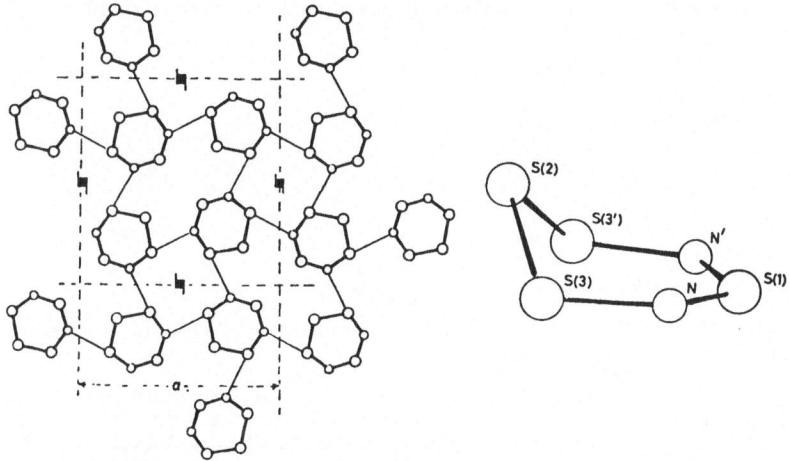

Fig. 1.   Structure of tetrasulphur dinitride.

SULPHUR - TETRASULPHUR TETRAIMIDE
$3S_8 \cdot S_4(NH)_4$

H. GARCIA-FERNANDEZ, M. GASPERIN and R. FREYMANN, 1981. C.R. Acad. Sci. Paris, 292, 1393-1396.

Monoclinic, P2/n, a = 8.440, b = 13.034, c = 8.203 Å, β = 112.49°, Z = 1. Mo radiation, R = 0.039 for 1557 reflexions.

The structure contains two independent molecules with 8-membered rings in crown conformations, with random distributions of S and N at the atomic sites. The arrangement is close to that in monoclinic γ-sulphur.

1,5-DICHLOROCYCLOTETRA(AZATHIENE)
$S_4N_4Cl_2$

Z. ŽÁK, 1981. Acta Cryst., B37, 23-26.

Monoclinic, $P2_1/c$, a = 9.077, b = 6.580, c = 13.311 Å, β = 108.46°, D_m = 2.25, Z = 4. Cu radiation, R = 0.094 for 1235 reflexions (films, densitometer intensities).

The structure contains discrete molecules with approximate C_S symmetry (Fig. 1). S-N = 1.51-1.68, S-Cl = 2.18 Å, angles at N = 117-121, angles at S = 99-112°.

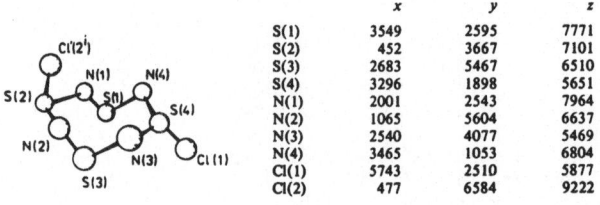

	x	y	z
S(1)	3549	2595	7771
S(2)	452	3667	7101
S(3)	2683	5467	6510
S(4)	3296	1898	5651
N(1)	2001	2543	7964
N(2)	1065	5604	6637
N(3)	2540	4077	5469
N(4)	3465	1053	6804
Cl(1)	5743	2510	5877
Cl(2)	477	6584	9222

Fig. 1.   The $S_4N_4Cl_2$ molecule, and atomic positional parameters (x 10^4).

TETRASULPHURTETRANITRIDE(2+) SALTS

$S_4N_4(Sb_3F_{14})(SbF_6)$ (I)

$S_4N_4(SO_3F)_2$ (II)

$S_4N_4(AsF_6)_2 \cdot SO_2$ (III)

$S_4N_4(AlCl_4)_2$ (IV)

$S_4N_4(SbCl_6)_2$ (V)

R.J. GILLESPIE, J.P. KENT, J.F. SAWYER, D.R. SLIM and J.D. TYRER, 1981. Inorg. Chem., 20, 3799-3812.

I.   Monoclinic, $P2_1/n$, a = 16.292, b = 16.162, c = 8.446 Å, β = 109.52°, Z = 4.
II. Monoclinic, $P2_1/c$, a = 6.121, b = 13.540, c = 7.005 Å, β = 108.89°, Z = 2.
III. Monoclinic, C2/c, a = 13.404, b = 7.958, c = 15.341 Å, β = 100.05°, Z = 4.
IV. Monoclinic, $P2_1/n$, a = 6.908, b = 16.117, c = 9.005 Å, β=122.24°, Z = 2.
V.   Orthorhombic, Pbcn, a = 12.775, b = 12.543, c = 13.550 Å, Z = 4.

Mo radiation, R = 0.040, 0.034, 0.063, 0.057, 0.051 for 4312, 1125, 1180, 1594, 1551 reflexions.

All the structures contain the $S_4N_4{}^{2+}$ cation, which consists of a planar D_{4h} ring, except in (V) where the ring is boat-shaped (Fig. 1); most of the structures exhibit some disorder. The $F_5Sb(V)-F-Sb(III)F_2-F-Sb(V)F_5^-$ anion has octahedral coordination for Sb(V) and trigonal-bipyramidal SbF_4E (E = lone-pair) coordination for Sb(III), with four longer interionic Sb(III)...F interactions. The other anions have normal geometries.

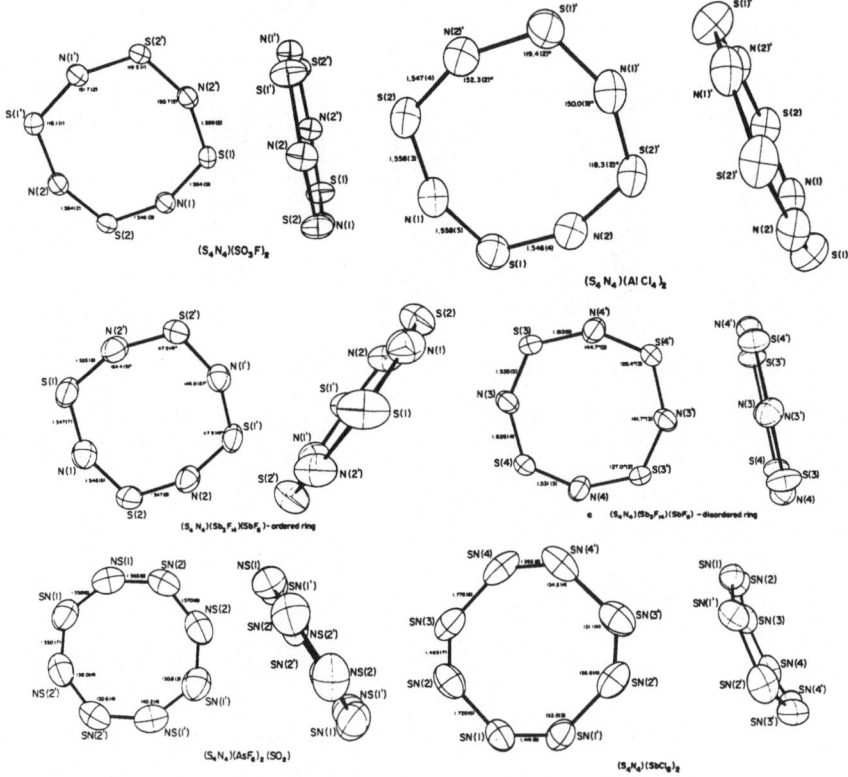

Fig. 1. The $S_4N_4^{2+}$ cation.

TETRASULPHUR TETRANITRIDE - IRON(III) CHLORIDE
β-S_4N_4·$FeCl_3$

U. THEWALT, 1981. Z. anorg. Chem., <u>476</u>, 105-108.

Monoclinic, $P2_1/c$, a = 6.803, b = 11.312, c = 13.784 Å, β = 95.02°, Z = 4. Mo radiation, R = 0.066 for 1153 reflexions.

The molecular structure (Fig. 1) is similar to that in the α-form (<u>1</u>), with an eight-membered S_4N_4 ring and a tetrahedrally-coordinated Fe atom bonded to a ring N atom. N(1)-S = 1.63, 1.68, other N-S = 1.48-1.58, Fe-N = 1.96, Fe-Cl = 2.166 Å, S-N(1)-S = 110, other S-N-S = 135-137, N-S-N = 111-120°.

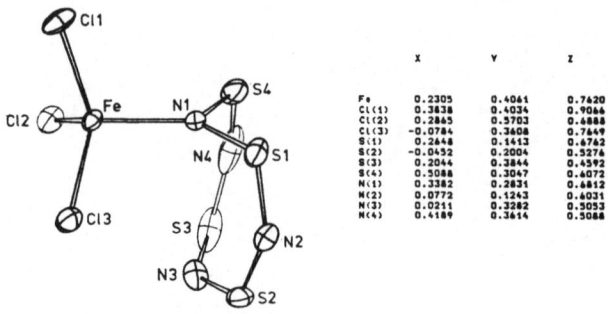

	x	y	z
Fe	0.2305	0.4061	0.7620
Cl(1)	0.3838	0.4034	0.9066
Cl(2)	0.2865	0.5703	0.6888
Cl(3)	-0.0784	0.3608	0.7649
S(1)	0.2444	0.1413	0.6762
S(2)	-0.0452	0.2004	0.5274
S(3)	0.2044	0.3844	0.4592
S(4)	0.5088	0.3047	0.6072
N(1)	0.3382	0.2831	0.6812
N(2)	0.0772	0.1243	0.6031
N(3)	0.0211	0.3282	0.5053
N(4)	0.4189	0.3614	0.5088

Fig. 1.   Structure of β-$S_4N_4.FeCl_3$.

1. Structure Reports, 46A, 148.

THIODITHIAZYL HEXAFLUOROARSENATE
$S_3N_2AsF_6$ (I)

BIS(THIODITHIAZYL) FLUOROTHIOSULPHATE
$S_6N_4(S_2O_2F)_2$ (II)

BIS(THIODITHIAZYL) FLUOROSULPHATE
$S_6N_4(SO_3F)_2$ (III)

R.J. GILLESPIE, J.P. KENT and J.F. SAWYER, 1981. Inorg. Chem., 20, 3784-3799.

I.   Triclinic, $P\bar{1}$, a = 9.471, b = 5.645, c = 7.843 Å, α = 74.78, β = 90.53, γ = 97.99°, Z = 2. Mo radiation, R = 0.064 for 1190 reflexions.

II. Monoclinic, C2/c, a = 12.108, b = 10.350, c = 12.045 Å, β = 104.82°, Z = 4. Mo radiation, R = 0.036 for 1089 reflexions.

III. Monoclinic, $P2_1/n$ [given as $P2_1/c$ in abstract], a = 6.099, b = 11.714, c = 9.765 Å, β = 102.44°, Z = 2. Mo radiation, R = 0.029 for 1239 reflexions.

The $S_3N_2^+$ cation contains a five-membered ring, linked to octahedral AsF_6^- anions by S...F interactions (Fig. 1); S-S = 2.14, S-N = 1.56-1.61(1), S...F = 2.77-3.22 Å. The material previously described as $S_3N_2AsF_6$ (1) is in fact dimeric, and contains the $S_6N_4^{2+}$ dimer.

The $S_6N_4^+$ cation in II and III is similar to that previously observed (2), and contains two S_3N_2 five-membered rings (S-S = 2.14, S-N = 1.56-1.61 Å) linked by longer S-S bonds, 3.03 and 2.99 Å. The cations are linked to tetrahedral SO_3F^- and disordered $S_2O_2F^-$ anions (Fig. 2).

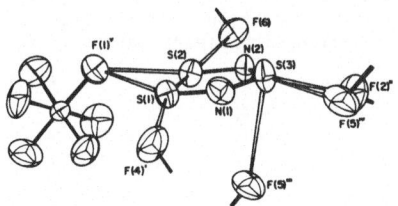

Fig. 1.   Structure of $S_3N_2^+AsF_6^-$.

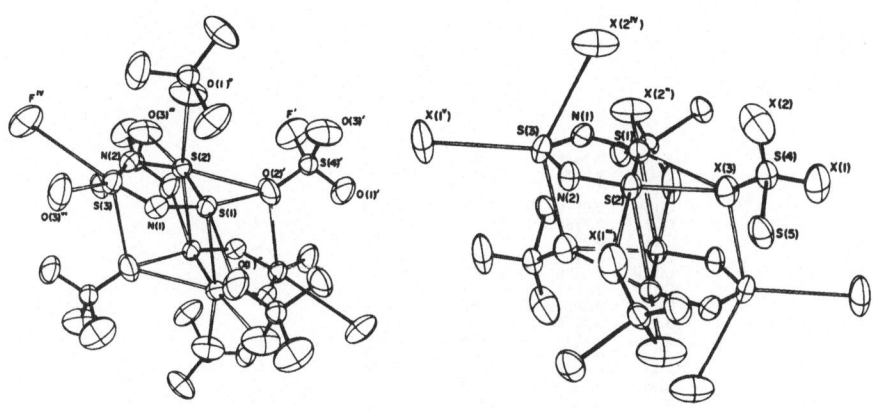

Fig. 2.   Structures of $S_6N_4^{2+}(SO_3F^-)_2$ (left) and $S_6N_4^{2+}(S_2O_2F^-)_2$ (right).

1. Structure Reports, 41A, 150.
2. Ibid., 40A, 295; 46A, 146 [see 46B].

BIS(THIODISELENAZYL) HEXAFLUOROARSENATE and HEXAFLUOROANTIMONATE
$Se_4S_2N_4(AsF_6)_2$ $Se_4S_2N_4(SbF_6)_2$

R.J. GILLESPIE, J.P. KENT and J.F. SAWYER, 1981. Inorg. Chem., 20, 4053-4060.

Monoclinic, P2$_1$/n, a = 10.358, 10.619, b = 16.002, 16.523, c = 9.748, 9.873 Å, β = 98.16, 97.70°, at -30, 22°C, Z = 4. Mo radiation, R = 0.079, 0.069, 0.073 for 2129, 2248, 3287 reflexions for the arsenate at 22, -30°C, and the antimonate at 22°C.

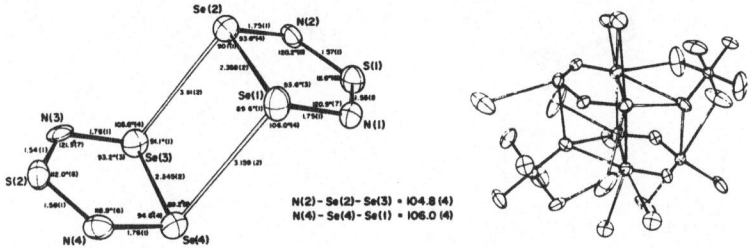

Fig. 1.   Structure of $Se_4S_2N_4(AsF_6)_2$.

Both materials contain the $Se_4S_2N_4^{2+}$ cation (Fig. 1), which is similar to the $S_6N_4^{2+}$ ion (1), containing two five-membered rings linked by longer Se...Se inter-actions, 3.12-3.18 Å; there are also significant cation-anion interactions.

1. Preceding report.

THIODITHIAZYL TETRACHLOROFERRATE(III)
$S_6N_4(FeCl_4)_2$

U. THEWALT and M. BURGER, 1981. Z. Naturforsch., 36B, 293-296.

Monoclinic, $P2_1/c$, a = 12.090, b = 11.745, c = 13.918 Å, β = 101.25°, Z = 4. Mo radiation, R = 0.044 for 3095 reflexions.

Atomic positions

	x	y	z
Fe(1)	0,4438	0,1964	0,2317
Fe(2)	0,0354	0,1453	0,2370
Cl(1)	0,4288	0,2360	0,0753
Cl(2)	0,6270	0,1843	0,2970
Cl(3)	0,3746	0,3350	0,3047
Cl(4)	0,3697	0,0312	0,2510
Cl(5)	—0,1456	0,1959	0,1915
Cl(6)	0,0328	—0,0397	0,2260
Cl(7)	0,1243	0,2241	0,1326
Cl(8)	0,0980	0,1941	0,3893
S(1)	0,2275	1,0261	—0,0513
S(2)	0,2015	0,8589	0,0811
S(3)	0,3613	0,8496	0,0328
S(4)	0,2624	0,4645	0,0738
S(5)	0,2830	0,6195	—0,0700
S(6)	0,1257	0,6349	—0,0193
N(1)	0,1538	0,9687	0,0162
N(2)	0,3420	0,9577	—0,0380
N(3)	0,3327	0,5138	—0,0007
N(4)	0,1487	0,5338	0,0604

The structure (Fig. 1) contains tetrahedral $FeCl_4^-$ anions, and $S_6N_4^{2+}$ cations with dimensions and conformation similar to those in other salts (1). There are some short interionic S...Cl contacts, 3.04, 3.16 Å.

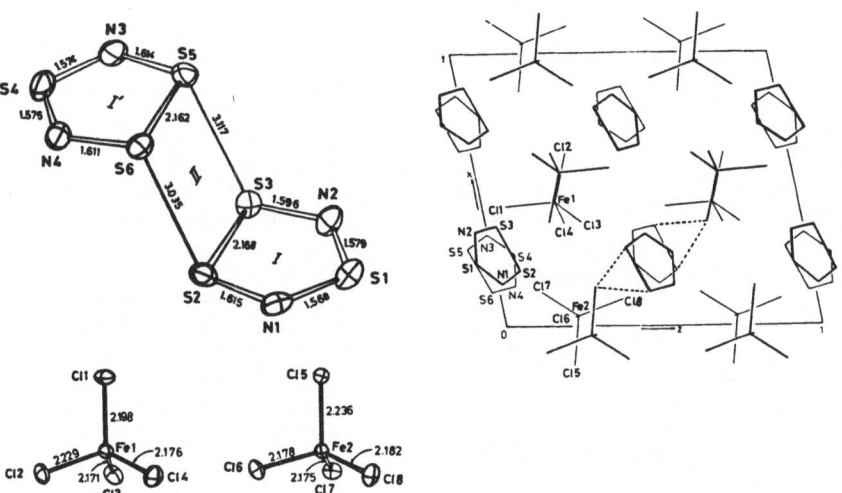

Fig. 1.   Structure of $S_6N_4(FeCl_4)_2$.

1. Structure Reports, 40A, 295; 41A, 150; 46A, 146; preceding two reports.

HYDROGEN FLUORIDE MONOHYDRATE
HF.H₂O $H_3O^+F^-$

HYDROGEN FLUORIDE HEMIHYDRATE
HF.0·5H₂O $0·5 \times H_3O^+FHF^-$

D. MOOTZ, U. OHMS and W. POLL, 1981. Z. anorg. Chem., 479, 75-83.

Monohydrate, orthorhombic, Pnma, a = 6.216, b = 4.184, c = 6.233 Å, at -62°C, Z = 4.
Mo radiation, R = 0.060 for 447 reflexions.

Hemihydrate, monoclinic, $P2_1/c$, a = 3.478, b = 6.039, c = 11.415 Å, β = 96.57°, Z =
8. Mo radiation, R = 0.066 for 837 reflexions.

    The structures (Fig. 1) contain H_3O^+ ions and F^- or FHF^- ions, linked into
puckered sheets by O-H...F hydrogen bonds.

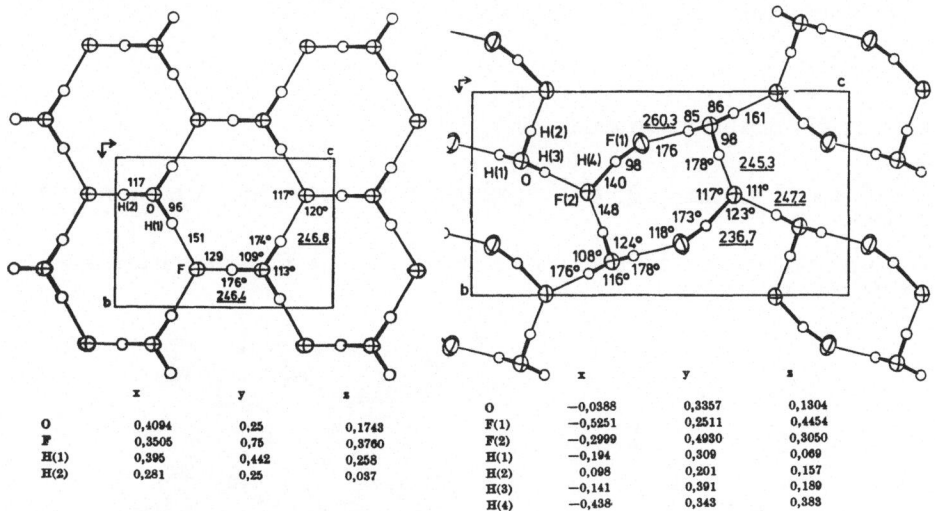

	x	y	z
O	0,4094	0,25	0,1743
F	0,3505	0,75	0,3760
H(1)	0,395	0,442	0,258
H(2)	0,281	0,25	0,037

	x	y	z
O	−0,0388	0,3357	0,1304
F(1)	−0,5251	0,2511	0,4454
F(2)	−0,2999	0,4930	0,3050
H(1)	−0,194	0,309	0,069
H(2)	0,098	0,201	0,157
H(3)	−0,141	0,391	0,189
H(4)	−0,438	0,343	0,383

Fig. 1.   Structures of HF.H₂O (left) and HF.0·5H₂O (right).

BORON TRIFLUORIDE DIHYDRATE
BF₃.2H₂O

D. MOOTZ and M. STEFFEN, 1981. Acta Cryst., B37, 1110-1112.

Monoclinic, $P2_1/c$, a = 5.562, b = 7.334, c = 8.746 Å, β = 90.30°, at 173K, Z = 4.
Mo radiation, R = 0.056 for 1194 reflexions. Preliminary study in 1.

The structure (Fig. 1) is as previously described (2), location of the H atoms
establishing the formulation as hydroxytrifluoroboric acid monohydrate, $F_3BOH_2.H_2O$.

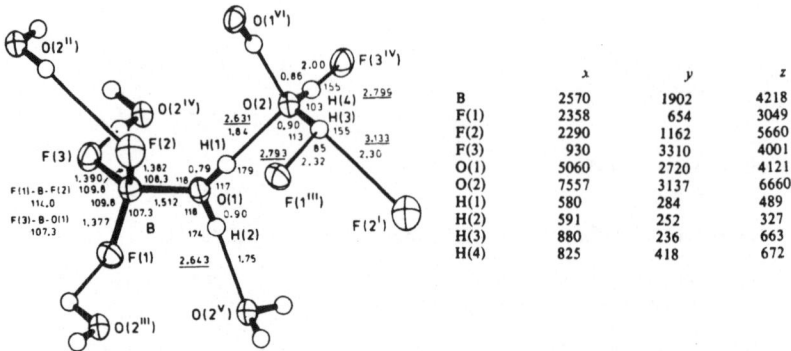

	x	y	z
B	2570	1902	4218
F(1)	2358	654	3049
F(2)	2290	1162	5660
F(3)	930	3310	4001
O(1)	5060	2720	4121
O(2)	7557	3137	6660
H(1)	580	284	489
H(2)	591	252	327
H(3)	880	236	663
H(4)	825	418	672

Fig. 1.   Structure of BF$_3$.2H$_2$O, and atomic positional parameters
(x 10^4 for non-H, x 10^3 for H).

1. D. MOOTZ and M. STEFFEN, 1979. Z. Kristallogr., 149, 139.
2. Structure Reports, 29, 252.

ZIRCONIUM(IV) FLUORIDE MONOHYDRATE
ZrF$_4$.H$_2$O

B. KOJIĆ-PRODIĆ, F. GABELA, Ž. RUŽIĆ-TOROŠ and M. ŠLJUKIĆ, 1981. Acta Cryst., B37, 1963-1965.

Tetragonal, I$\bar{4}$2d, a = 7.724, c = 11.678 Å, D$_m$ = 3.50, Z = 8. Mo radiation, R = 0.026 for 414 reflexions.

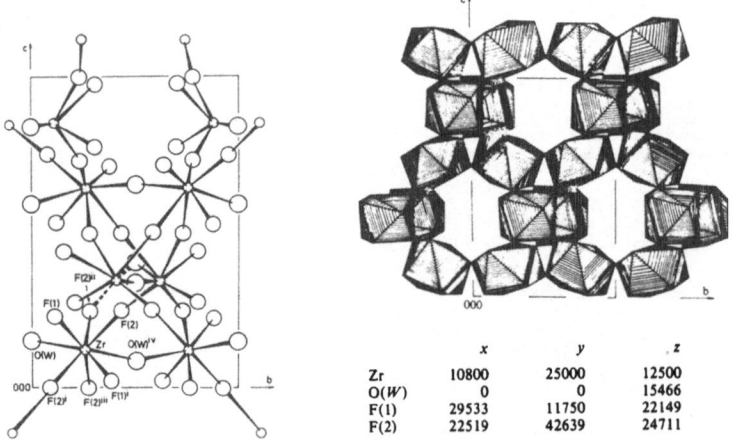

	x	y	z
Zr	10800	25000	12500
O(W)	0	0	15466
F(1)	29533	11750	22149
F(2)	22519	42639	24711

Fig. 1.   Structure of ZrF$_4$.H$_2$O, and atomic positional parameters (x 10^5).

The structure (Fig. 1) contains ZrF$_6$(H$_2$O)$_2$ dodecahedra, which share six corners with adjacent polyhedra to form an infinite three-dimensional network. Zr-F = 2.058, 2.100, 2.170(2), Zr-O = 2.132(1), O-H...F = 2.479-2.595 Å.

PALLADIUM(II) FLUORIDE (HIGH-PRESSURE)
PdF_2

A. TRESSAUD, J.L. SOUBEYROUX, H. TOUHARA, G. DEMAZEAU and F. LANGLAIS, 1981. Mater. Res. Bull., 16, 207-214.

Cubic, $P2_13$, a = 5.322 Å, Z = 4. Neutron powder data. Atoms in 4(a): x,x,x, x = 0, 0.344, 0.658 for Pd, F(1), F(2), (Pa3 is also possible, except for possible weak 110 and 310 reflexions; Pd in 4(a): 0,0,0; F in 8(c): x,x,x, x = 0.3431).

   Distorted fluorite-type structure (fluorite has x(F) = 1/4). Pd-6F = 2.18, Pd-2F = 3.17 Å.

AMMONIUM TETRAFLUOROBERYLLATE (PARAELECTRIC)
$(NH_4)_2BeF_4$

A. ONEDERA and Y. SHIOZAKI, 1981. Ferroelectrics, 31, 27-36.

Orthorhombic, Pnam, a = 7.646, b = 10.430, c = 5.918 Å, Z = 4. Cu radiation, R = 0.037 for 446 reflexions.

   β-K_2SO_4-type structure (1), as previously described (2). Be-F = 1.548 Å (corrected for libration).

1. Strukturbericht, 2, 86.
2. Structure Reports, 3, 94, 436; 44A, 141; 45A, 146, 393; 46A, 154.

OXONIUM TETRAFLUOROBORATES
H_3OBF_4 $HBF_4 \cdot H_2O$
$H_5O_2BF_4$ $HBF_4 \cdot 2H_2O$

D. MOOTZ and M. STEFFEN, 1981. Z. anorg. Chem., 482, 193-200.

H_3OBF_4, triclinic, $P\bar{1}$, a = 4.758, b = 6.047, c = 6.352 Å, α = 80.40, β = 79.48, γ = 88.25°, at -26°C, Z = 2. Mo radiation, R = 0.076 for 750 reflexions.

$H_5O_2BF_4$, monoclinic, $P2_1/c$, a = 6.584, b = 9.725, c = 7.084 Å, β = 95.15°, at -100°C, Z = 4. Mo radiation, R = 0.052 for 1382 reflexions.

   The structures contain tetrahedral BF_4^- anions and H_3O^+ or H_2O-H-OH_2^+ anions, linked by systems of O-H...F hydrogen bonds; B-F = 1.37-1.42, O-H...F = 2.58-2.87 Å. The O-H...O hydrogen bond in the $H_5O_2^+$ ion is short but asymmetric, O...O = 2.41, O-H = 1.18, 1.26 Å.

TIN FLUORIDE TETRAFLUOROBORATES
$Sn_2F_3BF_4$ $Sn_2F_5BF_4$

J. BÖNISCH and G. BERGERHOFF, 1981. Z. anorg. Chem., 473, 35-41.

$Sn_2F_3BF_4$, monoclinic, $P2_1/c$, a = 5.395, b = 8.978, c = 14.894 Å, β = 111.31°, Z = 4. Mo radiation, R = 0.069 for 501 reflexions.

$Sn_2F_5BF_4$, monoclinic, $P2_1/c$, a = 10.772, b = 7.638, c = 10.328 Å, β = 98.24°, Z = 4. Mo radiation, R = 0.036 for 677 reflexions.

The $Sn_2F_3BF_4$ structure (Fig. 1) contains $Sn_2F_3^+$ bands of SnF_3 trigonal pyramids sharing F atoms, and tetrahedral BF_4^- ions. Sn-3F = 2.09 Å (next F at 3.0 Å), F-Sn-F = 84°. In the $Sn_2F_5BF_4$ structure (Fig. 1) SnF_3 trigonal pyramids are linked into layers containing connected Sn_6F_{10} groups with 12-membered rings; BF_4^- ions lie between the layers. Sn-F = 2.06-2.21 Å (further F at 2.43, 2.8 Å).

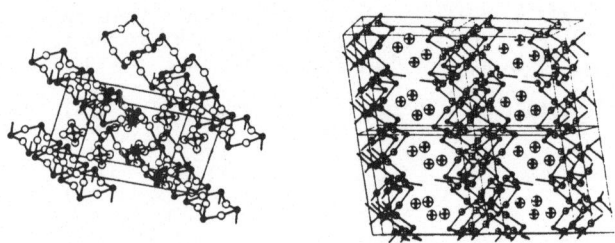

Fig. 1.   Structures of $Sn_2F_3BF_4$ (left) and $Sn_2F_5BF_4$ (right).

CHIOLITE
$Na_5Al_3F_{14}$

C. JACOBONI, A. LEBLE and J.J. ROUSSEAU, 1981.   J. Solid State Chem., 36, 297-304.

Tetragonal, P4/mnc, a = 7.014, c = 10.402 Å, Z = 2. Mo radiation, R = 0.021 for 466 reflexions.

The structure contains $[Al_3F_{14}^{5-}]_n$ octahedral sheets (Fig. 1), as previously described (1); the sheets are linked by the sodium ions.

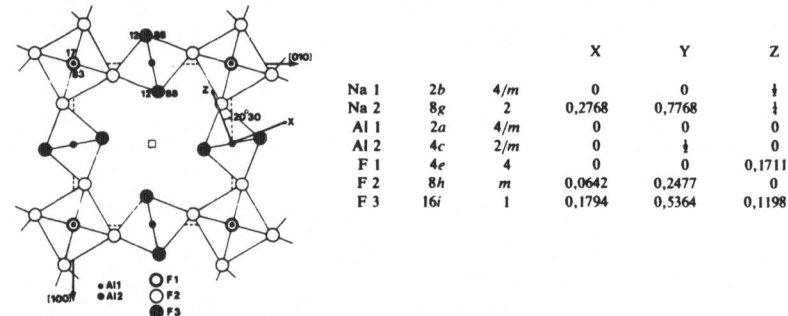

			X	Y	Z
Na 1	2b	4/m	0	0	¼
Na 2	8g	2	0,2768	0,7768	¼
Al 1	2a	4/m	0	0	0
Al 2	4c	2/m	0	½	0
F 1	4e	4	0	0	0,1711
F 2	8h	m	0,0642	0,2477	0
F 3	16i	1	0,1794	0,5364	0,1198

Fig. 1.   Structure of chiolite.

1.   Strukturbericht, 6, 31, 121.

POTASSIUM TETRAFLUOROALUMINATE
$KAlF_4$

J. NOUET, J. PANNETIER and J.L. FOURQUET, 1981. Acta Cryst., B37, 32-34.

Tetragonal, P4/mbm, a = 5.043, c = 6.164 Å, Z = 2. Neutron radiation, R = 0.030 for 152 reflexions. K in 2(c); Al in 2(a); F(eq) in 4(g): x = 0.2989; F(ax) in 4(e): z = 0.2842.

The structure (Fig. 1) is derived from that of TlAlF₄ (P4/mmm, $\underline{1}$), and contains layers of corner-sharing compressed AlF₆ octahedra parallel to (001), with K ions between the layers. Al-F = 1.817 (x 4), 1.752(1) (x 2), K-F(ax) = 2.848 (x 8) Å.

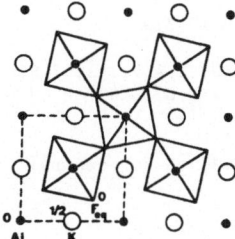

Fig. 1.   Structure of KAlF₄.

$\underline{1}$. Strukturbericht, $\underline{5}$, 17, 97; $\underline{45A}$, 148.

RUBIDIUM PENTAFLUOROALUMINATE MONOHYDRATE
Rb₂AlF₅.H₂O

J.-L. FOURQUET, F. PLET and R. DE PAPE, 1981. Rev. Chim. Minér., $\underline{18}$, 19-26.

Orthorhombic, Cmcm, a = 9.604, b = 8.379, c = 7.542 Å, D_m = 3.40, Z = 4. Mo radiation, R = 0.042 for 744 reflexions. Isostructural with Rb₂MnF₅.H₂O ($\underline{1}$); the compound described previously as Tl₂AlF₅ ($\underline{2}$) is probably a monohydrate.

The structure (Fig. 1) contains chains along $\underline{c}$ of AlF₆ octahedra sharing trans corners; Al-F = 1.77, 1.79, 1.89 Å. The chains are linked by 12-coordinate Rb ions, Rb-F = 2.80-3.30, Rb-O = 3.03, 3.37 Å, and O-H...F hydrogen bonds, O...F = 2.73 Å.

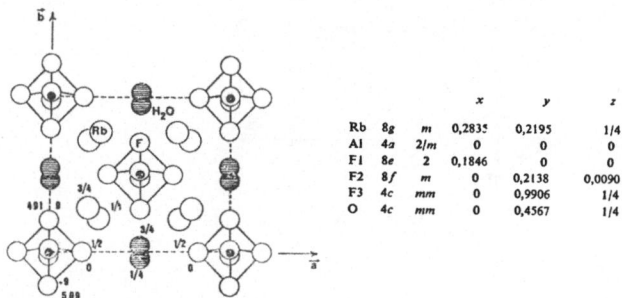

			x	y	z
Rb	8g	m	0,2835	0,2195	1/4
Al	4a	2/m	0	0	0
F1	8e	2	0,1846	0	0
F2	8f	m	0	0,2138	0,0090
F3	4c	mm	0	0,9906	1/4
O	4c	mm	0	0,4567	1/4

Fig. 1.   Structure of Rb₂AlF₅.H₂O.

$\underline{1}$. Structure Reports, $\underline{44A}$, 147.
$\underline{2}$. Strukturbericht, $\underline{5}$, 104.

CAESIUM SILVER(II) HEXAFLUOROALUMINATE
CsAgAlF$_6$

CAESIUM SILVER(II) HEXAFLUOROFERRATE(III)
CsAgFeF$_6$

CAESIUM SILVER(II) HEXAFLUOROALUMINATE-HEXAFLUOROFERRATE
CsAg(Al,Fe)F$_6$

B.G. MÜLLER, 1981. J. Fluor. Chem., <u>17</u>, 317-329.

Orthorhombic, Pnma, a = 7.380, 7.338, 7.19, b = 7.241, 7.564, 7.39, c = 10.352,
10.554, 10.32 Å, Z = 4. Mo radiation, R = 0.073, 0.069, 0.070 for 808, 966, 799
reflexions.

Atomic positions [for CsAgAlF$_6$, similar values for others]

			x	y	z
Cs	in	4(c)	0.4787	1/4	.0.1327
Ag		4(c)	0.2835	1/4	0.7549
Al		4(a)	0	0	0
F(1)		4(c)	0.0356	1/4	0.6590
F(2)		4(c)	0.4621	1/4	0.4928
F(3)		8(d)	0.1307	0.0104	0.1439
F(4)		8(d)	0.1981	0.0332	0.9044

The structure (Fig. 1) is related to that of cubic RbNiCrF$_6$ (<u>1</u>), but with
ordered distribution of the M(II) and M(III) cations. Al and Fe have octahedral
coordinations, Al-F = 1.78-1.83, Fe-F = 1.90-1.95, Al/Fe-F = 1.87-1.90; Ag has
distorted octahedral coordination with two short (2.05-2.08 Å) and four longer
(2.28-2.32 Å) bonds; Cs ions have 9-coordination. Several other Rb/Cs, Ga/Fe
compounds are isostructural.

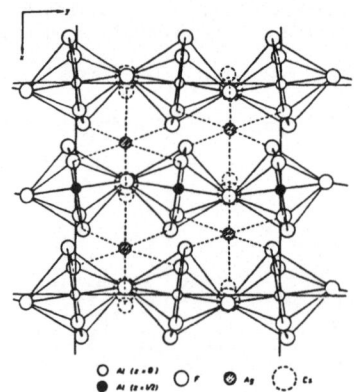

Fig. 1.   Structure of CsAgAlF$_6$.

<u>1</u>. Structure Reports, <u>38A</u>, 205.

MERCURY(I) PENTAFLUOROALUMINATE DIHYDRATE
Hg$_2$AlF$_5$.2H$_2$O

J.L. FOURQUET, F. PLET and R. DE PAPE, 1981. Acta Cryst., B<u>37</u>, 2136-2138.

Tetragonal, I4cm, a = 9.353, c = 7.241 Å, Z = 4.   Mo radiation, R = 0.037 for 383
reflexions.

The structure (Fig. 1) contains AlF_6 octahedra, which share trans corners to
form chains along c; mean Al-F = 1.80 Å. Between the chains are linear $[H_2O-Hg-Hg-OH_2]^{2+}$ cations, Hg-Hg = 2.511(1), Hg-O = 2.14(1) Å; four F complete octahedral
geometry at Hg, Hg-F = 2.83, 2.89 Å. The shortest O...F distances, 2.56-2.86 Å, are
probably hydrogen bonds.

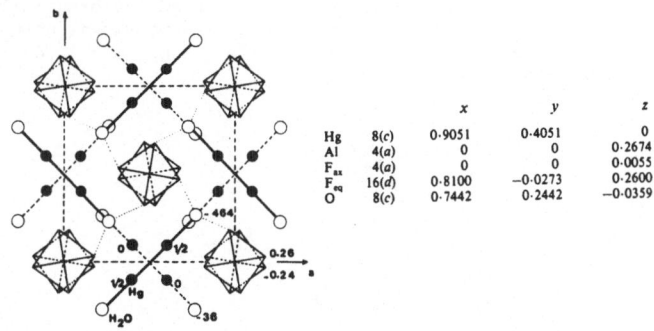

		x	y	z
Hg	8(c)	0.9051	0.4051	0
Al	4(a)	0	0	0.2674
F_{ax}	4(a)	0	0	0.0055
F_{eq}	16(d)	0.8100	−0.0273	0.2600
O	8(c)	0.7442	0.2442	−0.0359

Fig. 1.   Structure of $Hg_2AlF_5.H_2O$.

IRON(II) HEXAFLUOROSILICATE HEXAHYDRATE
$FeSiF_6.6H_2O$

G. CHEVRIER, A. HARDY and G. JÉHANNO, 1981. Acta Cryst., A37, 578-584.

The structure contains two anion orientations, A and B. Below 240K these are ordered
in space group $P2_1/c$. Above 240K, there is a periodic antiphase (Fig. 1), and the
structure can be described in P3̄, a = 9.616, c = 9.676 Å, Z = 3 (compare 1); Mo
radiation, R = 0.080 for 78 reflexions. The Mg compound has been described previously
in similar terms (2).

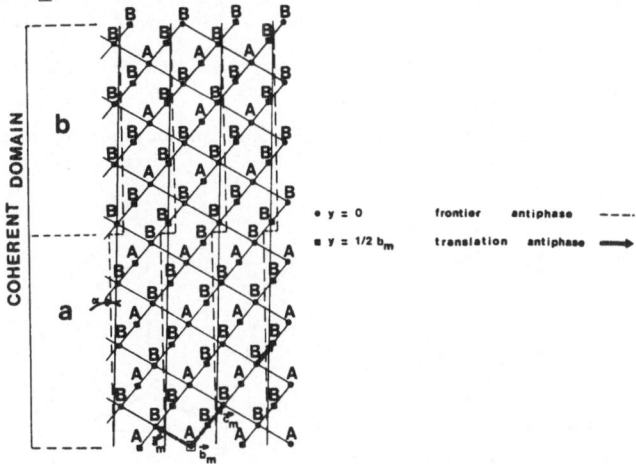

Fig. 1.   Model of the $FeSiF_6.6H_2O$ structure above 240K.

1. Structure Reports, 27, 461.
2. G. CHEVRIER and G. JÉHANNO, 1979. Acta Cryst., A35, 912.

HYDRAZINIUM(2+) HEXAFLUOROGERMANATE(IV) MONOHYDRATE
$N_2H_6GeF_6 \cdot H_2O$

B. FRLEC, D. GANTAR, L. GOLIČ and I. LEBAN, 1981. Acta Cryst., B37, 666-668.

Orthorhombic, Pnma, a = 8.869, b = 9.292, c = 7.400 Å, Z = 4. Mo radiation, R = 0.018 for 794 reflexions.

The structure (Fig. 1) contains $N_2H_6^{2+}$ ions (staggered conformation), GeF_6^{2-} octahedra, and H_2O molecules, interconnected by a three-dimensional network of hydrogen bonds. N-N = 1.428(2), Ge-F = 1.762-1.842(2), N-H...O = 2.76, N-H...F = 2.70-2.88, O-H...F = 2.73, 2.88 Å.

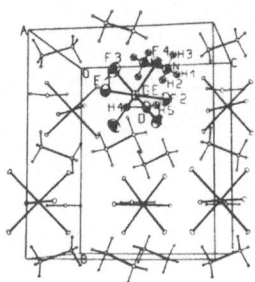

Fig. 1. Structure of $N_2H_6GeF_6 \cdot H_2O$.

TRIARSENICTETRASULPHUR HEXAFLUOROARSENATE(V)
TRIARSENICTETRASULPHUR HEXAFLUOROANTIMONATE(V)
TRIARSENICTETRASELENIUM HEXAFLUOROANTIMONATE(V)
$(As_3S_4)(AsF_6)$ $(As_3S_4)(SbF_6)$ $(As_3Se_4)(SbF_6)$

B.H. CHRISTIAN, R.J. GILLESPIE and J.F. SAWYER, 1981. Inorg. Chem., 20, 3410-3420.

Sulphur compounds, orthorhombic, Pcam, a = 19.962, 20.453, b = 5.930, 5.990, c = 9.441, 9.609 Å, Z = 4. Mo radiation, R = 0.043, 0.036 for 546, 699 reflexions.

Selenium compound, monoclinic, $P2_1/m$, a = 6.224, b = 9.564, c = 10.643 Å, β = 92.65°, Z = 2. Mo radiation, R = 0.064 for 620 reflexions.

The structures contain $As_3X_4^+$ cations (Fig. 1) and octahedral anions, with some short interionic contacts.

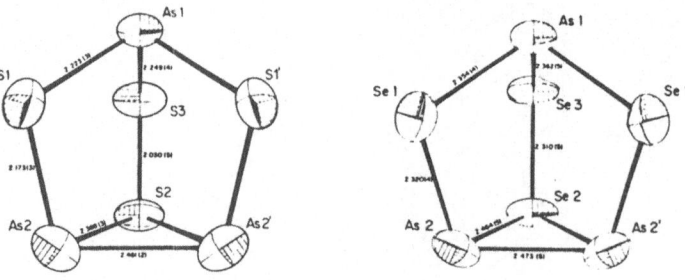

Fig. 1.   The $As_3X_4^+$ cations (X = S, Se).

IODOCYCLOHEPTASULPHUR HEXAFLUOROANTIMONATE(V) and HEXAFLUOROARSENATE(V)
S_7ISbF_6 S_7IAsF_6

J. PASSMORE, G. SUTHERLAND, P. TAYLOR, T.K. WHIDDEN and P.S. WHITE, 1981. Inorg.
Chem., 20, 3839-3845.

S_7ISbF_6, orthorhombic, $P2_12_12_1$, a = 11.786, b = 9.187, c = 12.400 Å, Z = 4. Mo radiation, R = 0.046 for 1628 reflexions.

S_7IAsF_6, triclinic, $P\bar{1}$, a = 15.516, b = 11.813, c = 11.650 Å, α = 107.30, β = 74.71, γ = 104.62°, Z = 6. Mo radiation, R = 0.14 for 2833 reflexions.

The structures (Fig. 1) contain octahedral anions and S_7I^+ cations, with one inter-cationic I...S interaction.

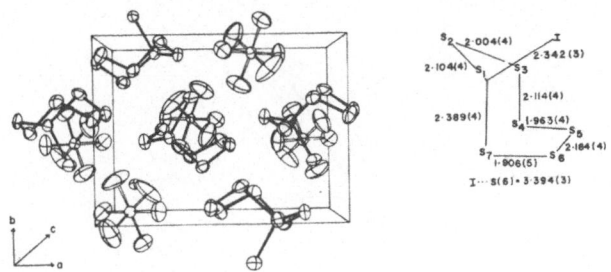

Fig. 1.   Structure of S_7ISbF_6.

TRIIODINE(1+) HEXAFLUOROARSENATE(V)
I_3AsF_6

J. PASSMORE, G. SUTHERLAND and P.S. WHITE, 1981. Inorg. Chem., 20, 2169-2171.

Triclinic, $P\bar{1}$, a = 8.054, b = 5.942, c = 10.503 Å, α = 103.08, β = 88.95, γ = 100.35°, Z = 2. Mo radiation, R = 0.091 for 1268 reflexions.

The structure (Fig. 1) contains bent I_3^+ cations and octahedral AsF_6^- anions, with some cation-anion interactions.

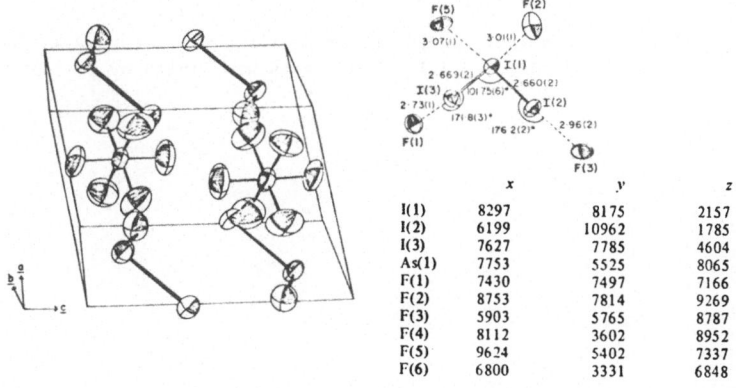

	x	y	z
I(1)	8297	8175	2157
I(2)	6199	10962	1785
I(3)	7627	7785	4604
As(1)	7753	5525	8065
F(1)	7430	7497	7166
F(2)	8753	7814	9269
F(3)	5903	5765	8787
F(4)	8112	3602	8952
F(5)	9624	5402	7337
F(6)	6800	3331	6848

Fig. 1.   Structure of I_3AsF_6, and atomic positional parameters
            $(\times 10^4)$; As-F = 1.68-1.72(1) Å.

DIOXYGENYL FLUOROTITANATE
$[O_2^+]_2Ti_7F_{30}^{2-}$

B.G. MULLER, 1981. J. Fluor. Chem., **17**, 489-499.

Trigonal, $P\bar{3}$, a = 10.192, c = 6.500 Å, at -120°C, Z = 1. Mo radiation, R = 0.086
for 748 reflexions.

The structure (Fig. 1) contains columns of corner-sharing TiF_6 octahedra, loosely
connected by disordered O_2^+ cations; Ti(1)-F = 1.87, Ti(2)-F = 1.74-2.10, O-O = 0.96 Å.

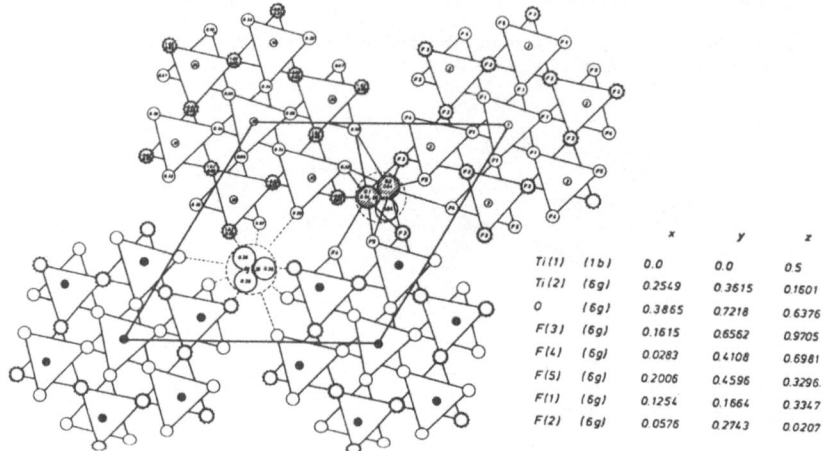

		x	y	z
Ti(1)	(1b)	0.0	0.0	0.5
Ti(2)	(6g)	0.2549	0.3615	0.1501
O	(6g)	0.3865	0.7218	0.6376
F(3)	(6g)	0.1615	0.6562	0.9705
F(4)	(6g)	0.0283	0.4108	0.6981
F(5)	(6g)	0.2006	0.4596	0.3296
F(1)	(6g)	0.1254	0.1664	0.3347
F(2)	(6g)	0.0578	0.2743	0.0207

Fig. 1.   Structure of $[O_2]_2Ti_7F_{30}$ (O has occupancy 0.62).

THALLIUM(I) PENTAFLUOROZIRCONATE(IV)
$TlZrF_5$

D. AVIGNANT, I. MANSOURI, R. CHEVALIER and J.C. COUSSEINS, 1981. J. Solid State Chem.,
38, 121-127.

Monoclinic, $P2_1/c$, a = 8.112, b = 7.927, c = 7.929 Å, β = 123.99°, D_m = 6.12, Z = 4.
Mo radiation, R = 0.057 for 1140 reflexions.

The structure (Fig. 1) contains sheets of edge- and corner-sharing ZrF_8 bicapped
trigonal prisms, Zr-F = 2.00-2.21(2) Å. The sheets are linked by 12-coordinate Tl
ions, Tl-F = 2.71-3.48(2) Å.

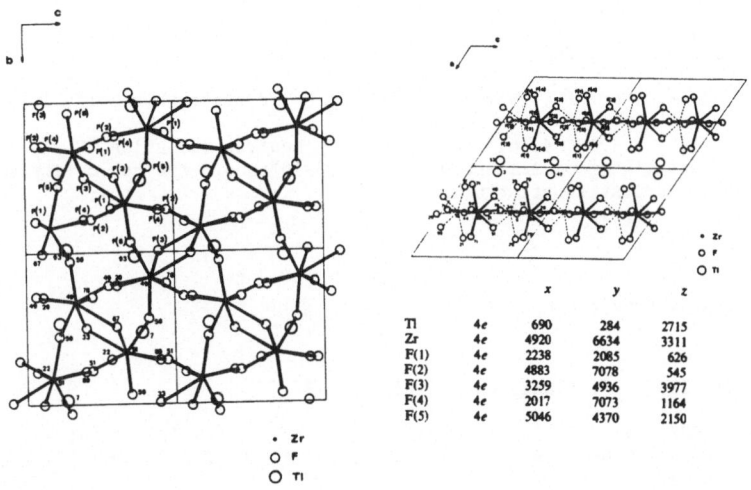

		x	y	z
Tl	4e	690	284	2715
Zr	4e	4920	6634	3311
F(1)	4e	2238	2085	626
F(2)	4e	4883	7078	545
F(3)	4e	3259	4936	3977
F(4)	4e	2017	7073	1164
F(5)	4e	5046	4370	2150

Fig. 1.   Structure of TlZrF$_5$ and atomic positional parameters (x 10^4).

CADMIUM OCTAFLUOROZIRCONATE HEXAHYDRATE
Cd$_2$ZrF$_8$.6H$_2$O

M.F. EIBERMAN, T.A. KAIDALOVA, R.L. DAVIDOVIČ, T.F. LEVČIŠINA and B.V. BUKVECKIJ, 1980.   Koordin. Khim., 6, 1885-1890.

Monoclinic, B2/b, a = 11.464, b = 13.732, c = 8.223 Å, γ = 119.82°, D$_m$ = 3.30, Z = 4.   R = 0.034.

The structure contains ZrF$_8$ Thompson cubes, linked by CdF$_4$O$_3$ pentagonal bipyramids with 4 F and 1 H$_2$O equatorial.

YTTERBIUM ZIRCONIUM HEPTAFLUORIDE
YbZrF$_7$

M. POULAIN and B.C. TOFIELD, 1981.   J. Solid State Chem., 39, 314-328.

Cubic, Pm3m, a = 4.067 Å, Z = 1/2.   Mo radiation, R = 0.03 for 83 reflexions, and neutron powder data.   1 Yb/Zr in 6(e): 0,0,0.0437; 2.5 F(1) in 12(j): 1/2,0.0496, 0.0496; 1 F(2) in 24(ℓ): 1/2,0.310,0.062.

The disordered structure is similar to that of (Yb,Zr)(F,O)$_{3.5}$ (1), with seven-coordinate metal atoms.

1.   Structure Reports, 45A, 197.

POTASSIUM MANGANESE(II) FLUORIDE
RUBIDIUM CALCIUM FLUORIDE
CAESIUM LEAD(II) CHLORIDE
KMnF$_3$ RbCaF$_3$ CsPbCl$_3$

J. HUTTON and R.J. NELMES, 1981. J. Phys. C: Solid State Phys., 14, 1713-1736.

Cubic, [Pm3m], a = 4.19, 4.45, 5.61 Å, Z = 1. Neutron radiation, R = 0.02-0.03 for
92-273 reflexions at room temperature and at 5-10° above the cubic-tetragonal
transitions. Perovskite structures.

DIOXYGENYL NONAFLUORODIMANGANATE(IV)

O$_2$$^+Mn_2F_9$$^-$

B.G. MÜLLER, 1981. J. Fluor. Chem., 17, 409-421.

Monoclinic, C2/c, a = 17.552, b = 8.373, c = 9.101 Å, β = 102.3°, at -150°C, Z = 8.
Mo radiation, R = 0.053 for 1619 reflexions at -150°C (0.268 for 1853 reflexions at
room temperature).

The structure contains double chains of corner-sharing MnF$_6$ octahedra (Fig. 1)
stacked in layers parallel to (100), with O$_2$$^+$ cations between the layers; O-O =
1.10 Å.

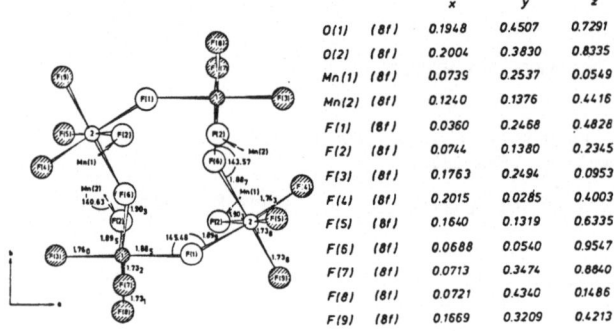

		x	y	z
O(1)	(8f)	0.1948	0.4507	0.7291
O(2)	(8f)	0.2004	0.3830	0.8335
Mn(1)	(8f)	0.0739	0.2537	0.0549
Mn(2)	(8f)	0.1240	0.1376	0.4416
F(1)	(8f)	0.0360	0.2468	0.4828
F(2)	(8f)	0.0744	0.1380	0.2345
F(3)	(8f)	0.1763	0.2494	0.0953
F(4)	(8f)	0.2015	0.0285	0.4003
F(5)	(8f)	0.1640	0.1319	0.6335
F(6)	(8f)	0.0688	0.0540	0.9547
F(7)	(8f)	0.0713	0.3474	0.8840
F(8)	(8f)	0.0721	0.4340	0.1486
F(9)	(8f)	0.1669	0.3209	0.4213

Fig. 1.   Structure of dioxygenyl nonafluorodimanganate(IV).

BARIUM MANGANESE(II) IRON(III) FLUORIDE
BaMnFeF$_7$

H. HOLLER, D. BABEL, M. SAMOUËL and A. de KOZAK, 1981. J. Solid State Chem., 39,
345-350.

Monoclinic, P2$_1$/c, a = 5.532, b = 10.980, c = 9.183 Å, β = 94.67°, D$_m$ = 4.65, Z = 4.
Mo radiation, R = 0.033 for 1771 reflexions.

The structure (Fig. 1) contains pairs of edge-sharing MnF$_6$ octahedra linked by
corner-sharing via FeF$_6$ octahedra; mean Mn-F = 2.12, Fe-F = 1.93 Å. Ba has 12-
coordination, mean Ba-F = 2.91 Å.

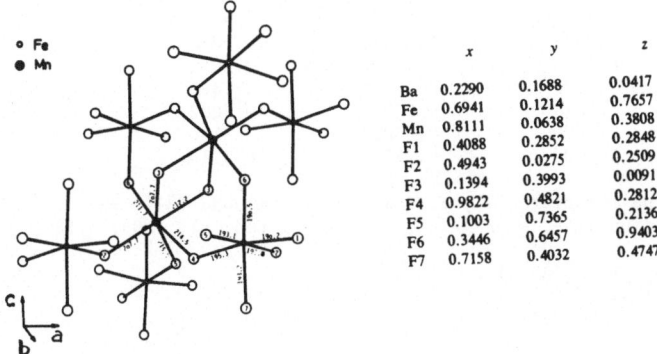

	x	y	z
Ba	0.2290	0.1688	0.0417
Fe	0.6941	0.1214	0.7657
Mn	0.8111	0.0638	0.3808
F1	0.4088	0.2852	0.2848
F2	0.4943	0.0275	0.2509
F3	0.1394	0.3993	0.0091
F4	0.9822	0.4821	0.2812
F5	0.1003	0.7365	0.2136
F6	0.3446	0.6457	0.9403
F7	0.7158	0.4032	0.4747

Fig. 1.   Structure of $BaMnFeF_7$.

AMMONIUM IRON FLUORIDE
$NH_4Fe_2F_6$ $NH_4Fe(II)Fe(III)F_6$

G. FEREY, M. LeBLANC and R. de PAPE, 1981. J. Solid State Chem., <u>40</u>, 1-7.

Orthorhombic, Pnma, a = 7.045, b = 7.454, c = 10.116 Å, D_m = 3.1, Z = 4. Mo radiation, R = 0.024 for 798 reflexions.

The structure (Fig. 1) is an ordered pyrochlore, with chains of corner-sharing
$Fe(II)F_6$ octahedra along <u>a</u> and $Fe(III)F_6$ octahedra along <u>b</u>; further corner sharing
builds up the three-dimensional structure. Fe(II)-F = 1.97-2.13, Fe(III)-F = 1.91-
1.95 Å. N has 13 F neighbours at 2.89-3.77 Å.

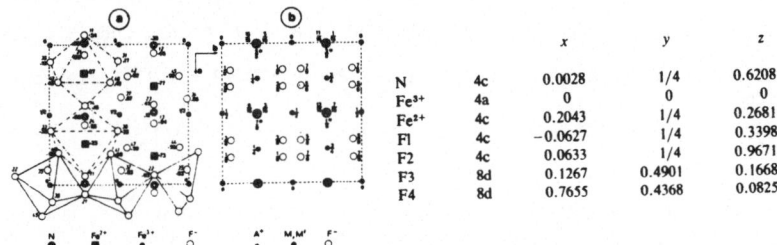

		x	y	z
N	4c	0.0028	1/4	0.6208
Fe^{3+}	4a	0	0	0
Fe^{2+}	4c	0.2043	1/4	0.2681
F1	4c	−0.0627	1/4	0.3398
F2	4c	0.0633	1/4	0.9671
F3	8d	0.1267	0.4901	0.1668
F4	8d	0.7655	0.4368	0.0825

Fig. 1.   The $NH_4Fe_2F_6$ structure (a) and positional parameters,
            and the ideal pyrochlore structure (b).

SODIUM TETRAFLUOROFERRATE(III)
$NaFeF_4$

J.-M. DANCE, A. TRESSAUD, W. MASSA and D. BABEL, 1981.   J. Chem. Res., (S), 202-203;
(M), 2282-2296.

Monoclinic, $P2_1/c$, a = 7.908, b = 5.351, c = 7.531 Å, β = 101.92°, D_m = 3.28, Z = 4.
Mo radiation, R = 0.035 for 642 reflexions.

Atomic positions

	x	y	z
Na	0.3711	0.7771	0.1065
Fe	0.1725	0.2557	0.2985
F(1)	0.1679	0.1518	0.0481
F(2)	0.0319	0.5492	0.2092
F(3)	0.3670	0.4648	0.3140
F(4)	0.3190	0.9855	0.3710

$NaNbO_2F_2$-type structure (1), as for the Cr compound (2). Fe-F = 1.86-1.96 (octahedral coordination), Na-$\overline{F}$ = 2.23-2.60 (6 distances), 2.88, 3.18 Å.

1. Structure Reports, 34A, 217.
2. Ibid., 45A, 156.

POTASSIUM TRIFLUOROCOBALTATE(II)
$KCoF_3$

N. KIJIMA, K. TANAKA and F. MARUMO, 1981. Acta Cryst., B37, 545-548.

Cubic, Pm3m, a = 4.0688 Å, at 293K, Z = 1. Mo radiation, R = 0.009 for 156 reflexions. Co in 1(a); K in 1(b); F in 3(d).

Ideal perovskite (1). Electron density distribution is studied.

1. Strukturbericht, 1, 300.

TETRAFLUOROAMMONIUM HEXAFLUORONICKELATE(IV)
$(NF_4)_2NiF_6$

P. CHARPIN, M. LANCE, T. BUI HUY and R. BOUGON, 1981. J. Fluor. Chem., 17, 479-484.

Tetragonal, I4/m, a = 6.828, c = 9.270 Å, Z = 2. R = 0.12 for 47 reflexions (films, visual data).

Atomic positions

			x	y	z
Ni	in	2(a)	0	0	0
N		4(d)	0	1/2	1/4
F(1)		8(h)	0.131	0.223	0
F(2)		4(e)	0	0	0.185
F(3)		16(i)	0.144	0.567	0.166

The structure is a distortion of the K_2PtCl_6 type, containing octahedral NiF_6^{2-} and tetrahedral NF_4^+ ions. Ni-F(1,2) = 1.76, 1.72(6), N-F(3) = 1.3-1.4 Å.

POTASSIUM TETRAFLUOROCUPRATE(II)
K_2CuF_4

POTASSIUM HEPTAFLUORODICUPRATE(II)
$K_3Cu_2F_7$

E. HERDTWECK and D. BABEL, 1981. Z. anorg. Chem., $\underline{474}$, 113-122.

K_2CuF_4, tetragonal, I4/mmm, a = 4.147, c = 12.73 Å, Z = 2. Mo radiation, R = 0.029 for 234 reflexions. [Another form has been described ($\underline{1}$).]

$K_3Cu_2F_7$, tetragonal, I4/mmm, a = 4.156, c = 20.52 Å, Z = 2. Mo radiation, R = 0.025 for 209 reflexions.

Atomic positions

K_2CuF_4

			x	y	z
K	in	4(e)	0	0	0.3569
Cu		2(a)	0	0	0
F(1)		4(e)	0	0	0.1523
0.5 F(2)		8(i)	0.4603	0	0

$K_3Cu_2F_7$

			x	y	z
K(1)	in	4(e)	0	0	0.3171
K(2)		2(b)	0	0	1/2
Cu		4(e)	0	0	0.0959
F(1)		4(e)	0	0	0.1898
0.5 F(2)		16(n)	0.4571	0	0.0942
F(3)		2(a)	0	0	0

Isostructural with K_2NiF_4 ($\underline{2}$) and $Sr_3Ti_2O_7$ ($\underline{3}$), respectively, but with disorder of bridging F atoms resulting from Jahn-Teller distortion of CuF_6 octahedra, Cu-F = 1.91, 1.94, 2.24 Å and 1.90, 1.95, 2.26 Å.

$\underline{1}$. Structure Reports, $\underline{40A}$, 290; following report.
$\underline{2}$. Ibid., $\underline{17}$, 332; $\underline{19}$, $\overline{323}$.
$\underline{3}$. Ibid., $\underline{22}$, 308; $\underline{24}$, 440.

POTASSIUM TETRAFLUOROCUPRATE(II)
K_2CuF_4

D. REINEN and S. KRAUSE, 1981. Inorg. Chem., $\underline{20}$, 2750-2759.

Tetragonal, I$\overline{4}$c2, a = 5.854, c = 25.42 Å, Z = 8. Neutron powder data.

Atomic positions

			x	y	z
Cu(1)	in	4(a)	0	0	1/4
Cu(2)		4(d)	0	1/2	0
K(1)		8(f)	0	0	0.0730
K(2)		8(g)	0	1/2	0.1818
F(1)		8(f)	0	0	0.1733
F(2)		8(g)	0	1/2	0.0767
F(3)		8(e)	0.2252	0.2252	1/4
F(4)		8(h)	0.2248	0.7248	0

Distorted K_2NiF_4-type structure, as previously described ($\underline{1}$), a = $\sqrt{2}$a(K_2NiF_4), c = 2c(K_2NiF_4). Cu has distorted octahedral coordination, Cu-F = 1.86, 1.95, 2.28 Å.

$\underline{1}$. Structure Reports, $\underline{40A}$, 290.

CALCIUM CHLORIDE AMMINES
$CaCl_2(NH_3)_8$ $CaCl_2(NH_3)_2$

CALCIUM CHLORIDE HYDROXIDE
CaClOH

S. WESTMAN, P.-E. WERNER, T. SCHULER and W. RALDOW, 1981.   Acta Chem. Scand., A35,
467-472.

$CaCl_2(NH_3)_8$, orthorhombic, Pnma, a = 12.114, b = 7.308, c = 15.083 Å, Z = 4.   Cu
radiation, powder data.

$CaCl_2(NH_3)_2$, orthorhombic, Abm2, a = 6.004, b = 7.825, c = 12.349 Å, Z = 4.   Cu
radiation, powder data.

CaClOH, hexagonal, $P6_3mc$, a = 3.864, c = 9.904 Å, Z = 2.   Cu radiation, powder data.

Atomic positions

$CaCl_2(NH_3)_8$

		x	y	z
Ca	4(c)	0.746	1/4	0.133
Cl1	4(c)	0.144	1/4	0.034
Cl2	4(c)	0.047	1/4	0.681
N1	8(d)	0.147	0.493	0.457
N2	8(d)	0.434	0.049	0.362
N3	8(d)	0.204	0.015	0.250
N4	8(d)	0.191	0.016	0.638

$CaCl_2(NH_3)_2$

		x	y	z
Ca	4(c)	0.275	1/4	0.095
Cl1	4(a)	0	0	0
Cl2	4(b)	1/2	0	0.205
N1	4(c)	0.494	1/4	0.440
N2	4(c)	−0.073	1/4	0.239

CaClOH

		x	y	z
Ca	2(b)	1/3	2/3	0.680
Cl	2(b)	1/3	2/3	0
O	2(a)	0	0	0.237

In the ammines Ca ions have octahedral coordination to 6 NH_3 and to 4 Cl and
2 NH_3, in the octaammine and diammine, respectively (Fig. 1); the additional two NH_3
in the octaammine connect the Cl^- ions and the Ca octahedra, probably by hydrogen
bonding. CaClOH is isostructural with CdClOH (1).

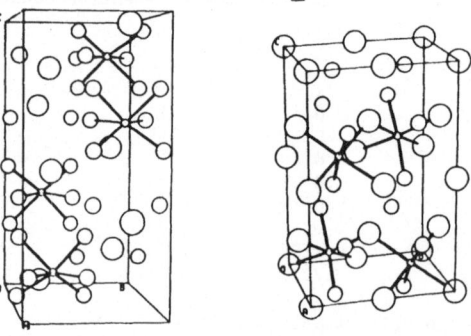

Fig. 1.   Structures of $CaCl_2(NH_3)_8$ (left) and $CaCl_2(NH_3)_2$ (right).

1. Strukturbericht, 3, 65, 373.

INDIUM SESQUICHLORIDE   (INDIUM(I) HEXACHLOROINDATE(III))
In_2Cl_3 $0.5 \times In_3[InCl_6]$

G. MEYER, 1981. Z. anorg. Chem., 478, 39-51.

Orthorhombic, Pnma, a = 12.614, b = 25.238, c = 14.562 Å, Z = 32. Mo radiation, R = 0.086 for 1289 reflexions.

    Isostructural with α-Tl_2Cl_3 (1). The structure contains $In(III)Cl_6^{3-}$ octahedra linked by In^+ ions with 7- to 11-coordinations; In(III)-Cl = 2.45-2.57, In(I)-Cl = 3.08-4.08 Å.

1. Structure Reports, 46A, 169.

LEAD(II) CHLORIDE IODIDE
PbClI

LEAD(II) BROMIDE IODIDE
$PbBr_{1.2}I_{0.8}$

L.H. BRIXNER, H.-Y. CHEN and C.M. FORIS, 1981.   J. Solid State Chem., 40, 336-343.

Orthorhombic, Pbnm, a = 9.669, 10.452, b = 8.200, 8.639, c = 4.605, 4.427 Å, Z = 4. Mo radiation, R = 0.063, 0.035 for 486, 501 reflexions.

Atomic positions

	PbClI			$PbBr_{1.2}I_{0.8}$		
	x	y	z	x	y	z
Pb	0.1240	0.2082	1/4	0.1580	0.1622	1/4
Cl or Br	0.0596	0.859	1/4	0.0534	0.8414	1/4
I or X	0.8305	0.4703	1/4	0.8286	0.5014	1/4

X = 0.8 I + 0.2 Br

    $PbCl_2$-type structure (1, 2), with ordered halogen positions. The Pb positional parameters differ to maintain normal Pb-halogen distances.

1. Strukturbericht, 2, 16.
2. Structure Reports, 34A, 176.

SELENIUM TETRACHLORIDE
α-$SeCl_4$

I.   R. KNIEP, L. KORTE and D. MOOTZ, 1981.   Z. Naturforsch., 36B, 1660-1662.

Cubic, P$\overline{4}$3m, a = 16.433 Å, D_m = 2.68, Z = 32. Mo radiation, R = 0.071 for 642 reflexions.

    The structure (Fig. 1) contains tetrameric cubane-like Se_4Cl_{16} molecules, packed in a β-tungsten type arrangement.

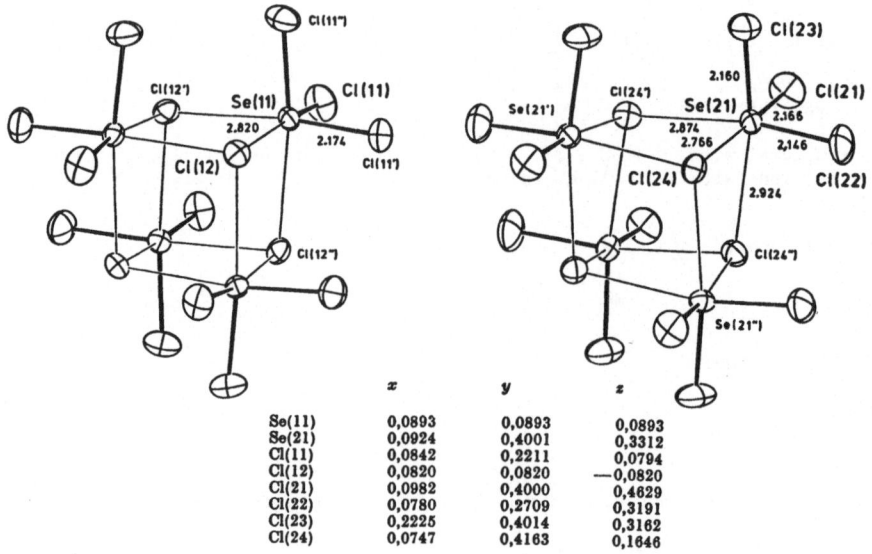

	x	y	z
Se(11)	0,0893	0,0893	0,0893
Se(21)	0,0924	0,4001	0,3312
Cl(11)	0,0842	0,2211	0,0794
Cl(12)	0,0820	0,0820	—0,0820
Cl(21)	0,0982	0,4000	0,4629
Cl(22)	0,0780	0,2709	0,3191
Cl(23)	0,2225	0,4014	0,3162
Cl(24)	0,0747	0,4163	0,1646

Fig 1.   Structure of α-SeCl₄.

β-SeCl$_4$

II.   P. BORN, R. KNIEP, D. MOOTZ, M. HEIN and B. KREBS, 1981. Z. Naturforsch, $\underline{36B}$, 1516-1519.

Monoclinic, C2/c, a = 16.548, b = 9.810, c = 15.029 Å, β = 116.95°, D_m = 2.70, Z = 16. Mo radiation, R = 0.064 for 2396 reflexions. Metastable form.

Isostructural with TeCl₄ ($\underline{1}$), the structure containing tetrameric cubane-like Se₄Cl₁₆ molecules (Fig. 2); Se-$\overline{\text{C}}$l = 2.15-2.19 (terminal), 2.76-2.86 Å (bridging).

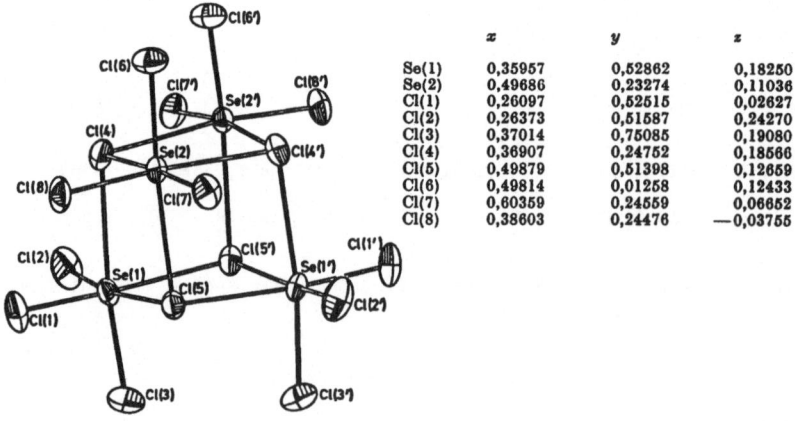

	x	y	z
Se(1)	0,35957	0,52862	0,18250
Se(2)	0,49686	0,23274	0,11036
Cl(1)	0,26097	0,52515	0,02627
Cl(2)	0,26373	0,51587	0,24270
Cl(3)	0,37014	0,75085	0,19080
Cl(4)	0,36907	0,24752	0,18566
Cl(5)	0,49879	0,51398	0,12659
Cl(6)	0,49814	0,01258	0,12433
Cl(7)	0,60359	0,24559	0,06652
Cl(8)	0,38603	0,24476	—0,03755

Fig. 2.   Structure of β-SeCl₄.

<u>1</u>. Structure Reports, <u>37</u>A, 194; <u>46</u>A, 401.

TITANIUM CHLORIDE   TITANIUM BROMIDE
Ti_7Cl_{16} Ti_7Br_{16}

I.   H. SCHÄFER and R. LAUMANNS, 1981. Z. anorg. Chem., <u>474</u>, 135-148.
II. B. KREBS and G. HENKEL, 1981. Ibid., <u>474</u>, 149-156.

Orthorhombic, Pnmm, a = 14.421, 15.228, b = 9.987, 10.577, c = 6.890, 7.276 Å, D_m =
2.96, 4.63, Z = 2. Mo radiation, R = 0.029, 0.063 for 1185, 1390 reflexions.

The structure (Fig. 1) contains Ti_3X_{13} octahedral clusters linked to each other
and to isolated TiX_6 octahedra. Ti-Ti interactions in the cluster displace the Ti
atoms from the centres of the octahedra. Ti-Ti = 2.95, 3.09, Ti-Cl = 2.36-2.42 and
2.51-2.64, Ti-Br = 2.51-2.57 and 2.66-2.83 Å.

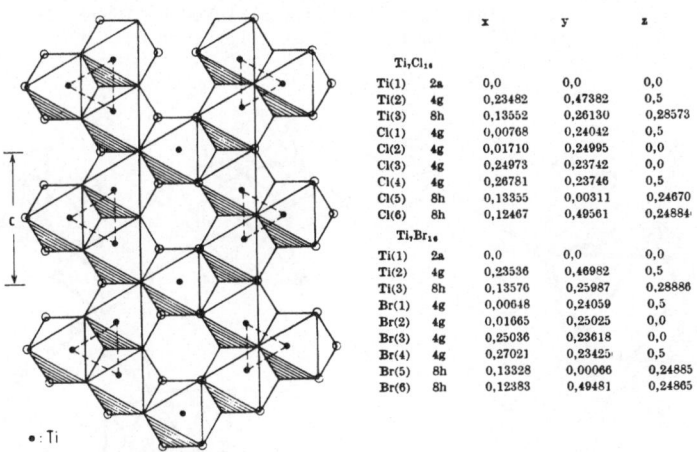

		x	y	z
Ti_7Cl_{16}				
Ti(1)	2a	0,0	0,0	0,0
Ti(2)	4g	0,23482	0,47382	0,5
Ti(3)	8h	0,13552	0,26130	0,28573
Cl(1)	4g	0,00768	0,24042	0,5
Cl(2)	4g	0,01710	0,24995	0,0
Cl(3)	4g	0,24973	0,23742	0,0
Cl(4)	4g	0,26781	0,23746	0,5
Cl(5)	8h	0,13355	0,00311	0,24670
Cl(6)	8h	0,12467	0,49561	0,24884
Ti_7Br_{16}				
Ti(1)	2a	0,0	0,0	0,0
Ti(2)	4g	0,23536	0,46982	0,5
Ti(3)	8h	0,13576	0,25987	0,28886
Br(1)	4g	0,00648	0,24059	0,5
Br(2)	4g	0,01665	0,25025	0,0
Br(3)	4g	0,25036	0,23618	0,0
Br(4)	4g	0,27021	0,23425	0,5
Br(5)	8h	0,13328	0,00066	0,24885
Br(6)	8h	0,12383	0,49481	0,24865

• : Ti

Fig. 1.   Structure of Ti_7Cl_{16} and Ti_7Br_{16}.

ZIRCONIUM CHLORIDE   ZIRCONIUM BROMIDE
Zr_6Cl_{12} Zr_6Br_{12}

POTASSIUM ZIRCONIUM CHLORIDE
$K_2Zr_7Cl_{18}$ $Zr_6Cl_{12}.K_2ZrCl_6$

H. IMOTO, J.D. CORBETT and A. CISAR, 1981. Inorg. Chem., <u>20</u>, 145-151.

Zr_6Cl_{12}, Zr_6Br_{12}, rhombohedral, R$\bar{3}$, a = 12.973, 13.577, c = 8.782, 9.287 Å, Z = 3.
Powder data.

$K_2Zr_7Cl_{18}$, rhombohedral, R$\bar{3}$, a = 9.499, c = 25.880 Å, Z = 3. Mo radiation, R =
0.037 for 736 reflexions.

Atomic positions

		Zr_6Cl_{12}				Zr_6Br_{12}		
		x	y	z		x	y	z
Zr		0.159	0.043	0.149		0.153	0.043	0.140
X(1)		0.127	0.174	0.332		0.126	0.178	0.325
X(2)		0.310	0.228	0.000		0.312	0.231	0.001

		$K_2Zr_7Cl_{18}$		
		x	y	z
Zr(1)	in	18(f)		
		0.0509	0.2164	0.0498
Zr(2)		3(b)	0	0	1/2
K		6(c)	0	0	0.2241
Cl(1)		18(f)	0.2401	0.1843	0.1135
Cl(2)		18(f)	0.4284	0.1377	0.0024
Cl(3)		18(f)	0.1033	0.4761	0.1117

Zr_6Cl_{12} and Zr_6Br_{12} are isostructural with Zr_6I_{12} (1), the structures containing Zr_6X_{12} clusters. $K_2Zr_7Cl_{18}$ contains six such clusters linked by longer Zr-Cl bonds to an octahedral $ZrCl_6^{2-}$ anion (Fig. 1); K has 12 Cl neighbours at 3.47-3.66 Å.

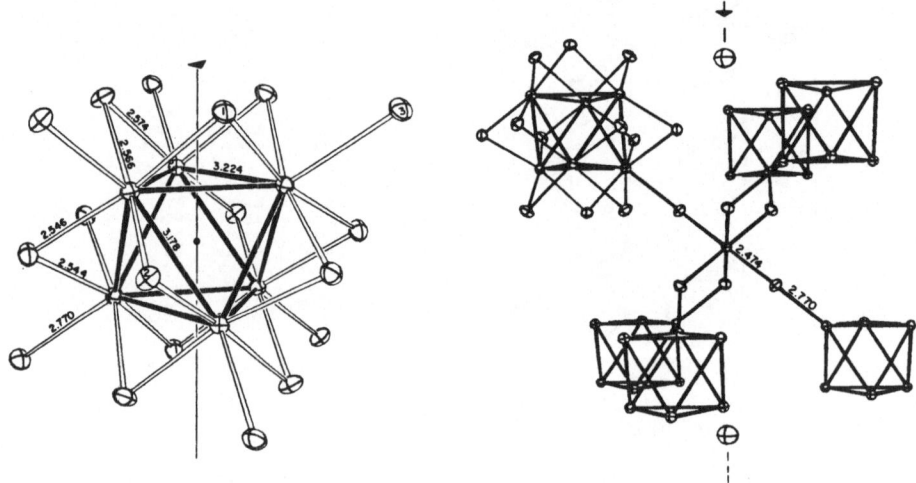

Fig. 1.   The Zr_6Cl_{12} cluster plus six neighbouring Cl (left) and the surroundings of the $ZrCl_6^{2-}$ anion (right) in $K_2Zr_7Cl_{18}$.

1. Structure Reports, 44A, 153.

DIAMMINEDICHLOROZINC(II)
$ZnCl_2(NH_3)_2$

T. YAMAGUCHI and O. LINDQVIST, 1981. Acta Chem. Scand., A35, 727-728.

Orthorhombic, Imam, a = 7.809, b = 8.512, c = 8.114 Å, Z = 4. Mo radiation, R = 0.029 for 1435 reflexions.

The structure (Fig. 1) is as previously determined (1), and contains tetrahedral molecules, linked by possible N-H...Cl hydrogen bonds. $\overline{Zn}$-Cl = 2.273(1), Zn-N = 2.024(2) Å, Cl-Zn-Cl = 109.2, N-Zn-N = 111.9, Cl-Zn-N = 108.9°, N...Cl = 3.49, 3.60, 3.62 Å.

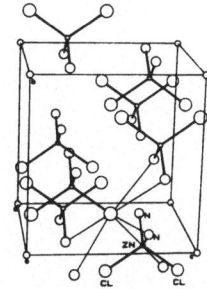

	x	y	z
Zn	0	0.38583	1/4
Cl	0	0.23111	0.02170
N	0.21467	0.51902	1/4

Fig. 1.   Structure of $ZnCl_2(NH_3)_2$.

1. Strukturbericht, 4, 31, 144.

LANTHANUM MONOCHLORIDE   LANTHANUM SESQUICHLORIDE
LaCl La_2Cl_3

R.E. ARAUJO and J.D. CORBETT, 1981. Inorg. Chem., 20, 3082-3086.

LaCl, rhombohedral, $R\overline{3}m$, a = 4.1026, c = 27.596 Å, Z = 6. Mo radiation, R = 0.086 for 150 reflexions. La, Cl in 6(c): 0,0,z, z = 0.22101, 0.3843. Isostructural with SrCl and ZrBr (1), La-La = 3.823, 4.103, La-Cl = 2.911(3) Å.

La_2Cl_3, monoclinic, [Cm], a = 15.890, b = 4.404, c = 10.231 Å, β = 119.14°, [Z = 4]. Powder data, atomic positional parameters not determined. Isostructural with Gd_2Cl_3 (2).

1. Structure Reports, 43A, 132, 139.
2. Ibid., 35A, 150; 39A, 166.

RUBIDIUM MAGNESIUM CHLORIDE   (6H)
$RbMgCl_3$

K.O. DEVANEY, M.R. FREEDMAN, G.L. McPHERSON and J.L. ATWOOD, 1981. Inorg. Chem., 20, 140-145.

Hexagonal, $P6_3/mmc$, a = 7.095, c = 17.578 Å, D_m = 2.77, Z = 6. Mo radiation, R = 0.070 for 249 reflexions.

Atomic positions

			x	y	z
Rb(1)	in	2(b)	0	0	1/4
Rb(2)		4(f)	1/3	2/3	0.0896
Mg(1)		2(a)	0	0	0
Mg(2)		4(f)	1/3	2/3	0.8403
Cl(1)		6(h)	0.5090	1.0172	1/4
Cl(2)		12(k)	0.8339	1.6677	0.0816

The structure is very similar to those of CsCdCl$_3$ (1) and RbMnCl$_3$ (2), and contains face- and corner-sharing MgCl$_6$ octahedra, Mg-Cl = 2.47-2.51 Å; Rb-Cl = 3.55-3.60(1) Å. Doping with Ni produces 4H (3) and 9R polytypes.

1. Structure Reports, 29, 280; 41A, 178.
2. Ibid., 43A, 137.
3. Ibid., 41A, 171.

BARIUM PHOSPHIDE CHLORIDE
Ba$_2$P$_7$Cl

H.G. von SCHNERING and G. MENGE, 1981. Z. anorg. Chem., 481, 33-40.

Monoclinic, P2$_1$/m, a = 11.726, b = 6.829, c = 6.337 Å, β = 95.27°, Z = 2. Mo radiation, R = 0.035 for 964 reflexions.

The structure (Fig. 1) contains an NaCl-type arrangement of cations and anions, half the anions being P$_7^{3-}$.

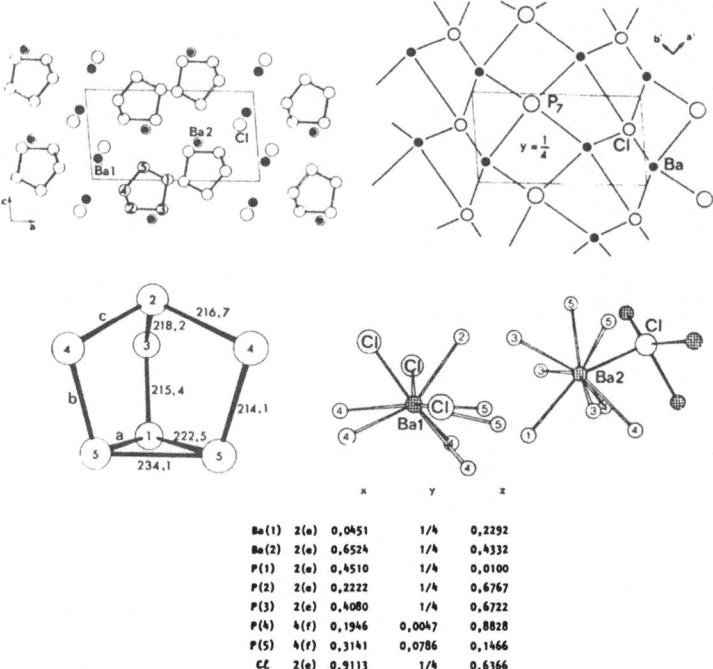

		x	y	z
Ba(1)	2(e)	0,0451	1/4	0,2292
Ba(2)	2(e)	0,6524	1/4	0,4332
P(1)	2(e)	0,4510	1/4	0,0100
P(2)	2(e)	0,2222	1/4	0,6767
P(3)	2(e)	0,4080	1/4	0,6722
P(4)	4(f)	0,1946	0,0047	0,8828
P(5)	4(f)	0,3141	0,0786	0,1466
Cl	2(e)	0,9113	1/4	0,6366

Fig. 1.   Structure of Ba$_2$P$_7$Cl.

GALLIUM TELLURIUM CHLORIDE
GaTeCl

A. WILMS and R. KNIEP, 1981. Z. Naturforsch., 36B, 1658-1659.

Orthorhombic, Pnnm, a = 5.852, b = 14.469, c = 4.082 Å, Z = 4. Mo radiation, R = 0.082 for 675 reflexions.

The structure (Fig. 1) contains layers of $GaTe_3Cl$ tetrahedra which share Te atoms with six neighbouring tetrahedra, Cl atoms being terminal; the arrangement is similar to that in black phosphorus. Ga-Te = 2.625-2.638(1), Ga-Cl = 2.182(4) Å, angles at Ga = 101.4-116.7, at Te = 99.4, 101.4°.

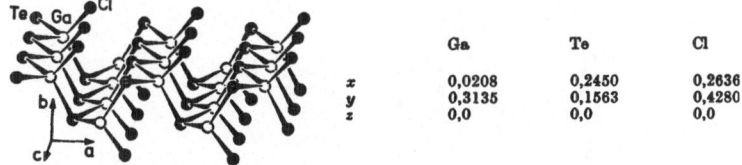

	Ga	Te	Cl
x	0,0208	0,2450	0,2636
y	0,3135	0,1563	0,4280
z	0,0	0,0	0,0

Fig. 1.   Structure of GaTeCl.

DICHLOROIODINE(1+) HEXAFLUOROANTIMONATE(V)
$[ICl_2]^+[SbF_6^-]$

T. BIRCHALL and R.D. MYERS, 1981. Inorg. Chem., <u>20</u>, 2207-2211.

Orthorhombic, Cmca, a = 10.751, b = 12.087, c = 12.982 Å, Z = 8. Mo radiation, R = 0.031 for 788 reflexions.

The structure (Fig. 1) contains chains of alternating bent ICl_2^+ cations and octahedral SbF_6^- anions, with two strong I...F interactions completing fourfold planar coordination at I. I-Cl = 2.268(2), I...F = 2.650(6), Sb-F = 1.894 (bridging), 1.851-1.867(8) Å (terminal), Cl-I-Cl = 97.2, F...I...F = 91.2, I...F-Sb = 155.3°.

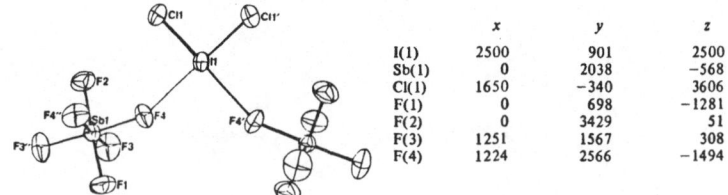

	x	y	z
I(1)	2500	901	2500
Sb(1)	0	2038	−568
Cl(1)	1650	−340	3606
F(1)	0	698	−1281
F(2)	0	3429	51
F(3)	1251	1567	308
F(4)	1224	2566	−1494

Fig. 1.   Structure of $ICl_2.SbF_6$, and atomic positional parameters (x 10^4).

DICHLOROTRIIODINE HEXACHLOROANTIMONATE
$I_3Cl_2SbCl_6$ (I)

DICHLOROTRIIODINE TETRACHLOROALUMINATE
$I_3Cl_2AlCl_4$ (II)

DIBROMOTRIIODINE HEXACHLOROANTIMONATE
$I_3Br_2SbCl_6$ (III)

S. POHL and W. SAAK, 1981. Z. Naturforsch., <u>36B</u>, 283-288.

I, III, triclinic, $P\bar{1}$, a = 7.027, 7.132, b = 7.050, 7.220, c = 7.722, 7.752 Å, α = 88.37, 87.98, β = 82.83, 82.99, γ = 81.58, 81.13°, at 143K, Z = 1. Mo radiation, R = 0.037, 0.074 for 1036, 1040 reflexions.

II, monoclinic, C2/c, a = 27.53, b = 7.120, c = 14.188 Å, β = 110.36°, at 143K, Z = 8. Mo radiation, R = 0.079 for 1645 reflexions.

Atomic positions

		x	y	z
1:	I1	0,65577	0,23330	0,16952
	I2	1,00000	0,00000	0,00000
	Sb	1,00000	0,50000	0,50000
	Cl1	1,10925	0,46646	0,20112
	Cl2	0,71984	0,70372	0,43366
	Cl3	0,83411	0,22296	0,49312
	Cl4	0,50023	0,24321	−0,08313
2:	I1	0,11567	0,04584	0,21555
	I2	0,20691	−0,15762	0,33120
	I3	0,02624	0,27364	0,09705
	Cl1	0,24619	−0,15312	0,20780
	Cl2	−0,03867	0,07079	0,10566
	Al	0,13200	0,45113	−0,03561
	Cl3	0,10870	0,53349	0,09028
	Cl4	0,16719	0,17400	−0,00129
	Cl5	0,18823	0,64113	−0,04800
	Cl6	0,06639	0,41941	−0,16726
3:	I1	0,66146	0,23133	0,16842
	I2	1,00000	0,00000	0,00000
	Br	0,49223	0,24783	−0,09309
	Sb	1,00000	0,50000	0,50000
	Cl1	1,09859	0,47391	0,19884
	Cl2	0,71297	0,69571	0,44137
	Cl3	0,84648	0,22630	0,49574

The structures (Fig. 1) contain $I_3Cl_2^+$ and $I_3Br_2^+$ cations, and $SbCl_6^-$ and $AlCl_4^-$ anions, linked into infinite chains by weaker I...Cl interactions. Cation bond distances are shown in Fig. 1. I-I-I = 180, 177, I-I-Cl/Br = 92-98°, Sb-Cl = 2.35-2.42, Al-Cl = 2.11-2.18 Å.

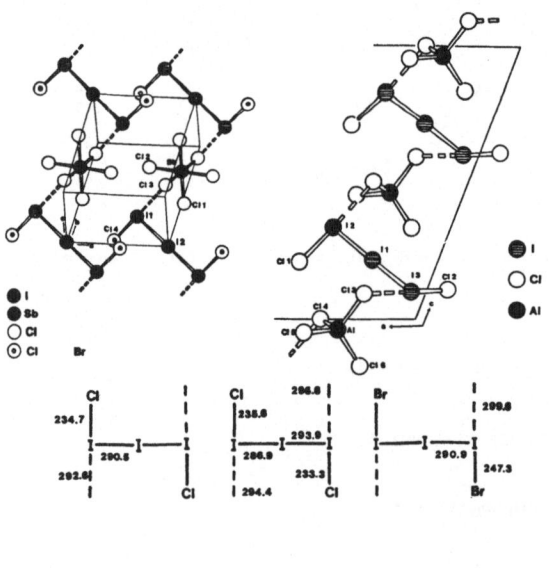

$I_3Cl_2SbCl_6$ $I_3Cl_2AlCl_4$· $I_3Br_2SbCl_6$

Fig. 1.   Structures of $I_3Cl_2SbCl_6$ and $I_3Br_2SbCl_6$ (left) and of $I_3Cl_2AlCl_4$ (right), and distances involving the cations (Å x 10^2).

LITHIUM VANADIUM(II) CHLORIDE
Li_6VCl_8 $6LiCl.VCl_2$

L. HANEBALI, T. MACHEJ, C. CROS and P. HAGENMULLER, 1981. Mater. Res. Bull., 16, 887-901.

Cubic, Fm3m, a = 10.294 Å, D_m = 2.28, Z = 4. Powder data. V in 4(a); ☐ in 4(b); Li in 24(d); Cl(1) in 8(c); Cl(2) in 24(e): x = 0.245. Li may also be distributed over the 4(b) sites. Other phases are found at different compositions.

Structure as in $6NaCl.CdCl_2$ (1).

1. Structure Reports, 26, 316.

CAESIUM CHROMIUM(II) TRICHLORIDE
CAESIUM CHROMIUM(II) TRIIODIDE
α-$CsCrX_3$ (X = Cl, I)

W.J. CRAMA and H.W. ZANDBERGEN, 1981. Acta Cryst., B37, 1027-1031.

Hexagonal, $P6_3/mmc$, a = 7.257, 8.107, c = 6.238, 6.917 Å, Z = 2. Mo radiation, R = 0.034, 0.027 for 421, 289 reflexions. Previous studies (1) are in space group $P6_3mc$.

Atomic positions (for two possible models, A and B)

		x	y	z	Model			x	y	z	Model
Cs	2(d)	$\frac{1}{3}$	$\frac{2}{3}$	$\frac{1}{4}$	A.B	Cs	2(d)	$\frac{1}{3}$	$\frac{2}{3}$	$\frac{1}{4}$	A.B
Cr	2(a)	0	0	0	A.B	Cr	2(a)	0	0	0	A.B
Cl(1)	12(k)	0·1613	−0·1613	0·2873	A	I(1)	12(k)	0·1677	−0·1677	0·2840	A
	12(k)	0·1647	−0·1647	0·2978	B		12(k)	0·1707	−0·1707	0·2946	B
Cl(2)	6(h)	0·1501	−0·1501	0·25	A	I(2)	6(h)	0·1555	−0·1555	0·25	A
	12(k)	0·1543	−0·1543	0·2688	B		12(k)	0·1597	−0·1597	0·2672	B

These high-temperature phases have a slightly-distorted $BaNiO_3$ structure, the distortion resulting from a local Jahn-Teller effect which leads to elongated octahedra with randomly distributed elongation directions.

1. Structure Reports, 39A, 173; 45A, 184.

DICAESIUM DICHLOROTETRAAQUOCHROMIUM(III) CHLORIDE
$Cs_2CrCl_5.4H_2O$ $Cs_2[CrCl_2(H_2O)_4]Cl_3$

TRICAESIUM DICHLOROTETRAAQUOVANADIUM(III) CHLORIDE
$Cs_3VCl_6.4H_2O$ $Cs_3[VCl_2(H_2O)_4]Cl_4$

I.   P.J. McCARTHY, J.C. LAUFFENBURGER, P.M. SKONEZNY and D.C. ROHRER, 1981. Inorg. Chem., 20, 1566-1570.
II. P.J. McCARTHY, J.C. LAUFFENBURGER, M.M. SCHREINER and D.C. ROHRER, 1981. Ibid., 20, 1571-1576.

$Cs_2CrCl_5.4H_2O$
Monoclinic, C2/m, a = 17.604, b = 6.140, c = 6.979 Å, β = 106.04°, Z = 2. Mo radiation, R = 0.083 for 1140 reflexions.

$Cs_3VCl_6.4H_2O$
Orthorhombic, Immm, a = 8.354, b = 17.666, c = 6.044 Å, D_m = 2.74, Z = 2. Mo radiation, R = 0.079 for 929 reflexions.

Both structures (Fig. 1) contain trans-octahedral complex cations; Cr-Cl = 2.302, Cr-O = 1.925, V-Cl = 2.377, V-O = 1.969 Å. Cs ions have eight-coordinations to bound and free Cl, Cs-Cl = 3.43-3.77 Å, and O-H...Cl hydrogen bonds are present, O...Cl = 3.01-3.09 Å.

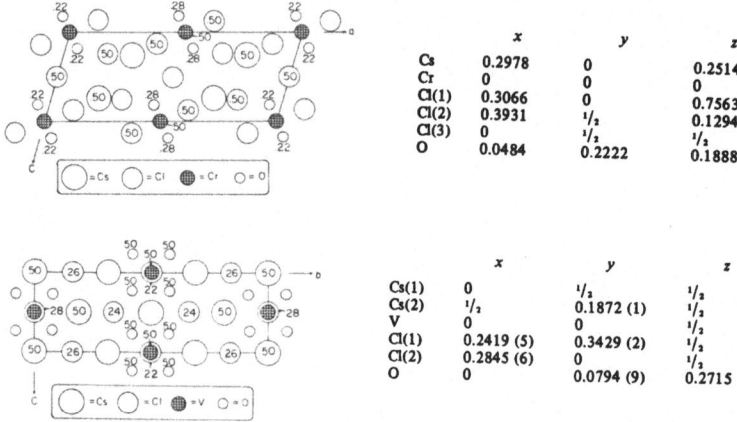

	x	y	z
Cs	0.2978	0	0.2514
Cr	0	0	0
Cl(1)	0.3066	0	0.7563
Cl(2)	0.3931	1/2	0.1294
Cl(3)	0	1/2	1/2
O	0.0484	0.2222	0.1888

	x	y	z
Cs(1)	0	1/2	1/2
Cs(2)	1/2	0.1872 (1)	1/2
V	0	0	1/2
Cl(1)	0.2419 (5)	0.3429 (2)	1/2
Cl(2)	0.2845 (6)	0	1/2
O	0	0.0794 (9)	0.2715

Fig. 1.   Structures of $Cs_2CrCl_5.4H_2O$ (top) and $Cs_3VCl_6.4H_2O$ (bottom).

OXONIUM AQUOPENTACHLOROFERRATE(III)
$(H_3O)_2[FeCl_5(H_2O)]$ $H_2FeCl_5.3H_2O$

I. SØTOFTE and K. NIELSEN, 1981. Acta Chem. Scand., A35, 821-822.

Orthorhombic, Pcmn, a = 7.038, b = 9.926, c = 13.720 Å, Z = 4. Mo radiation, R = 0.031 for 965 reflexions.

Isostructural with the ammonium compound (1), the structure (Fig. 1) containing H_3O^+ cations and distorted octahedral $FeCl_5(H_2O)^{2-}$ anions; Fe-Cl = 2.328-2.394(1), Fe-O = 2.107(4) Å. The ions are linked by hydrogen bonding, which is quite weak for cation-anion interactions, O-H...Cl = 3.30-3.36 Å, but stronger for anion-anion interactions, O-H...Cl = 3.19 Å.

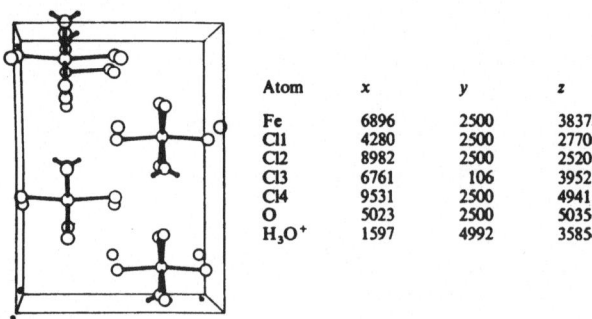

Atom	x	y	z
Fe	6896	2500	3837
Cl1	4280	2500	2770
Cl2	8982	2500	2520
Cl3	6761	106	3952
Cl4	9531	2500	4941
O	5023	2500	5035
H_3O^+	1597	4992	3585

Fig. 1.   Structure of oxonium aquopentachloroferrate(III), and atomic positional parameters (x 10^4).

1. Structure Reports, 11, 417.

CAESIUM ENNEACHLORODIRUTHENATE(III)

$Cs_3Ru_2Cl_9$

J. DARRIET, 1981. Rev. Chim. Minér., 18, 27-32.

Hexagonal, $P6_3$/mmc, a = 7.221, c = 17.556 Å, Z = 2. Mo radiation, R = 0.035 for 309 reflexions.

Atomic positions

			x	y	z
Cs(1)	in	2(b)	0	0	1/4
Cs(2)		4(f)	1/3	2/3	0.0746
Ru		4(f)	1/3	2/3	0.8276
Cl(1)		6(h)	0.5096	1.0192	1/4
Cl(2)		12(k)	0.8193	1.6386	0.0961

Isostructural with the Cr compound (1). The structure contains $Ru_2Cl_9{}^{3-}$ ions which consist of two face-sharing octahedra, linked by Cs ions. Ru-Ru = 2.725 Å.

1. Structure Reports, 21, 277.

CAESIUM TETRACHLOROCOBALTATE(II) CHLORIDE

Cs_3CoCl_5

P.A. REYNOLDS, B.N. FIGGIS and A.H. WHITE, 1981. Acta Cryst., B37, 508-513.

Tetragonal, I4/mcm, a = 9.232, c = 14.554 Å, Z = 4. Mo radiation, various refinements.

Structure as previously described (1). Ionic charges and 3d-electron populations are determined.

1. Strukturbericht, 3, 134, 498; 29, 277; 46A, 184, 399.

AMMONIUM AQUOPENTACHLORORHODATE(III)

$(NH_4)_2[RhCl_5(H_2O)]$

G. BUGLI and C. POTVIN, 1981. Acta Cryst., B37, 1394-1396.

Orthorhombic, Pnma, a = 13.767, b = 9.787, c = 7.059 Å, D_m = 2.33, Z = 4. Mo radiation, R = 0.035 for 1581 reflexions.

The structure (Fig. 1) contains octahedral anions, linked into chains along b by O-H...Cl hydrogen bonds. The chains are linked by 8-coordinate ammonium ions. Rh-O = 2.090(6), Rh-Cl = 2.304 (trans to O), 2.337-2.356(2) (cis to O), O-H...Cl = 3.194, N...Cl = 3.28-3.50 Å (the shorter contacts may be hydrogen bonds).

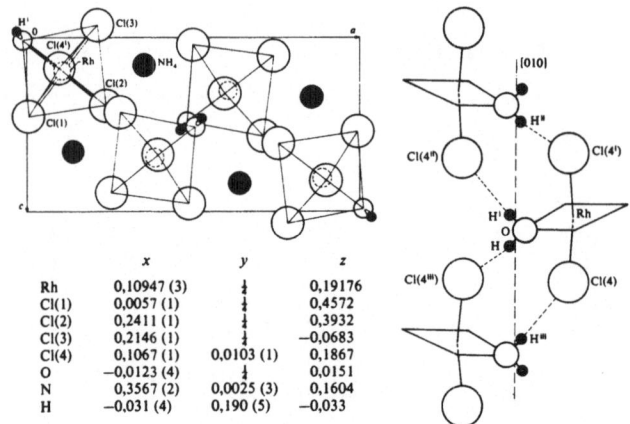

	x	y	z
Rh	0,10947 (3)	¼	0,19176
Cl(1)	0,0057 (1)	¼	0,4572
Cl(2)	0,2411 (1)	¼	0,3932
Cl(3)	0,2146 (1)	¼	−0,0683
Cl(4)	0,1067 (1)	0,0103 (1)	0,1867
O	−0,0123 (4)	¼	0,0151
N	0,3567 (2)	0,0025 (3)	0,1604
H	−0,031 (4)	0,190 (5)	−0,033

Fig. 1.   Structure of ammonium aquopentachlororhodate(III).

PALLADIUM ALUMINUM CHLORIDE
PdAl$_2$Cl$_8$

W. LENHARD, H. SCHÄFER, H.-U. HÜRTER and B. KREBS, 1981. Z. anorg. Chem., **482**, 19-26.

Monoclinic, P2$_1$/c, a = 6.583, b = 7.342, c = 13.033 Å, β = 96.40°, D$_m$ = 2.35, Z = 2.
Mo radiation, R = 0.030 for 1153 reflexions.

Atomic positions

	x	y	z
Pd	0	0	0
Al	0.3367	−0.2802	0.1020
Cl(1)	0.1400	−0.2722	−0.0470
Cl(2)	0.2300	−0.0094	0.1479
Cl(3)	0.2508	−0.4786	0.2015
Cl(4)	0.6431	−0.2850	0.0768

The structure contains centrosymmetric Cl$_2$AlCl$_2$PdCl$_2$AlCl$_2$ molecules in which Pd has square-planar and Al tetrahedral coordinations; Pd-Cl = 2.313, Al-Cl = 2.212 (bridging), 2.075(2) Å (terminal).

AMMONIUM TRICHLOROCUPRATE(II)
NH$_4$CuCl$_3$

G. O'BANNON and R.D. WILLETT, 1981. Inorg. Chim. Acta, **53**, L131-L132.

Monoclinic, P2$_1$/c, a = 4.034, b = 14.206, c = 8.992 Å, β = 96.43°, Z = 4. Mo radiation, R = 0.067 for 1214 reflexions.

The structure (Fig. 1) is as previously described (1), containing $Cu_2Cl_6{}^{2-}$ anions with square-planar coordination at Cu, linked by 8-coordinate ammonium ions; longer Cu...Cl contacts complete (4+2)-coordination at Cu. Cu-Cl = 2.26-2.31, Cu... Cl = 2.95, 3.17, N...Cl = 3.18-3.51 Å.

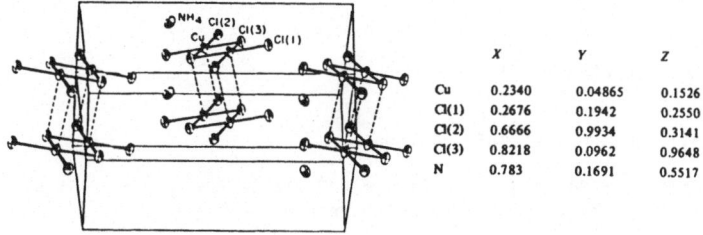

	X	Y	Z
Cu	0.2340	0.04865	0.1526
Cl(1)	0.2676	0.1942	0.2550
Cl(2)	0.6666	0.9934	0.3141
Cl(3)	0.8218	0.0962	0.9648
N	0.783	0.1691	0.5517

Fig. 1.   Structure of NH_4CuCl_3.

1. Structure Reports, 28, 85.

RUBIDIUM COPPER(II) CHLORIDE
$RbCuCl_3$

W.J. CRAMA, 1981. J. Solid State Chem., 39, 168-172.

α-Form (above 339K), $P6_3/mmc$, a = 6.978, c = 6.156 Å, at 390K, Z = 2. X-ray powder data; atomic positional parameters not determined.

β-Form (262-339K), orthorhombic, Pcan, a = 11.929, b = 6.971, c = 6.164 Å, at 295K, Z = 4. Neutron powder data.

γ-Form (below 262K), monoclinic, C2, a = 11.932, b = 6.844, c = 12.244 Å, β = 91.93°, at 4.2K, Z = 8 [not 2]. Neutron powder data.

Atomic positions

β-$RbCuCl_3$ AT 295 K

		x	y	z
Rb	4c	0.3313	0	¼
Cu	4a	0	0	0
Cl(1)	8d	0.0831	0.2427	0.1972
Cl(2)	4c	0.1509	0	¼

γ-$RbCuCl_3$ AT 4.2 K

		x	y	z
Rb(a)	4c	0.342	-0.029	0.381
Rb(b)	4c	0.324	0.007	0.868
Cu(1a)	2a	0	0	0
Cu(1b)	2b	0	-0.051	¼
Cu(2)	4c	-0.001	-0.016	0.245
Cl(1a)	4c	0.168	-0.019	0.158
Cl(1b)	4c	0.157	-0.060	0.659
Cl(2a)	4c	0.093	0.158	0.392
Cl(2b)	4c	0.087	0.254	0.896
Cl(3a)	4c	0.073	-0.302	0.394
Cl(3b)	4c	0.068	-0.241	0.880

All three structures are distortions of the $CsMgCl_3$ type (1). The α-form is probably isostructural with α-$CsCrCl_3$ (2), the β-form is isostructural with β-$CsCrI_3$ (3), and the γ-form with γ-$RbCrCl_3$ (4). The β and γ structures contain chains of face-sharing $CuCl_6$ octahedra, with Jahn-Teller distortions, Cu-Cl = 2.26-2.41 (4 distances), 2.66-2.80 Å (2 distances).

1. Structure Reports, 35A, 160.
2. Ibid., 39A, 173.
3. Ibid., 46A, 204.
4. Ibid., 45A, 171.

RUBIDIUM HEPTACHLORODICUPRATE(II)
$Rb_3Cu_2Cl_7$

W.J. CRAMA, 1981. Acta Cryst., B37, 662-664.

Orthorhombic, Ccca, a = 24.843, b = 7.216, c = 7.216 Å, Z = 4. Mo radiation, R = 0.028 for 5448 reflexions [includes symmetry-related reflexions].

The structure (Fig. 1) can be derived from that of $Rb_3Mn_2Cl_7$ (1). It contains corner-sharing, Jahn-Teller distorted $CuCl_6$ octahedra; Cu-Cl = 2.234-2.381 (4 distances), 2.870 (x 2) Å.

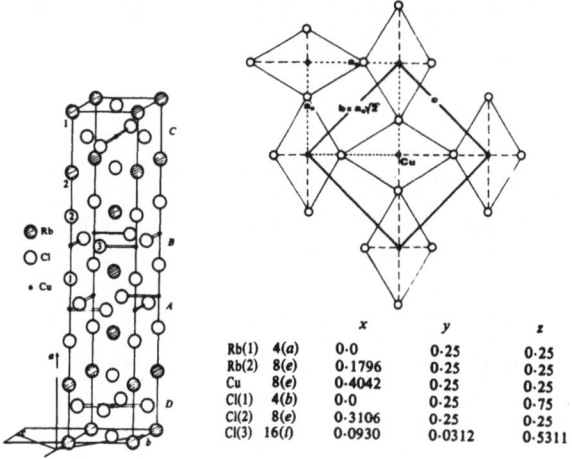

		x	y	z
Rb(1)	4(a)	0·0	0·25	0·25
Rb(2)	8(e)	0·1796	0·25	0·25
Cu	8(e)	0·4042	0·25	0·25
Cl(1)	4(b)	0·0	0·25	0·75
Cl(2)	8(e)	0·3106	0·25	0·25
Cl(3)	16(f)	0·0930	0·0312	0·5311

Fig. 1.   Structure of $Rb_3Cu_2Cl_7$ (a_0 is the a-axis of the $Rb_3Mn_2Cl_7$ structure (1)).

1. Structure Reports, 38A, 214.

CAESIUM TRICHLOROCUPRATE(II)
α-$CsCuCl_3$

W.J. CRAMA, 1981. Acta Cryst., B37, 2133-2136.

Hexagonal, $P6_3/mmc$, a = 7.212, 7.229, c = 6.141, 6.149 Å, at 430, 470K, Z = 2. Mo radiation, R = 0.022, 0.019 for 148, 208 reflexions.

Atomic positions (430K)

			x	y	z
2 Cs	in	2(d)	1/3	2/3	3/4
2 Cu		2(a)	0	0	0
4 Cl(1)		12(k)	0.1603	-0.1603	0.2998
2 Cl(2)		6(h)	0.1407	-0.1407	1/4

Slightly distorted $BaNiO_3$-type structure, with disorder of Jahn-Teller elongated CuO_6 octahedra.

AMMONIUM TETRACHLOROZINCATE CHLORIDE
$(NH_4)_3ZnCl_5$

D. SCHMITZ, 1981. Acta Cryst., B**37**, 518-525.

Orthorhombic, Pnma, a = 8.716, b = 9.887, c = 12.625 Å, Z = 4. Mo radiation, R = 0.044 for 934 reflexions.

The structure (Fig. 1) is as previously described (**1**). It contains $ZnCl_4^{2-}$ tetrahedra, NH_4^+, and Cl^- ions; Zn-Cl = 2.245-2.278(3) Å. Structural relationships with Ba_3SiS_5 are described.

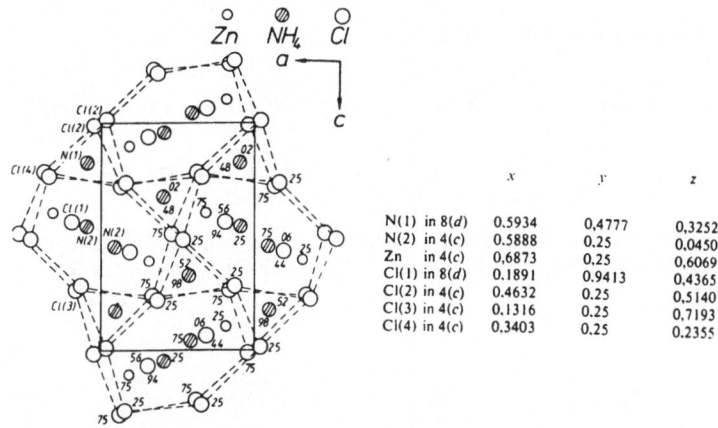

	x	y	z
N(1) in 8(d)	0.5934	0.4777	0.3252
N(2) in 4(c)	0.5888	0.25	0.0450
Zn in 4(c)	0.6873	0.25	0.6069
Cl(1) in 8(d)	0.1891	0.9413	0.4365
Cl(2) in 4(c)	0.4632	0.25	0.5140
Cl(3) in 4(c)	0.1316	0.25	0.7193
Cl(4) in 4(c)	0.3403	0.25	0.2355

Fig. 1.   Structure of $(NH_4)_3ZnCl_5$.

1. Structure Reports, **9**, 202.

MAGNESIUM HEXACHLOROCADMATE DODECAHYDRATE
$Mg_2CdCl_6 \cdot 12H_2O$

M. LEDESERT and J.C. MONIER, 1981. Acta Cryst., B**37**, 652-654.

Trigonal, P31c, a = 9.981, c = 11.556 Å, D_m = 1.967, Z = 2. Mo radiation, R = 0.030 for 1292 reflexions.

The structure (Fig. 1) contains $CdCl_6^{4-}$ and $Mg(OH_2)_6^{2+}$ octahedra, linked by hydrogen bonds. Cd-Cl = 2.636, Mg-O = 2.07, O-H...O = 2.94, O-H...Cl = 3.09-3.33 Å.

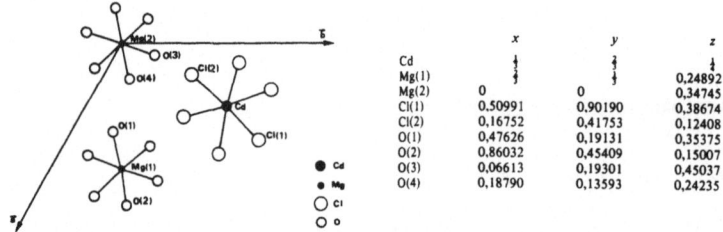

	x	y	z
Cd	$\frac{1}{4}$	$\frac{1}{4}$	$\frac{1}{4}$
Mg(1)	$\frac{1}{4}$	$\frac{1}{4}$	0,24892
Mg(2)	0	0	0,34745
Cl(1)	0,50991	0,90190	0,38674
Cl(2)	0,16752	0,41753	0,12408
O(1)	0,47626	0,19131	0,35375
O(2)	0,86032	0,45409	0,15007
O(3)	0,06613	0,19301	0,45037
O(4)	0,18790	0,13593	0,24235

Fig. 1.   Structure of $Mg_2CdCl_6.12H_2O$.

HEXAAMMINECOBALT(III) PENTACHLOROMERCURATE(II)
$[Co(NH_3)_6][HgCl_5]$

A.W. HERLINGER, J.N. BROWN, M.A. DWYER and S.F. PAVKOVIC, 1981. Inorg. Chem., 20, 2366-2371.

Orthorhombic, Pnma, a = 21.033, b = 7.591, c = 8.377 Å, D_m = 2.67, Z = 4. Mo radiation, R = 0.048 for 1015 reflexions.

The cation is octahedral, mean Co-N = 1.96(1) Å, and the $HgCl_5^{3-}$ anion is a severely distorted trigonal bipyramid (Fig. 1; Cl(axial)-Hg-Cl(axial) = 159°).

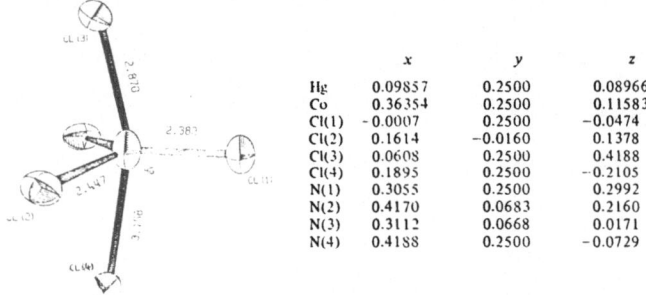

	x	y	z
Hg	0.09857	0.2500	0.08966
Co	0.36354	0.2500	0.11583
Cl(1)	-0.0007	0.2500	-0.0474
Cl(2)	0.1614	-0.0160	0.1378
Cl(3)	0.0608	0.2500	0.4188
Cl(4)	0.1895	0.2500	-0.2105
N(1)	0.3055	0.2500	0.2992
N(2)	0.4170	0.0683	0.2160
N(3)	0.3112	0.0668	0.0171
N(4)	0.4188	0.2500	-0.0729

Fig. 1.   The $HgCl_5^{3-}$ ion and the atomic positional parameters in $[Co(NH_3)_6][HgCl_5]$.

trans-DIAMMINEDIBROMOPLATINUM(II)-trans-DIAMMINETETRABROMOPLATINUM(IV)
$[PtBr_2(NH_3)_2][PtBr_4(NH_3)_2]$ $Pt_2Br_6(NH_3)_4$

H.J. KELLER, B. KEPPLER, G. LEDEZMA-SANCHEZ and W. STEIGER, 1981. Acta Cryst., B37, 674-675.

Orthorhombic, Immm, a = 7.760, b = 8.234, c = 5.553 Å, Z = 1. Mo radiation, R = 0.055 for 584 reflexions. Previous studies in 1.

Atomic positions

			x	y	z
2 Pt	in	2(a)	0	0	0
2 Br(1)		4(i)	0	0	0.552
4 Br(2)		4(e)	0.297	0	0
4 N		4(g)	0	0.266	0

The structure is disordered (Fig. 1), with half-occupancy for Br(1). Pt-Br(1) = 2.485(4), Pt-Br(2) = 2.445(2), Pt-N = 2.07(2) Å.

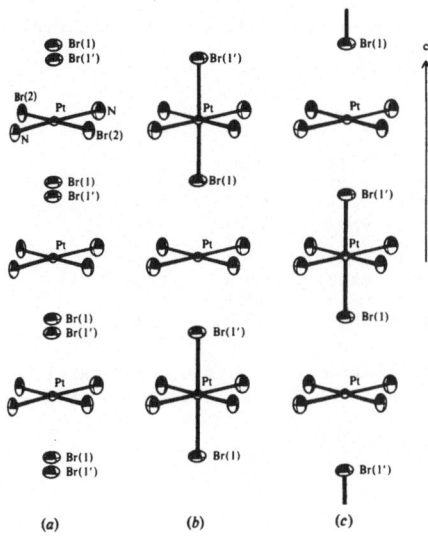

Fig. 1.   Structure of $Pt_2Br_6(NH_3)_4$. (a) is a superposition of (b) and (c).

1. Structure Reports, 11, 483; 22, 261; 27, 667.

THORIUM(IV) BROMIDE
β-ThBr$_4$

R. de KOUCHKOVSKY, M.F. LE CLOAREC and P. DELAMOYE, 1981. Mater. Res. Bull, 16, 1421-1427.

Tetragonal, I4$_1$/amd, a = 8.923, c = 7.920 Å, at 130K, Z = 4. Neutron powder data. Th in 4(a): 0,3/4,1/8; Br in 16(h): 0,0.564,0.798. Structure as previously described (1). Below 92K, the c axis triples, with an incommensurate modulation.

1. Structure Reports, 39A, 167.

THALLIUM(I) LEAD(II) BROMIDE
Tl$_3$PbBr$_5$

I.   H.-L. KELLER, 1981. Z. anorg. Chem., 482, 154-162.

Low-temperature form
Orthorhombic, $P2_12_12_1$, a = 15.399, b = 9.063, c = 8.532 Å, Z = 4. R = 0.105 for 940 reflexions.

Pb has pentagonal-bipyramidal 7-coordination, Pb-Br = 2.89-3.32 Å, and Tl ions have 8-, 8-, and 7-coordinations, Tl-Br = 3.21-3.72 Å; there is also one Tl...Tl contact of 3.63 Å.

II. H.-L. KELLER, 1981. J. Less-Common Metals, <u>78</u>, 281-286.

High-temperature form
Tetragonal, $P4_1$, a = 8.903, c = 15.486 Å (at room temperature), Z = 4. Mo radiation, R = 0.13 for 324 reflexions.

Atomic positions

	x	y	z
Tl(1)	0,989	0,807	0,801
Tl(2)	0,127	0,317	0,875
Tl(3)	0,623	0,181	0,816
Pb	0,494	0,694	0,898
Br(1)	0,337	0,161	0,219
Br(2)	0,688	0,006	0,357
Br(3)	0,028	0,848	0,207
Br(4)	0,513	0,651	0,501
Br(5)	0,493	0,813	0,113

Isostructural with the chloride (<u>1</u>), with ordered metal positions in $P4_1$. Metal ions have 7-coordinations, Tl-Br = 3.09-3.76, Pb-Br = 3.06-3.50 Å.

<u>1</u>. Structure Reports, <u>43A</u>, 136; <u>45A</u>, 169.

CADMIUM ARSENIC BROMIDE
CADMIUM PHOSPHORUS HALIDES
Cd_2As_3Br Cd_2P_3I Cd_2P_3Br Cd_2P_3Cl

A. REBBAH, J. YAZBECK, R. LANDÉ and A. DESCHANVRES, 1981. Mater. Res. Bull., <u>16</u>, 525-533.

Monoclinic, Cc, a = 8.286, 8.243, 8.077, 7.969, b = 9.408, 9.334, 9.088, 8.984, c = 7.987, 7.516, 7.534, 7.554 Å, β = 101.3, 99.8, 100.3, 100.8°, Z = 4. Powder data.

Atomic positions

	x	y	z			x	y	z
		Cd_2As_3Br					Cd_2P_3Br	
Cd(1)	0,0	0,153	0,0		Cd(1)	0,0	0,147	0,0
Cd(2)	0,502	0,142	0,086		Cd(2)	0,504	0,147	0,08
As(1)	0,121	0,047	0,317		P(1)	0,130	0,038	0,315
As(2)	-0,258	0,290	0,039		P(2)	-0,25	0,292	0,005
As(3)	0,370	0,047	-0,243		P(3)	0,378	0,040	-0,21
Br	0,253	0,370	0,040		Br	0,249	0,374	0,07
		Cd_2P_3I					Cd_2P_3Cl	
Cd(1)	0,0	0,141	0,0		Cd(1)	0,0	0,149	0,0
Cd(2)	0,500	0,145	0,069		Cd(2)	0,494	0,144	0,080
P(1)	0,147	0,043	0,324		P(1)	0,116	0,046	0,318
P(2)	-0,258	0,290	0,04		P(2)	-0,258	0,296	0,038
P(3)	0,388	0,044	-0,239		P(3)	0,366	0,045	-0,24
I	0,240	0,374	0,060		Cl	0,253	0,345	-0,004

Isostructural with Cd_2As_3I (<u>1</u>).

<u>1</u>. Structure Reports, <u>45</u>A, 188.

CADMIUM PHOSPHORUS ARSENIC BROMIDE
Cd_4PAsBr_3

A. REBBAH and A. DESCHANVRES, 1981. Ann. Chim., <u>6</u>, 585-590.

Cubic, Pa3, a = 12.521 Å, D_m = 5.42, Z = 8. Mo radiation, R = 0.058 for 619 reflexions.

The structure (Fig. 1) contains a very-distorted face-centred-cubic arrangement of Cd ions, with octahedral sites occupied by Br or a pair of A atoms (A = P or As); tetrahedral sites are occupied by the remaining A atoms. Cd(1) has distorted tetrahedral coordination to 3 Br at 2.76 Å and 1 A at 2.54 Å, Cd(2) has 5-coordination to 3 Br at 2.80-3.65 Å and 2 A at 2.52 Å. A(1) has 4 Cd neighbours, and A(2) has 3 Cd neighbours plus 1 A(1) at 2.35 Å.

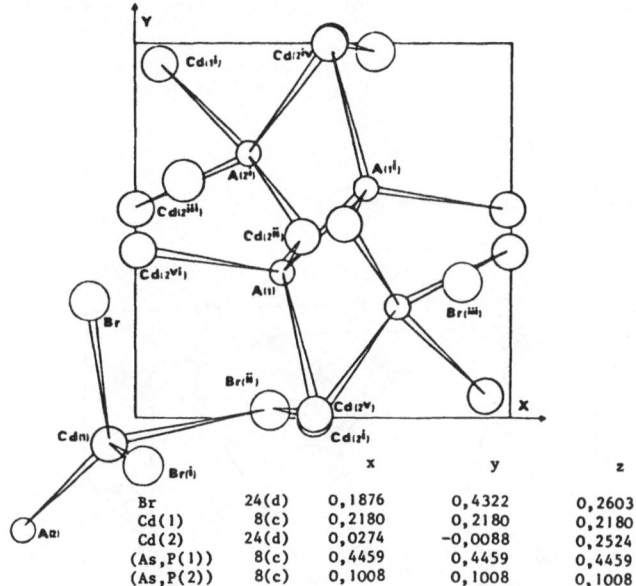

		x	y	z
Br	24(d)	0,1876	0,4322	0,2603
Cd(1)	8(c)	0,2180	0,2180	0,2180
Cd(2)	24(d)	0,0274	-0,0088	0,2524
(As,P(1))	8(c)	0,4459	0,4459	0,4459
(As,P(2))	8(c)	0,1008	0,1008	0,1008

Fig. 1.   Structure of Cd_4PAsBr_3.

CAESIUM AMMINEDIBROMONITROPLATINATE(II) AMMINETETRABROMONITROPLATINATE(IV)
$Cs_2[Pt(NO_2)(NH_3)Br_2][Pt(NO_2)(NH_3)Br_4]$

CAESIUM AMMINEDIIODONITROPLATINATE(II) AMMINETETRAIODONITROPLATINATE(IV)
$Cs_2[Pt(NO_2)(NH_3)I_2][Pt(NO_2)(NH_3)I_4]$

R.J.H. CLARK, M. KURMOO, A.M.R. GALAS and M.B. HURSTHOUSE, 1981. Inorg. Chem., <u>20</u>, 4206-4212.

Orthorhombic, Pnam, a = 15.769, 16.738, b = 10.150, 10.543, c = 5.820, 6.063 Å,
D_m = 4.50, 4.79, Z = 4. Mo radiation, R = 0.036, 0.038 for 839, 1399 reflexions.

There is only one type of Pt, the structures (Fig. 1) containing trans-square-planar PtN_2X_2 groups bridged by two X atoms, each disordered over two positions. $Pt-NH_3$ = 2.076, 2.067, $Pt-NO_2$ = 2.064, 2.033, Pt-Br = 2.443, 2.492, Pt-I = 2.631, 2.711 Å.

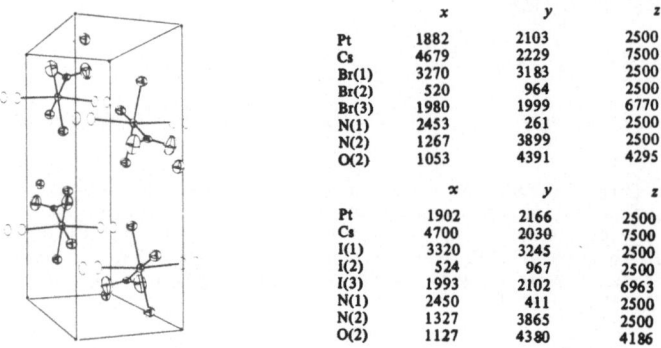

	x	y	z
Pt	1882	2103	2500
Cs	4679	2229	7500
Br(1)	3270	3183	2500
Br(2)	520	964	2500
Br(3)	1980	1999	6770
N(1)	2453	261	2500
N(2)	1267	3899	2500
O(2)	1053	4391	4295

	x	y	z
Pt	1902	2166	2500
Cs	4700	2030	7500
I(1)	3320	3245	2500
I(2)	524	967	2500
I(3)	1993	2102	6963
N(1)	2450	411	2500
N(2)	1327	3865	2500
O(2)	1127	4380	4186

Fig. 1.   Structure of the mixed-valence Pt compounds, and atomic positional parameters (x 10^4; Br(3) and I(3) have half-occupancy).

CAESIUM TETRABROMOAURATE(III) TRIBROMIDE
$Cs_3(AuBr_4)_2Br_3$

B. LEHNIS and J. STRÄHLE, 1981.   Z. Naturforsch., 36B, 1504-1508.

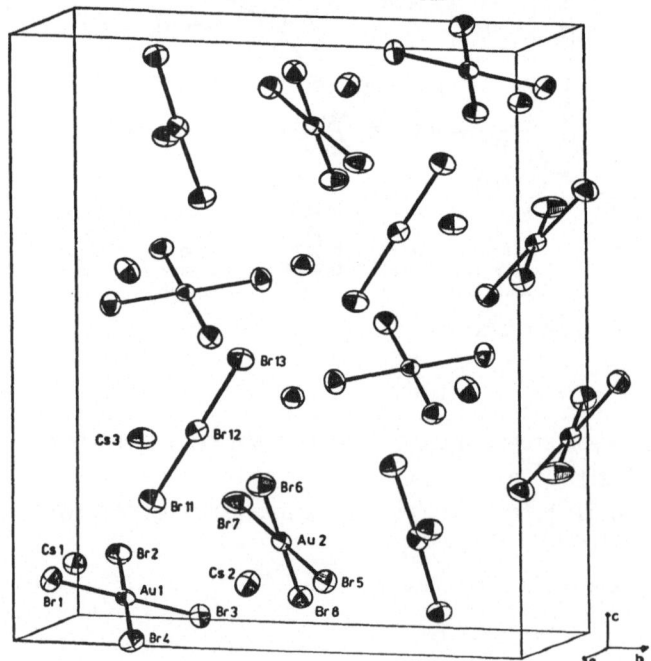

Fig. 1.   Structure of $Cs_3(AuBr_4)_2Br_3$.

Monoclinic, $P2_1/c$, a = 7.706, b = 16.063, c = 18.454 Å, β = 93.65°, Z = 4. Mo radiation, R = 0.045 for 2103 reflexions.

The structure (Fig. 1) contains square-planar $AuBr_4^-$ and linear Br_3^- anions, linked by 11-coordinate Cs^+ cations. Au-Br = 2.417-2.432(1), Br-Br = 2.540, 2.562(2), Cs-Br = 3.59-4.22 Å.

RUBIDIUM ZINC BROMIDE
Rb_3ZnBr_5

CAESIUM TETRABROMOZINCATE
Cs_2ZnBr_4

M. HEMING, G. LEHMANN, G. HENKEL and B. KREBS, 1981. Z. Naturforsch., 36A, 286-293.

Rb_3ZnBr_5, orthorhombic, Pnma, a = 9.079, b = 10.496, c = 13.443 Å, Z = 4. Mo radiation, R = 0.060 for 935 reflexions.

Cs_2ZnBr_4, orthorhombic, Pnma, a = 10.202, b = 7.738, c = 13.539 Å, Z = 4. Mo radiation, R = 0.066 for 963 reflexions.

Rb_3ZnBr_5 is isostructural with $(NH_4)_3ZnCl_5$ (1), containing $ZnBr_4$ tetrahedra, Zn-Br = 2.384-2.416(3) Å, a Br^- ion, and 8-coordinate Rb^+ ions, Rb-Br = 3.349-3.848(3) Å.

Cs_2ZnBr_4 is isostructural with Cs_2ZnCl_4 (2), with $ZnBr_4$ tetrahedra, Zn-Br = 2.394-2.401(3) Å, and 8- and 11-coordinate Cs^+ ions, Cs-Br = 3.594-4.272 Å.

1. Structure Reports, 9, 202; this volume p. 161.
2. Ibid., 23, 317.

SODIUM HEXABROMOURANATE(IV)
Na_2UBr_6

A. BOGACZ, J. BROS, M. GAUNE-ESCARD, A.W. HEWAT and J.C. TAYLOR, 1980. J. Phys. C: Solid State Phys., 13, 5273-5278.

Trigonal, $P\bar{3}m1$, a = 12.4368, c = 6.6653 Å, Z = 3. Neutron powder data.

Atomic positions

			x	y	z
Na	in	6(g)	0.3434	0	0
U(1)		1(a)	0	0	0
U(2)		2(d)	1/3	2/3	0.4799
Br(1)		6(i)	0.1084	2x	0.2448
Br(2)		6(i)	0.2300	2x	0.7302
Br(3)		6(i)	0.4365	2x	0.2449

Na_2SiF_6-like structure (1). U-Br = 2.72-2.85 (octahedral), Na-Br = 2.89-3.06 Å (octahedral).

1. Structure Reports, 19, 325; 29, 264.

MAGNESIUM OCTAIODIDE HEXAHYDRATE
$MgI_8 \cdot 6H_2O$

R. THOMAS and F.H. MOORE, 1981. Acta Cryst., B**31**, 2153-2155.

Triclinic, $P\bar{1}$, a = 8.909, b = 9.719, c = 7.725 Å, α = 114.46, β = 102.78, γ = 105.09°, D_m = 3.40, Z = 1. Neutron radiation, R = 0.086 for 561 reflexions.

The structure (Fig. 1) contains octahedral $Mg(OH_2)_6^{2+}$ cations and Z-shaped I_8^{2-} anions.

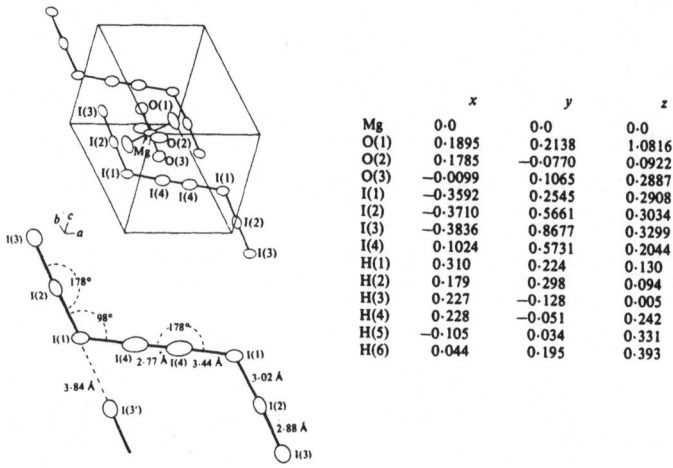

	x	y	z
Mg	0·0	0·0	0·0
O(1)	0·1895	0·2138	1·0816
O(2)	0·1785	−0·0770	0·0922
O(3)	−0·0099	0·1065	0·2887
I(1)	−0·3592	0·2545	0·2908
I(2)	−0·3710	0·5661	0·3034
I(3)	−0·3836	0·8677	0·3299
I(4)	0·1024	0·5731	0·2044
H(1)	0·310	0·224	0·130
H(2)	0·179	0·298	0·094
H(3)	0·227	−0·128	0·005
H(4)	0·228	−0·051	0·242
H(5)	−0·105	0·034	0·331
H(6)	0·044	0·195	0·393

Fig. 1.   Structure of magnesium octaiodide hexahydrate; Mg-O = 2.04-2.07 Å.

CALCIUM DECAIODIDE HEPTAHYDRATE
$CaI_{10} \cdot 7H_2O$

R. THOMAS and F.H. MOORE, 1981. Acta Cryst., B**37**, 2156-2159.

Triclinic, $P\bar{1}$, a = 9.830, b = 15.192, c = 9.572 Å, α = 91.42, β = 108.77, γ = 93.03°, D_m = 3.45, Z = 2. Neutron radiation, R = 0.093 for 1157 reflexions.

The structure (Fig. 1) contains a $Ca(OH_2)_7^{2+}$ cation (distorted capped trigonal prism) and two V-shaped I_5^- anions.

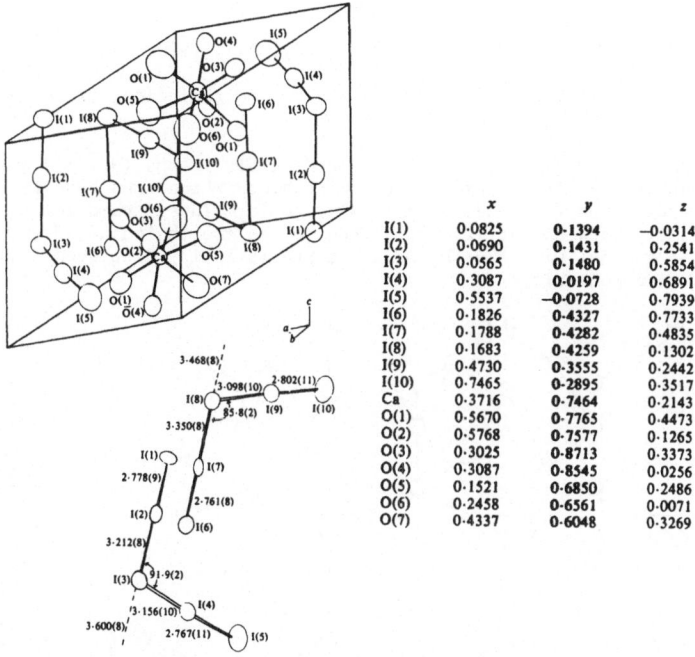

	x	y	z
I(1)	0·0825	0·1394	−0·0314
I(2)	0·0690	0·1431	0·2541
I(3)	0·0565	0·1480	0·5854
I(4)	0·3087	0·0197	0·6891
I(5)	0·5537	−0·0728	0·7939
I(6)	0·1826	0·4327	0·7733
I(7)	0·1788	0·4282	0·4835
I(8)	0·1683	0·4259	0·1302
I(9)	0·4730	0·3555	0·2442
I(10)	0·7465	0·2895	0·3517
Ca	0·3716	0·7464	0·2143
O(1)	0·5670	0·7765	0·4473
O(2)	0·5768	0·7577	0·1265
O(3)	0·3025	0·8713	0·3373
O(4)	0·3087	0·8545	0·0256
O(5)	0·1521	0·6850	0·2486
O(6)	0·2458	0·6561	0·0071
O(7)	0·4337	0·6048	0·3269

Fig. 1.   Structure of $CaI_{10}.7H_2O$; Ca-O = 2.42-2.49 Å.

BARIUM IODIDE
BaI_2

H.P. BECK, 1981. Z. Naturforsch., 36B, 1255-1260.

Normal-pressure form
Orthorhombic, Pbnm, a = 10.685, b = 8.904, c = 5.298 Å, Z = 4. Ba, I(1), I(2) in 4(c): (0.1215,0.2366,1/4), (0.4265,0.1393,1/4), (-0.1613,0.0290,1/4) (from 1).

High-pressure form
Hexagonal, $P\bar{6}2m$, a = 9.147, c = 5.173 Å, Z = 3. Powder data. Ba(1) in 1(b): 0,0,1/2; Ba(2) in 2(c): 1/3,2/3,0; I(1) in 3(f): 0.256,0,0; I(2) in 3(g): 0.594,0,1/2.

The normal-pressure form has a $PbCl_2$-type structure (1, 2), and the high-pressure form an anti-Fe_2P type structure (3).

1. Structure Reports, 28, 77.
2. Strukturbericht, 2, 16; Structure Reports, 42A, 184.
3. Structure Reports, 45A, 86.

NITROGEN IODIDE AMMONIA
I[NI$_4$].NH$_3$

M. PLEWA and K.-F. TEBBE, 1981. Z. anorg. Chem., <u>477</u>, 7-20.

Hexagonal, P6$_3$mc, a = 8.425, c = 8.765 Å, Z = 2. Mo radiation, R = 0.047 for 496 reflexions.

The structure (Fig. 1) contains tetrahedral NI$_4$ groups (N-I = 2.190 (x 3), 2.236 Å), connected by three corners to iodine atoms which thus have trigonal pyramidal coordination (I-I = 3.088 Å (x 3)), to form layers parallel to (00.1). NH$_3$ molecules lie between the layers, linked to the fourth iodine of the NI$_4$ group, I-N = 2.289 Å; a model with the NH$_3$ molecule statistically displaced from the three-fold axis is also possible.

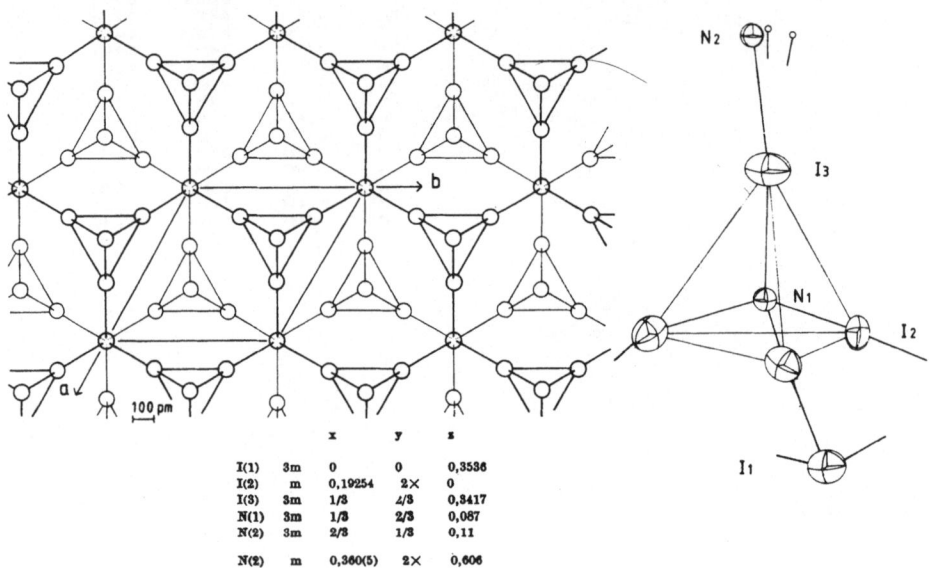

		x	y	z
I(1)	3m	0	0	0,3536
I(2)	m	0,19254	2×	0
I(3)	3m	1/3	2/3	0,3417
N(1)	3m	1/3	2/3	0,087
N(2)	3m	2/3	1/3	0,11
N(2)	m	0,360(5)	2×	0,606

Fig. 1.   Structure of I[NI$_4$].NH$_3$.

ARSENIC(III) IODIDE
AsI$_3$

A.S. KANIŠČEVA, Ju.N. MIKHAJLOV and A.P. ČERNOV, 1980. Izv. Akad. Nauk SSSR, Neorg. Mater., <u>16</u>, 1885-1886.

Rhombohedral, R$\overline{3}$, a = 7.215, c = 21.468 Å, Z = 6. R = 0.068. Structure as previously determined (<u>1</u>).

<u>1</u>. Structure Reports, <u>31</u>A, 239; <u>46</u>A, 197.

ZIRCONIUM(II) IODIDE
α-ZrI$_2$

D.H. GUTHRIE and J.D. CORBETT, 1981. J. Solid State Chem., $\underline{37}$, 256-263.

Monoclinic, P2$_1$/m, a = 6.821, b = 3.741, c = 14.937 Å, β = 95.66°, Z = 4. Mo radiation, R = 0.064 for 669 reflexions.

Isostructural with β-MoTe$_2$ ($\underline{1}$), a distorted CdI$_2$-type structure, with Zr atoms displaced 0.440 Å from octahedral centres along $\underline{a}$ to form infinite zigzag metal chains along $\underline{b}$ (Fig. 1), Zr-Zr = 3.182(3) Å. Another form contains Zr$_6$I$_{12}$ clusters ($\underline{2}$).

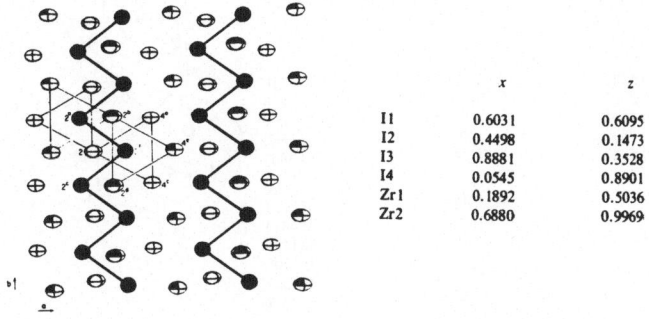

	x	z
I1	0.6031	0.6095
I2	0.4498	0.1473
I3	0.8881	0.3528
I4	0.0545	0.8901
Zr1	0.1892	0.5036
Zr2	0.6880	0.9969

Fig. 1.   Structure of α-ZrI$_2$, and atomic positional parameters (y = 1/4).

$\underline{1}$. Structure Reports, $\underline{31A}$, 69.
$\underline{2}$. Ibid., $\underline{44A}$, 153.

trans-TETRAAMMINEDIFLUOROCHROMIUM(III) IODIDE MONOHYDRATE
[Cr(NH$_3$)$_4$F$_2$]I.H$_2$O (I)

cis-TETRAAMMINEDIFLUOROCHROMIUM(III) PERCHLORATE
[Cr(NH$_3$)$_4$F$_2$]ClO$_4$ (II)

J.V. BRENČIČ, B. ČEH and I. LEBAN, 1981. Mh. Chem., $\underline{112}$, 1359-1368.

I.   Monoclinic, P2$_1$/m, a = 5.033, b = 16.333, c = 5.539 Å, β = 98.47°, D$_m$ = 2.20, Z = 2. Mo radiation, R = 0.065 for 884 reflexions.

II.   Tetragonal, I4$_1$md, a = 7.417, c = 16.610 Å, D$_m$ = 1.87, Z = 4. Mo radiation, R = 0.072 for 354 reflexions.

The structures (Fig. 1) contain octahedral cations, iodide or perchlorate anions, and water molecules (in (I)); there are no hydrogen bonds except for weak N-H...F interactions in (I). Cr-F = 1.89, Cr-N = 2.05-2.09, Cl-O = 1.40(1) Å.

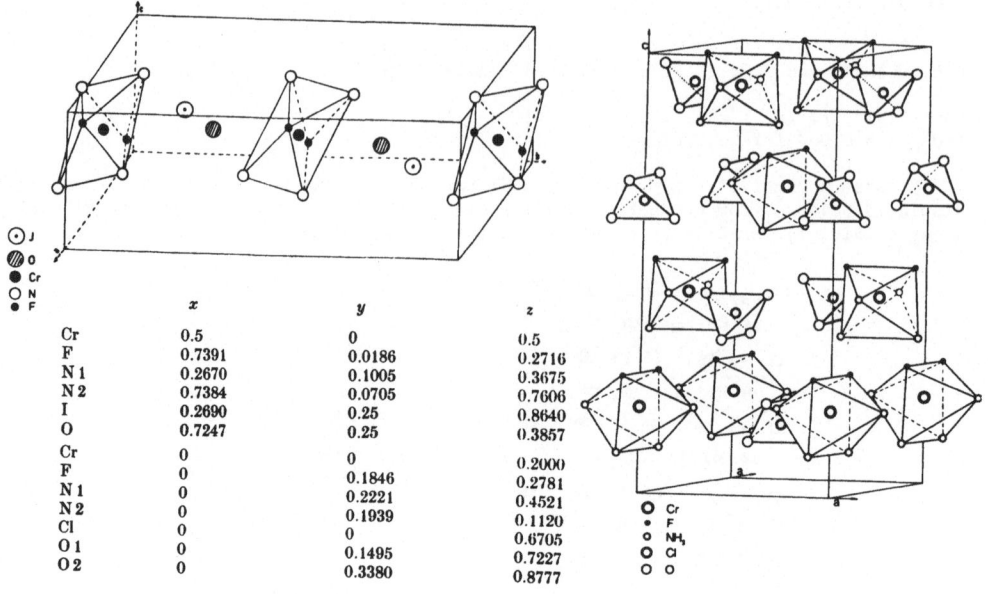

	x	y	z
Cr	0.5	0	0.5
F	0.7391	0.0186	0.2716
N 1	0.2670	0.1005	0.3675
N 2	0.7384	0.0705	0.7606
I	0.2690	0.25	0.8640
O	0.7247	0.25	0.3857
Cr	0	0	0.2000
F	0	0.1846	0.2781
N 1	0	0.2221	0.4521
N 2	0	0.1939	0.1120
Cl	0	0	0.6705
O 1	0	0.1495	0.7227
O 2	0	0.3380	0.8777

Fig. 1.   Structure of trans-[Cr(NH$_3$)$_4$F]I.H$_2$O (left) and
cis-[Cr(NH$_3$)$_4$F$_2$]ClO$_4$ (right), and atomic positional parameters.

TETRAAMMINEZINC(II) IODIDE
Zn(NH$_3$)$_4$I$_2$

T. YAMAGUCHI and O. LINDQVIST, 1981. Acta Chem. Scand., A35, 811-814.

Orthorhombic, Pmnb, a = 7.538, b = 10.396, c = 13.098 Å, Z = 4. Mo radiation, R = 0.034 for 1415 reflexions.

The structure (Fig. 1) contains tetrahedral cations, Zn-N = 2.00-2.03(1) Å, and iodide anions which have 9 N neighbours, I...N = 3.70-4.04 Å (possibly hydrogen bonds).

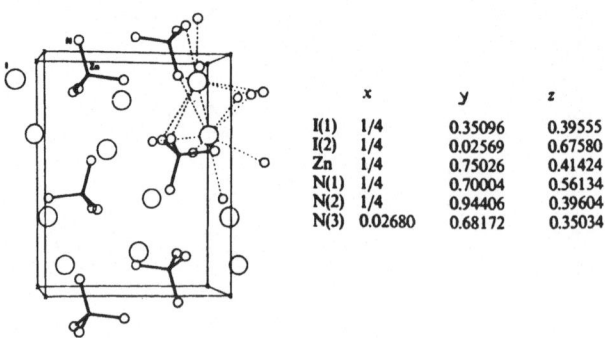

	x	y	z
I(1)	1/4	0.35096	0.39555
I(2)	1/4	0.02569	0.67580
Zn	1/4	0.75026	0.41424
N(1)	1/4	0.70004	0.56134
N(2)	1/4	0.94406	0.39604
N(3)	0.02680	0.68172	0.35034

Fig. 1.   Structure of tetraamminezinc (II) iodide.

CADMIUM ARSENIC IODIDE
CADMIUM PHOSPHORUS IODIDE
CADMIUM PHOSPHORUS CHLORIDE
Cd_3AsI_3 Cd_3PI_3 Cd_3PCl_3

A. REBBAH, J. YAZBECK and A. DESCHANVRES, 1981. Rev. Chim. Miner., <u>18</u>, 43-53.

Cd_3AsI_3, Cd_3PI_3
Hexagonal, $P6_3mc$, a = 4.422, 4.466, c = 7.24, 7.302 Å, D_m = 5.10, 5.24, Z = 1/2.
Cu radiation, powder data. 1.5 Cd + 0.5 □ in 2(b): 0,0,0; 1.5 I + 0.5 As or P in
2(b): z = 0,0,0.373 (origin shifted by 2/3,1/3,0).

Cd_3PCl_3
Trigonal, $P\overline{3}$, a = 7.633, c = 7.133 Å, D_m = 4.40, Z = 2. Cu radiation, powder data.
Cd(1) in 1(b): 0,0,1/2; Cd(2) in 3(e): 1/2,0,0; Cd(3) in 2(d): 1/3,2/3,0.4935; P in
2(d): 1/3,2/3,-0.1597; Cl in 6(g): 0.1669,0.3324,0.3316.

   Cd_3AsI_3 and Cd_3PI_3 have a wurtzite-type structure. The Cd_3PCl_3 structure is shown
in Fig. 1; this compound also exists in an orthorhombic modification, isostructural
with Cd_3AsCl_3 (<u>1</u>).

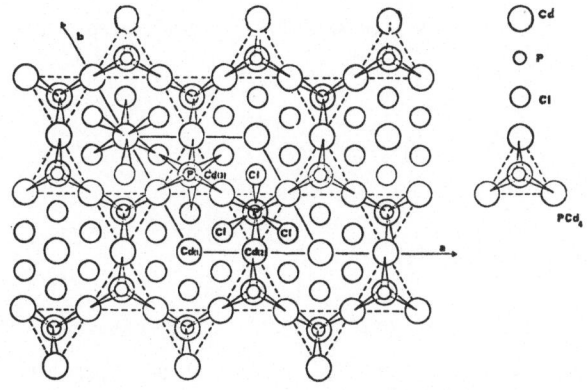

Fig. 1.   Structure of Cd_3PCl_3 ($P\overline{3}$).

<u>1</u>. Structure Reports, <u>46A</u>, 179.

TRIIODOTELLURIUM(IV) HEXAFLUOROARSENATE(V)
TeI_3AsF_6

J. PASSMORE, G. SUTHERLAND and P.S. WHITE, 1981. Canad. J. Chem., <u>59</u>, 2876-2878.

Monoclinic, $P2_1/c$, a = 8.243, b = 10.755, c = 12.660 Å, β = 100.91°, Z = 4. Mo
radiation, R = 0.075 for 1551 reflexions.

   Isostructural with $TeBr_3AsF_6$ (<u>1</u>), the structure (Fig. 1) containing trigonal
pyramidal TeI_3^+ and octahedral AsF_6^- ions, with some cation-anion interactions.

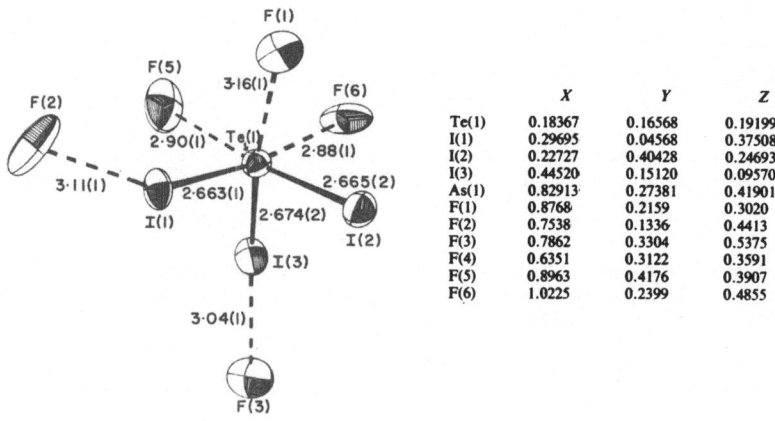

	X	Y	Z
Te(1)	0.18367	0.16568	0.19199
I(1)	0.29695	0.04568	0.37508
I(2)	0.22727	0.40428	0.24693
I(3)	0.44520	0.15120	0.09570
As(1)	0.82913	0.27381	0.41901
F(1)	0.8768	0.2159	0.3020
F(2)	0.7538	0.1336	0.4413
F(3)	0.7862	0.3304	0.5375
F(4)	0.6351	0.3122	0.3591
F(5)	0.8963	0.4176	0.3907
F(6)	1.0225	0.2399	0.4855

Fig. 1.   Structure of TeI_3AsF_6 (mean I-Te-I = 99.9°, mean As-F = 1.70 Å).

1. Structure Reports, 46A, 195.

CAESIUM ENNEAIODODIYTTRATE(III)
$Cs_3Y_2I_9$

CAESIUM ENNEAIODODIZIRCONATE(III)
$Cs_3Zr_2I_9$

D.H. GUTHRIE, G. MEYER and J.D. CORBETT, 1981. Inorg. Chem., 20, 1192-1196.

Hexagonal, $P6_3/mmc$, a = 8.406, 8.269, c = 21.280, 19.908 Å, Z = 2. Mo radiation, R = 0.050, 0.082 for 321, 610 reflexions.

Isostructural with $Cs_3Cr_2Cl_9$ (1), the transition from Y^{3+} (d^0) to Zr^{3+} (d^1) being accompanied by a decrease in the M-M distance as a result of Zr-Zr bond formation (Fig. 1).

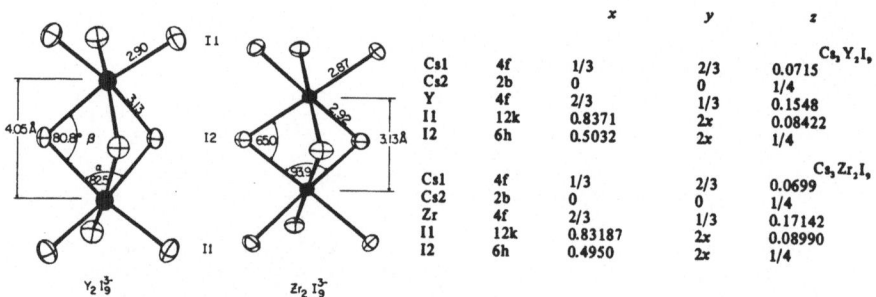

		x	y	z	
					$Cs_3Y_2I_9$
Cs1	4f	1/3	2/3	0.0715	
Cs2	2b	0	0	1/4	
Y	4f	2/3	1/3	0.1548	
I1	12k	0.8371	2x	0.08422	
I2	6h	0.5032	2x	1/4	
					$Cs_3Zr_2I_9$
Cs1	4f	1/3	2/3	0.0699	
Cs2	2b	0	0	1/4	
Zr	4f	2/3	1/3	0.17142	
I1	12k	0.83187	2x	0.08990	
I2	6h	0.4950	2x	1/4	

Fig. 1.   Structures of $Cs_3Y_2I_9$ and $Cs_3Zr_2I_9$.

1. Structure Reports, 21, 277.

CAESIUM TRIIODOVANADATE(II)
RUBIDIUM TRIIODOVANADATE(II)
RUBIDIUM TRIIODOTITANATE(II)
$CsVI_3$ $RbVI_3$ $RbTiI_3$

H.W. ZANDBERGEN, 1981. J. Solid State Chem., $\underline{37}$, 308-317.

$CsVI_3$, hexagonal, $P6_3/mmc$, a = 8.124, 8.062, c = 6.774, 6.704 Å at 293, 1.2K, Z = 2.
Neutron powder data.

$RbVI_3$, $RbTiI_3$, hexagonal, $P6_3cm$ or $P\bar{3}c1$, a = 13.863, 13.750, 14.024, c = 6.807, 6.761,
6.796 Å, for $RbVI_3$ at 293, 1.2K, $RbTiI_3$ at 293K, Z = 6. Neutron powder data.

Atomic positions

$CsVI_3$

293 K
a = 8.124(1) Å c = 6.774(1) Å

	x	y	z
Cs	$\frac{1}{3}$	$\frac{2}{3}$	$\frac{1}{4}$
V	0	0	0
I	0.169	−0.169	$\frac{1}{4}$

1.2K
a = 8.062(1) Å c = 6.704(1) Å

Cs	$\frac{1}{3}$	$\frac{2}{3}$	$\frac{1}{4}$
V	0	0	0
I	0.1734	−0.1734	$\frac{1}{4}$

$RbTiI_3$
AT 293

a = 14.024(3) Å c = 6.796(2) Å

	x	y	z
Rb	0.327	0.327	0.280
Ti(1)	0	0	0
Ti(2)	$\frac{1}{3}$	$\frac{2}{3}$	0.086
I(1)	0.172	0	$\frac{1}{4}$
I(2)	0.518	0.166	0.336

$RbVI_3$

293 K
$P6_3cm$ a = 13.863(2) Å c = 6.807(1) Å

	x	y	z
Rb	0.330	0.330	0.294
V(1)	0	0	0
V(2)	$\frac{1}{3}$	$\frac{2}{3}$	0.096
I(1)	0.166	0	$\frac{1}{4}$
I(2)	0.503	0.171	0.346

$P\bar{3}c1$ a = 13.863(2) Å c = 6.807(1) Å

Rb	0.330	0.330	$\frac{1}{4}$
V(1)	0	0	0
V(2)	$\frac{1}{3}$	$\frac{2}{3}$	0.057
I(1)	0.169	0	$\frac{1}{4}$
I(2)	0.506	0.170	0.193

1.2 K
$P6_3cm$ a = 13750(2) Å c = 6.761(2) Å

Rb	0.333	0.333	0.277
V(1)	0	0	0
V(2)	1.3	$\frac{2}{3}$	0.101
I(1)	0.161	0	$\frac{1}{4}$
I(2)	0.505	0.168	0.351

$P\bar{3}c1$ a = 13.750(2) Å c = 6.761(2) Å

Rb	0.334	0.334	$\frac{1}{4}$
V(1)	0	0	0
V(2)	$\frac{1}{3}$	$\frac{2}{3}$	0.063
I(1)	0.161	0	$\frac{1}{4}$
I(2)	0.510	0.168	0.187

$CsVI_3$ has the $BaNiO_3$ structure ($\underline{1}$). $RbVI_3$ and $RbTiI_3$ have distorted $BaNiO_3$
structures, with displacement of VI_3 chains along $\underline{c}$. V-I = 2.83-2.92 Å.

$\underline{1}$. Structure Reports, $\underline{42A}$, 287.

RUBIDIUM CHROMIUM(II) IODIDE
$RbCrI_3$

H.W. ZANDBERGEN and D.J.W. IJDO, 1981. J. Solid State Chem., $\underline{38}$, 199-210.

β-Form, 293K, monoclinic, C2/m, a = 13.772, b = 8.000, c = 7.069 Å, β = 95.85°, Z =
4. Neutron powder data.

γ-Form, 1.2K, monoclinic, C2, a = 13.586, b = 7.923, c = 14.094 Å, β = 96.88°, Z = 8.
Neutron powder data.

Atomic positions

	x	y	z

β-RbCrI₄ (293 K)
$a = 13.772(3), b = 8.000(2),$
$c = 7.069(2)$ Å, $\beta = 95.85(1)°$

	x	y	z
Rb	−0.3384	0	0.7430
Cr(1)	0	0	0
Cr(2)	0	0	0.5
I(1)	0.1686	0	0.1820
I(2)	0.0839	0.2446	0.7120
I(2a)	0.0857	0.2576	0.6856
I(2b)	0.0827	0.2331	0.7384

γ-RbCrI₄ (1.2 K)

$a = 13.586(2), b = 7.923(2),$
$c = 14.094(3)$ Å, $\beta = 96.88(1)°$

	x	y	z
Rb(a)	0.350	0.036	0.391
Rb(b)	0.326	0.015	0.876
Cr(1a)	0	0	0
Cr(1b)	0	0.063	0.5
Cr(2)	−0.002	0.035	0.249
I(1a)	0.166	0.023	0.164
I(1b)	0.176	0.061	0.667
I(2a)	0.085	0.310	0.401
I(2b)	0.089	0.252	0.909
I(3a)	0.085	−0.185	0.391
I(3b)	0.082	−0.254	0.881

Isostructural with the corresponding chloride phases (1), all having Jahn-Teller-distorted $BaNiO_3$ structures, with elongated face-sharing CrX_6 octahedra (observation of one flattened octahedron in the β-form suggests disorder of I(2)). Cr-I = 2.70-2.93 (4 distances), 3.00-3.18 Å (2 distances).

1. Structure Reports, 44A, 158; 45A, 171.

THALLIUM(I) CHROMIUM(II) IODIDE
Tl₄CrI₆

H.W. ZANDBERGEN, 1981. J. Solid State Chem., 38, 239-245.

α-Form, 293K, tetragonal, P4/mnc, a = 9.132, c = 9.661 Å, Z = 2. Neutron powder data at 293 and 77K.

β-Form, 4.2K, orthorhombic, Cccm, a = 12.941, b = 12.596, c = 9.602 Å, Z = 4. Neutron powder data at 4.2 and 1.2K.

Atomic positions

	α-Form				β-Form		
	x	y	z			4.2 K	
					$a = 12.941(3), b = 12.596(3), c = 9.602(2)$ Å,		
	293 K						
	$a = 9.132(1), c = 9.661(1)$ Å						
Tl	0.1449	0.3551	0.25	Tl(1)	0.1406	0	0.25
Cr	0	0	0	Tl(2)	0	0.3524	0.25
I(1)	0	0	0.2839	Cr	0.25	0.25	0
I(2)	0.3049	0.1371	0	I(1)	0.25	0.25	0.2896
I(2a)	0.3219	0.1436	0	I(2)	0.3268	0.0345	0
I(2b)	0.2879	0.1307	0	I(3)	0.0191	0.1692	0
	77K					1.2 K	
	$a = 9.013(1), b = 9.580(1)$ Å				$a = 12.927(3), b = 12.584(3), c = 9.593(2)$ Å,		
Tl	0.1445	0.3555	0.25	Tl(1)	0.1414	0	0.25
Cr	0	0	0	Tl(2)	0	0.3522	0.25
I(1)	0	0	0.2871	Cr	0.25	0.25	0
I(2)	0.3032	0.1390	0	I(1)	0.25	0.25	0.2901
I(2a)	0.3233	0.1517	0	I(2)	0.3257	0.0348	0
I(2b)	0.2831	0.1263	0	I(3)	0.0189	0.1692	0

The α-form structure is as previously described (1), with a random distribution of Jahn-Teller-distorted octahedra, Cr-I = 2.74-2.88 (4 distances), 3.22 Å (2 distances). In the β-form the octahedra are ordered, Cr-I = 2.78, 2.89, 3.16 (each x 2) Å.

1. Structure Reports, 45A, 185.

THALLIUM(I) TRIIODOMANGANATE(II)
THALLIUM(I) TRIIODOFERRATE(II)
TlMI$_3$ (M = Mn, Fe)

H.W. ZANDBERGEN, 1981. J. Solid State Chem., 37, 189-203.

Orthorhombic, Pnma, a = 10.074, 9.991, 9.967, 9.884, b = 4.2967, 4.2701, 4.2407, 4.2134, c = 16.172, 16.020, 15.981, 15.823 Å, for Mn compound at 293, 4.2K and Fe compound at 293, 4.2K, respectively, Z = 4. Neutron powder data.

Atomic positions

TlMnI$_3$, AT 293 AND 4.2 K

293 K (a = 10.074(1), b = 4.2967(5), c = 16.172(2) Å)

	x	y	z
Tl	0.4418(5)	0.25	0.1753
Mn	0.1628(9)	0.25	0.9421
I(1)	0.2810(6)	0.25	0.7884
I(2)	0.1692(5)	0.25	0.5079
I(3)	0.0245(6)	0.25	0.1052

4.2 K (a = 9.991(2), b = 4.2701(8), c = 16.020(3) Å)

	x	y	z
Tl	0.4454	0.25	0.1745
Mn	0.1626	0.25	0.9439
I(1)	0.2834	0.25	0.7857
I(2)	0.1670	0.25	0.5078
I(3)	0.0216	0.25	0.1049

TlFeI$_3$, AT 293 AND 4.2 K

	x	y	z

293 K (a = 9.967(1), b = 4.2407(5), c = 15.981(1) Å)

	x	y	z
Tl	0.4442(5)	0.25	0.1734
Fe	0.1619(4)	0.25	0.9458
I(1)	0.2772(6)	0.25	0.7885
I(2)	0.1714(6)	0.25	0.5068
I(3)	0.0222(6)	0.25	0.1043

4.2 K (a = 9.884(1), b = 4.2134(5), c = 15.823(1) Å)

	x	y	z
Tl	0.4458	0.25	0.1732
Fe	0.1628	0.25	0.9472
I(1)	0.2763	0.25	0.7871
I(2)	0.1721	0.25	0.5059
I(3)	0.0200	0.25	0.1034

Isostructural with NH$_4$CdCl$_3$ (1). The magnetic structures are described.

1. Strukturbericht, 6, 13, 79; 7, 19, 115.

SODIUM PENTAFLUOROPEROXONIOBATE(V) DIHYDRATE
Na$_2$[NbF$_5$(O$_2$)].2H$_2$O

R. STOMBERG, 1981. Acta Chem. Scand., A35, 489-495.

Orthorhombic, C2cb, a = 6.890, b = 10.263, c = 10.361 Å, Z = 4. Mo radiation, R = 0.035 for 708 reflexions.

The structure (Fig. 1) contains a pentagonal bipyramidal anion, with 3 F and the peroxo group equatorial; the anion is disordered in two orientations related by 90° rotation about a. Na has 7 F and O neighbours at 2.25-2.50 Å, and there is no significant hydrogen bonding.

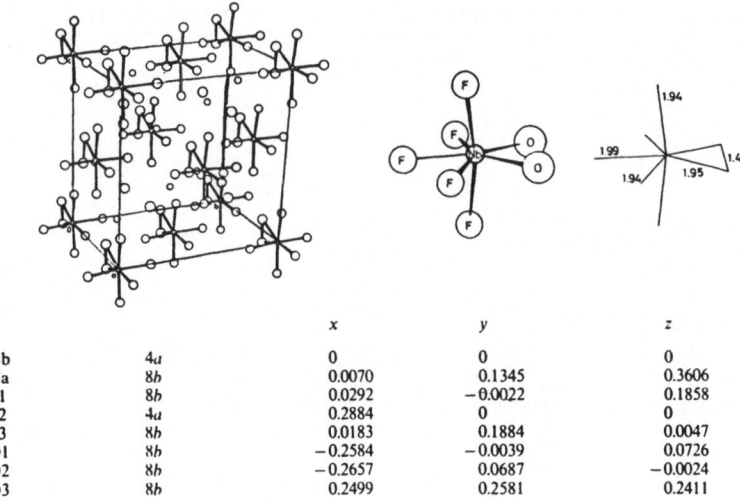

		x	y	z
Nb	4a	0	0	0
Na	8h	0.0070	0.1345	0.3606
F1	8h	0.0292	−0.0022	0.1858
F2	4a	0.2884	0	0
F3	8h	0.0183	0.1884	0.0047
O1	8h	−0.2584	−0.0039	0.0726
O2	8h	−0.2657	0.0687	−0.0024
O3	8h	0.2499	0.2581	0.2411

Fig. 1.   Structure of $Na_2[NbF_5(O_2)].2H_2O$. Only one of the two anion
           orientations is shown; O(1) and O(2) have occupancy = 0.5.

SODIUM PENTAFLUOROPEROXONIOBATE(V) HYDROGENDIFLUORIDE
$Na_3[NbF_5(O_2)][HF_2]$

R. STOMBERG, 1981. Acta Chem. Scand., A35, 389-394.

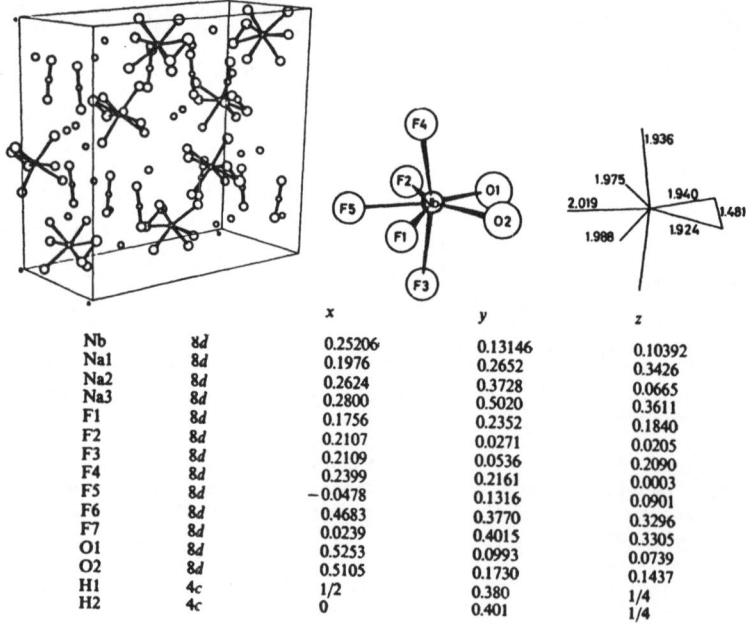

		x	y	z
Nb	8d	0.25206	0.13146	0.10392
Na1	8d	0.1976	0.2652	0.3426
Na2	8d	0.2624	0.3728	0.0665
Na3	8d	0.2800	0.5020	0.3611
F1	8d	0.1756	0.2352	0.1840
F2	8d	0.2107	0.0271	0.0205
F3	8d	0.2109	0.0536	0.2090
F4	8d	0.2399	0.2161	0.0003
F5	8d	−0.0478	0.1316	0.0901
F6	8d	0.4683	0.3770	0.3296
F7	8d	0.0239	0.4015	0.3305
O1	8d	0.5253	0.0993	0.0739
O2	8d	0.5105	0.1730	0.1437
H1	4c	1/2	0.380	1/4
H2	4c	0	0.401	1/4

Fig. 1.   Structure of $Na_3[NbF_5(O_2)][HF_2]$.

Orthorhombic, Pbcn, a = 6.703, b = 14.982, c = 14.092 Å, Z = 8. Mo radiation, R = 0.022 for 1694 reflexions.

The structure (Fig. 1) contains pentagonal-bipyramidal $NbF_5O_2^{2-}$ ions, linear FHF^- ions, and Na^+ ions with 6- and 7-coordinations to F and O. F...H...F = 2.29 Å.

RUBIDIUM POTASSIUM FLUOROMOLYBDATE
$Rb_2KMoO_3F_3$

S.C. ABRAHAMS, J.L. BERNSTEIN and J. RAVEZ, 1981. Acta Cryst., B37, 1332-1336.

Above 328K, cubic, Fm3m, a = 8.945 Å, at 328K, D_m (298K) = 3.81, Z = 4. Mo radiation, R = 0.032, 0.056 for 97, 81 reflexions at 343, 473K, respectively. Rb in 8(c): 1/4, 1/4,1/4; K in 4(b): 1/2,1/2,1/2; Mo in 4(a): 0,0,0; O/F in 24(e): x,0,0, x = 0.2105, 0.2119.

The structure (Fig. 1) is similar to that of K_2PtCl_6, but with additional K in site 4(b); O and F randomly occupy the 24(e) site. Mo-O/F = 1.90(1), 1.93(1) Å at 343, 473K (corrected for thermal motion), Rb-O/F = 3.19, 3.20 (12-coordination), K-O/F = 2.59, 2.59 Å (6-coordination).

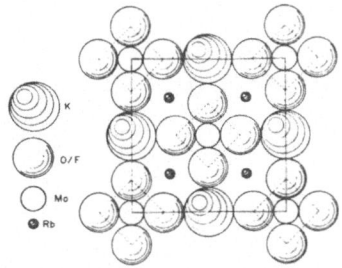

Fig. 1. (001) Plane of $Rb_2KMoO_3F_3$ structure (Rb radius is about 12% greater than that of K).

MOLYBDENUM TETRAFLUORIDE OXIDE - ANTIMONY PENTAFLUORIDE
RHENIUM TETRAFLUORIDE OXIDE - ANTIMONY PENTAFLUORIDE
$MF_4O.SbF_5$ (M = Mo, Re)

J. FAWCETT, J.H. HOLLOWAY and D.R. RUSSELL, 1981. J. Chem. Soc., Dalton, 1212-1218.

M = Mo, monoclinic, $P2_1/n$, a = 7.470, b = 10.40, c = 9.606 Å, β = 93.13°, Z = 4. Mo radiation, R = 0.041 for 1123 reflexions.

M = Re, monoclinic, $P2_1/c$, a = 5.561, b = 10.198, c = 12.622 Å, β = 99.37°, Z = 4. Mo radiation, R = 0.058 for 1075 reflexions.

The Mo compound contains a polymeric zigzag chain, whereas the Re compound contains dimeric molecules with an eight-membered ring (Fig. 1). Bond distances in the Mo compound are in Fig. 1; Mo-F-Sb = 162°. For the Re compound, Re-F = 2.08, 2.23 (bridging), 1.73, 1.83, 1.85 (terminal), Re-O = 1.66, Sb-F = 1.96, 2.04 (bridging), 1.80-1.86 Å (terminal), Re-F-Sb = 138, 148°.

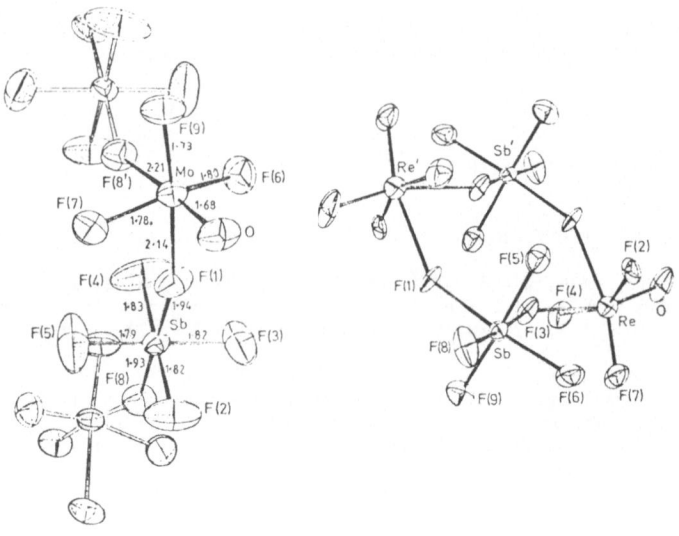

Fig. 1. $MF_4O.SbF_5$, M = Mo (left) and Re (right).

SODIUM URANYL FLUORIDE HYDRATE
$Na_3(UO_2)_2F_7.6H_2O$

NGUYEN QUY DAO, S. CHOUROU and J. HECKLY, 1981. J. Inorg. Nucl. Chem., **43**, 1835-1839.

Triclinic, $P\bar{1}$, a = 6.997, b = 7.176, c = 8.630 Å, α = 77.84, β = 113.30, γ = 104.95°, D_m = 4.01, Z = 1. Mo radiation, R = 0.039 for 2592 reflexions.

	x	y	z
U	0,00042	0,01014	0,26664
Na 1	0,7845	0,5231	0,0225
Na 2	0	0,5	0,5
F 1	0	0	0
F 2	-0,0175	-0,3047	0,2474
F 3	0,0221	0,3159	0,1471
F 4	0,0097	0,1820	0,4738
O 1	-0,283	-0,025	0,182
O 2	0,280	0,039	0,352
O 3	0,367	0,671	0,220
O 4	0,375	0,598	0,583
O 5	0,497	0,301	0,090

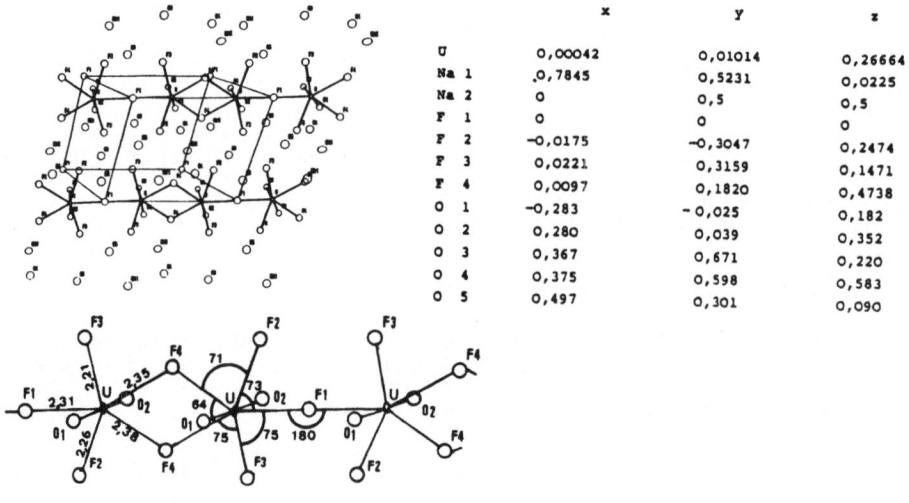

Fig. 1.   Structure of $Na_3(UO_2)_2F_7.6H_2O$.

The structure (Fig. 1) contains $[(UO_2)_2F_6F_{2/2}{}^{3-}]_\infty$ chains, linked by six-coordinate Na^+ ions and water molecules.

AMMONIUM CAESIUM DIURANYL HEPTAFLUORIDE
$NH_4Cs_2(UO_2)_2F_7$

S.B. IVANOV, Ju.N. MIKHAJLOV, V.G. KUZNECOV and R.L. DAVIDOVIČ, 1980. Koordin. Khim., 6, 1746-1750.

Orthorhombic, Immm, a = 6.526, b = 8.553, c = 12.434 Å, D_m = 4.62, Z = 2. R = 0.068.

The structure contains $UO_2F_5{}^{3-}$ pentagonal bipyramids, which share edges and corners to form $[(UO_2)_2F_7{}^{3-}]_n$ chains.

NICKEL URANYL FLUORIDE HYDRATE
$Ni_3[(UO_2)_2F_7]_2 \cdot 18H_2O$

S.B. IVANOV, Ju.N. MIKHAJLOV, V.G. KUZNECOV and R.L. DAVIDOVIČ, 1981. Ž. Strukt. Khim., 22, No. 2, 188-191 [J. Struct. Chem., 22, 302-304].

Monoclinic, $P2_1/b$, a = 9.131, b = 16.925, c = 12.500 Å, γ = 114.62°, Z = 2. Mo radiation, R = 0.067 for 2372 reflexions.

The structure contains UO_2F_5 pentagonal bipyramids, sharing an F...F edge to form pairs, which are joined into chains along b by sharing a F corner. $Ni(H_2O)_6{}^{2+}$ octahedra lie between the chains, with O-H...F hydrogen bonds being formed. U-O = 1.75-1.79 (apical), U-F = 2.23-2.39 (equatorial), Ni-O = 2.02-2.09, O-H...F = 2.61-2.82 Å.

TIN(II) CHLORIDE HYDROXIDE
$Sn_{21}Cl_{16}(OH)_{14}O_6$

H.G. von SCHNERING, R. NESPER and H. PELSHENKE, 1981. Z. Naturforsch., 36B, 1551-1560.

Rhombohedral, R32, a = 10.018, c = 44.030 Å, D_m = 4.42, Z = 3 (rhombohedral cell has a = 15.775 Å, α = 37.09°, Z = 1). Mo radiation, R = 0.071, 0.045 for 463, 452 reflexions from two twinned crystals.

The structure (Fig. 1) contains two types of layer, M1 = $[Sn_5(OH)_6Cl_6{}^{2-}]_\infty$ and M2 = $[Sn_8O_3(OH)_4Cl_3{}^{3+}]_\infty$, stacked along c and intercalated with Cl^- ions. M1 consists of condensed $Sn(OH)_3$ and $Sn(OH)_2Cl_2$ groups; M2 contains condensed $Sn(OH)_3$ and $Sn(OH)_2Cl$ groups and also a cubane-type $Sn_4O_3(OH)$ group linked by $Sn(OH)_3$ and SnCl fragments. One Sn has trigonal bipyramidal $Sn(OH)_2Cl_2E$ coordination (E = lone-pair) and the other Sn have trigonal-pyramidal coordination with lone-pairs completing tetrahedral geometry.

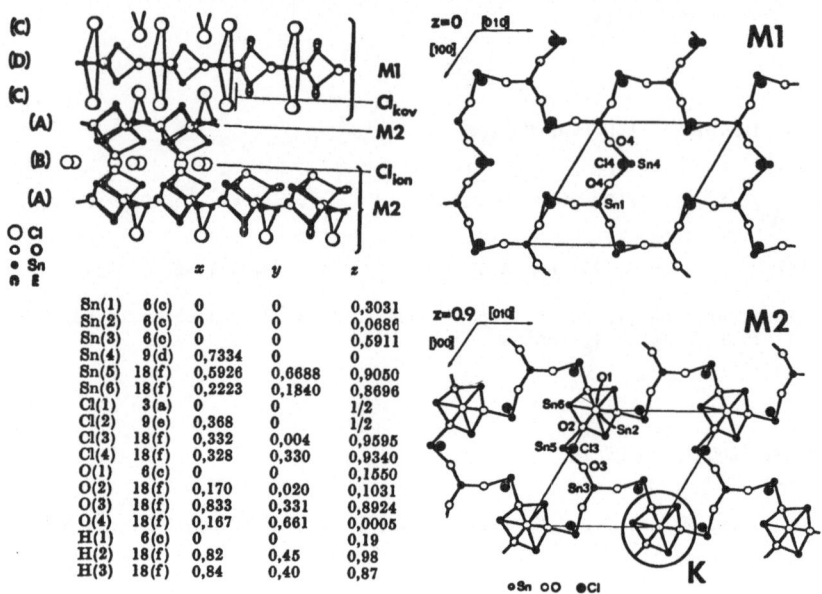

Fig. 1.   Structure of $Sn_{21}Cl_{16}(OH)_{14}O_6$.

HEMATOPHANITE

$Pb_4Fe_3O_8Cl$

J. PANNETIER and P. BATAIL, 1981.   J. Solid State Chem., **39**, 15-21.

Tetragonal, P4/mmm, a = 3.9097, c = 15.2873 Å, Z = 1.   Mo radiation, R = 0.061 for 365 reflexions (synthetic material).

Atomic positions

		x	y	z
Pb(1)	2h	½	½	0.1201
Pb(2)	2h	½	½	0.3802
Fe(1)	2g	0	0	0.2440
Fe(2)	1b	0	0	½
O(1)	4i	0	½	0.212
O(2)	2g	0	0	0.370
O(3)	2e	0	½	½
Cl	1a	0	0	0

The structure is essentially as previously described (1), but in a higher-symmetry space group. It contains incomplete perovskite-like $Pb_4Fe_3O_8$ sheets, and sheets of Cl^- ions. Fe(1) has fivefold square-pyramidal coordination to oxygen (Fe-O = 1.93 (x 4), 2.01 Å), and Fe(2) has octahedral coordination to oxygen (Fe-O = 1.95 (x 4), 1.98 (x 2) Å). Pb(1) has distorted square antiprismatic coordination to 4 O at 2.41 and 4 Cl at 3.32 Å, and Pb(2) has 12-coordination, Pb-O = 2.68, 2.77, 3.23 (each x 4) Å.

1. Structure Reports, **39A**, 194.

TELLURIUM OXYCHLORIDE
$Te_6O_{11}Cl_2$

I.   W. ABRIEL, 1981. Z. Naturforsch., <u>36B</u>, 405-409.

II. Idem, 1981. Z. Kristallogr., <u>156</u>, 8-9.

Monoclinic, $P2_1/m$, a = 6.844, b = 6.800, c = 15.227 Å, γ = 120.19°, Z = 2. Mo
radiation, R = 0.067 for 1179 reflexions.

The structure contains $(Te_{12}O_{22}^{4+})_\infty$ chains (Fig. 1). Te(1) has tetrahedral
coordination (3 O at 1.87-1.95 Å, plus the lone electron pair), Te(2) and Te(3) have
trigonal bipyramidal geometries (4 O at 1.88-2.16 Å, plus an equatorial lone pair).
Chains are linked by ionic Te...Cl interactions, 2.99-3.45 Å.

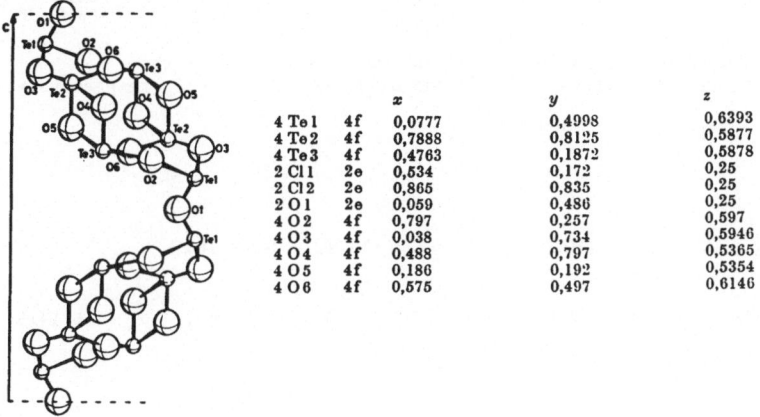

			x	y	z
4	Te 1	4f	0,0777	0,4998	0,6393
4	Te 2	4f	0,7888	0,8125	0,5877
4	Te 3	4f	0,4763	0,1872	0,5878
2	Cl 1	2e	0,534	0,172	0,25
2	Cl 2	2e	0,865	0,835	0,25
2	O 1	2e	0,059	0,486	0,25
4	O 2	4f	0,797	0,257	0,597
4	O 3	4f	0,038	0,734	0,5946
4	O 4	4f	0,488	0,797	0,5365
4	O 5	4f	0,186	0,192	0,5354
4	O 6	4f	0,575	0,497	0,6146

Fig. 1.   Structure of $Te_6O_{11}Cl_2$.

VANADYL CHLORIDE and BROMIDE
$VOCl_2$ $VOBr_2$

H.J. SEIFERT and J. UEBACH, 1981. Z. anorg. Chem., <u>479</u>, 32-40.

Orthorhombic, Immm, a = 3.360, 3.542, b = 11.750, 12.696, c = 3.838, 3.827 Å, Z = 2.
Powder data. V in 2(a): 0,0,0; O in 2(c): 1/2,1/2,0; Cl or Br in 4(h): 0,y,1/2, y =
0.360, 0.365 for Cl, Br, respectively.

The structures contain corner-sharing $trans$-VO_2X_4 octahedra, V-O = 1.92, V-Cl =
2.36, V-Br = 2.47 Å.

trans-AQUOTETRAAMMINECHROMIUM(III)-μ-HYDROXO-PENTAAMMINECHROMIUM(III) CHLORIDE
  TRIHYDRATE
$[(NH_3)_4(H_2O)Cr(OH)Cr(NH_3)_5]Cl_5 \cdot 3H_2O$

S.J. CLINE, J. GLERUP, D.J. HODGSON, G.S. JENSEN and E. PEDERSEN, 1981. Inorg. Chem.,
<u>20</u>, 2229-2233.

Monoclinic, C2/c, a = 23.855, b = 7.387, c = 16.763 Å, β = 129.8°, D_m = 1.53, Z = 4.
Mo radiation, R = 0.049 for 1638 reflexions.

The complex cation lies on a twofold axis (as for the decaammine complex, 1) and hence the trans aquo and ammine ligands are disordered (Fig. 1); the lattice water molecules exhibit partial occupancy. Cr-OH = 1.983(1), Cr-N = 2.062-2.084(4), Cr-O/N = 2.071(4) Å, Cr-O-Cr = 155°. The cation, chloride anions, and water molecules are linked by a system of hydrogen bonds.

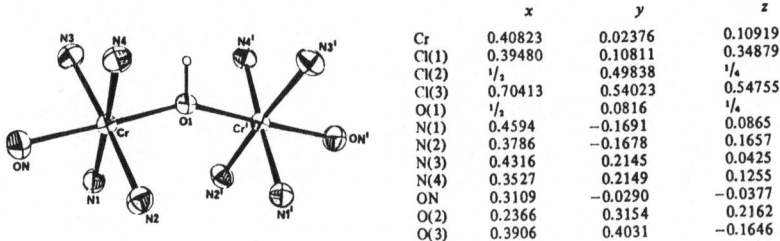

	x	y	z
Cr	0.40823	0.02376	0.10919
Cl(1)	0.39480	0.10811	0.34879
Cl(2)	¹/₂	0.49838	¹/₄
Cl(3)	0.70413	0.54023	0.54755
O(1)	¹/₂	0.0816	¹/₄
N(1)	0.4594	−0.1691	0.0865
N(2)	0.3786	−0.1678	0.1657
N(3)	0.4316	0.2145	0.0425
N(4)	0.3527	0.2149	0.1255
ON	0.3109	−0.0290	−0.0377
O(2)	0.2366	0.3154	0.2162
O(3)	0.3906	0.4031	−0.1646

Fig. 1.   The binuclear cation and atomic positional parameters; occupancy factors for O(2) and O(3) are 0.9 and 0.6, respectively.

1. Structure Reports, 39A, 193.

MOLYBDENUM(VI) OXIDE TETRACHLORIDE (HYDROGEN CHLORIDE SOLVATE)
MoOCl₄.HCl

M. MERCER, K.W. MUIR and D.W.A. SHARP, 1981. Z. Naturforsch., 36B, 1416-1418.

Monoclinic, P2₁/c, a = 6.635, b = 9.148, c = 23.849 Å, β = 93.09°, Z = 8. Mo radiation, R = 0.022 for 2018 reflexions.

The structure (Fig. 1) contains two independent square-pyramidal MoOCl₄ molecules, Mo-O = 1.64, Mo-Cl = 2.35-2.39(1) Å, linked by Cl bridging, Mo...Cl = 3.08, 2.99 Å, which completes distorted octahedral geometry at each Mo. Spherical holes probably contain HCl molecules.

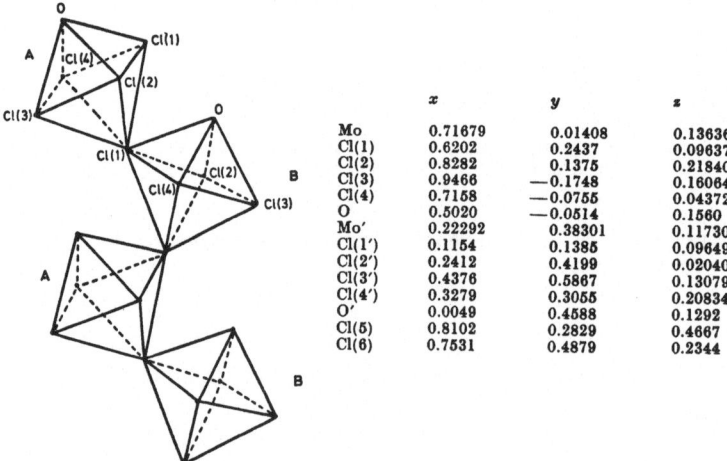

	x	y	z
Mo	0.71679	0.01408	0.13636
Cl(1)	0.6202	0.2437	0.09637
Cl(2)	0.8282	0.1375	0.21840
Cl(3)	0.9466	−0.1748	0.16064
Cl(4)	0.7158	−0.0755	0.04372
O	0.5020	−0.0514	0.1560
Mo′	0.22292	0.38301	0.11730
Cl(1′)	0.1154	0.1385	0.09649
Cl(2′)	0.2412	0.4199	0.02040
Cl(3′)	0.4376	0.5867	0.13079
Cl(4′)	0.3279	0.3055	0.20834
O′	0.0049	0.4588	0.1292
Cl(5)	0.8102	0.2829	0.4667
Cl(6)	0.7531	0.4879	0.2344

Fig. 1.   Structure of MoOCl₄ (solvated).

DI-μ-CHLORO-BIS[DICHLORO-OXO(PHOSPHORUS OXYTRICHLORIDE)MOLYBEDNUM]
[MoOCl$_3$.POCl$_3$]$_2$

J. STRÄHLE, G. BEYENDORFF, A. LIEBELT and K. DEHNICKE, 1981. Z. anorg. Chem., <u>474</u>, 171-181.

Monoclinic, P2$_1$/c, a = 6.264, b = 13.341, c = 12.089 Å, β = 100.27°, Z = 2 dimers. Mo radiation, R = 0.07 for 2327 reflexions.

The structure contains centrosymmetric dimers (Fig. 1). Mo-Cl = 2.468, 2.475 (bridging), 2.284, 2.307 (terminal), Mo-O(oxo) = 1.63 (triple bond), Mo-OP = 2.27, O-P = 1.44, P-Cl = 1.945-1.955 Å, Mo-O-P = 156°.

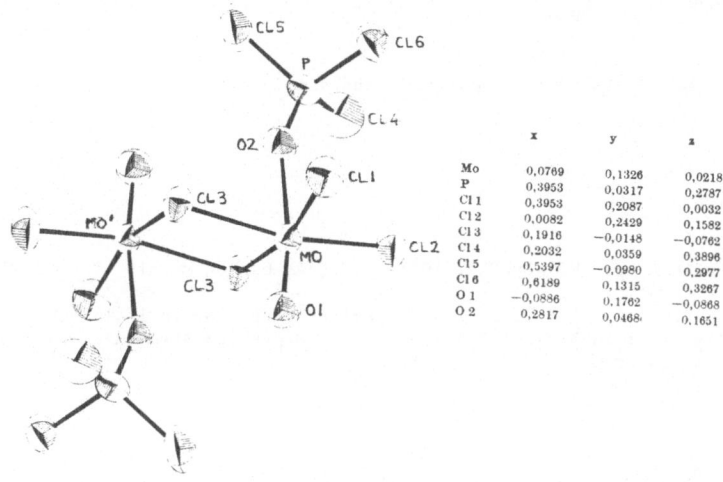

	x	y	z
Mo	0,0769	0,1326	0,0218
P	0,3953	0,0317	0,2787
Cl1	0,3953	0,2087	0,0032
Cl2	0,0082	0,2429	0,1582
Cl3	0,1916	-0,0148	-0,0762
Cl4	0,2032	0,0359	0,3896
Cl5	0,5397	-0,0980	0,2977
Cl6	0,6189	0,1315	0,3267
O1	-0,0886	0,1762	-0,0868
O2	0,2817	0,0468	0,1651

Fig. 1.   The [MoCl$_3$.POCl$_3$]$_2$ molecule, and atomic positional parameters.

DI-μ-CHLORO-BIS[DICHLORONITROSO(PHOSPHORUS OXYTRICHLORIDE)MOLYBDENUM]
[Mo(NO)Cl$_3$.POCl$_3$]$_2$

K. DEHNICKE, A. LIEBELT and F. WELLER, 1981. Z. anorg. Chem., <u>474</u>, 83-95.

Monoclinic, P2$_1$/c, a = 8.169, b = 11.045, c = 12.413 Å, β = 109.28°, Z = 2 dimers. Mo radiation, R = 0.040 for 1391 reflexions.

The structure contains centrosymmetric dimers (Fig. 1). Mo-Cl = 2.480, 2.510 (bridging), 2.273, 2.280 (terminal), Mo-N = 1.81 (double bond), N-O = 1.08 (very short), Mo-O = 2.21, O-P = 1.44, P-Cl = 1.940-1.945 Å, Mo-N-O = 178, Mo-O-P = 150°.

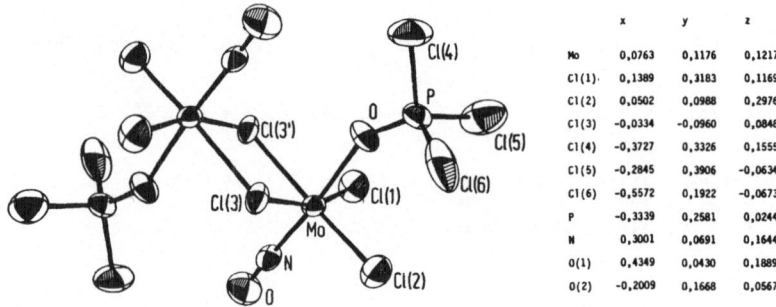

	x	y	z
Mo	0,0763	0,1176	0,1217
Cl(1)	0,1389	0,3183	0,1169
Cl(2)	0,0502	0,0988	0,2976
Cl(3)	-0,0334	-0,0960	0,0848
Cl(4)	-0,3727	0,3326	0,1555
Cl(5)	-0,2845	0,3906	-0,0634
Cl(6)	-0,5572	0,1922	-0,0673
P	-0,3339	0,2581	0,0244
N	0,3001	0,0691	0,1644
O(1)	0,4349	0,0430	0,1889
O(2)	-0,2009	0,1668	0,0567

Fig. 1.   The $[Mo(NO)Cl_3.POCl_3]_2$ molecule, and atomic positional parameters.

LITHIUM HYDROXIDE BROMIDE
$Li_2(OH)Br$

P. HARTWIG, A. RABENAU and W. WEPPNER, 1981.   J. Less-Common Metals, **78**, 227-233.

Cubic, Pm3m, a = 4.056 Å, D_m = 2.72, Z = 1.   Powder data.   Br in 1(a): 0,0,0; OH in
1(b): 1/2,1/2,1/2; 2 Li in 3(c): 1/2,1/2,0.   Anti-perovskite structure, with Li
sites only two-thirds occupied.

GERMANIUM SULPHIDE BROMIDE
$Ge_4S_6Br_4$

GERMANIUM SULPHIDE IODIDE
$Ge_4S_6I_4$

S. POHL, U. SEYER and B. KREBS, 1981.   Z. Naturforsch., **36B**, 1432-1436.

Bromide, triclinic, P$\bar{1}$, a = 8.806, b = 9.934, c = 10.106 Å, α = 86.05, β = 64.00,
γ = 89.87°, Z = 2.   Mo radiation, R = 0.078 for 2416 reflexions.

Iodide, trigonal, P$\bar{3}$, a = 10.640, c = 9.461 Å, Z = 2.   Mo radiation, R = 0.036 for
1052 reflexions.

    Both structures (Fig. 1) contain isolated adamantane-like molecules.   Ge-S =
2.20-2.22, Ge-Br = 2.27-2.28, Ge-I = 2.48, 2.49, S-Ge-S = 113, Ge-S-Ge = 101°.

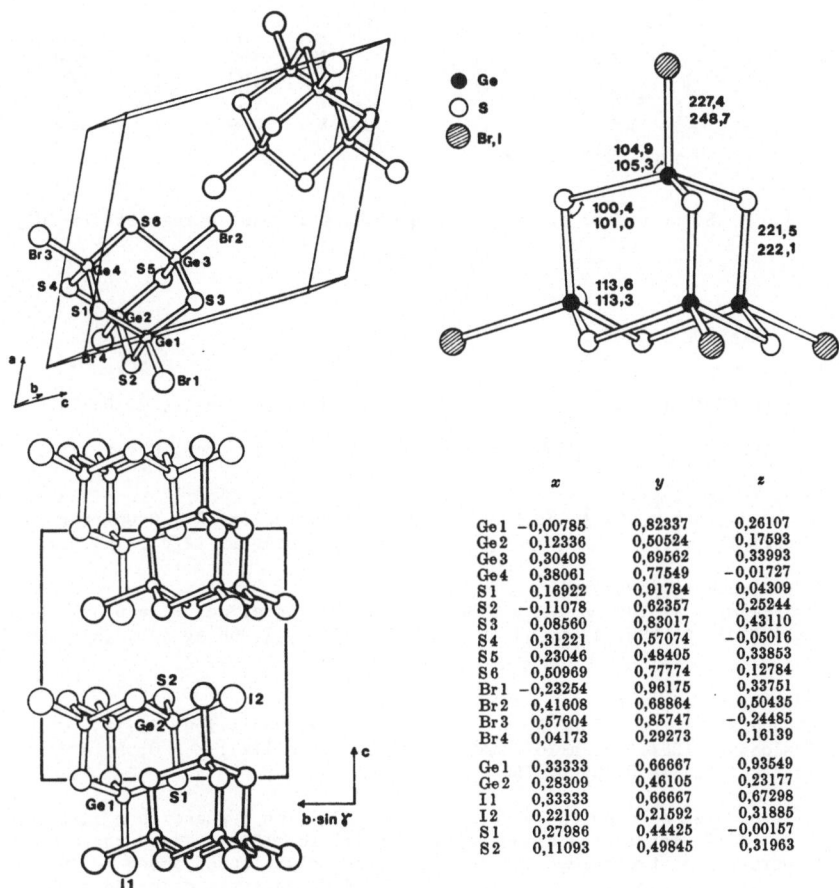

	x	y	z
Ge1	−0,00785	0,82337	0,26107
Ge2	0,12336	0,50524	0,17593
Ge3	0,30408	0,69562	0,33993
Ge4	0,38061	0,77549	−0,01727
S1	0,16922	0,91784	0,04309
S2	−0,11078	0,62357	0,25244
S3	0,08560	0,83017	0,43110
S4	0,31221	0,57074	−0,05016
S5	0,23046	0,48405	0,33853
S6	0,50969	0,77774	0,12784
Br1	−0,23254	0,96175	0,33751
Br2	0,41608	0,68864	0,50435
Br3	0,57604	0,85747	−0,24485
Br4	0,04173	0,29273	0,16139
Ge1	0,33333	0,66667	0,93549
Ge2	0,28309	0,46105	0,23177
I1	0,33333	0,66667	0,67298
I2	0,22100	0,21592	0,31885
S1	0,27986	0,44425	−0,00157
S2	0,11093	0,49845	0,31963

Fig. 1.   Structures of $Ge_4S_6Br_4$ and $Ge_4S_6I_4$.

NEODYMIUM THIOBROMIDE
$Nd_4Br_6S_3$

N. RYSANEK, A. MAZURIER, P. LARUELLE and C. DAGRON, 1980. Acta Cryst., B36, 2930-2932.

Monoclinic, B2/b, a = 29.346, b = 7.156, c = 6.837 Å, γ = 94.65°, D_m = 5.35, Z = 4. Mo radiation, R = 0.090 for 560 reflexions.

The structure (Fig. 1) is built up from Nd_4S tetrahedra linked into ribbons parallel to b, and surrounded by Br atoms. Nd atoms have 7-coordination, Nd-2 or 4S = 2.78-2.84(2), Nd-5 or 3Br = 2.96-3.29 Å.

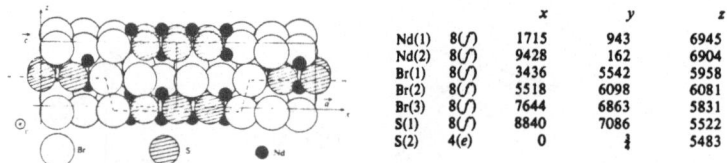

		x	y	z
Nd(1)	8(f)	1715	943	6945
Nd(2)	8(f)	9428	162	6904
Br(1)	8(f)	3436	5542	5958
Br(2)	8(f)	5518	6098	6081
Br(3)	8(f)	7644	6863	5831
S(1)	8(f)	8840	7086	5522
S(2)	4(e)	0	¼	5483

Fig. 1.   Structure of $Nd_4Br_6S_3$ and atomic positional parameters (x 10^4).

SILVER SULPHIDE IODIDE
Ag_3SI

I.   E. PERENTHALER and H. SCHULZ, 1981. Solid State Ionics, 2, 43-46.

II. E. PERENTHALER, H. SCHULZ and H.U. BEYELER, 1981. Acta Cryst., B37, 1017-1023.

α-Phase (high-temperature)
Cubic, Im3m, a = 4.994 Å, at 573K, Z = 1. Ag radiation, R = 0.007 for 30 reflexions.
I/S in 2(a): 0,0,0; 2.22Ag in 6(b): 0,1/2,1/2; 0.66Ag in 12(d): 1/4,0,1/2.

β-Phase (room-temperature)
Cubic, Pm3m, a = 4.897 Å, D_m = 6.86, Z = 1. Mo radiation, R_w = 0.041, 0.041 for 49,
76 reflexions at 295, 170K. I in 1(a); S in 1(b); 2.868 Ag in 12(h): x = 0.405,
0.395.

γ-Phase (low-temperature)
Rhombohedral, R3, a = 4.88 Å, γ = 90°, Z = 1. M_Q radiation, R_w(I) = 0.16 for 55
reflexions at 135K, for a 16-domain crystal. I in 1(a): x = 0; S in 1(a): x = 0.536;
Ag in 3(b): (0.515,0.399,0.016).

    Structures as previously described (1, 2). The α-phase is a fast ionic conductor,
with Ag^+ ions distributed along bands parallel to <100>. The β-phase has a modified
anti-perovskite structure, with three Ag^+ ions distributed among twelve (I,S) tetrahedra. The γ-phase crystal is composed of 16 differently-oriented domains with fully-
occupied Ag sites.

1. Structure Reports, 26, 423; 30A, 289.
2. S. HOSHINO, T. SAKUMA and Y. FUJII, 1979. J. Phys. Soc. Japan, 47, 1252.

POTASSIUM HEXACYANOCHROMATE(III) (TRICLINIC FORM)
$K_3Cr(CN)_6$

B.N. FIGGIS, P.A. REYNOLDS and G.A. WILLIAMS, 1981. Acta Cryst., B37, 504-508.

Triclinic, PĪ, a = 8.450, b = 10.576, c = 13.557 Å, α = β = γ = 90.0°, at 4.2K, Z =
4. Neutron radiation, R = 0.053 for 1575 reflexions, for a crystal consisting of
four twins with a stacking disorder.

    The structure is derived from that of the room temperature Pcan form (1) by
translations, but not rotations, of the $Cr(CN)_6^{3-}$ and K^+ ions. The anion appears
to be distorted from ideal octahedral geometry; Cr-C = 2.07, C-N = 1.16 Å, C-Cr-C
= 87.1-93.0°.

1. Structure Reports, 40A, 170.

THALLIUM(I) TETRACYANOPLATINATE(II) CARBONATE
$Tl_4[Pt(CN)_4]CO_3$

M.A. BENO, F.J. ROTELLA, J.D. JORGENSEN and J.M. WILLIAMS, 1981. Inorg. Chem., 20, 1802-1806.

Tetragonal, P4/mcc, a = 9.911, c = 6.490 Å, D_m = 6.0, Z = 2. Neutron radiation, R = 0.064 for 202 reflexions, and neutron time-of-flight powder data.

The structure (Fig. 1) contains square-planar $Pt(CN)_4^{2-}$ anions stacked along c, Tl^+ ions, and disordered carbonate ions. Pt-C = 1.98, C-N = 1.16, Tl-O = 2.52-2.97, Tl-N = 2.97, 3.06 Å.

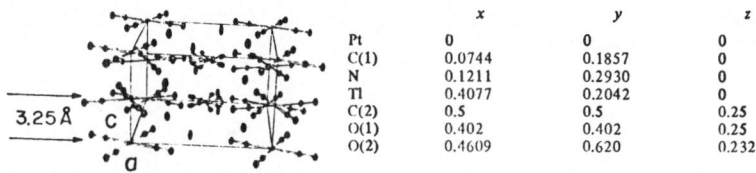

	x	y	z
Pt	0	0	0
C(1)	0.0744	0.1857	0
N	0.1211	0.2930	0
Tl	0.4077	0.2042	0
C(2)	0.5	0.5	0.25
O(1)	0.402	0.402	0.25
O(2)	0.4609	0.620	0.232

Fig. 1.   Structure of $Tl_4[Pt(CN)_4]CO_3$ [O(1) and O(2) occupancies are 0.5].

POTASSIUM COBALT(II) DICYANOAURATE(I)
$KCo[Au(CN)_2]_3$

S.C. ABRAHAMS, J.L. BERNSTEIN, R. LIMINGA and E.T. EISENMANN, 1980. J. Chem. Phys., 73, 4585-4590.

Trigonal, P312, a = 6.82802, c = 7.80807 Å, at 298K, D_m = 4.48, Z = 1. Mo radiation, R = 0.058 for 1130 reflexions. K in 1(d): 1/3,2/3,1/2; Co in 1(b): 0,0,1/2; Au in 3(j): 0.9918,0.4959,0; C in 6(ℓ): 0.987,0.650,0.2149; N in 6(ℓ): 0.995,0.752,0.3341.

The structure (Fig. 1) contains nearly linear $Au(CN)_2^-$ ions, linked by octa-hedrally-coordinated Co ions, and by K ions with 6 N and 6 C neighbours; a site with environment similar to that of the K ion is empty. Au-C = 1.99(1), C-N = 1.15(1) Å, C-Au-C = 178, Au-C-N = 175°, Co-N = 2.12(1), K-N = 2.95(2), K-C = 3.22(2) Å.

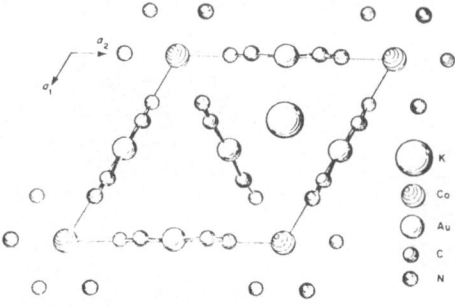

Fig. 1.   Structure of $KCo[Au(CN)_2]_3$.

SODIUM DICYANOMERCURATE(II) CHLORIDE MONOHYDRATE
POTASSIUM DICYANOMERCURATE(II) CHLORIDE MONOHYDRATE
RUBIDIUM DICYANOMERCURATE(II) CHLORIDE MONOHYDRATE
$MHg(CN)_2Cl.H_2O$ (M = Na, K, Rb)

G. THIELE, K. BRODERSEN and H. FROHRING, 1981. Z. Naturforsch., $\underline{36}$B, 180-187.

M = Na, orthorhombic, Pbca, a = 11.779, b = 6.060, c = 18.380 Å, D_m = 3.32, Z = 8.
R = 0.067 for 1164 reflexions.

M = K, orthorhombic, Pnma, a = 8.971, b = 4.221, c = 18.855 Å, D_m = 3.15, Z = 4.   R = 0.062 for 586 reflexions.

M = Rb, orthorhombic, Cmcm, a = 4.416, b = 19.035, c = 8.838 Å, D_m = 3.39, Z = 4.
R = 0.072 for 391 reflexions.

Atomic positions

	M = Na, atoms in 8(c)			M = K, atoms in 4(c)		
	x	y	z	x	y	z
Hg	0.1133	0.1645	0.2405	0.5182	1/4	0.2486
Cl	0.3562	0.1629	0.2148	0.2646	3/4	0.2389
Na/K	0.3394	0.0349	0.4847	0.2792	3/4	0.4790
C(1)	0.0582	0.1596	0.1356	0.5497	1/4	0.1409
N(1)	0.0262	0.1522	0.0770	0.5644	1/4	0.0811
C(2)	0.1495	0.1734	0.3505	0.4856	1/4	0.3572
N(2)	0.1699	0.1837	0.4093	0.4674	1/4	0.4169
O	0.3275	0.3797	0.5454	0.2263	1/4	0.5705

M = Rb

			x	y	z
Hg	in	4(a)	0	0	0
Cl		4(c)	0	0.1045	1/4
Rb		4(c)	0	0.7251	1/4
C		8(f)	0	0.1045	0.0244
N		8(f)	0	0.1667	0.0356
O		4(c)	0	0.3263	1/4

All three structures (Fig. 1) contain linear $Hg(CN)_2^-$ ions, M^+ and Cl^- ions, and water molecules. Hg-C = 2.00-2.07(2) Å; Hg atoms al   have Cl neighbours at 2.88-3.13 Å. M ions have 6-8 N or O neighbours at 2.37-2.57 (Na), 2.77-3.63 (K), 2.95-3.25 Å (Rb).

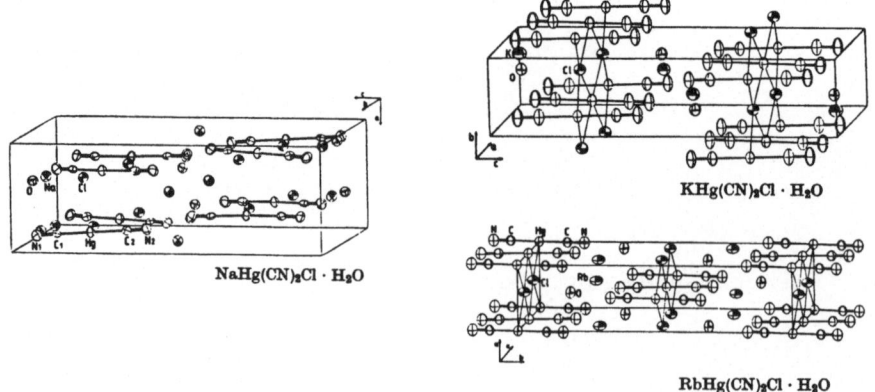

NaHg(CN)₂Cl · H₂O

KHg(CN)₂Cl · H₂O

RbHg(CN)₂Cl · H₂O

Fig. 1.   Structures of the dicyanomercurates.

RUBIDIUM TETRACYANOMERCURATE(II)
α-Rb$_2$Hg(CN)$_4$

P. KLÜFERS, H. FUESS and S. HAUSSÜHL, 1981. Z. Kristallogr., 156, 255-263.

Rhombohedral, R$\bar{3}$c, a = 9.076, c = 46.050 Å, D$_m$ = 2.881, Z = 12. Mo radiation, R = 0.088 for 457 reflexions; neutron radiation, R = 0.039 for 559 reflexions.

Atomic positions (neutron data)

			x	y	z
Rb(1)	in	6(b)	0	0	0
Rb(2)		18(e)	0.4355	0	1/4
Hg		12(c)	0	0	0.1854
C(1)		12(c)	0	0	0.1384
N(1)		12(c)	0	0	0.1135
C(2)		36(f)	0.1710	0.2560	0.2023
N(2)		36(f)	0.2555	0.3893	0.2113

The structure deviates only slightly from that of the cubic high-temperature spinel β-phase (Fig. 1), the Hg(CN)$_4$ tetrahedra being rotated 10.6° about the trigonal axis. Both Rb ions are nearly octahedrally surrounded by CN groups. Hg-C = 2.191 (x 3), 2.169, Rb-N = 3.019-3.115(2) Å.

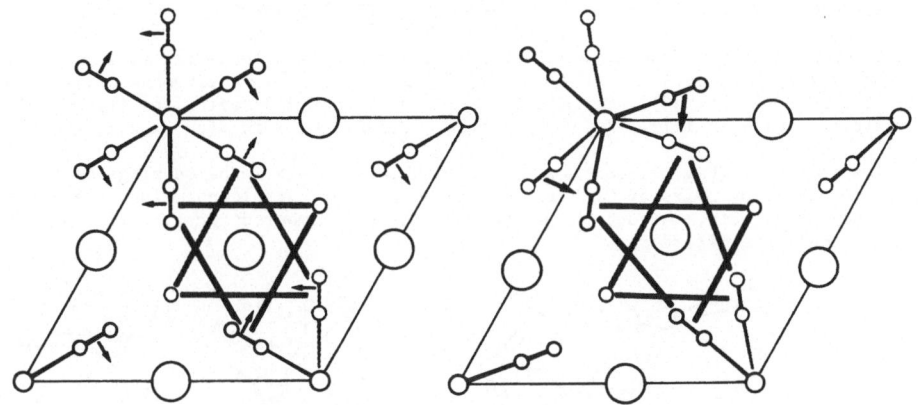

Fig. 1.   Structures of β- (left) and α- (right) Rb$_2$Hg(CN)$_4$.

COPPER(I) THIOCYANATE
β-CuNCS

D.L. SMITH and V.I. SAUNDERS, 1981. Acta Cryst., B37, 1807-1812.

The crystals consist of 3R and 2H polytypes in syntactic coalescence, along with some disorder and twinning. Space groups are R3m and P6$_3$mc, a = 3.856 Å for both, c = 16.453 and 10.97 Å (32.905 Å for an expanded cell used for data collection), Z = 3 and 2 (6 for the expanded cell). Mo radiation, R = 0.016-0.135 for various subsets of reflexions. Cu, N, C, S all in 0,0,z, z = 0, 0.0585, 0.0934, 0.1445 (relative to the expanded cell).

The structures (Fig. 1) contain (001) layers of close-packed CuNCS cylinders; the layers are linked by strong Cu-S bonds, so that both Cu and S have tetrahedral coordination. Cu-S = 2.343(1), Cu-N = 1.92(1), N-C = 1.15(2), C-S = 1.68(1) Å. An orthorhombic α-form is also known (1).

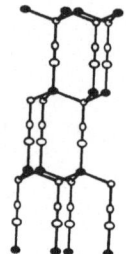

Fig. 1.   Structure of β-CuNCS 3R polytype.

<u>1</u>. Structure Reports, <u>42</u>A, 227.

ICE-IV
H_2O

H. ENGELHARDT and B. KAMB, 1981. J. Chem. Phys., <u>75</u>, 5887-5899.

Rhombohedral, R$\bar{3}$c, a = 7.60 Å, α = 70.1°, at 1 atm and 110K (material is a metastable high-pressure phase, obtained at 4-5.5kb), Z = 16, D_X = 1.272 (hexagonal cell has a = 8.74, c = 17.05 Å, Z = 48). Mo radiation, R = 0.067 for 213 reflexions (films, visual intensities). O(1) in 12(f): (0.3804,-0.1109,-0.2396); O(2) in 4(c): x,x,x, x = 0.0855.

    The structure (Fig. 1) contains puckered sheets of six-membered rings of O(1) atoms, connected by pairs of O(2) atoms; one-quarter of these connections are between second nearest sheets, through the centres of the six-membered rings. All the hydrogen bonds are disordered.

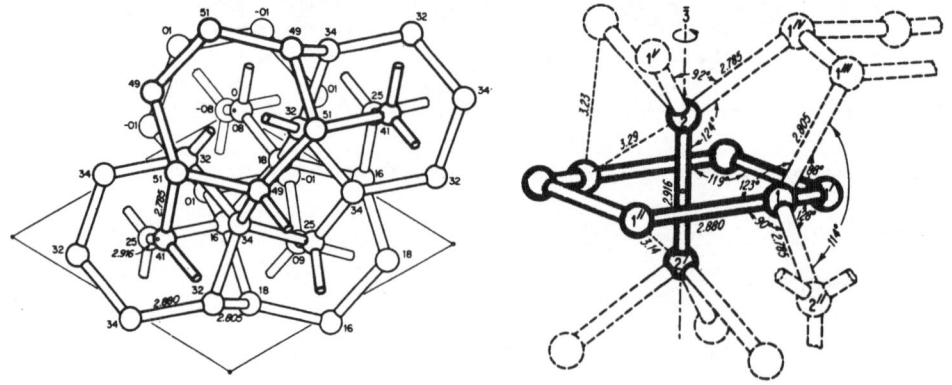

Fig. 1.   Structure of ice-IV.

TIN(II) OXIDE (TETRAGONAL)
SnO

F. IZUMI, 1981. J. Solid State Chem., <u>38</u>, 381-385.

Tetragonal, P4/nmm, a = 3.799, c = 4.841 Å, Z = 2.   Cu radiation, powder data.   Sn in
2(c): 1/2,0,z, z = 0.2369; O in 2(a): 0,0,0.

Isostructural with low-temperature PbO (1), as previously reported (2).   Sn has
four oxygen neighbours in a square, all on one side, Sn-O = 2.22 Å, with a stereochemically-active lone pair completing a square pyramid.

1. Strukturbericht, 1, 89; Structure Reports, 11, 237.
2. Structure Reports, 11, 238.

PHOSPHORUS(III,V) OXIDE
P_4O_7

K.H. JOST and M. SCHNEIDER, 1981.   Acta Cryst., B37, 222-224

Monoclinic, P2/n, a = 9.808, b = 9.966, c = 6.852 Å, β = 96.81°, Z = 4.   Mo radiation,
R = 0.08 for 1822 reflexions.

The structure (Fig. 1) contains molecules similar to those in other phosphorus
oxides (1).   P(1)-O(11) = 1.44, other P(1)-O = 1.58-1.59, other P-O = 1.63-1.69(1) Å,
O(11)-P($\bar{1}$)-O = 115-116, other O-P(1)-O = 103-104, other O-P-O = 97-100, P-O-P = 124-
129°.

	x	y	z
P(1)	−0·1176	0·3903	0·6749
P(2)	0·1692	0·3941	0·8196
P(3)	0·0506	0·1660	0·5906
P(4)	−0·0211	0·2100	0·9883
O(12)	0·0227	0·4678	0·7279
O(13)	−0·0772	0·2753	0·5327
O(14)	−0·1372	0·3145	0·8698
O(23)	0·1716	0·2730	0·6616
O(24)	0·1109	0·3136	0·9988
O(34)	0·0086	0·1186	0·8014
O(11)	−0·2304	0·4726	0·5943

Fig. 1.   Structure of P_4O_7.

1. Structure Reports, 8, 145; 29, 300, 360; 31A, 120; 32A, 252; 34A, 240.

TANTALUM OXIDE
δ'-Ta_2O_5

V.I. KHITROVA and V.V. KLEČKOVSKAJA, 1980.   Kristallografija, 25, 1169-1175 [Soviet
Physics - Crystallography, 25, 669-672].

Hexagonal, P6/mmm, a = 7.16, c = 11.52 Å.   Electron diffraction patterns for three
specimens with somewhat different reflexion intensities.   The structure contains
mixed Ta/O layers alternating with purely O layers, with variable composition.

TUNGSTEN OXIDE
$W_{18}O_{49}$

K. VISWANATHAN, K. BRANDT and E. SALJE, 1981.   J. Solid State Chem., 36, 45-51.

Monoclinic, P2/m, a = 18.334, b = 3.786, c = 14.044 Å, β = 115.20°, Z = 1. Mo
radiation, R = 0.065 for 1202 reflexions (twinned crystal).

The structure is essentially as previously described (1), but with changes of up
to 0.04 in x for two oxygens, which results in 7-coordination for W(5). The structure
(Fig. 1) contains edge- and corner-sharing WO_6 octahedra and one WO_7 polyhedron, with
empty hexagonal tunnels.

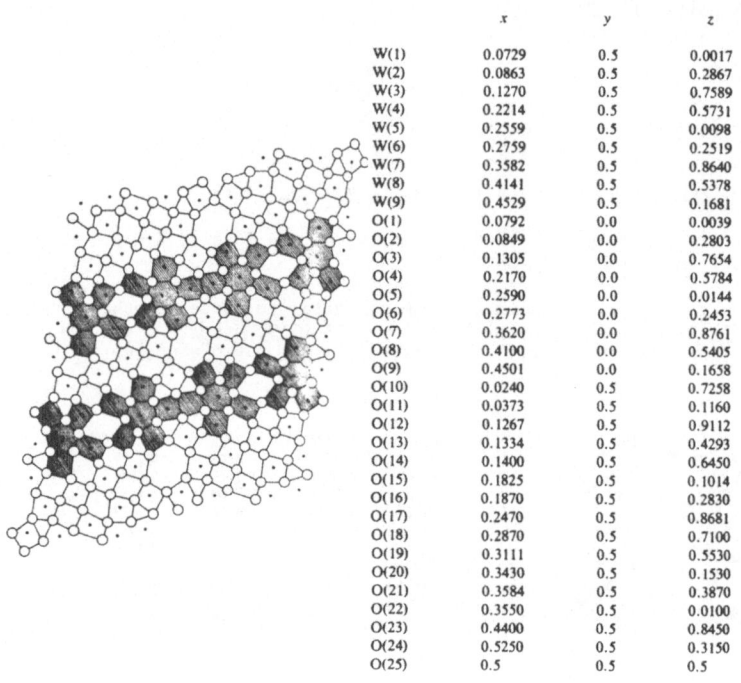

	x	y	z
W(1)	0.0729	0.5	0.0017
W(2)	0.0863	0.5	0.2867
W(3)	0.1270	0.5	0.7589
W(4)	0.2214	0.5	0.5731
W(5)	0.2559	0.5	0.0098
W(6)	0.2759	0.5	0.2519
W(7)	0.3582	0.5	0.8640
W(8)	0.4141	0.5	0.5378
W(9)	0.4529	0.5	0.1681
O(1)	0.0792	0.0	0.0039
O(2)	0.0849	0.0	0.2803
O(3)	0.1305	0.0	0.7654
O(4)	0.2170	0.0	0.5784
O(5)	0.2590	0.0	0.0144
O(6)	0.2773	0.0	0.2453
O(7)	0.3620	0.0	0.8761
O(8)	0.4100	0.0	0.5405
O(9)	0.4501	0.0	0.1658
O(10)	0.0240	0.5	0.7258
O(11)	0.0373	0.5	0.1160
O(12)	0.1267	0.5	0.9112
O(13)	0.1334	0.5	0.4293
O(14)	0.1400	0.5	0.6450
O(15)	0.1825	0.5	0.1014
O(16)	0.1870	0.5	0.2830
O(17)	0.2470	0.5	0.8681
O(18)	0.2870	0.5	0.7100
O(19)	0.3111	0.5	0.5530
O(20)	0.3430	0.5	0.1530
O(21)	0.3584	0.5	0.3870
O(22)	0.3550	0.5	0.0100
O(23)	0.4400	0.5	0.8450
O(24)	0.5250	0.5	0.3150
O(25)	0.5	0.5	0.5

Fig. 1.   Structure of $W_{18}O_{49}$.

1. Structure Reports, 12, 185.

TUNGSTEN TRIOXIDE HYDRATE
$WO_3 \cdot 1/3 H_2O$

B. GERAND, G. NOWOGROCKI and M. FIGLARZ, 1981. J. Solid State Chem., 38, 312-320.

Orthorhombic, Fmm2, a = 7.359, b = 12.513, c = 7.704 Å, Z = 12. Powder data (only
20 lines).

The structure derived from the limited data contains a layer of corner-sharing
WO_6 octahedra with two such layers stacked along c (Fig. 1); $W-OH_2$ = 2.1, W-O =
1.8-1.9 Å. The unit cell is closely related to that of hexagonal WO_3.

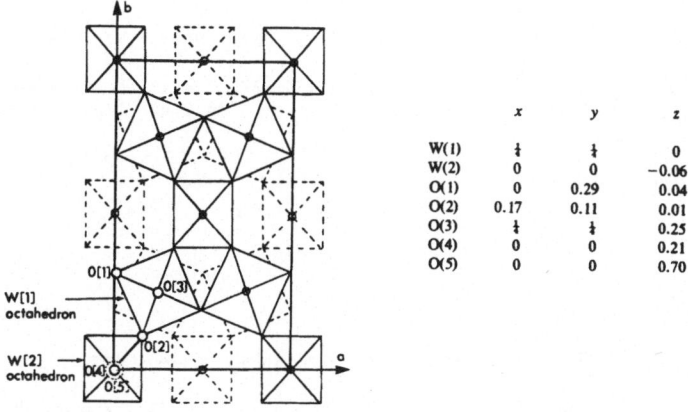

	x	y	z
W(1)	¼	¼	0
W(2)	0	0	−0.06
O(1)	0	0.29	0.04
O(2)	0.17	0.11	0.01
O(3)	¼	¼	0.25
O(4)	0	0	0.21
O(5)	0	0	0.70

Fig. 1.   Structure proposed for $WO_3.1/3H_2O$; O(4) is probably H_2O.

MAGNETITE
Fe_3O_4

M.E. FLEET, 1981. Acta Cryst., B**37**, 917-920.

Cubic, Fd3m, a = 8.394 Å, Z = 8. Mo radiation, R = 0.024 for 147 reflexions. 8 Fe^{3+} in 8(a): 1/8,1/8,1/8; 8 Fe^{3+} + 8 Fe^{2+} in 16(d): 1/2,1/2,1/2; 32 O in 32(e): x,x,x, x = 0.2549.

Inverse spinel, as previously described (1). Fe^{3+}-O = 1.888(2) (tetrahedral), Fe^{3+}/Fe^{2+}-O = 2.058(1) Å (octahedral).

1. Strukturbericht, **1**, 352, 417, 418; Structure Reports, **22**, 300; **33A**, 267.

RHODIUM(III) OXIDE (HIGH-TEMPERATURE, HIGH-PRESSURE)
Rh_2O_3

K.R. POEPPELMEIER and G.B. ANSELL, 1981. J. Cryst. Growth, **51**, 587-588.

Orthorhombic, Pbna (non-standard setting of Pbcn), a = 5.159, b = 5.381, c = 7.241 Å, Z = 4. Mo radiation, R = 0.074.

Atomic positions

		x	y	z
Rh	in	8(d)	0.7495	0.0317	0.1056
O(1)		8(d)	0.6080	0.1148	0.8482
O(2)		4(c)	0.0514	1/4	0

Structure as previously described (1).

1. Structure Reports, **35A**, 211.

PLUTONIUM(III) OXIDE (HEXAGONAL)
β-Pu$_2$O$_3$

B. McCART, G.H. LANDER and A.T. ALDRED, 1981. J. Chem. Phys., 74, 5263-5268.

Trigonal, P$\bar{3}$m1, a = 3.838, c = 5.917 Å, at 40 and 13K, Z = 1. Neutron powder data.
Pu in 2(d): 1/2,2/3,0.2422; O(1) in 1(a): 0,0,0; O(2) in 2(d): 1/2,2/3,0.6489.
La$_2$O$_3$-type structure (1).

1. Strukturbericht, 1, 744.

STRONTIUM ALUMINATE
SrAl$_2$O$_4$

A.-R. SCHULZE and H. MÜLLER-BUSCHBAUM, 1981. Z. anorg. Chem., 475, 205-210.

Monoclinic, P2$_1$, a = 8.447, b = 8.816, c = 5.163 Å, β = 93.42°, Z = 4. R = 0.096
for 1188 reflexions.

The structure (Fig. 1) is of stuffed tridymite type, with a framework of corner-sharing AlO$_4$ tetrahedra (Al-O = 1.73-1.77 Å) and 9-coordinated Sr ions (Sr-O = 2.51-3.55 Å).

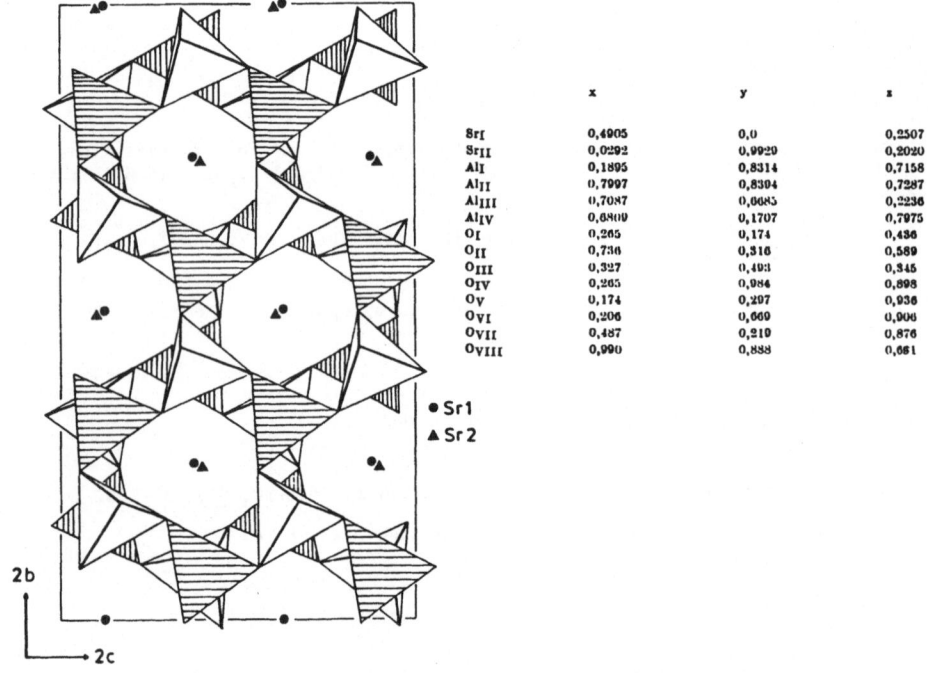

	x	y	z
Sr$_I$	0,4905	0,0	0,2507
Sr$_{II}$	0,0292	0,9929	0,2020
Al$_I$	0,1895	0,8314	0,7158
Al$_{II}$	0,7997	0,8394	0,7287
Al$_{III}$	0,7087	0,6685	0,2236
Al$_{IV}$	0,6809	0,1707	0,7975
O$_I$	0,265	0,174	0,436
O$_{II}$	0,736	0,516	0,589
O$_{III}$	0,327	0,493	0,345
O$_{IV}$	0,265	0,984	0,898
O$_V$	0,174	0,297	0,936
O$_{VI}$	0,206	0,669	0,906
O$_{VII}$	0,487	0,210	0,876
O$_{VIII}$	0,990	0,888	0,661

• Sr 1
▲ Sr 2

2b

2c

Fig. 1.   Structure of SrAl$_2$O$_4$.

COPPER(I) ALUMINATE
CuAlO$_2$

T. ISHIGURO, A. KITAZAWA, N. MIZUTANI and M. KATO, 1981. J. Solid State Chem., **40**, 170-174.

Rhombohedral, R$\overline{3}$m, a = 2.8604, b = 16.953 Å, D$_m$ = 5.06, Z = 3. Mo radiation, R = 0.038 for 333 reflexions. Cu in 3(a): 0,0,0; Al in 3(b): 0,0,1/2; O in 6(c): 0,0, 0.1098.

Delafossite-type structure (1). The structure contains sheets of edge-sharing AlO$_6$ octahedra, linked by linearly-coordinated Cu. Al-O = 1.912, Cu-O = 1.861(1) Å.

1. Strukturbericht, **3**, 75; Structure Reports, **37A**, 262.

COPPER(I) TETRAALUMINATE
Cu$_2$Al$_4$O$_7$

H. MEYER and H. MÜLLER-BUSCHBAUM, 1981. Mh. Chem., **112**, 51-57.

Cubic, F$\overline{4}$3m, a = 8.090 Å [in text, 9.080 Å in abstract], Z = 4. R = 0.10 for 72 reflexions.

The structure (Fig. 1) contains AlO$_4$ tetrahedra (Al-O = 1.75 (x 3), 1.88 Å), Cu(1)O$_6$ octahedra (Cu-O = 1.97 Å), and Cu(2)O$_{12}$ cuboctahedra (Cu-O = 2.86 Å).

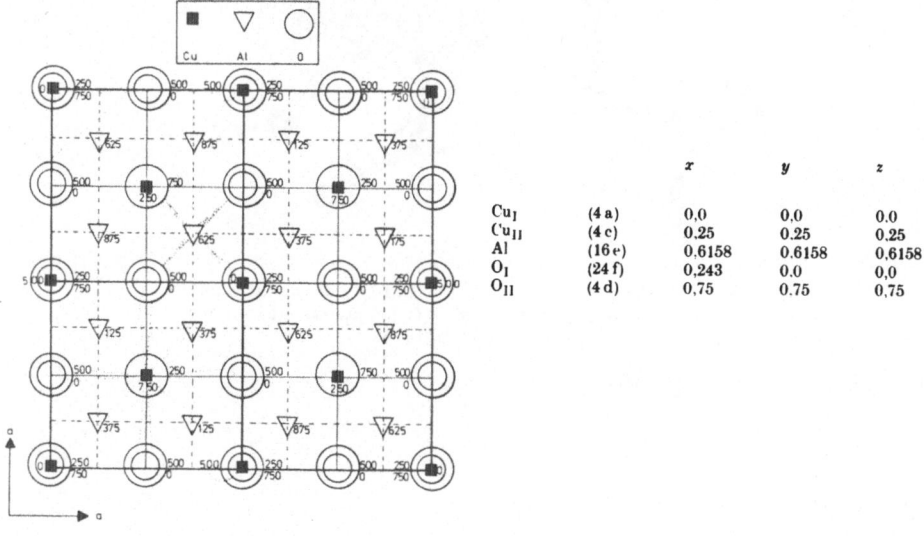

		x	y	z
Cu$_I$	(4 a)	0,0	0,0	0,0
Cu$_{II}$	(4 c)	0,25	0,25	0,25
Al	(16 e)	0,6158	0,6158	0,6158
O$_I$	(24 f)	0,243	0,0	0,0
O$_{II}$	(4 d)	0,75	0,75	0,75

Fig. 1.   Structure of Cu$_2$Al$_4$O$_7$.

LITHIUM METAGALLATE OCTAHYDRATE
LiGaO$_2$.8H$_2$O

C. CARANONI, L. CAPELLA, R. HASER and G. PÈPE, 1981. Acta Cryst., B**37**, 15-19.

Trigonal, P3c1, a = 6.804, c = 25.60 Å, Z = 4.   Cu radiation, R = 0.061 for 569 reflexions.

The metal ions are disordered (Ga occupancies 50, 50, 25, 75% for Ga,Li(1), (2), (3), (4), respectively) and each has tetrahedral coordination (Fig. 1).   The structure contains a system of hydrogen bonds.   The hexahydrate is described in 1.

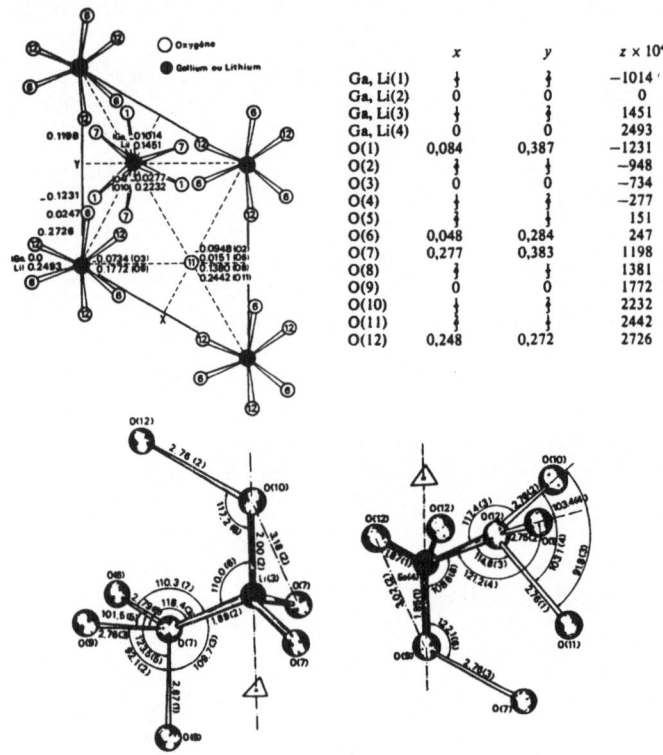

	x	y	$z \times 10^4$
Ga, Li(1)	⅓	⅔	−1014
Ga, Li(2)	0	0	0
Ga, Li(3)	⅓	⅔	1451
Ga, Li(4)	0	0	2493
O(1)	0,084	0,387	−1231
O(2)	⅓	⅔	−948
O(3)	0	0	−734
O(4)	⅓	⅔	−277
O(5)	⅓	⅔	151
O(6)	0,048	0,284	247
O(7)	0,277	0,383	1198
O(8)	⅓	⅔	1381
O(9)	0	0	1772
O(10)	0	0	2232
O(11)	⅓	⅔	2442
O(12)	0,248	0,272	2726

Fig. 1.   Structure of LiGaO$_2$.8H$_2$O, Li(3) and Ga(4) coordinations, and atomic positional parameters (oxygen atoms belong to water molecules, except O(4) and O(10)).

1. Structure Reports, 44A, 190.

CALCIUM TETRAGALLATE
Ca$_3$Ga$_4$O$_9$

A.-R. SCHULZE and H. MÜLLER-BUSCHBAUM, 1981.   Mh. Chem., 112, 149-156.

Orthorhombic, Cmm2, a = 14.358, b = 16.825, c = 5.321 Å, Z = 6.   R = 0.062 for 1217 reflexions.

The structure (Fig. 1) contains rings of 4 and 5 GaO$_4$ tetrahedra; Ca ions have 6- and 8-coordinations. Ga-O = 1.80-1.95, Ca-O = 2.26-2.97 Å.

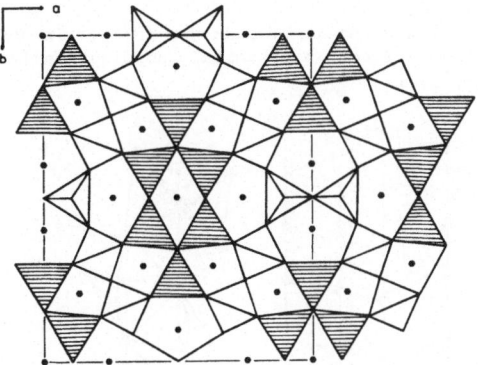

Fig. 1.   Structure of Ca$_3$Ga$_4$O$_9$.

STRONTIUM GALLATE
β-SrGa$_2$O$_4$

A.-R. SCHULZE and H. MÜLLER-BUSCHBAUM, 1981.   Z. Naturforsch., 36B, 892-893.

Monoclinic, P2$_1$/c, a = 8.392, b = 9.018, c = 10.697 Å, β = 93.9°, Z = 8.   R = 0.08 for 1784 reflexions.

Atomic positions

	x	y	z
Ga$_I$	0,2941	0,3923	0,1202
Ga$_{II}$	0,6959	0,4052	0,1100
Ga$_{III}$	0,8825	0,5561	0,3604
Ga$_{IV}$	0,2192	0,5647	0,3690
Sr$_I$	0,5342	0,2393	0,3498
Sr$_{II}$	0,0042	0,2129	0,3695
O$_I$	0,348	0,441	0,285
O$_{II}$	0,742	0,414	0,461
O$_{III}$	0,153	0,244	0,076
O$_{IV}$	0,490	0,339	0,068
O$_V$	0,217	0,565	0,041
O$_{VI}$	0,715	0,577	0,202
O$_{VII}$	0,774	0,249	0,209
O$_{VIII}$	0,020	0,491	0,321

The material has a stuffed-tridymite structure, and is isostructural with monoclinic CaGa$_2$O$_4$ (1).   Ga-O = 1.83-1.87 (tetrahedral), Sr-O = 2.47-3.06 Å (6- and 7-coordinations).

1.   Structure Reports, 39A, 212.

COPPER GALLIUM INDIUM OXIDE
CuGaInO$_4$

A. ROESLER and D. REINEN, 1981.   Z. anorg. Chem., 479, 119-124.

Rhombohedral, R$\bar{3}$m, a = 3.354, c = 24.81 Å, Z = 3.   Mo radiation, R = 0.055 for 380 reflexions.   In in 3(a): 0,0,0; Cu/Ga, O(1), O(2) in 6(c): 0,0,z, z = 0.2143, 0.2917, 0.1294, respectively.

The anions have a 12-layer packing, sequence $(31)_3$.   In has compressed-octahedral coordination and Cu/Ga has trigonal-bipyramidal coordination (Fig. 1); In-O = 2.20, Cu/Ga-O = 1.90, 1.95 (x 3), 2.12 Å.

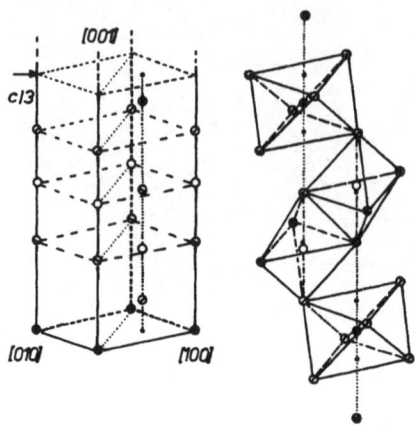

● = In; ○ = (Cu,Ga); ◕ = O(I); ◑ = O(II)

Fig. 1.   Structure of CuGaInO$_4$.

STRONTIUM METAGERMANATE (TRICLINIC)
SrGeO$_3$

I.   T.N. NADEŽINA, E.A. POBEDIMSKAJA, V.V. ILJUKHIN and N.V. BELOV, 1981.
     Kristallografija, <u>26</u>, 54-62 [Soviet Physics - Crystallography, <u>26</u>, 27-32].

II. Idem, 1981.   Ibid., <u>26</u>, 473-479 [Ibid., <u>26</u>, 268-272].

Triclinic, P$\bar{1}$, a = 8.699 [in I, 8.688 in II], b = 9.935, c = 11.148 Å, α = 106.04, β = 89.97, γ = 102.11°, D$_m$ = 4.47, Z = 12.   Mo radiation, R = 0.07 for 2615 reflexions.

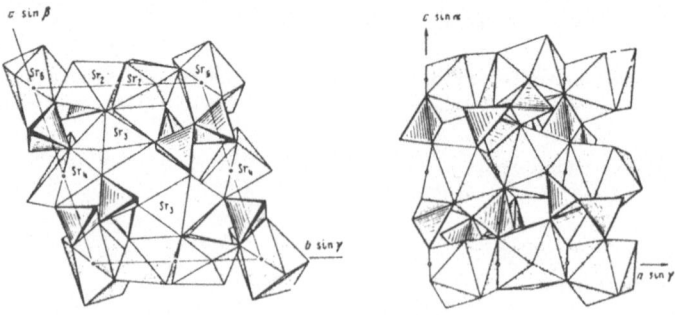

Fig. 1.   Structure of triclinic strontium germanate.

The structure (Fig. 1) contains Ge_3O_9 groups of three corner-sharing tetrahedra, Ge-O = 1.70-1.84 Å, linked by 6- and 8-coordinate Sr ions. Another form has been described previously (1).

1. Structure Reports, 27, 501.

SODIUM ALUMINUM GERMANATE CARBONATE HYDRATE
$Na_8Al_6Ge_6O_{24}(CO_3) \cdot 2H_2O$

Ju.A. SOKOLOV, B.A. MAKSIMOV, R.V. GALIULIN, V.V. ILJUKHIN and N.V. BELOV, 1981. Kristallografija, 26, 287-292 [Soviet Physics - Crystallography, 26, 161-164].

Trigonal, P31c (pseudo P6$_3$mc), a = 12.913, c = 10.493 Å, Z = 2. R = 0.056. Preliminary study in 1.

Atomic positions

	x	y	z
Ge$_1$	0.0002	0.2523	0.0000
Ge$_2$	0.4177	0.3336	0.7511
Al$_1$	0.7456	0.0011	0.0000
Al$_2$	0.4173	0.0845	0.7511
Na$_1$	0	0	0.112
Na$_2$	$^1/_3$	$^1/_3$	0.609
Na$_3$	0.552	0.102	0.046
Na$_4$	0.386	0.192	0.253
O$_1$	0.099	0.201	0.473
O$_2$	0.035	0.376	0.408
O$_3$	0.284	-0.005	0.654
O$_4$	0.278	0.140	0.447
O$_5$	0.474	0.239	0.732
O$_6$	0.287	0.294	0.653
O$_7$	0.342	0.380	0.407
O$_8$	0.524	0.052	0.694
O$_9$	0.553	0.277	0.138
O$_{10}$ (H$_2$O)	0.615	0.202	0.404
O$_{11}$ (H$_2$O)	0.017	0.058	0.198
O$_{12}$ (H$_2$O)	0	0	0.207
C	$^1/_3$	$^1/_3$	0.154

The structure is cancrinite-like (2), with an Al/Ge tetrahedral framework of rings of six tetrahedra.

1. Structure Reports, 43A, 336; 45A, 390.
2. Strukturbericht, 3, 150, 524; Structure Reports, 19, 480; 30A, 428; 35A, 458.

LEAD(II) TETRAGERMANATE
α-PbGe$_4$O$_9$

A.Ju. ŠAŠKOV, V.A. EFREMOV, I. MATSIČEK, N.V. RANNEV, Ju.N. VENEVCEV and V.K. TRUNOV, 1981. Ž. Neorg. Khim., 26, 583-587.

Trigonal, P321, a = 11.420, c = 4.753 Å, Z = 3. R = 0.039.

The structure contains Ge_3O_9 groups with a ring of three GeO$_4$ tetrahedra, linked by GeO$_6$ octahedra; Pb has 10-coordination.

COPPER(II) GERMANATE
Cu_2GeO_4

W. HEGENBART, F. RAU and K.-J. RANGE, 1981. Mater. Res. Bull., 16, 413-417.

Tetragonal, $I4_1/amd$, a = 5.593, c = 9.396 Å, Z = 4. Mo radiation, R = 0.045. Ge in
4(a): 0,0,0; Cu in 8(d): 0,1/4,5/8; O in 16(h): 0,0.241,0.359.

Hausmannite (Mn_3O_4) type structure (1), but with greater Jahn-Teller distortion
of the CuO_6 octahedron. Cu-O = 1.94 (x 4), 2.50 (x 2), Ge-O (tetrahedral) = 1.78 Å.

1. Strukturbericht, 1, 418; Structure Reports, 23, 331; 26, 407.

CADMIUM GERMANATE
Cd_2GeO_4

CADMIUM DIGERMANATE
$Cd_2Ge_2O_6$

M.A. SIMONOV, E.L. BELOKONEVA and N.V. BELOV, 1981. Ž. Strukt. Khim., 22, No. 3, 199-
200 [J. Struct. Chem., 22, 478-479].

Cd_2GeO_4, orthorhombic, Pmcn, a = 6.584, b = 5.211, c = 11.160 Å, D_m = 6.4, Z = 4. Mo
radiation, R = 0.077 for 1458 reflexions. Previous study in 1.

$Cd_2Ge_2O_6$, monoclinic, B2/b, a = 10.204, b = 5.385, c = 9.669 Å, γ = 102.23°, D_m = 5.7,
Z = 4. Mo radiation, R = 0.088 for 1975 reflexions.

Atomic positions

	x	y	z
		$Cd_2[GeO_4]$	
Cd_1	0	0	0
Cd_2	0,25	0,4888(2)	0,21948
Ge	0,75	0,4330(2)	0,0981
O_1	0,25	0,270(2)	0,4046
O_2	0,75	0,210(2)	0,4541
O_3	0,540	0,281(1)	0,1647
		$Cd_2[Ge_2O_6]$	
Cd_1	0	0,25	0,0958
Cd_2	0	0,25	0,7112
Ge	0,2045	0,2755	0,4108
O_1	0,3797	0,347	0,407
O_2	0,3517	0,439	0,053
O_3	0,375	0,373	0,765

Olivine (2) and diopside (3) structures. Ge-O = 1.73-1.81, Cd-O = 2.17-2.51 Å.

1. Structure Reports, 38A, 251.
2. Strukturbericht, 1, 352.
3. Ibid., 2, 130.

MERCURY(II) GERMANATE
Hg_2GeO_4

K.-F. HESSE and W. EYSEL, 1981. Acta Cryst., B$\underline{37}$, 429-431.

Orthorhombic, Fddd, a = 6.603, b = 10.596, c = 11.485 Å, Z = 8. Mo radiation, R = 0.076 for 683 reflexions.

Isostructural with thenardite ($\underline{1}$). The structure (Fig. 1) contains isolated GeO_4 tetrahedra (Ge-O = 1.75 Å, O-Ge-O = 99-115°), linked by Hg^{2+} ions with irregular six-coordination (Hg-O = 2.08, 2.64, 2.66 Å (each x 2)).

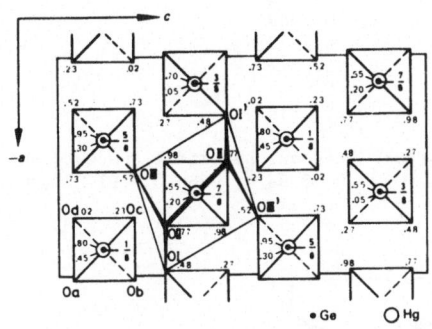

		x	y	z
Hg in	16(f)	1/8	0.4472	1/8
Ge	8(a)	1/8	1/8	1/8
O	32(h)	-0.0171	0.2320	0.0420

(origin at $\bar{1}$)

Fig. 1.   Structure of mercury(II) germanate.

$\underline{1}$. Strukturbericht, $\underline{2}$, 88; Structure Reports, $\underline{41A}$, 343.

POTASSIUM DISTANNATE(II)
$K_2Sn_2O_3$

R.M. BRAUN and R. HOPPE, 1981. Z. anorg. Chem., $\underline{478}$, 7-12.

Rhombohedral, R$\bar{3}$m, a = 6.001, c = 14.327 Å, D_m = 3.98, Z = 3. Mo radiation, R = 0.021 for 219 reflexions. K(1) in 3(b): 0,0,1/2; K(2) in 3(a): 0,0,0; Sn in 6(c): 0,0,0.2401; O in 9(d): 1/2,0,1/2.

The structure (Fig. 1) contains a tin-oxygen framework, Sn having trigonal pyramidal coordination, Sn-O = 2.03 Å, O-Sn-O = 95.4°. K ions have six-coordinations, K-O = 3.00, 2.95 Å.

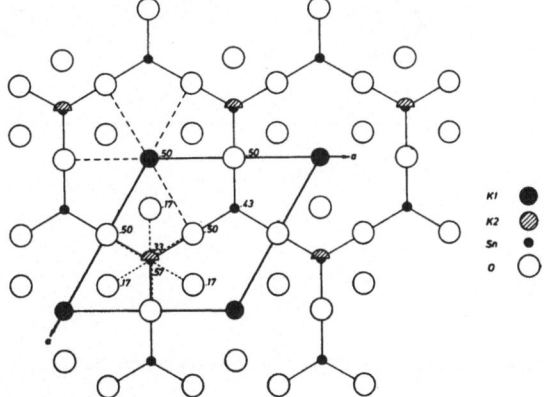

Fig. 1.   Structure of $K_2Sn_2O_3$.

CAESIUM DISTANNATE(II)
$Cs_2Sn_2O_3$

R.M. BRAUN and R. HOPPE, 1981. Z. anorg. Chem., <u>480</u>, 81-89.

Orthorhombic, Pnma, a = 13.708, b = 6.098, c = 8.921 Å, D_m = 4.84, Z = 4. Mo radiation,
R = 0.082 for 955 reflexions.

Atomic positions

	x	y	z
Cs(1)	0.3703	1/4	0.5343
Cs(2)	0.0949	1/4	0.3369
Sn(1)	0.1355	1/4	0.7823
Sn(2)	0.3033	1/4	0.0813
O(1)	0.1600	1/4	0.0053
O(2)	0.2729	0.0007	0.2319

The structure (Fig. 1) contains layers, with trigonal pyramidal Sn coordinations
and Cs ions with 6- and 3-coordinations. Sn-O = 2.02-2.08 Å, O-Sn-O = 91-97°, Cs-O =
3.02-3.37 Å.

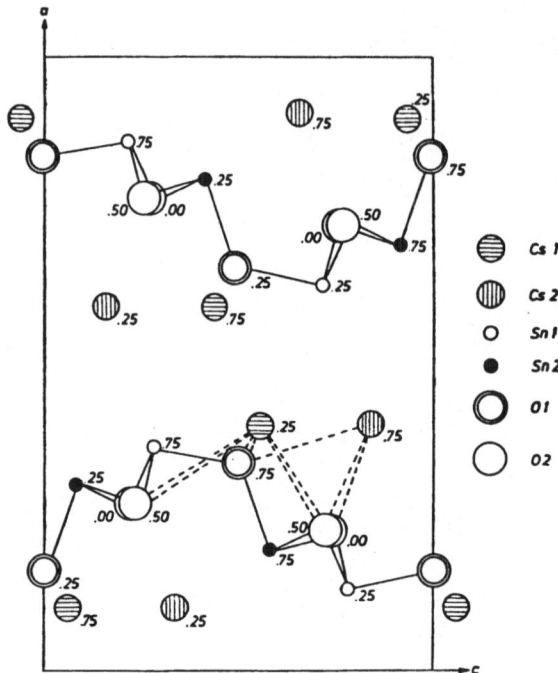

Fig. 1.   Structure of $Cs_2Sn_2O_3$.

STRONTIUM TIN OXIDE
Sr_3SnO

A. WIDERA and H. SCHÄFER, 1981. J. Less-Common Metals, <u>77</u>, 29-36.

Cubic, Pm3m, a = 5.12 Å, D_m = 4.78, Z = 1.   Mo radiation, R = 0.053 for 72 reflexions.
Sr in 3(c); Sn in 1(a); O in 1(b).   Perovskite structure.

TIN(IV) ANTIMONY(III) OXIDE
(Sn,Sb)O_2 $Sn_{0.9}Sb_{0.1}O_2$

F.J. BERRY and C. GREAVES, 1981.   J. Chem. Soc., Dalton, 2447-2451.

Tetragonal, $P4_2/mnm$, a = 4.737, c = 3.182 Å, Z = 2.   Neutron powder data.   1.8 Sn in
2(a): (0,0,0); 0.2 Sb in 16(k): (-0.055,0.283,0.326); 4 O in 4(f): (0.3046,0.3046,0).

Rutile structure (1) with Sb(III) in interstitial positions and deficiency of
Sn(IV).   Sb is displaced by 1.2 Å from the octahedral interstice to a position which
is 1.0 Å from an oxygen site and 1.7 Å from a tin site; in the presence of Sb, the O
and Sn sites must be vacant.   Sb coordination suggests a stereochemically active lone
electron-pair.

1.   Strukturbericht, 1, 155.

ALUMINUM LEAD(II) OXIDE
$Al_8Pb_9O_{21}$

K.-B. PLÖTZ and H. MÜLLER-BUSCHBAUM, 1981.   Z. anorg. Chem., 480, 149-152.

Cubic, Pa3, a = 13.26 Å, Z = 4.   R = 0.098 for 866 reflexions.

The structure contains a framework of AlO_4 tetrahedra, with large cavities
containing Pb_9O_2 groups (Fig. 1).   Al-O = 1.69-1.83, Pb-O = 2.26-3.30 Å (6- and 8-
coordinations).

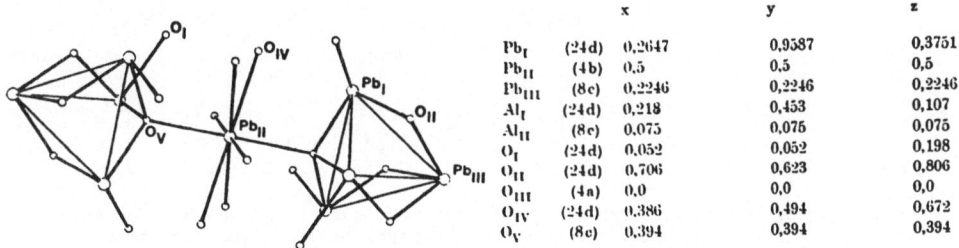

		x	y	z
Pb_I	(24d)	0.2647	0.9587	0.3751
Pb_{II}	(4b)	0.5	0.5	0.5
Pb_{III}	(8c)	0.2246	0.2246	0.2246
Al_I	(24d)	0.218	0.453	0.107
Al_{II}	(8c)	0.075	0.075	0.075
O_I	(24d)	0.052	0.052	0.198
O_{II}	(24d)	0.706	0.623	0.806
O_{III}	(4a)	0.0	0.0	0.0
O_{IV}	(24d)	0.386	0.494	0.672
O_V	(8c)	0.394	0.394	0.394

Fig. 1.   Pb coordinations and atomic positional parameters in $Al_8Pb_9O_{21}$.

TIN(IV) DILEAD(II) OXIDE
$SnPb_2O_4$

J.R. GAVARRI, J.P. VIGOUROUX, G. CALVARIN and A.W. HEWAT, 1981.   J. Solid State Chem.,
36, 81-90.

Orthorhombic, Pbam (pseudo-$P4_2/mbc$), a = 8.715-8.722, b = 8.701-8.709, c = 6.287-6.293 Å, at 5-300K, Z = 4. X-ray and neutron powder data at 5, 100, 200, 300K.

Atomic positions (5K)

			x	y	z
Sn	in	4(e)	0	0	0.250
Pb(1)		4(g)	0.1424	0.1582	0
Pb(2)		4(h)	0.1613	0.8586	1/2
O(1)		8(i)	0.6650	0.1670	0.250
O(2c)		4(g)	0.0959	0.6253	0
O(2a)		4(h)	0.1246	0.5969	1/2

The structure is as previously described (1), but with a distortion from tetragonal symmetry, as for low-temperature Pb_3O_4 (2). Sn^{4+} and Pb^{2+} ions have octahedral coordinations, Sn-O = 2.045, 2.087, Pb-O = 2.194, 2.300, 2.798(1) Å.

1. Structure Reports, 9, 174.
2. Ibid., 41A, 212; 44A, 183.

HYDROXONIUM ANTIMONATE
$(H_3O)_{12}Sb_{12}O_{36}$ $[H(H_2O)_n]_{12}Sb_{12}O_{36}$, n = 1

$[H(H_2O)_{0.26}]_{12}Sb_{12}O_{36}$ n = 0.26

H. WATELET, J.-P. PICARD, G. BAUD, J.-P. BESSE and R. CHEVALIER, 1981. Mater. Res. Bull., 16, 1131-1137.

Cubic, Im3, a = 9.497, 9.470 Å, Z = 1. Mo radiation, R = 0.030, 0.047 for 409, 322 reflexions.

Atomic positions

			n = 1				n = 0.26			
			x	y	z	Occ.	x	y	z	Occ.
Sb	in	12(e)	0.836	0	1/2	1	0.834	0	1/2	1
O(1)		12(d)	0.367	0	0	1	0.369	0	0	1
O(2)		24(g)	0	0.341	0.292	1	0	0.340	0.295	1
$H_2O(1)$		16(f)	0.086	0.086	0.086	0.30	0.090	0.090	0.090	0.17
$H_2O(2)$		16(f)	0.172	0.172	0.172	0.45	0.175	0.175	0.175	0.03

The structure contains a framework of edge- and corner-sharing SbO_6 octahedra, as in $KSbO_3$ (1) and related compounds (2), with partially-occupied H_3O^+ sites along [111].

1. Structure Reports, 11, 443.
2. Ibid., 45A, 265; 46A, 402 [$Ag_{14}Sb_{12}O_{36}F_2$].

POTASSIUM SODIUM ANTIMONATE
$K_2NaSb_3O_9$

H. WATELET, J.-P. PICARD, G. BAUD, J.-P. BESSE and R. CHEVALIER, 1981. Mater. Res. Bull., 16, 877-882.

Cubic, Pn3, a = 9.515 Å, Z = 4. Mo radiation, R = 0.059 for 563 reflexions. K in
8(e): x = 0.399; 0.5 Na in 8(e): x = 0.018; Sb in 12(g): x = 0.590; O(1) in 12(f):
x = 0.613; O(2) in 24(h): 0.250,0.592,0.542 (origin at centre).

Cubic $KSbO_3$-type (1). Sb-6 O = 1.96-2.01; Na-6 O = 2.51, 2.66, K-9 O = 2.68-
3.38 Å.

1. Structure Reports, 11, 443.

BARIUM ANTIMONATE
$BaSb_3O_5(OH)$

F. THUILLIER-CHEVIN, P. MARAINE and G. PÉREZ, 1981. Acta Cryst., B37, 11-15.

Monoclinic, $P2_1/c$, a = 8.974, b = 5.714, c = 14.195 Å, β = 100.69°, D_m = 5.54, Z = 4.
Mo radiation, R = 0.042 for 2878 reflexions.

The structure (Fig. 1) contains $(Sb_3O_6^{3-})_n$ chains, linked to form tunnels into
which the Sb(III) lone pairs project. Each Sb has SbO_4E trigonal bipyramidal coordination (E = equatorial lone pair), Sb-O = 1.93-2.28 Å; Ba has 7-coordination, Ba-O
= 2.78-2.95 Å. The H atom was not located.

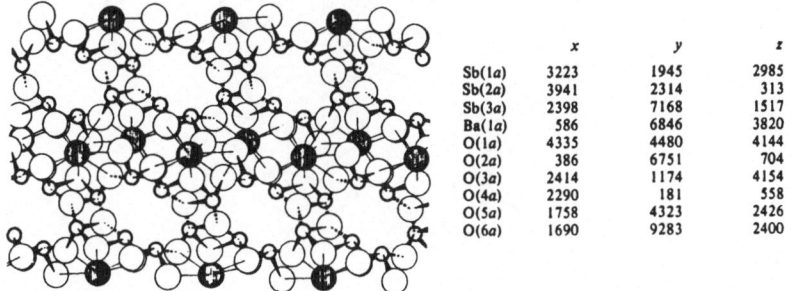

	x	y	z
Sb(1a)	3223	1945	2985
Sb(2a)	3941	2314	313
Sb(3a)	2398	7168	1517
Ba(1a)	586	6846	3820
O(1a)	4335	4480	4144
O(2a)	386	6751	704
O(3a)	2414	1174	4154
O(4a)	2290	181	558
O(5a)	1758	4323	2426
O(6a)	1690	9283	2400

Fig. 1.   Structure of $BaSb_3O_5(OH)$, and atomic positional parameters (x 10^4).

POTASSIUM TITANIUM ANTIMONATE
$K_{2.38}Sb_{3.62}Ti_{0.38}O_{11}$

Y. PIFFARD, R. MARCHAND and M. TOURNOUX, 1981. Ann. Chim., 6, 419-427.

Monoclinic, $P2_1/a$, a = 21.055, b = 10.233, c = 10.121 Å, β = 110.48°, Z = 8. Mo
radiation, R = 0.048 for 3808 reflexions.

The structure (Fig. 1) is related to that of $K_2Sb_4O_{11}$ (1), both containing
M_8O_{22} units, built up from MO_6 octahedra. Ti partially occupies one of eight M
positions. K ions have 8- to 11-coordinations.

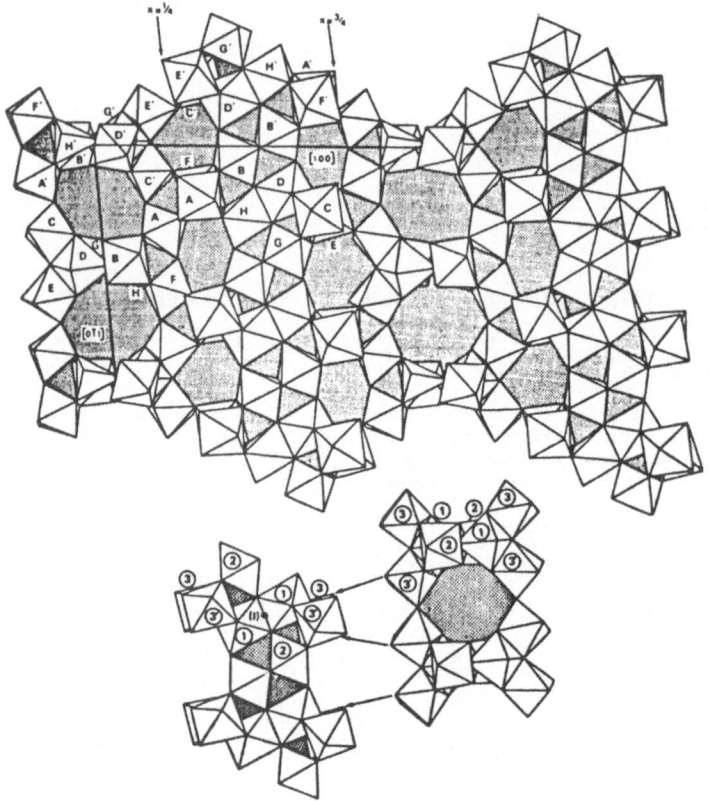

Fig. 1.   Structure of $K_{2.38}Sb_{3.62}Ti_{0.38}O_{11}$ viewed along [011], and the
M_8O_{22} units viewed along $\underline{b}$ and $\underline{a}$.

1. Structure Reports, 40A, 183.

CHROMIUM(III) ANTIMONATE(V)
IRON(III) ANTIMONATE(V)
$MSbO_4$ (M = Cr, Fe)

J. AMADOR and I. RASINES, 1981. J. Appl. Cryst., 14, 348-349.

Tetragonal, $P4_2/mnm$, a = 4.5899, 4.6388, c = 3.0525, 3.0773 Å, D_m = 5.31, 6.00, Z = 1.
Powder data. M/Sb in 2(a); O in 4(f): x = 0.318, 0.315. Rutile structure, with
statistical distribution of M and Sb.

BARIUM MANGANESE ANTIMONY OXIDE
BARIUM MANGANESE BISMUTH OXIDE
$Ba_2Mn_2Sb_2O$ $Ba_2Mn_2Bi_2O$

E. BRECHTEL, G. CORDIER and H. SCHÄFER, 1981. Z. Naturforsch., $\underline{36B}$, 27-30.

Hexagonal, $P6_3/mmc$, a = 4.71, 4.803, c = 20.04, 20.097 Å, D_m = 6.56 (Bi), Z = 2.
Mo radiation, R = 0.055, 0.078 for 199, 263 reflexions. Ba(1) in 2(b); Ba(2) in 2(a);
Mn in 4(f): z = 0.6495, 0.6499; Sb or Bi in 4(f): z = 0.3873, 0.3883; O in 2(d).

Mn and Sb (or Bi) atoms form corrugated hexagonal nets, which are bridged by
Mn-O-Mn linkages to form double layers, separated by Ba atoms.

NICKEL ANTIMONATE   ZINC ANTIMONATE
$NiSb_2O_4$ $ZnSb_2O_4$

J.-R. GAVARRI, 1981. C.R. Acad. Sci. Paris, $\underline{292}$, 895-898.

Tetragonal, $P4_2/mbc$, a = 8.372-8.360, 8.501-8.482, c = 5.908-5.906, 5.921-5.922 Å, at
300-5K, Z = 4. Neutron radiation, powder data. Ni or Zn in 4(d): 0,1/2,1/4; Sb in
8(h): x = 0.1718, 0.1775, y = 0.1627, 0.1613; O(1) in 8(g): x = 0.6756, 0.6802, O(2)
in 8(h): x = 0.0983, 0.0948, y = 0.6386, 0.6412 (parameters at 5K, similar values at
100, 200, 300K).

STRONTIUM TITANATE
$SrTiO_3$

I.   J. HUTTON, R.J. NELMES and H.J. SCHEEL, 1981. Acta Cryst., A$\underline{37}$, 916-920.

II. J. HUTTON and R.J. NELMES, 1981. J. Phys. C: Solid State Phys., $\underline{14}$, 1713-1736.

Cubic, [Pm3m], a = 3.90 Å, at 112K, Z = 1. Neutron radiation, R_W = 0.030 for 452
reflexions. Perovskite structure.

BARIUM TITANATE (RHOMBOHEDRAL)
$BaTiO_3$

W. SCHILDKAMP and K. FISCHER, 1981. Z. Kristallogr., $\underline{155}$, 217-226.

Rhombohedral, R3m, unit cell parameters not given [see $\underline{1}$], Z = 1. Neutron radiation,
R = 0.017, 0.021 for 950, 295 reflexions at 132, 196K. Ba in 1(a): x,x,x, x = 0; Ti
in 1(a): x,x,x, x = 0.4889; O in 3(b): x,x,z, x = 0.5110, z = 0.0180.

Structure as previously described ($\underline{1}$).

$\underline{1}$. Structure Reports, $\underline{40A}$, 185.

BARIUM TITANATE (TETRAGONAL, HIGH-PRESSURE)
$BaTiO_3$

R. WÄSCHE, W. DENNER and H. SCHULZ, 1981. Mater. Res. Bull., 16, 497-500.

Below 3.4 GPa, tetragonal, P4mm, a = 3.99-3.97, c = 4.03-3.97 Å, Z = 1. Ag radiation, no details of data. Ba in 1(a): 0,0,0; Ti in 1(b): 1/2,1/2,z, z = 0.510-0.507; O(1) in 1(b): z = -0.028 to -0.020; O(2) in 2(c): z = 0.479-0.488.

Above 3.4 GPa, lattice becomes cubic, a = 3.97 Å, but structure remains tetragonal; at 4.4 GPa z = 0.505, -0.011, 0.492 for Ti, O(1), O(2).

Structure as previously described (1).

1. Structure Reports, 11, 446; 12, 200; 15, 198; 17, 432; 19, 382; 21, 319; 26, 370; 40A, 302.

BARIUM ALUMINUM TITANATE
$Ba_4Al_2Ti_{10}O_{27}$

J. SCHMACHTEL and H. MÜLLER-BUSCHBAUM, 1981. Z. anorg. Chem., 472, 89-94.

Monoclinic, C2/m, a = 19.737, b = 11.349, c = 9.837 Å, β = 109.4°, Z = 4. R = 0.073 for 3402 reflexions.

The structure (Fig. 1) contains a framework of MO_6 distorted octahedra (M = 5/6 Ti + 1/6 Al), M-O = 1.81-2.15 (and one distance of 2.53) Å. Ba ions have 11- and 12-coordinations, Ba-O = 2.68-3.22 Å.

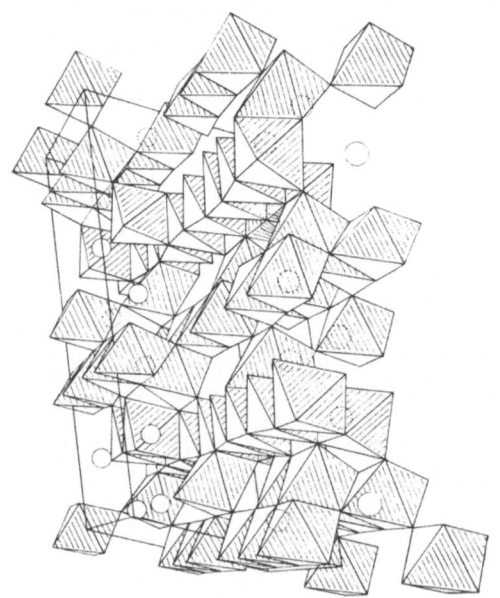

Fig. 1.   Structure of $Ba_4Al_2Ti_{10}O_{27}$.

ZIRCONOLITE
$CaZrTi_2O_7$

B.M. GATEHOUSE, I.E. GREY, R.J. HILL and H.J. ROSSELL, 1981. Acta Cryst., B37, 306-312.

$CaZr_xTi_{3-x}O_7$, monoclinic, C2/c, a = 12.445, 12.444, b = 7.288, 7.266, c = 11.487, 11.341 Å, β = 100.39, 100.59°, for x = 1.30, 0.85, Z = 8. Mo radiation, R = 0.045, 0.054 for 2011, 1126 reflexions.

The structure (Fig. 1) contains (001) sheets of (Ti,Zr)O_6 octahedra sharing corners in an arrangement of three- and six-membered rings. The sheets are linked by eight-coordinate Ca and seven-coordinate (Zr,Ti). One Ti is disordered over two positions close to the centre of a six-membered ring of octahedra, and has trigonal bipyramidal coordination.

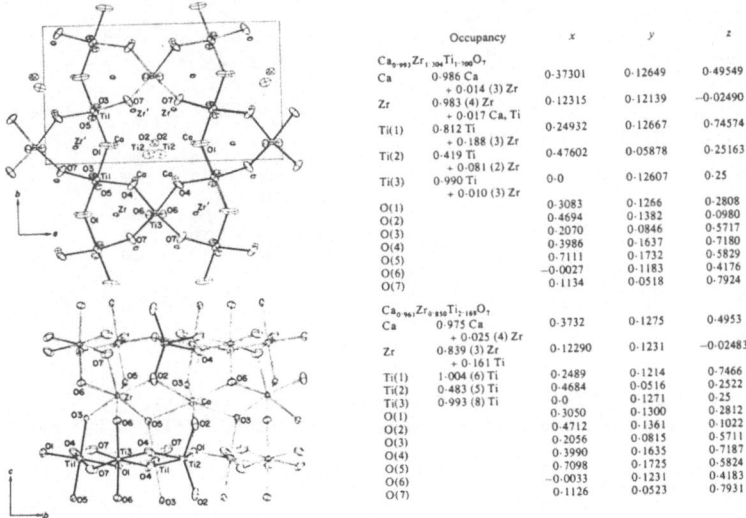

	Occupancy	x	y	z
$Ca_{0.993}Zr_{1.304}Ti_{1.700}O_7$				
Ca	0.986 Ca + 0.014 (3) Zr	0.37301	0.12649	0.49549
Zr	0.983 (4) Zr + 0.017 Ca, Ti	0.12315	0.12139	−0.02490
Ti(1)	0.812 Ti + 0.188 (3) Zr	0.24932	0.12667	0.74574
Ti(2)	0.419 Ti + 0.081 (2) Zr	0.47602	0.05878	0.25163
Ti(3)	0.990 Ti + 0.010 (3) Zr	0.0	0.12607	0.25
O(1)		0.3083	0.1266	0.2808
O(2)		0.4694	0.1382	0.0980
O(3)		0.2070	0.0846	0.5717
O(4)		0.3986	0.1637	0.7180
O(5)		0.7111	0.1732	0.5829
O(6)		−0.0027	0.1183	0.4176
O(7)		0.1134	0.0518	0.7924
$Ca_{0.961}Zr_{0.850}Ti_{2.189}O_7$				
Ca	0.975 Ca + 0.025 (4) Zr	0.3732	0.1275	0.4953
Zr	0.839 (3) Zr + 0.161 Ti	0.12290	0.1231	−0.02483
Ti(1)	1.004 (6) Ti	0.2489	0.1214	0.7466
Ti(2)	0.483 (5) Ti	0.4684	0.0516	0.2522
Ti(3)	0.993 (8) Ti	0.0	0.1271	0.25
O(1)		0.3050	0.1300	0.2812
O(2)		0.4712	0.1361	0.1022
O(3)		0.2056	0.0815	0.5711
O(4)		0.3990	0.1635	0.7187
O(5)		0.7098	0.1725	0.5824
O(6)		−0.0033	0.1231	0.4183
O(7)		0.1126	0.0523	0.7931

Fig. 1.   Structure of zirconolite.

POTASSIUM TITANIUM TANTALATES
KTi_3TaO_9 $K_3TiTa_7O_{21}$

B.M. GATEHOUSE and M.C. NESBIT, 1981. J. Solid State Chem., 39, 1-6.

KTi_3TaO_9, orthorhombic, Pnmm, a = 6.392, b = 3.793, c = 14.877 Å, Z = 2. Mo radiation, R = 0.055 for 283 reflexions.

$K_3TiTa_7O_{21}$, hexagonal, $P6_3/mcm$, a = 9.095, c = 12.063 Å, Z = 2. Mo radiation, R = 0.035 for 323 reflexions.

KTi$_3$TaO$_9$ is isostructural with KTi$_3$NbO$_9$ (1), the structure (Fig. 1) containing pairs of edge-shared octahedra joined by further edge sharing to form double zigzag units; further corner sharing produces tunnels which contain 12-coordinate K ions. Ta/Ti are disordered in the octahedral sites, 42/58 for B1 and 8/92 for B2 positions.

K$_3$TiTa$_7$O$_{21}$ is isostructural with the non-stoichiometric phase K$_{5.5}$Ta$_{15.7}$O$_{42}$ (2), as previously reported (3). Ti substitutes for Ta only in position B1 (Fig. 1).

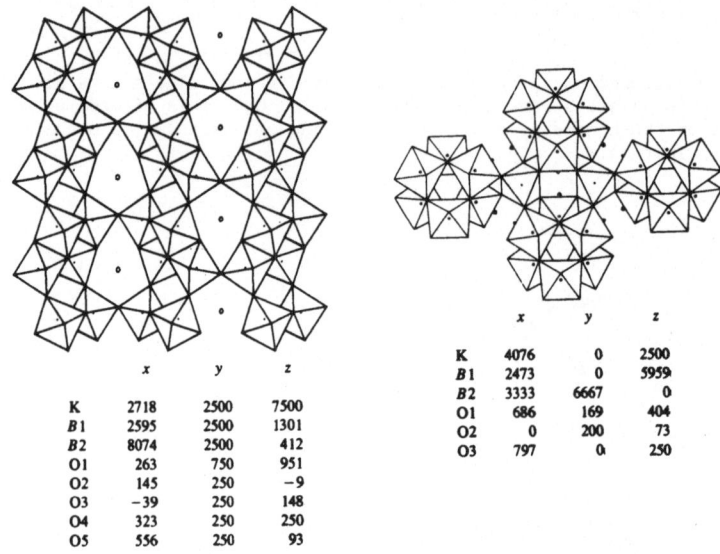

	x	y	z
K	2718	2500	7500
B1	2595	2500	1301
B2	8074	2500	412
O1	263	750	951
O2	145	250	−9
O3	−39	250	148
O4	323	250	250
O5	556	250	93

	x	y	z
K	4076	0	2500
B1	2473	0	5959
B2	3333	6667	0
O1	686	169	404
O2	0	200	73
O3	797	0	250

Fig. 1.   Structures of KTi$_3$TaO$_9$ (left) and K$_3$TiTa$_7$O$_{21}$ (right), and atomic positional parameters (x 10^4 for K, Ta/Ti, x 10^3 for O).

1. Structure Reports, 29, 327.
2. Ibid., 45A, 247.
3. Ibid., 43A, 197.

IRON TITANIUM OXIDE (MONOCLINIC)
Fe$_2$TiO$_5$

I.   M. DROFENIK, L. GOLIČ, D. HANŽEL, V. KRAŠEVEC, A. PRODAN, M. BAKKER and D. KOLAR, 1981. J. Solid State Chem., 40, 47-51.

II. V. KRAŠEVEC, A. PRODAN, M. BAKKER, M. DROFENIK, L. GOLIČ, D. HANŽEL and D. KOLAR, 1981. Ibid., 40, 52-58.

Monoclinic, C2/c, a = 10.101, b = 5.037, c = 7.024 Å, β = 110.9°, D$_m$ = 4.81, Z = 4. Mo radiation, R = 0.11 for 477 reflexions.

The structure (Fig. 1) is of V$_3$O$_5$-type (1) and contains slabs of TiO$_6$ and face-sharing pairs of FeO$_6$ octahedra. Ti-O = 2.00-2.03, Fe-O = 1.85-2.26 Å. There is probably some Ti/Fe disorder, and the structure contains stacking faults, which are investigated by electron microscopy and diffraction. Pseudo-brookite (Fe$_2$TiO$_5$) and rutile (TiO$_2$) phases are also found in the Fe$_2$O$_3$-TiO$_2$ system.

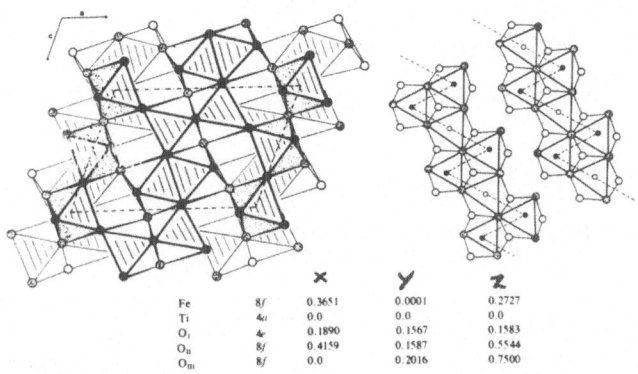

Fe	8f	0.3651	0.0001	0.2727
Ti	4a	0.0	0.0	0.0
O_I	4c	0.1890	0.1567	0.1583
O_{II}	8f	0.4159	0.1587	0.5544
O_{III}	8f	0.0	0.2016	0.7500

Fig. 1.   Structure of monoclinic Fe_2TiO_5.

1. Structure Reports, 23, 337; 42A, 235.

TITANIUM IRON OXIDE
Ti_4Fe_2O

TITANIUM IRON OXIDE HYDRIDE
$Ti_4Fe_2OD_{2.2}$

C. STIOUI, D. FRUCHART, A. ROUAULT, R. FRUCHART, E. ROUDAUT and J. REBIÉRE, 1981.
Mater. Res. Bull., 16, 869-876.

Cubic, Fd3m, a = 11.28 Å, Z = 16.   Neutron powder data.   Ti(1) in 48(f): x = 0.940;
Ti(2) in 16(d); Fe in 32(e): x = 0.293 (oxide), 0.298 (hydride); O in 16(c).   In the
hydride, 8 D(1) in 8(a); 19 D(2) in 192(i): 0.543,0.028,0.334; 9 D(3) in 96(g): x =
0.972, z = 0.142 (origin at centre). [Previous study in 1.]

   D(1) is located in a Ti_6 octahedron, and D(2) and D(3) form a cluster arrangement.

1. Structure Reports, 15, 52.

NICKEL TITANIUM OXIDE
$Ni_{2.62}Ti_{0.69}O_4$

NICKEL TITANIUM SILICON OXIDE
$Ni_{2.42}Ti_{0.74}Si_{0.05}O_4$

G.A. LAGER, T. ARMBRUSTER, F.K. ROSS, F.J. ROTELLA and J.D. JORGENSEN, 1981. J. Appl.
Cryst., 14, 261-264.

Cubic, Fd3m, a = 8.3416, 8.3222 Å, D_m = 5.70, 5.58, Z = 8.   Neutron powder data.   16
Ni in 16(d); 4.5, 3.2 Ni in 16(c); Ti and Ti/Si in 8(a); O in 32(e): x = 0.2519, 0.2515
[origin at centre]. Defect spinels, with Ti in tetrahedral sites, and excess Ni in
the octahedral 16(c) site.

BARIUM TITANIUM PLATINUM OXIDES
$Ba(Ti_{0.88}Pt_{0.12})O_3$ $Ba_4(Ti_{1.81}Pt_{0.19})PtO_{10}$

R. FISCHER AND E. TILLMANNS, 1981. Z. Kristallogr., 157, 69-81.

$Ba(Ti_{0.88}Pt_{0.12})O_3$, hexagonal, $P6_3/mmc$, a = 5.723, c = 14.023 Å, Z = 6. Mo radiation, R = 0.021 for 309 reflexions.

$Ba_4(Ti_{1.81}Pt_{0.19})PtO_{10}$, orthorhombic, Cmca, a = 5.783, b = 13.368, c = 13.129 Å, Z = 4. Mo radiation, R = 0.037 for 1026 reflexions.

The structures (Fig. 1) contain hexagonal close-packed Ba and O with some positions not occupied, and Ti and Pt in octahedral holes.

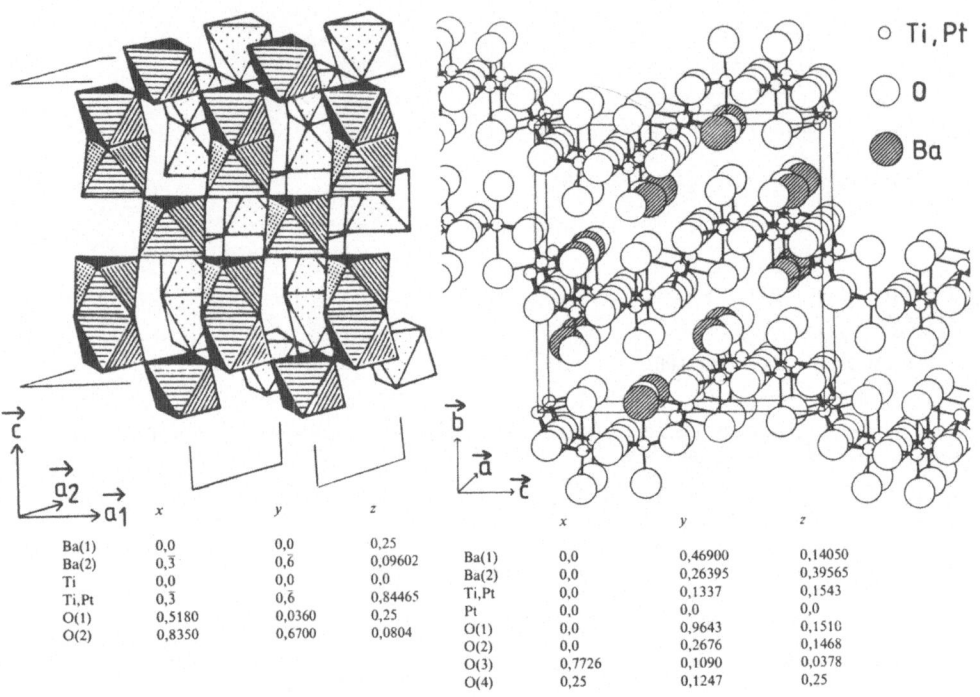

	x	y	z
Ba(1)	0,0	0,0	0,25
Ba(2)	0,$\bar{3}$	0,$\bar{6}$	0,09602
Ti	0,0	0,0	0,0
Ti,Pt	0,$\bar{3}$	0,$\bar{6}$	0,84465
O(1)	0,5180	0,0360	0,25
O(2)	0,8350	0,6700	0,0804

	x	y	z
Ba(1)	0,0	0,46900	0,14050
Ba(2)	0,0	0,26395	0,39565
Ti,Pt	0,0	0,1337	0,1543
Pt	0,0	0,0	0,0
O(1)	0,0	0,9643	0,1510
O(2)	0,0	0,2676	0,1468
O(3)	0,7726	0,1090	0,0378
O(4)	0,25	0,1247	0,25

○ Ti,Pt
◯ O
◓ Ba

Fig. 1.   Structures of $Ba(Ti_{0.88}Pt_{0.12})O_3$ (left) and
            $Ba_4(Ti_{1.81}Pt_{0.19})PtO_{10}$ (right).

LEAD TITANATE ZIRCONATE
$Pb(Zr,Ti)O_3$ $PbZr_{0.6}Ti_{0.4}O_3$

A. AMIN, R.E. NEWNHAM, L.E. CROSS and D.E. COX, 1981. J. Solid State Chem., 37, 248-255.

295K, rhombohedral, R3m, a = 5.755, c = 14.214 Å (double pseudo-cubic cell has 2a = 8.162 Å, α = 89.68°); 9K, rhombohedral, R3c, a = 5.760, c = 14.251 Å (2a = 8.173 Å, α = 89.61°). Neutron powder data.

Distorted perovskites (<u>1</u>). The phase transition is of a diffuse type, possibly involving short-range ordering of Zr and Ti.

<u>1</u>. Structure Reports, <u>44</u>A, 197.

LITHIUM VANADIUM OXIDE (HIGH-PRESSURE)
$LiVO_2$

C. CHIEH, B.L. CHAMBERLAND and A.F. WELLS, 1981. Acta Cryst., B<u>37</u>, 1813-1816.

Cubic, Fd3m, a = 8.227 Å, D_m = 4.23, Z = 16. Mo radiation, R = 0.072 for 136 reflexions. 2/3V + 1/3Li in 16(c): 1/8,1/8,1/8; 1/3V + 2/3Li in 16(d): 5/8,5/8,5/8; O in 32(e): x,x,x, x = -0.1236.

The structure is a 2 x 2 x 2 NaCl superstructure, with partial or short-range ordering of Li and V in the octahedral holes of a slightly-disorted close-packed O lattice.

SODIUM LITHIUM METAVANADATE
$NaLiV_2O_6$

R.S. BUBNOVA, S.K. FILATOV, V.S. GRUNIN and Z.N. ZONN, 1980. Kristallografija, <u>25</u>, 1287-1289 [Soviet Phyics - Crystallography, <u>25</u>, 734-736].

Monoclinic, C2/c, a = 10.187, b = 9.074, c = 5.840 Å, β = 108.95°, Z = 4. Mo radiation, R = 0.022 for 712 reflexions.

Atomic positions

	x	y	z
Na	1/2	0.2066	1/4
Li	1/2	0.4121	3/4
V	0.2889	0.0922	0.7448
O(1)	0.1171	0.0947	0.6476
O(2)	0.3505	0.2606	0.7963
O(3)	0.3563	-0.0018	0.0317

Diopside-type structure (Fig. 1) as for $NaVO_3$ (<u>1</u>) and $LiVO_3$ (<u>2</u>), with ordered Na/Li distribution.

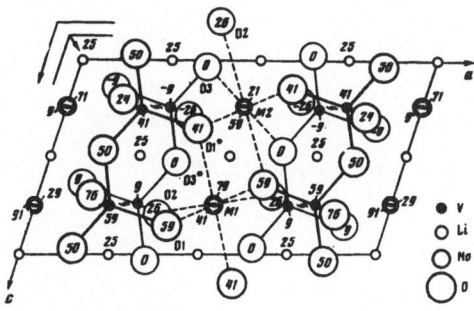

Fig. 1.   Structure of $NaLiV_2O_6$.

1. Structure Reports, 9, 180; 40A, 188.
2. Ibid., 39A, 222.

STRONTIUM DECAVANADATE
$SrV_{10}O_{15}$ $SrV(III)_8V(II)_2O_{15}$

D. CHALES de BEAULIEU and H. MÜLLER-BUSCHBAUM, 1981. Z. anorg. Chem., 472, 33-37.

Orthorhombic, Ccmb, a = 9.915, b = 11.574, c = 9.324 Å, Z = 4. R = 0.08 for 753 reflexions.

    Isostructural with $BaV_{10}O_{15}$ (1), the structure containing a $V_{12}O_{50}$ octahedral framework with a large cavity (Fig. 1). V-O = 1.95-2.19, Sr-O (12-coordination) = 2.73-2.95 Å.

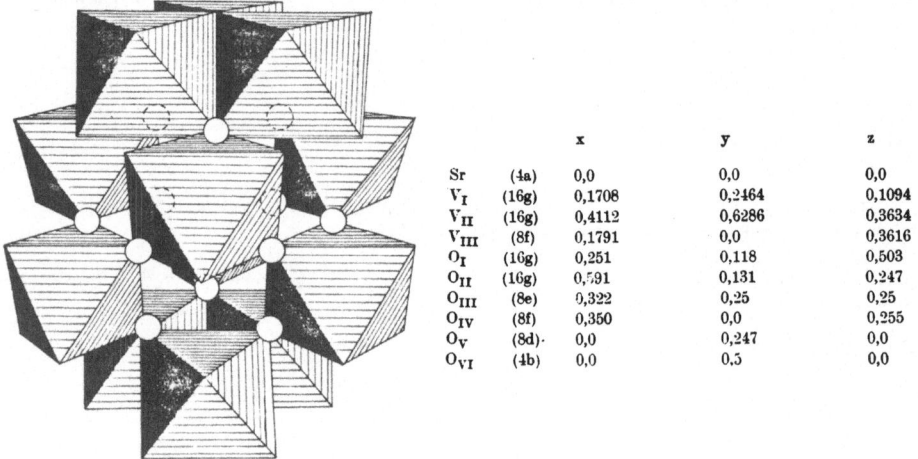

		x	y	z
Sr	(4a)	0,0	0,0	0,0
V_I	(16g)	0,1708	0,2464	0,1094
V_{II}	(16g)	0,4112	0,6286	0,3634
V_{III}	(8f)	0,1791	0,0	0,3616
O_I	(16g)	0,251	0,118	0,503
O_{II}	(16g)	0,591	0,131	0,247
O_{III}	(8e)	0,322	0,25	0,25
O_{IV}	(8f)	0,350	0,0	0,255
O_V	(8d)	0,0	0,247	0,0
O_{VI}	(4b)	0,0	0,5	0,0

Fig. 1.   The $V_{12}O_{50}$ grouping in $SrV_{10}O_{15}$, and atomic positional parameters.

1. Structure Reports, 46A, 247.

INDIUM LITHIUM VANADATE
$In_{0.6}Li_{1.2}VO_4$

M. TOUBOUL and P. TOLEDANO, 1981. J. Solid State Chem., 38, 386-393.

Orthorhombic, Cmcm, a = 5.763, b = 8.742, c = 6.385 Å, Z = 4. Mo radiation, R = 0.019 for 726 reflexions.

Atomic positions

			x	y	z
0.6 In	in	4(a)	0	0	0
V		4(c)	0	0.3574	1/4
O(1)		8(g)	0.2441	0.4792	1/4
O(2)		8(f)	0	0.2466	0.0348
n* Li(1)		4(a)	0	0	0
(1.2-n) Li(2)		4(c)	1/2	0.165	1/4

* Three models give the same R value: n = 0.2, 0.3, 0.4.

The structure (Fig. 1) contains a three-dimensional framework of VO_4 tetrahedra and InO_6 octahedra, as in $InVO_4$ (1). Li(1) substitutes for In, with additional Li(2) in tetrahedral sites. V-O = 1.68, 1.76, In-O = 2.17, 2.18, Li(2)-O = 1.98, 2.15 Å.

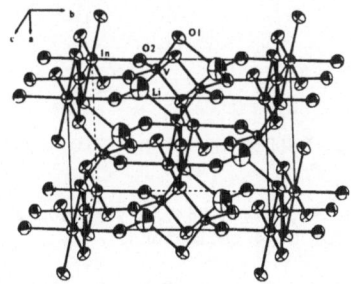

Fig. 1.   Structure of $In_{0.6}Li_{1.2}VO_4$.

1. Structure Reports, 46A, 247.

DREYERITE
$BiVO_4$

G. DREYER and E. TILLMANNS, 1981. Neues Jb. Miner. Mh., 151-154.

Tetragonal, $I4_1/amd$, a = 7.303, c = 6.584 Å, Z = 4. Mo radiation, R = 0.06 for 154 reflexions. Bi in 4(a): 0,3/4,1/8; V in 4(b): 0,1/4,3/8; O in 16(h): 0,0.072,0.208 (origin at centre).

Zircon structure (1). V-O = 1.69 (tetrahedral), Bi-O = 2.41, 2.51(2) Å (each x 4).

1. Strukturbericht, 1, 345, 408; Structure Reports, 22, 314; 29, 411; 37A, 349; 41A, 402.

MANGANESE MOLYBDATE VANADATE
$Mn(Mo,V)_2O_6$

R. KOZŁOWSKI and K. STADNICKA, 1981. J. Solid State Chem., 39, 271-276.

Monoclinic, C2, a = 9.412, b = 3.643, c = 6.767 Å, β = 112.00°, for composition $Mn_{0.47}Mo_{1.06}V_{0.94}O_6$, Z = 2. Mo radiation, R = 0.029 for 712 reflexions.

Atomic positions

			x	y	z
Mn	in	2(a)	0	0	0
Mo,V		4(c)	0.3139	0.5018	0.3502
O(1)		4(c)	0.1705	0.4444	0.1170
O(2)		4(c)	0.4749	0.5237	0.2950
O(3)		4(c)	0.3084	0.0042	0.4375

Brannerite-type structure ($\underline{1}$), with Mo/V disorder, and partial occupancy of the Mn site. Mn has octahedral coordination, Mn-O = 2.10, 2.21, 2.52 Å, and Mo/V has very-distorted octahedral coordination, Mo/V-O = 1.66, 1.69, 1.91, 1.93, 2.15, 2.48 Å.

$\underline{1}$. Structure Reports, $\underline{31}$A, 139; $\underline{33}$A, 313, 336; $\underline{38}$A, 255.

IRON VANADIUM OXIDE

$(Fe,V)_{18}O_{35}$ $Fe_{6.5}V_{11.5}O_{35}$

I.E. GREY, M. ANNE, A. COLLOMB, J. MULLER and M. MAREZIO, 1981. J. Solid State Chem., $\underline{37}$, 219-227.

Triclinic, P$\overline{1}$, a = 10.209, b = 9.387, c = 6.564 Å, α = 100.52, β = 94.35, γ = 98.85°, Z = 1. Ag radiation, R = 0.053 for 5654 reflexions.

The structure (Fig. 1) contains cubic close-packed oxygens, with cations ordered into tetrahedral (V^{5+}), square-pyramidal (V^{4+}), and octahedral (Fe^{3+}, V^{3+}, or V^{4+}) sites. Zigzag chains of edge-sharing octahedra are cross-linked by corner-sharing with tetrahedra and square pyramids, and by edge-sharing with square pyramids and octahedra.

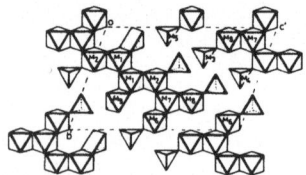

Fig. 1.   Structure of $(Fe,V)_{18}O_{35}$.

CADMIUM METAVANADATE TETRAHYDRATE

$Cd(VO_3)_2 \cdot 4H_2O$

L. ULICKA, 1980. Acta Fac. Rerum Nat. Univ. Comenianae Chim., no. 28, 153-156.

Monoclinic, Cc, a = 13.606, b = 10.496, c = 7.182 Å, β = 111.72°, Z = 4. R = 0.069 for 814 reflexions.

The structure contains chains parallel to $\underline{c}$ of edge-sharing VO_5 square-pyramids, $CdO_4(H_2O)_2$ octahedra, and additional water molecules; the water molecules are involved in hydrogen bonding.

CAESIUM TETRANIOBATE
$Cs_2Nb_4O_{11}$

M. GASPERIN, 1981. Acta Cryst., B37, 641-643.

Orthorhombic, P2nn, a = 10.484, b = 28.898, c = 7.464 Å, D_m = 4.785, Z = 8. Mo radiation, R = 0.042 for 3461 reflexions.

    The structure (Fig. 1) is derived from that of pyrochlore, and contains a three-dimensional framework of NbO_4 tetrahedra and NbO_6 octahedra, with Cs ions in inter-linked tunnels. Nb-O = 1.74-1.93, mean 1.83 (tetrahedra), 1.79-2.25, mean 2.00 (octahedra), Cs-O = 2.89-3.43 Å (4 and 7 oxygen neighbours).

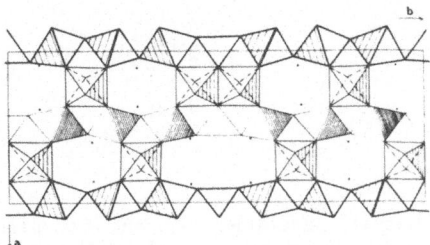

Fig. 1.   Structure of $Cs_2Nb_4O_{11}$.

STRONTIUM DINIOBATE
$SrNb_2O_6$

V.K. TRUNOV, I.M. AVERINA and Ju.A. VELIKODNIJ, 1981. Kristallografija, 26, 390-391 [Soviet Physics - Crystallography, 26, 222-223].

Monoclinic, $P2_1/c$, a = 7.722, b = 5.592, c = 10.989 Å, β = 90.36°, Z = 4. R = 0.052 for 1488 reflexions.

Atomic positions

	x	y	z
Nb(1)	0.0144	0.0271	0.1438
Nb(2)	0.5254	0.0358	0.1423
Sr	0.2500	0.5340	0.0387
O(1)	0.0329	0.2185	-0.0256
O(2)	0.4676	0.2199	-0.0255
O(3)	0.0474	0.3648	0.2122
O(4)	0.4493	0.3581	0.2133
O(5)	0.2500	-0.0308	0.1475
O(6)	-0.2485	0.1390	0.1294

    The structure is a slight distortion of the orthorhombic $CaTa_2O_6$ structure (1). Nb atoms have octahedral coordinations, Nb-O = 1.85-2.16 Å, and Sr has 8-coordination, Sr-O = 2.53-2.71 Å.

1. Structure Reports, 28, 201.

ALUMINUM NIOBATE
$AlNbO_4$

V.A. EFREMOV, V.K. TRUNOV and A.A. EVDOKIMOV, 1981. Kristallografija, 26, 305-311
[Soviet Physics - Crystallography, 26, 172-176].

Monoclinic, C2/m, a = 12.151, b = 3.735, c = 6.486 Å, β = 107.63°, Z = 4. Mo radiation, R = 0.028 for 708 reflexions.

Atomic positions

	x	y	z
M(1)	0.1024	0	0.2332
M(2)	0.1938	0	0.2009
O(1)	0.1376	0	0.5169
O(2)	-0.0579	0	0.1359
O(3)	-0.3592	0	0.2022
O(4)	0.2636	0	0.1466

M(1) = 0.81 Nb + 0.19 Al
M(2) = 0.19 Nb + 0.81 Al

The structure is as previously determined (1), but with partial disordering of
cations in the octahedral sites. M(1)-O = 1.76-2.29, M(2)-O = 1.75-2.16 Å.

1. Structure Reports, 27, 523.

ANTIMONY(III) NIOBATE(V)
$SbNbO_4$

V.I. PONOMAREV, O.S. FILIPENKO, L.O. ATOVMJAN, N.V. RANNEV, S.A. IVANOV and Ju.N.
VENEVCEV, 1981. Kristallografija, 26, 341-348 [Soviet Physics - Crystallography,
26, 194-198].

300-683K, orthorhombic, $Pna2_1$, a = 5.557, b = 4.932, c = 11.795 Å, at 300K, Z = 4.
Mo radiation, R = 0.044, 0.062, 0.069 for 1700, 1037, 334 reflexions, at 300, 573,
623K.

683-963K, orthorhombic, Pnan, a = 5.613, b = 4.955, c = 11.860 Å, at 963K, Z = 4.
Mo radiation, R = 0.078, 0.072, 0.072, 0.10 for 898, 868, 840, 754 reflexions, at
723, 773, 873, 963K.

Atomic positions

	300K			963K		
	x	y	z	x	y	z
Sb	0.4745	0.0516	0.2484	1/2	0.0534	1/4
Nb	0.1055	-0.0083	0	0.1072	0	0
O(1)	0.1779	-0.1655	-0.1505	0.1632	-0.1680	-0.1547
O(2)	0.4182	-0.2112	0.0582	0.3911	-0.2455	0.0511
O(3)	0.1529	0.1676	0.1603			
O(4)	0.3697	0.2767	-0.0443			

In the room-temperature structure, Nb has octahedral coordination, Nb-O = 1.82-
2.12 Å, and Sb also has 6-coordination but with a stereochemically-active lone electron
pair, Sb-O = 2.00-2.34 (all on one side), 2.61, 3.06 Å. The high-temperature form
has a very similar but more symmetrical arrangement.

FERGUSONITE (MONOCLINIC)
YNbO$_4$

V.K. TRUNOV, V.A. EFREMOV, Ju.A. VELIKODNIJ and I.M. AVERINA, 1981. Kristallografija,
26, 67-71 [Soviet Physics - Crystallography, **26**, 35-37].

Monoclinic, C2/c, a = 7.037, b = 10.945, c = 5.298 Å, β = 134.07°, Z = 4. R = 0.077
for 1232 reflexions, for a synthetic sample. Previous study in **1**.

Atomic positions

	x	y	z
Nb	0	0.1445	1/4
Y	0	0.6212	1/4
O(1)	0.2442	0.0418	0.3374
O(2)	0.2919	0.2819	0.2968

The structure is related to that of scheelite, both containing 6- and 8-coordinate
cation polyhedra. Nb-O = 1.82-2.42 (6-coordination), Y-O = 2.32-2.41 Å (8-coordination).

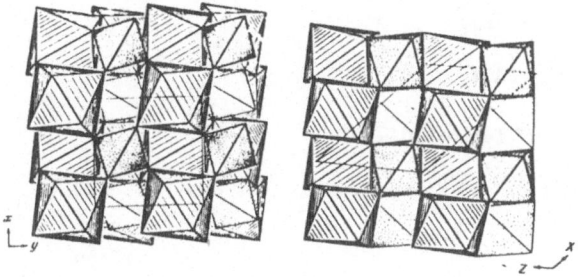

Fig. 1.   Structures of scheelite and fergusonite.

1. Structure Reports, **23**, 389.

BARIUM COBALT NIOBATE
Ba$_6$CoNb$_9$O$_{30}$

U. LEHMANN and H. MÜLLER-BUSCHBAUM, 1981. Z. anorg. Chem., **481**, 7-12.

Tetragonal, P4bm, a = 12.589, c = 4.009 Å, Z = 1. Mo radiation, R = 0.076 for 767
reflexions.

Tetragonal bronze type structure (**1**). Nb and Co/Nb have octahedral coordinations
(Fig. 1), Nb-O = 1.89-2.12, Co/Nb-O = 1.97-2.04 Å. Ba(1) has 12-coordination, Ba-O =
2.80, 2.81 Å, and Ba(2) has irregular coordination with 15 oxygen neighbours within
3.5 Å.

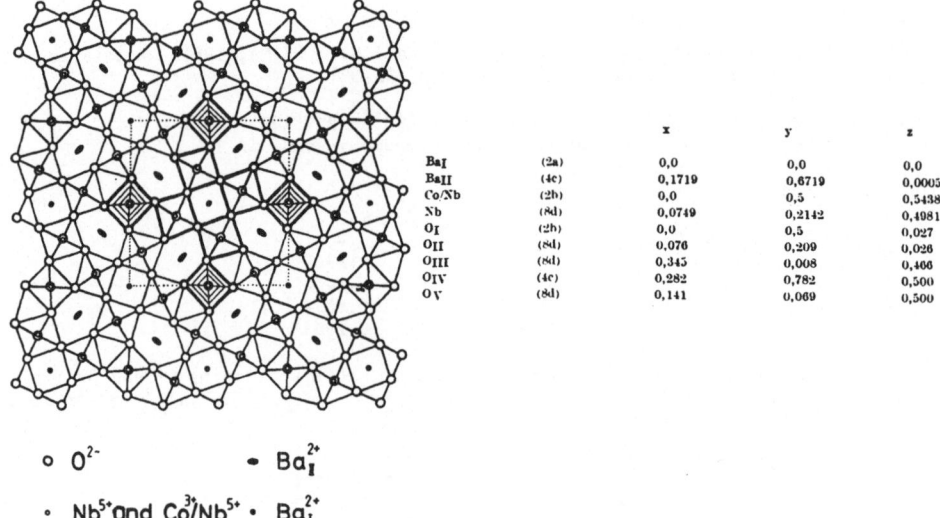

		x	y	z
Ba_I	(2a)	0,0	0,0	0,0
Ba_II	(4c)	0,1719	0,6719	0,0005
Co/Nb	(2b)	0,0	0,5	0,5438
Nb	(8d)	0,0749	0,2142	0,4981
O_I	(2b)	0,0	0,5	0,027
O_II	(8d)	0,076	0,209	0,026
O_III	(8d)	0,345	0,008	0,466
O_IV	(4c)	0,282	0,782	0,500
O_V	(8d)	0,141	0,069	0,500

○ O^{2-} ● Ba_I^{2+}

● Nb^{5+} and Co/Nb^{5+} ● Ba_I^{2+}

Fig. 1.   Structure of $Ba_6CoNb_9O_{30}$.

1. Structure Reports, 30A, 346.

LANTHANUM NIOBATE
$La_{0.33}NbO_3$

V.K. TRUNOV, I.M. AVERINA, A.A. EVDOKIMOV and A.M. FROLOV, 1981. Kristallografija, 26, 189-191 [Soviet Physics - Crystallography, 26, 104-105].

Orthorhombic, Pmmm, a = 3.909, b = 3.917, c = 7.910 Å, Z = 2. Mo radiation, R = 0.053 for 601 reflexions.

Atomic positions

	x	y	z
2/3 La	0	0	0
Nb	1/2	1/2	0.2624
O(1)	1/2	1/2	0
O(2)	1/2	1/2	1/2
O(3)	1/2	0	0.2314
O(4)	0	1/2	0.2335

      Orthorhombically-distorted perovskite structure. Nb-O = 1.88-2.08 (octahedral), La-O = 2.68-2.77 Å (12-coordination).

STRONTIUM TANTALATE NIOBATE
$Sr_2(Ta_{1-x}Nb_x)_2O_7$ (x ∼ 0.12)

STRONTIUM DITANTALATE
$Sr_2Ta_2O_7$

N. ISHIZAWA, F. MARUMO and S. IWAI, 1981. Acta Cryst., B37, 26-31.

Crystal data for $Sr_2(Ta_{1-x}Nb_x)_2O_7$ ($x \simeq 0.12$)

Temperature	300 K	573 K	773 K	1073 K
Crystal system	Orthorhombic	Orthorhombic	Orthorhombic	Orthorhombic
Space group	$Cmc2_1$	$Cmc2_1$	$Cmcm$	$Cmcm$
Cell dimensions				
a	3·961 Å	3·967 Å	3·967 Å	3·975 Å
b	27·110	27·247	27·325	27·393
c	5·687	5·705	5·723	5·739
z	4	4	4	4

Crystal data for $Sr_2Ta_2O_7$

Temperature	123 K	Room temperature
Crystal system	Orthorhombic	Orthorhombic
Space group	$Cmc2_1$	$Cmcm$
Cell dimensions		
a	3·940 Å	3·937 Å
b	27·15	27·198
c	5·692	5·692
z	4	4

Mo radiation, R = 0.047, 0.070, 0.087, 0.097, and 0.033 for 907, 748, 663, 598, and 520 reflexions for $Sr_2(Ta,Nb)_2O_7$ at 300, 573, 773, 1073K, and $Sr_2Ta_2O_7$ at 123K (Curie points are 675 and 166K, respectively).

Above the Curie points the structures are as previously described (1). Below the Curie points there is a slight movement of the metal atoms along c, and a small rotation of $(Ta,Nb)O_6$ octahedra around axes parallel to a (Fig. 1).

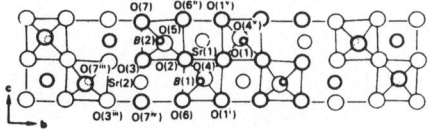

Fig. 1.   Structure of $Sr_2(Ta,Nb)_2O_7$ at 300K.

1. Structure Reports, 41A, 240; 42A, 267.

LANTHANUM TANTALATE (HIGH-TEMPERATURE)
$LaTaO_4$

R.J. CAVA and R.S. ROTH, 1981.   J. Solid State Chem., 36, 139-147.

Orthorhombic, $A2_1am$, a = 5.6643, b = 14.6411, c = 3.9457 Å, at 300°C, Z = 4. Neutron powder data.

Isostructural with room-temperature $BaMnF_4$ (1). The structure (Fig. 1) contains sheets perpendicular to b of corner-sharing TaO_6 octahedra; the sheets are linked by 9-coordinate La ions. Ta-O = 1.88-2.10, La-O = 2.39-2.89 Å. Below 175°C the structure is monoclinic (2, 3) with reoriented TaO_6 octahedra resulting in 8-coordination for La.

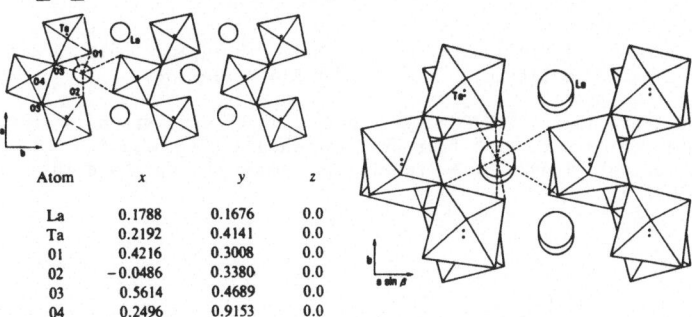

Atom	x	y	z
La	0.1788	0.1676	0.0
Ta	0.2192	0.4141	0.0
O1	0.4216	0.3008	0.0
O2	−0.0486	0.3380	0.0
O3	0.5614	0.4689	0.0
O4	0.2496	0.9153	0.0

Fig. 1.   Structure of orthorhombic $LaTaO_4$ at 300°C (left) and the monoclinic form at 25°C (2, right).

1. Structure Reports, 35A, 164.
2. T.A. KUROVA and V.B. ALEKSANDROV, 1971. Dokl. Akad. Nauk SSSR, 201, 1095.
3. Structure Reports, 46A, 256 [CeNbO$_4$].

SODIUM CHROMATE (FORM II)
Na$_2$CrO$_4$

J.K. NIMMO, 1981. Acta Cryst., B37, 431-433.

Orthorhombic, Cmcm, a = 5.862, b = 9.251, c = 7.145 Å, Z = 4. Neutron radiation,
R = 0.03 for 298 reflexions.

The structure (Fig. 1) is as previously described (1). It contains CrO$_4$
tetrahedra, linked by six-coordinate Na ions. Cr-O = 1.635, 1.656(1) Å, O-Cr-O =
108.8-109.8°, Na-O = 2.314-3.270 Å (two further O at 3.76, 3.88 Å).

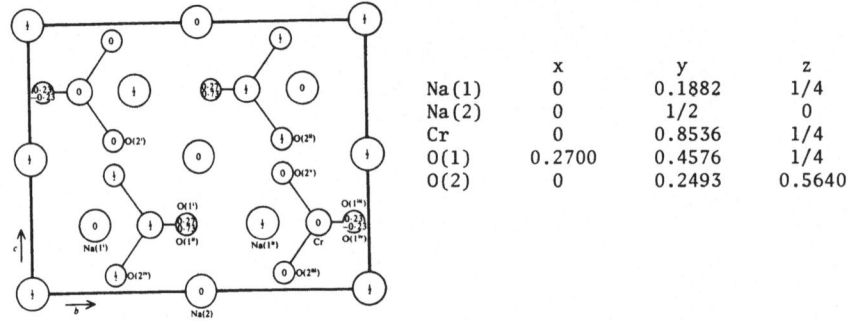

	x	y	z
Na(1)	0	0.1882	1/4
Na(2)	0	1/2	0
Cr	0	0.8536	1/4
O(1)	0.2700	0.4576	1/4
O(2)	0	0.2493	0.5640

Fig. 1.   Structure of sodium chromate.

1. Structure Reports, 18, 474.

AMMONIUM PHOSPHOTETRACHROMATE
(NH$_4$)$_3$PCr$_4$O$_{16}$

M.T. AVERBUCH-POUCHOT, A. DURIF and J.C. GUITEL, 1981. J. Solid State Chem., 36,
381-384.

Rhombohedral, R3m, a = 7.710 Å, α = 102.59°, Z = 1 (hexagonal cell, a = 12.033, c =
10.032 Å, Z = 3). Ag radiation, R = 0.054 for 516 reflexions.

The structure contains a P(OCrO$_3$)$_4^{3-}$ anion, which consists of a central PO$_4$
tetrahedron sharing corners with four CrO$_4$ tetrahedra (Fig. 1); P-O = 1.49, Cr-O =
1.79, 1.84 (bridging), 1.44-1.61 Å (terminal). Ammonium ions have 10-coordination,
N...O = 2.84-3.46 Å.

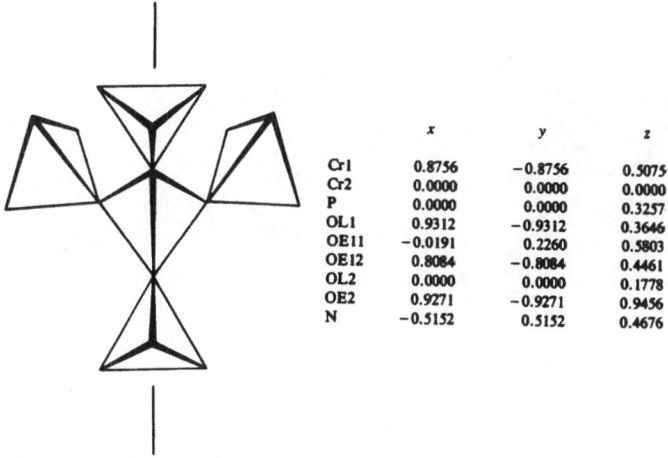

	x	y	z
Cr1	0.8756	−0.8756	0.5075
Cr2	0.0000	0.0000	0.0000
P	0.0000	0.0000	0.3257
OL1	0.9312	−0.9312	0.3646
OE11	−0.0191	0.2260	0.5803
OE12	0.8084	−0.8084	0.4461
OL2	0.0000	0.0000	0.1778
OE2	0.9271	−0.9271	0.9456
N	−0.5152	0.5152	0.4676

Fig. 1.   The $PCr_4O_{16}^{3-}$ ion and atomic positional parameters in $(NH_4)_3PCr_4O_{16}$.

POTASSIUM PHOSPHOTETRACHROMATE
$K_3PCr_4O_{16}$

M.T. AVERBUCH-POUCHOT, A. DURIF and J.C. GUITEL, 1981. J. Solid State Chem., **38**, 253-258.

Monoclinic, Cc, a = 9.512, b = 11.74, c = 14.74 Å, β = 106.13°, Z = 4. Ag radiation, R = 0.055 for 1011 reflexions. The unit cell is closely related to the rhombohedral cell of the ammonium compound (<u>1</u>).

   The structure contains a $P(OCrO_3)_4^{3-}$ anion with a central PO_4 tetrahedron. P-O = 1.50-1.66, Cr-O = 1.69-1.84 (bridging), 1.48-1.60(1) Å (terminal); K-O = 2.60-3.50 Å.

<u>1</u>. Preceding report.

URANYL CHROMATE HYDRATE
$UO_2CrO_4.5 \cdot 5H_2O$

V.N. SEREŽKIN and V.K. TRUNOV, 1981. Kristallografija, **26**, 301-304 [Soviet Physics - Crystallography, **26**, 169-172].

Monoclinic, P2$_1$/c, a = 11.179, b = 7.119, c = 26.49 Å, β = 94.19°, Z = 8. Mo radiation, R = 0.072 for 2747 reflexions.

   The structure (Fig. 1) contains CrO_4 tetrahedra linked by linear uranyl ions and by hydrogen bonding via the water molecules; U has pentagonal bipyramidal coordination. Cr-O = 1.59-1.67, U-O = 1.73-1.77 (uranyl), 2.32-2.52 Å.

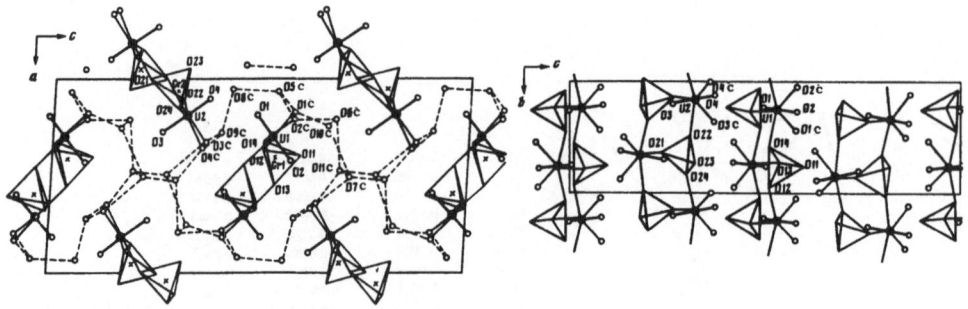

Fig. 1.   Structure of uranyl chromate hydrate.

PEROXOMOLYBDATES

$(NH_4)_4[Mo_3O_7(O_2)_4] \cdot 2H_2O$

$K_5[Mo_7O_{21}(O_2)_2(OH)] \cdot 6H_2O$

$(NH_4)_4[Mo_8O_{24}(O_2)_2(H_2O)_2] \cdot 4H_2O$

L. TRYSBERG and R. STOMBERG, 1981. Acta Chem. Scand., A35, 823-825.

Preliminary account of the structures [atomic positional parameters not listed].
The anion structures are shown in Fig. 1.

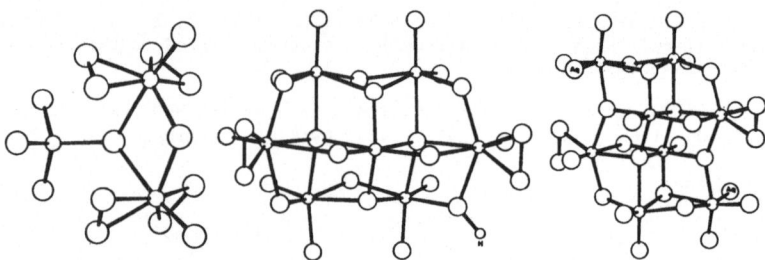

Fig. 1.   The peroxomolybdate anions.

SODIUM TETRAMOLYBDATE
$NaMo_4O_6$

C. C. TORARDI and R.E. McCARLEY, 1979. J. Amer. Chem. Soc., 101, 3963-3964.

Tetragonal, P4/mbm, a = 9.559, c = 2.860 Å, Z = 2. R = 0.046 for 216 reflexions
(atomic positional parameters not listed).

The structure (Fig. 1) contains chains of Mo_6O_{12} clusters sharing an Mo-Mo edge; the chains are cross-linked by sharing an oxygen atom, to give channels containing eight-coordinate Na ions. Mo-Mo = 2.751, 2.778, 2.860(3), Mo-O = 2.01-2.07(1), Na-O = 2.74(1) Å.

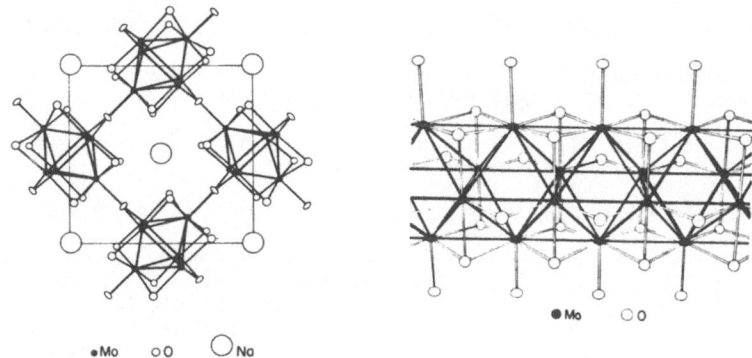

●Mo   o O   ◯ Na   ● Mo   ◯ O

Fig. 1.   Structure of $NaMo_4O_6$.

SODIUM MAGNESIUM MOLYBDATE
$Na_2Mg_5(MoO_4)_6$

R.F. KLEVCOVA, V.G. KIM and P.V. KLEVCOV, 1980. Kristallografija, _25_, 1148-1154 [Soviet Physics - Crystallography, _25_, 657-660].

Triclinic, $P\bar{1}$, a = 10.575, b = 8.617, c = 6.951 Å, α = 103.42, β = 102.67, γ = 112.37°, D_m = 3.55, Z = 1. Mo radiation, R = 0.051 for 3538 reflexions.

The structure (Fig. 1) contains MoO_4 tetrahedra, linked by MgO_6 and NaO_6 distorted octahedra and by one five-coordinate polyhedron, which is occupied statistically by Na and Mg; the formula is thus $Na(Na,Mg)_2Mg_4(MoO_4)_6$.

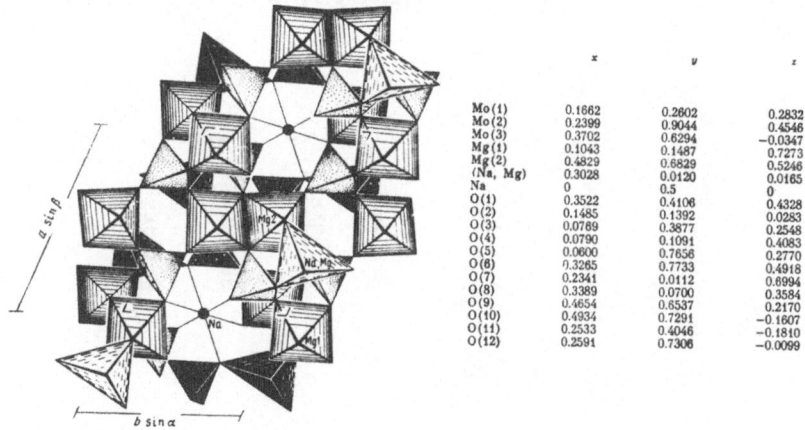

	x	y	z
Mo(1)	0.1662	0.2602	0.2832
Mo(2)	0.2399	0.9044	0.4546
Mo(3)	0.3702	0.6294	−0.0347
Mg(1)	0.1043	0.1487	0.7273
Mg(2)	0.4829	0.6829	0.5246
(Na, Mg)	0.3028	0.0120	0.0165
Na	0	0.5	0
O(1)	0.3522	0.4106	0.4328
O(2)	0.1485	0.1392	0.0283
O(3)	0.0789	0.3877	0.2548
O(4)	0.0790	0.1091	0.4083
O(5)	0.0600	0.7656	0.2770
O(6)	0.3265	0.7733	0.4918
O(7)	0.2341	0.0112	0.6994
O(8)	0.3389	0.0700	0.3584
O(9)	0.4654	0.6537	0.2170
O(10)	0.4934	0.7291	−0.1607
O(11)	0.2533	0.4046	−0.1810
O(12)	0.2591	0.7306	−0.0099

Fig. 1.   Structure of sodium magnesium molybdate.

BARIUM MOLYBDATES

$Ba_{0.62}Mo_4O_6$ $Ba_{1.13}Mo_8O_{16}$

LITHIUM ZINC MOLYBDATE

$LiZn_2Mo_3O_8$

C.C. TORARDI and R.E. McCARLEY, 1981. J. Solid State Chem., 37, 393-397.

$Ba_{0.62}Mo_4O_6$, orthorhombic, Pbam, a = 9.509, b = 9.825, c = 2.853 Å, Z = 1 (subcell; true cell has c' = 8c). R = 0.057 for 265 reflexions.

$Ba_{1.13}Mo_8O_{16}$, triclinic, P$\overline{1}$, a = 7.311, b = 7.453, c = 5.726 Å, α = 101.49, β = 99.60, γ = 89.31°, Z = 1. R = 0.036 for 1024 reflexions.

$LiZn_2Mo_3O_8$, rhombohedral, R$\overline{3}$m, a = 5.812, c = 30.013 Å, Z = 6. R = 0.043.

No atomic positional parameters are given. The $Ba_{0.62}Mo_4O_6$ structure is related to that of tetragonal $NaMo_4O_6$ (1); it contains chains of edge-sharing Mo_6 octahedra, linked by corner-sharing to form four-sided tunnels in which the Ba ions are located. $Ba_{1.13}Mo_8O_{16}$ is a low-symmetry metal-metal bonded variant of the hollandite structure, with two infinite chains of Mo_4O_8 units linked to form again four-sided tunnels which contain the Ba ions. The $LiZn_2Mo_3O_8$ structure is related to that of $Zn_2Mo_3O_8$ (2), each containing Mo_3O_8 clusters.

1. This volume, p. 226.
2. Structure Reports, 21, 333; 31A, 155.

SCANDIUM MOLYBDATE

$Sc_2(MoO_4)_3$

V.A. EFREMOV, B.I. LAZORJAK and V.K. TRUNOV, 1981. Kristallografija, 26, 72-81 [Soviet Physics - Crystallography, 26, 38-43].

Orthorhombic, Pbcn, a = 13.242, b = 9.544, c = 9.637 Å, Z = 4. Mo radiation, R = 0.035 for 2325 reflexions.

The structure (Fig. 1) contains a framework of corner-sharing MoO_4 tetrahedra and ScO_6 octahedra. Mo-O = 1.74-1.76, Sc-O = 2.06-2.11 Å.

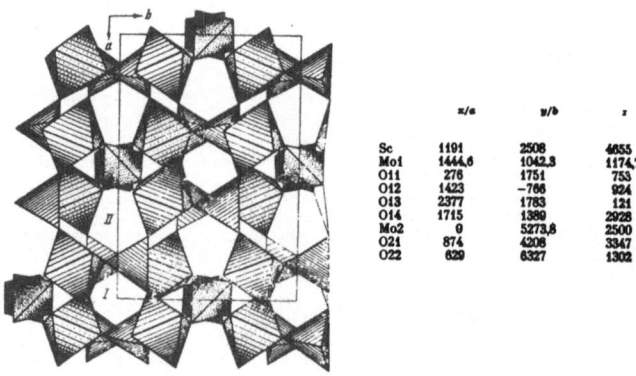

	x/a	y/b	z
Sc	1191	2508	4655
Mo1	1444.6	1042.3	1174.7
O11	276	1751	753
O12	1423	-766	924
O13	2377	1783	121
O14	1715	1389	2928
Mo2	0	5273.8	2500
O21	874	4208	3347
O22	629	6327	1302

Fig. 1.   Structure of scandium molybdate, and atomic positional parameters (x 10^4).

YTTRIUM TUNGSTATE (HIGH-TEMPERATURE FORM)

$\varepsilon - Y_2WO_6$ $Y_2O_3 \cdot WO_3$

O. BEAURY, M. FAUCHER and G. TESTE de SAGEY, 1981. Acta Cryst., B$\underline{37}$, 1166-1170.

Orthorhombic, $P2_12_12_1$, a = 8.591, b = 20.840, c = 5.233 Å, D_m = 6.38, Z = 8. Cu radiation, R = 0.067 for 454 reflexions.

The structure (Fig. 1) contains WO_6 octahedra, linked by 8- and 7-coordinate Y atoms. W-O = 1.76-2.04, Y-O in Fig. 1.

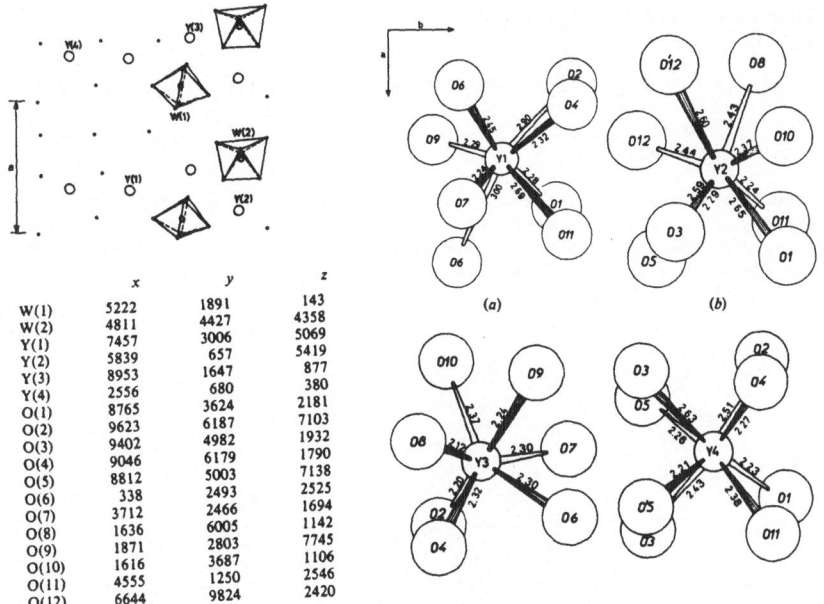

	x	y	z
W(1)	5222	1891	143
W(2)	4811	4427	4358
Y(1)	7457	3006	5069
Y(2)	5839	657	5419
Y(3)	8953	1647	877
Y(4)	2556	680	380
O(1)	8765	3624	2181
O(2)	9623	6187	7103
O(3)	9402	4982	1932
O(4)	9046	6179	1790
O(5)	8812	5003	7138
O(6)	338	2493	2525
O(7)	3712	2466	1694
O(8)	1636	6005	1142
O(9)	1871	2803	7745
O(10)	1616	3687	1106
O(11)	4555	1250	2546
O(12)	6644	9824	2420

Fig. 1. WO_6 octahedra, Y coordinations, and atomic positional parameters (x10^4) in $\varepsilon - Y_2WO_6$.

POTASSIUM YTTRIUM MOLYBDATE

$\alpha - K_5Y(MoO_4)_4$

B.I. LAZORJAK and V.A. EFREMOV, 1981. Kristallografija, $\underline{26}$, 464-472 [Soviet Physics - Crystallography, $\underline{26}$, 263-267].

Trigonal, $P\overline{3}1m$ or $P\overline{3}$, a = 10.453, c = 41.04 Å, Z = 9 (subcell, rhombohedral, $R\overline{3}m$, a = 6.035, c = 20.52 Å, Z = 3/2). Mo radiation, R = 0.045 for 367 sub-cell reflexions.

Atomic positions (subcell, $R\overline{3}m$)

			x	y	z
3M*	in	3(a)	0	0	0
6Mo		18(h)	-0.0005	0.0005	0.3981
6K		18(h)	0.0090	-0.0090	0.1934
6 O(1)		18(h)	-0.0489	0.0489	0.3197
9 O(2a)		18(h)	-0.1690	0.1690	0.4086
9 O(2b)		18(h)	-0.1413	0.1413	0.4436

* M = 0.5K + 0.5Y

The material is a high-temperature modification; the structure is related to
that of palmierite (1), but with displacement of atoms from ideal positions giving
rise to a supercell [superstructure not determined].

<u>1</u>. Structure Reports, <u>17</u>, 499.

AMMONIUM DICUPRO-18-MOLYBDODISILICATE HYDRATE
$(NH_4)_{12}[Cu_2Si_2Mo_{18}O_{66}].14H_2O$

H.F. FUKUSHIMA, A. KOBAYASHI and Y. SASAKI, 1981. Acta Cryst., B<u>37</u>, 1613-1615.

Triclinic, $P\bar{1}$, a = 12.19, b = 15.78, c = 10.81 Å, α = 91.3, β = 74.6, γ = 75.8°,
D_m = 2.91, Z = 1. Mo radiation, R = 0.048 for 4602 reflexions.

The structure contains a centrosymmetric polyanion (Fig. 1), which consists of
two edge-sharing CuO_6 octahedra attached to two SiO_4 tetrahedra; each tetrahedron is
surrounded by nine MoO_6 octahedra. Cu-O = 1.95-1.99 (4 distances), 2.44 (2 distances)
(Jahn-Teller distortion), Si-O = 1.61-1.66 Å.

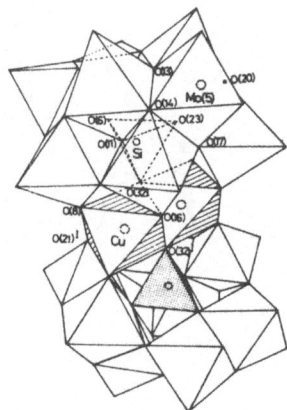

Fig. 1.   The $[Cu_2Si_2Mo_{18}O_{66}]^{12-}$ anion.

SILVER ZINC MOLYBDATE
$Ag_2Zn_2(MoO_4)_3$

C. GICQUEL-MAYER, M. MAYER and G. PEREZ, 1981. Acta Cryst., B<u>37</u>, 1035-1039.

Triclinic, $P\bar{1}$, a = 6.992, b = 8.712, c = 10.818 Å, α = 64.24, β = 66.51, γ = 76.27°,
Z = 2. Mo radiation, R = 0.042 for 1877 reflexions.

The structure (Fig. 1) contains MoO_4 tetrahedra and pairs of edge-sharing ZnO_6
octahedra, joined to form (θ10) layers which are linked by Ag ions. Mo-O = 1.73-
1.80, Zn-O = 1.98-2.19, Ag-O = 1.98-3.20 Å.

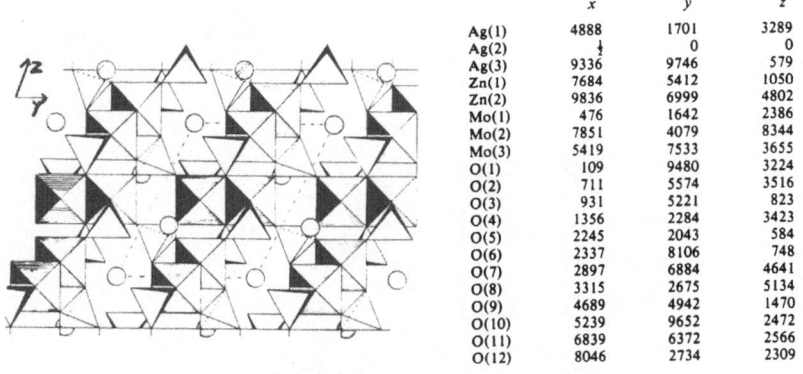

	x	y	z
Ag(1)	4888	1701	3289
Ag(2)	½	0	0
Ag(3)	9336	9746	579
Zn(1)	7684	5412	1050
Zn(2)	9836	6999	4802
Mo(1)	476	1642	2386
Mo(2)	7851	4079	8344
Mo(3)	5419	7533	3655
O(1)	109	9480	3224
O(2)	711	5574	3516
O(3)	931	5221	823
O(4)	1356	2284	3423
O(5)	2245	2043	584
O(6)	2337	8106	748
O(7)	2897	6884	4641
O(8)	3315	2675	5134
O(9)	4689	4942	1470
O(10)	5239	9652	2472
O(11)	6839	6372	2566
O(12)	8046	2734	2309

Fig. 1.   Structure of silver zinc molybdate, and atomic positional parameters
(x 10^4).

TETRAMETHYLAMMONIUM α-DODECATUNGSTOSILICATE
$[N(CH_3)_4]_4SiW_{12}O_{40}$ (I)

TETRABUTYLAMMONIUM β-DODECATUNGSTOSILICATE
$[N(C_4H_9)_4]_4SiW_{12}O_{40}$ (II)

J. FUCHS, A. THIELE and R. PALM, 1981. Z. Naturforsch., 36B, 161-171.

I.   Tetragonal, $I\bar{4}$, a = 14.642, c = 12.706 Å, Z = 2. Cu radiation, R = 0.057 for
1306 reflexions.

II.   Orthorhombic, $P2_12_12_1$, a = 29.277, b = 22.181, c = 15.381 Å, D_m = 2.5, Z = 4.
Cu radiation, R = 0.106 for 6367 reflexions.

The heteropolyanions (Fig. 1) have the well-known α-Keggin (1) and β-type (2)
structures.

Fig. 1.   The α- (left) and β-$SiW_{12}O_{40}^{4-}$ (right) anions.

1. Strukturbericht, 3, 114, 463.
2. Structure Reports, 41A, 256; 46A, 274.

SODIUM β-HYDROGENENNEATUNGSTOSILICATE HYDRATE
$Na_9(SiW_9O_{34}H).23H_2O$

F. ROBERT and A. TÉZÉ, 1981. Acta Cryst., B37, 318-322.

Triclinic, P$\overline{1}$, a = 13.166, b = 12.609, c = 18.350 Å, α = 69.53, β = 73.51, γ = 63.16°, Z = 2. Mo radiation, R = 0.059 for 4596 reflexions.

The structure (Fig. 1) contains a β-$SiW_9O_{34}H^{9-}$ ion, which is of Keggin type with three WO_6 octahedra missing; Si has tetrahedral coordination. W-O = 1.72-2.37, Si-O = 1.62-1.65 Å. Na ions have 4-, 5-, and 6-coordinations.

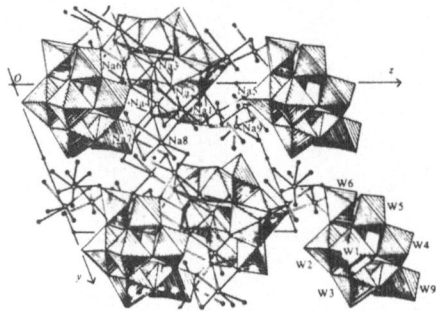

Fig. 1.   Structure of $Na_9(SiW_9O_{34}H).23H_2O$.

PHOSPHORUS TUNGSTEN OXIDE
$P_4W_8O_{32}$

J.P. GIROULT, M. GOREAUD, P. LABBÉ and B. RAVEAU, 1981. Acta Cryst., B37, 2319-2142.

Orthorhombic, $P2_12_12_1$, a = 5.285, b = 6.569, c = 17.351 Å, Z = 1. Mo radiation, R = 0.058 for 967 reflexions.

The structure (Fig. 1) is built up from corner sharing WO_6 octahedra and PO_4 tetrahedra, with pentagonal tunnels along a. W-O = 1.81-2.04, P-O = 1.49-1.52(3) Å.

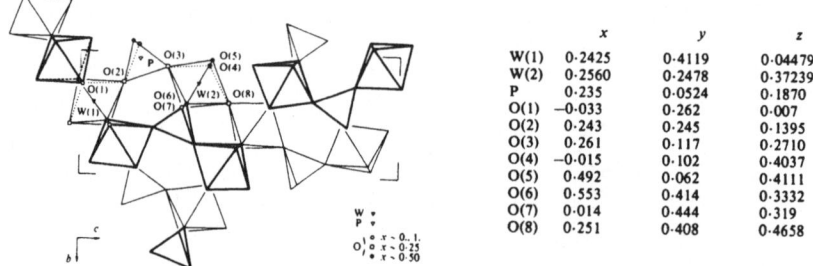

	x	y	z
W(1)	0·2425	0·4119	0·04479
W(2)	0·2560	0·2478	0·37239
P	0·235	0·0524	0·1870
O(1)	−0·033	0·262	0·007
O(2)	0·243	0·245	0·1395
O(3)	0·261	0·117	0·2710
O(4)	−0·015	0·102	0·4037
O(5)	0·492	0·062	0·4111
O(6)	0·553	0·414	0·3332
O(7)	0·014	0·444	0·319
O(8)	0·251	0·408	0·4658

Fig. 1.   Structure of $P_4W_8O_{12}$.

SODIUM TETRAMETHYLAMMONIUM α-UNDECATUNGSTOPHOSPHATE HYDRATE
$Na[N(CH_3)_4]_4HPW_{11}O_{39}.7H_2O$

J. FUCHS, A. THIELE and R. PALM, 1981. Z. Naturforsch., 36B, 544-550.

Monoclinic, C2/m, a = 23.895, b = 11.417, c = 28.930 Å, β = 126.0°, D_m = 3.2, Z = 4. Cu radiation, R = 0.116 for 3019 reflexions.

The compound contains a defect Keggin-type polyanion (Fig. 1).

Fig. 1.   The $HPW_{11}O_{39}^{6-}$ anion.

RUBIDIUM PHOSPHOTUNGSTATE

$Rb_{0.2}PW_3O_{11}$ $1/8 \times [Rb_{1.8}P_8W_{24}O_{88}]$

J.P. GIROULT, M. GOREAUD, Ph. LABBÉ and B. RAVEAU, 1981. Acta Cryst., B<u>37</u>, 1163-1166.

Monoclinic, P2/m, a = 13.991, b = 3.7650, c = 8.561 Å, β = 114.22°, Z = 2 (sub-cell; true cell is A2/m with doubled b and c). Mo radiation, R = 0.039 for 820 reflexions.

The material is one of a series of bronzes $Rb_xP_8W_{8n}O_{24n+16}$, with n = 3 (the n = 4 material having been described previously (<u>1</u>)). The structure (Figs. 1 and 2) contains three-octahedra-wide perovskite slabs, connected by P_2O_7 groups, with hexagonal tunnels along <u>b</u> which contain 8-coordinate Rb ions.

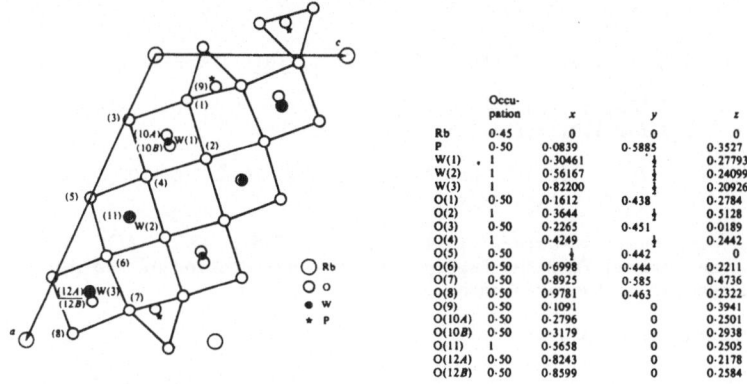

	Occupation	x	y	z
Rb	0·45	0	0	0
P	0·50	0·0839	0·5885	0·3527
W(1)	1	0·30461	¼	0·27793
W(2)	1	0·56167	¼	0·24099
W(3)	1	0·82200	¼	0·20926
O(1)	0·50	0·1612	0·438	0·2784
O(2)	1	0·3644	¼	0·5128
O(3)	0·50	0·2265	0·451	0·0189
O(4)	1	0·4249	¼	0·2442
O(5)	0·50	¼	0·442	0
O(6)	0·50	0·6998	0·444	0·2211
O(7)	0·50	0·8925	0·585	0·4736
O(8)	0·50	0·9781	0·463	0·2322
O(9)	0·50	0·1091	0	0·3941
O(10A)	0·50	0·2796	0	0·2501
O(10B)	0·50	0·3179	0	0·2938
O(11)	1	0·5658	0	0·2505
O(12A)	0·50	0·8243	0	0·2178
O(12B)	0·50	0·8599	0	0·2584

○ Rb
○ o
● w
✳ P

Fig. 1.   Mean structure of $Rb_{1.8}P_8W_{24}O_{88}$.

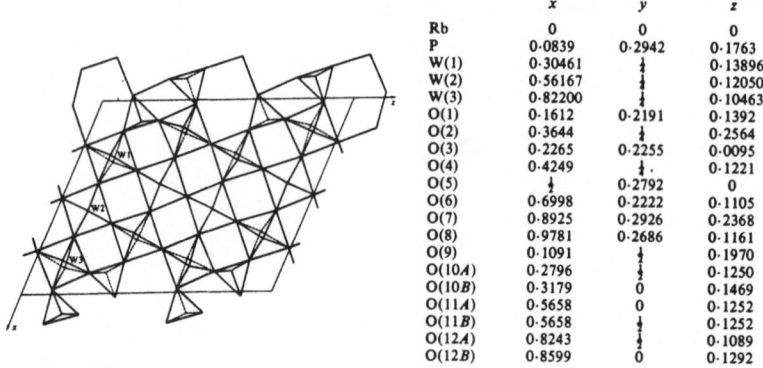

	x	y	z
Rb	0	0	0
P	0·0839	0·2942	0·1763
W(1)	0·30461	$\frac{1}{4}$	0·13896
W(2)	0·56167	$\frac{1}{4}$	0·12050
W(3)	0·82200	$\frac{1}{4}$	0·10463
O(1)	0·1612	0·2191	0·1392
O(2)	0·3644	$\frac{1}{4}$	0·2564
O(3)	0·2265	0·2255	0·0095
O(4)	0·4249	$\frac{1}{4}$	0·1221
O(5)	$\frac{1}{4}$	0·2792	0
O(6)	0·6998	0·2222	0·1105
O(7)	0·8925	0·2926	0·2368
O(8)	0·9781	0·2686	0·1161
O(9)	0·1091	$\frac{1}{4}$	0·1970
O(10A)	0·2796	$\frac{1}{4}$	0·1250
O(10B)	0·3179	0	0·1469
O(11A)	0·5658	0	0·1252
O(11B)	0·5658	$\frac{1}{4}$	0·1252
O(12A)	0·8243	$\frac{1}{4}$	0·1089
O(12B)	0·8599	0	0·1292

Fig. 2.   Actual structure of $Rb_{1.8}P_8W_{24}O_{88}$.

1. Structure Reports, 46A, 276.

CAESIUM PHOSPHOTUNGSTATES

$Cs_6W_5P_2O_{23}.7.5H_2O$ (I)

$Cs_7W_{10}PO_{36}.7H_2O$ (II)

W.H. KNOTH and R.L. HARLOW, 1981. J. Amer. Chem. Soc., 103, 1865-1867.

I.   Monoclinic, a = 13.009, b = 15.974, c = 17.087 Å, β = 98.93°, at -100°C, Z = 4.
Mo radiation, R = 0.038 for 5201 reflexions.

II.   Orthorhombic, $P2_12_12$, a = 12.401, b = 18.948, c = 9.636 Å, at -100°C, Z = 2.   Mo
radiation, R = 0.047 for 2665 reflexions.

The $W_5P_2O_{23}{}^{6-}$ ion has a structure similar to that of $Mo_5P_2O_{23}{}^{6-}$ (1).   The
$W_{10}PO_{36}{}^{7-}$ ion can be derived from the Keggin structure by rotation of two W_3O_{13}
sets by 60°, and removal of two octahedra.

1. Structure Reports, 39A, 292.

RUBIDIUM 21-TUNGSTODIARSENATE(III) HYDRATE

$Rb_4H_2[As_2W_{21}O_{69}(H_2O)].34H_2O$

Y. JEANNIN and J. MARTIN-FRÈRE, 1981. J. Amer. Chem. Soc., 103, 1664-1667.

Hexagonal, $P6_3/mmc$, a = 16.926, c = 18.767 Å, D_m = 4.40, Z = 2. Mo radiation, R =
0.059 for 1572 reflexions (one rubidium and 19 water molecules were not located).

The heteropolyanion (Fig. 1) consists of two AsW_9O_{33} units (1), joined asymmetrically by three W atoms, one of which has octahedral and two of which have square-
pyramidal coordination; there is some disorder of the W(3) positions. Rb ions have
ten-coordination.

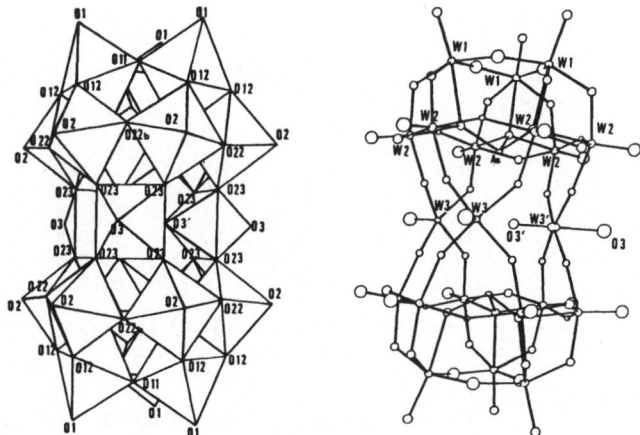

Fig. 1.   The $As_2W_{21}O_{69}(H_2O)^{6-}$ heteropolyanion.

<u>1</u>. Structure Reports, <u>45</u>A, 261.

TUNGSTEN MOLYBDENUM OXIDE
$(W_{12.64}Mo_{1.36})O_{41}$

K. VISWANATHAN and E. SALJE, 1981. Acta Cryst., A<u>37</u>, 449-456.

Monoclinic, P2/c, a = 23.939, b = 3.966, c = 16.729 Å, β = 106.74°, Z = 2. Mo
radiation, R = 0.10 for 1285 reflexions.

The structure (Fig. 1) contains MO_6 octahedra, with Mo segregated in crystallographic shear (CS) planes.

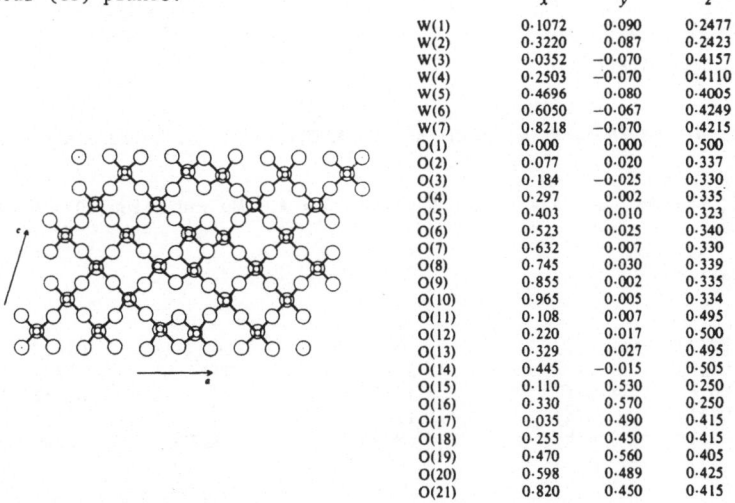

	x	y	z
W(1)	0·1072	0·090	0·2477
W(2)	0·3220	0·087	0·2423
W(3)	0·0352	−0·070	0·4157
W(4)	0·2503	−0·070	0·4110
W(5)	0·4696	0·080	0·4005
W(6)	0·6050	−0·067	0·4249
W(7)	0·8218	−0·070	0·4215
O(1)	0·000	0·000	0·500
O(2)	0·077	0·020	0·337
O(3)	0·184	−0·025	0·330
O(4)	0·297	0·002	0·335
O(5)	0·403	0·010	0·323
O(6)	0·523	0·025	0·340
O(7)	0·632	0·007	0·330
O(8)	0·745	0·030	0·339
O(9)	0·855	0·002	0·335
O(10)	0·965	0·005	0·334
O(11)	0·108	0·007	0·495
O(12)	0·220	0·017	0·500
O(13)	0·329	0·027	0·495
O(14)	0·445	−0·015	0·505
O(15)	0·110	0·530	0·250
O(16)	0·330	0·570	0·250
O(17)	0·035	0·490	0·415
O(18)	0·255	0·450	0·415
O(19)	0·470	0·560	0·405
O(20)	0·598	0·489	0·425
O(21)	0·820	0·450	0·415

Fig. 1.   Structure of $(W,Mo)_{14}O_{41}$ and atomic positional parameters;
          W(5) contains the Mo.

CADMIUM TUNGSTATE
$CdWO_4$

D.J. MORELL, J.S. CANTRELL and L.L.Y. CHANG, 1980. J. Amer. Ceram. Soc., 63, 261-264.

Monoclinic, P2/b, a = 5.013, b = 5.090, c = 5.866 Å, γ = 91.46°, D_m = 7.83, Z = 2.
Mo radiation, R = 0.076 for 696 reflexions. Previous study in 1.

Atomic positions

			x	y	z
W	in	2(e)	0	1/4	0.1784
Cd		2(f)	1/2	3/4	0.3018
O(1)		4(g)	0.2037	0.4474	0.9033
O(2)		4(g)	0.2429	0.2874	0.3719

Wolframite structure (2). W-O = 1.79-2.15, Cd-O = 2.18-2.42 Å. The Cd and Zn
compounds form a complete range of solid solutions.

1. Structure Reports, 31A, 158.
2. Strukturbericht, 2, 85; Structure Reports, 21, 305; 37A, 256.

CALCIUM MANGANESE OXIDE
$CaMn_7O_{12}$ $[CaMn_3](Mn_4)O_{12}$

B. BOCHU, J.L. BUEVOZ, J. CHENAVAS, A. COLLOMB, J.C. JOUBERT and M. MAREZIO, 1980.
Solid State Comm., 36, 133-138.

Rhombohedral, R$\bar{3}$, a = 10.442, c = 6.343 Å, Z = 3. Neutron powder data. Ca in 3(a);
Mn in 9(e), 9(d), and 3(b); O(1) and O(2) in 18(f): (0.2226,0.2731,0.0814) and (0.3422,
0.5221,0.3410).

Perovskite-like structure, as for $NaMn_7O_{12}$ (1),with Jahn-Teller distorted
Mn(III)O_6 octahedra.

1. Structure Reports, 39A, 247.

STRONTIUM MANGANATE(IV)
α-$SrMnO_3$

K. KURODA, N. ISHIZAWA, N. MIZUTANI and M. KATO, 1981. J. Solid State Chem., 38, 297-
299.

Hexagonal, P6$_3$/mmc, a = 5.454, c = 9.092 Å, Z = 4. Mo radiation, R = 0.043 for 203
reflexions.

Atomic positions

			x	y	z
0.5 Sr(1)	in	4(e)	0	0	0.0120
Sr(2)		2(c)	1/3	2/3	1/4
Mn		4(f)	1/3	2/3	0.6127
O(1)		6(g)	1/2	0	0
O(2)		6(h)	-0.8179	0.8179	3/4

Isostructural with high-temperature $BaMnO_3$ (1) as previously reported (2), except for disorder of $Sr(1)$. The structure contains four close-packed SrO_3 layers in an ABAC stacking sequence along c. Mn are in octahedral holes in the A layer, with pairs of face-sharing MnO_6 octahedra linked by further corner sharing. Mn-O = 1.88, Sr-O = 2.73-2.85 Å; Mn...Mn distances of 2.49 Å between face-sharing octahedra suggest metal-metal interaction.

1. Structure Reports, 27, 663.
2. Ibid., 42A, 283.

STRONTIUM MANGANATE(IV)
$\beta-Sr_2MnO_4$

J.-C. BOULOUX, J.-L. SOUBEYROUX, G. LE FLEM and P. HAGENMULLER, 1981. J. Solid State Chem., 38, 34-39.

Tetragonal, I4/mmm, a = 3.787, c = 12.496 Å, Z = 2. Neutron powder data at 300 and 4.2K. Sr in 4(e): z = 0.356; Mn in 2(a); O(1) in 4(c); O(2) in 4(e): z = 0.157.

K_2NiF_4-type structure (1), as previously suggested (2). Mn-O = 1.89 (x 4), 1.98 (x 2), Sr-O = 2.49-2.68 Å.

1. Structure Reports, 17, 332; 19, 323.
2. N. MIZUTANI, A. KITAZAWA, N. ONKUMA and M. KATO, 1970. Kogyo Kogaku Zasshi, 43, 1097.

BARIUM MANGANATE(II)
Ba_2MnO_3

K. SANDER and H. MÜLLER-BUSCHBAUM, 1981. Z. anorg. Chem., 478, 52-56.

Monoclinic, Cc, a = 5.846, b = 11.579, c = 12.707 Å, β = 93.74°, Z = 8. Mo radiation, R = 0.057 for 902 reflexions.

The structure (Fig. 1) contains chains of corner-sharing MnO_4 tetrahedra, linked by 7-coordinate Ba ions. Mn-O = 1.96-2.19, Ba-O = 2.11[?]-3.36 Å.

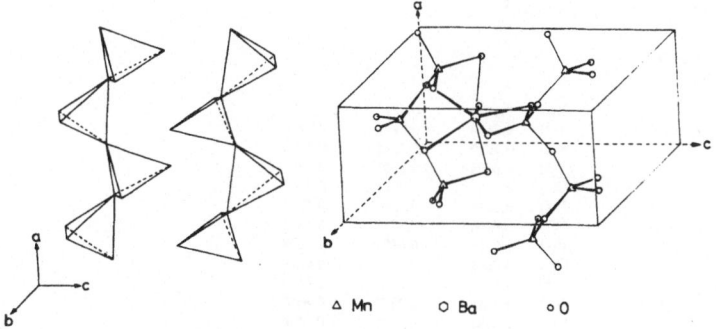

△ Mn   ○ Ba   ○ O

Fig. 1.   Structure of Ba_2MnO_3.

STRONTIUM LANTHANUM MANGANATE
$Sr_{1.50}La_{0.50}MnO_4$

J.-C. BOULOUX, J.-L. SOUBEYROUX, A. DAOUDI and G. LE FLEM, 1981. Mater. Res. Bull.,
16, 855-860.

Tetragonal, I4/mmm, a = 3.852, 3.853, 3.868, c = 12.39, 12.39, 12.45 Å, at 2, 160,
300K, Z = 4. Neutron powder data. Sr/La in 4(e): z = 0.358, 0.358, 0.357; Mn (Mn^{3+}
+ Mn^{4+}) in 2(a); O(1) in 4(e): z = 0.160, 0.160, 0.159; O(2) in 4(c).

    K_2NiF_4-type structure (1).

1. Structure Reports, **17**, 332; **19**, 323.

STRONTIUM NEODYMIUM MANGANATE(III)
STRONTIUM NEODYMIUM CHROMATE(III)
$SrNdMO_4$ (M = Mn, Cr)

K. SANDER, U. LEHMANN and H. MÜLLER-BUSCHBAUM, 1981. Z. anorg. Chem., **480**, 153-156.

Tetragonal, I4/mmm, a = 3.777, 3.842, c = 12.96, 12.338 Å, Z = 2. Mo radiation,
R = 0.051, 0.10 for 141, 145 reflexions. Sr/Nd in 4(e): z = 0.3557, 0.3594; Mn or
Cr in 2(a); O(1) in 4(c); O(2) in 4(e): z = 0.176, 0.162.

    K_2NiF_4-type structure (1), with elongated MO_6 octahedra, and disordered 9-
coordinate Sr/Nd. Mn-O = 1.$\overline{89}$ (x 4), 2.28 (x 2), Cr-O = 1.92, 2.00, Sr/Nd-O = 2.33-
2.73 Å.

1. Structure Reports, **17**, 332; **19**, 323.

YTTRIUM RHENIUM OXIDE
β-Y_3ReO_8

G. BAUD, J.-P. BESSE, R. CHEVALIER and M. GASPERIN, 1981. J. Solid State Chem., **38**,
186-191.

Monoclinic, $P2_1/a$, a = 14.391, b = 7.196, c = 6.045 Å, γ = 112°8' [or 112.8°?], Z =
4. R = 0.058 for 1341 reflexions.

Atomic positions

	x	y	z
Re	0,0874	0,8215	0,7671
Y(1)	0,1684	0,0917	0,3429
Y(2)	0,3029	0,4032	0,7553
Y(3)	0,0505	0,3507	0,7765
O(1)	0,3291	0,7032	0,5179
O(2)	0,1633	0,3596	0,5252
O(3)	0,9766	0,5831	0,8211
O(4)	0,2174	0,0403	0,7078
O(5)	0,0885	0,9014	0,0618
O(6)	0,1676	0,6734	0,8620
O(7)	0,0942	0,7293	0,4658
O(8)	0,0258	0,9911	0,6826

Isostructural with Sm_3ReO_8 (1). Re-O = 1.83-1.97 (octahedral), Y-O = 2.21-2.63(2) Å (7- and 8-coordinations). The material is synthesized at high pressure; α (cubic) and α' (trigonal) forms are also found.

1. Structure Reports, 42A, 290.

CALCIUM FERRITES
$Ca_2Fe_9O_{13}$ $Ca_2Fe_7O_{11}$

B. MALAMAN, H. ALEBOUYEH, F. JEANNOT, A. COURTOIS, R. GÉRARDIN and O. EVRARD, 1981.
Mater. Res. Bull., 16, 1139-1148.

Monoclinic, C2/m, a = 10.022, 9.96, b = 3.047, 3.03, c = 16.877, 15.77 Å, β = 99.20°, 118°, Z = 2. $Ca_2Fe_9O_{13}$, Mo radiation, R = 0.054 for 1106 reflexions; $Ca_2Fe_7O_{11}$, powder data.

The structures (Fig. 1) contain stackings of $CaFe_2O_4$ and FeO blocks, as in related compounds (1). Fe ions have octahedral coordinations, and Ca ions have 8-coordination.

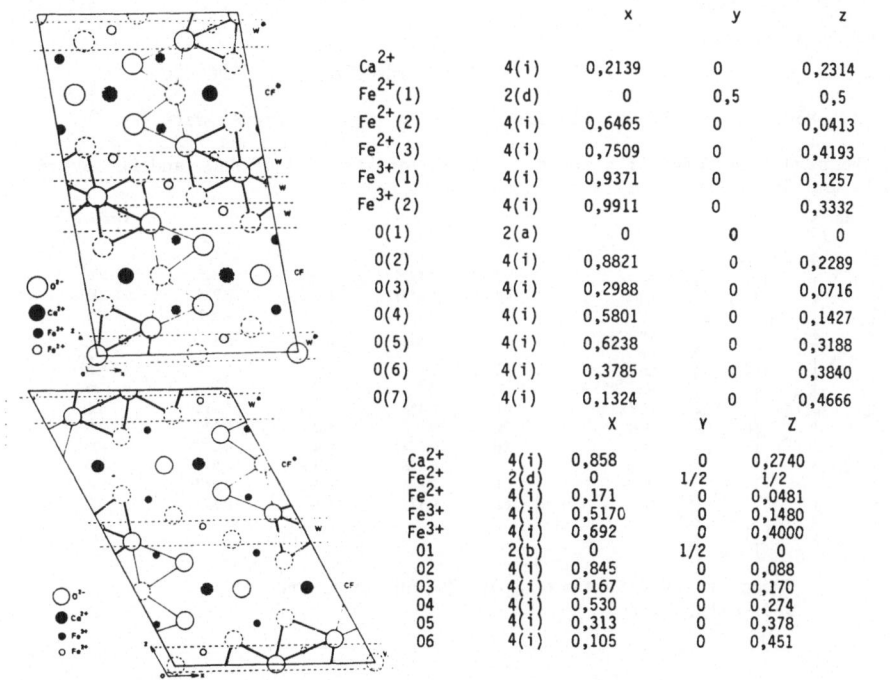

		x	y	z
Ca^{2+}	4(i)	0,2139	0	0,2314
$Fe^{2+}(1)$	2(d)	0	0,5	0,5
$Fe^{2+}(2)$	4(i)	0,6465	0	0,0413
$Fe^{2+}(3)$	4(i)	0,7509	0	0,4193
$Fe^{3+}(1)$	4(i)	0,9371	0	0,1257
$Fe^{3+}(2)$	4(i)	0,9911	0	0,3332
0(1)	2(a)	0	0	0
0(2)	4(i)	0,8821	0	0,2289
0(3)	4(i)	0,2988	0	0,0716
0(4)	4(i)	0,5801	0	0,1427
0(5)	4(i)	0,6238	0	0,3188
0(6)	4(i)	0,3785	0	0,3840
0(7)	4(i)	0,1324	0	0,4666

		X	Y	Z
Ca^{2+}	4(i)	0,858	0	0,2740
Fe^{2+}	2(d)	0	1/2	1/2
Fe^{2+}	4(i)	0,171	0	0,0481
Fe^{3+}	4(i)	0,5170	0	0,1480
Fe^{3+}	4(i)	0,692	0	0,4000
01	2(b)	0	1/2	0
02	4(i)	0,845	0	0,088
03	4(i)	0,167	0	0,170
04	4(i)	0,530	0	0,274
05	4(i)	0,313	0	0,378
06	4(i)	0,105	0	0,451

Fig. 1.   Structures of $Ca_2Fe_9O_{13}$ (top) and $Ca_2Fe_7O_{11}$ (bottom).

1. Structure Reports, 46A, 280.

LITHIUM IRON TIN OXIDE
LiFeSnO$_4$

J. CHOISNET, M. HERVIEU, B. RAVEAU and P. TARTE, 1981. J. Solid State Chem., 40, 344-351.

High-temperature form, orthorhombic, Pmcn, a = 3.066, b = 5.066, c = 9.874 Å, D$_m$ = 5.32, Z = 2. Powder data.

Low-temperature form, hexagonal, P6$_3$mc, a = 6.012, c = 9.776 Å, D$_m$ = 5.29, Z = 4. Powder data.

The high-temperature form has the γ-MnO$_2$ ramsdellite structure (1), with partial occupancy of Li sites and Fe/Sn disorder; the low-temperature form is isostructural with Li$_{1.6}$Zn$_{1.6}$Sn$_{2.8}$O$_8$ (2), with some Li/Fe/Sn disorder.

1. Structure Reports, 12, 152; 13, 188; 15, 184.
2. Ibid., 45A, 230.

BARIUM TIN IRON OXIDE
BaSn$_{0.9}$Fe$_{5.47}$O$_{11}$

M.C. CADÉE and D.J.W. IJDO, 1981. J. Solid State Chem., 40, 290-300.

Trigonal, P$\bar{3}$m1, a = 5.937, c = 14.336 Å, Z = 2. Neutron powder data.

The structure (Fig. 1) contains a six-layer stacking of BaO$_3$ and O$_4$ layers, with Fe and Sn in octahedral and tetrahedral sites.

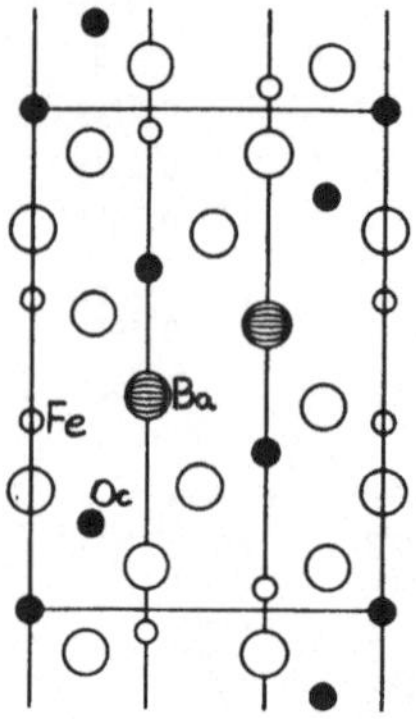

		x	y	z	Occupation (%)
Ba	2d	0.3333	0.6667	0.425	100
Fe(1)	2d	0.3333	0.6667	0.955	100
Fe(2)	2c	0.0	0.0	0.378	100
Sn / Fe Oc(1)	2d	0.3333	0.6667	0.680	65 / 35
Sn / Fe Oc(2)	1a	0.0	0.0	0.0	50 / 49
Fe Oc(3)	6i	0.170	0.341	0.173	96
O(1)	2c	0.0	0.0	0.239	100
O(2)	2d	0.3333	0.6667	0.087	100
O(3)	6i	0.158	0.316	0.915	100
O(4)	6i	0.486	0.972	0.247	100
O(5)	6i	0.173	0.346	0.593	100

Fig. 1.   Structure of BaSn$_{0.9}$Fe$_{5.47}$O$_{11}$ (occupancies are not well-established).

BARIUM LANTHANUM RUTHENATE
Ba$_2$LaRuO$_6$

CALCIUM LANTHANUM RUTHENATE
Ca$_2$LaRuO$_6$

P.D. BATTLE, 1981. Mater. Res. Bull., 16, 397-405.

Ba_2LaRuO_6, monoclinic, $P2_1/n$, a = 6.0285, b = 6.0430, c = 8.5409 Å, β = 90.44°, Z = 2. Neutron powder data.

Ca_2LaRuO_6, triclinic, $P\bar{1}$, a = 5.6179, b = 5.8350, c = 8.0667 Å, α = 90.0, β = 89.76, γ = 90.0°, Z = 2. Neutron powder data.

Both materials are distorted perovskites (Fig. 1), with some Ca/La disorder in the Ca compound. Ru-O = 1.91-2.06 Å.

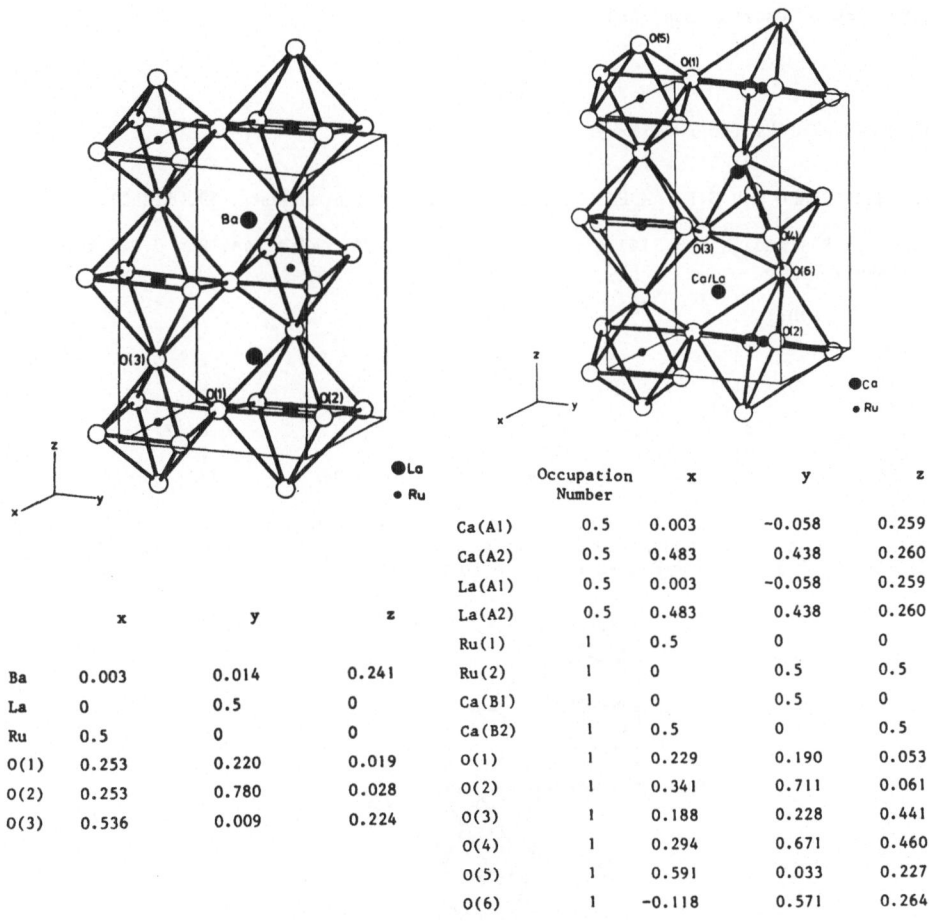

	x	y	z
Ba	0.003	0.014	0.241
La	0	0.5	0
Ru	0.5	0	0
O(1)	0.253	0.220	0.019
O(2)	0.253	0.780	0.028
O(3)	0.536	0.009	0.224

	Occupation Number	x	y	z
Ca(A1)	0.5	0.003	−0.058	0.259
Ca(A2)	0.5	0.483	0.438	0.260
La(A1)	0.5	0.003	−0.058	0.259
La(A2)	0.5	0.483	0.438	0.260
Ru(1)	1	0.5	0	0
Ru(2)	1	0	0.5	0.5
Ca(B1)	1	0	0.5	0
Ca(B2)	1	0.5	0	0.5
O(1)	1	0.229	0.190	0.053
O(2)	1	0.341	0.711	0.061
O(3)	1	0.188	0.228	0.441
O(4)	1	0.294	0.671	0.460
O(5)	1	0.591	0.033	0.227
O(6)	1	−0.118	0.571	0.264

Fig. 1.   Structures of Ba_2LaRuO_6 (left) and Ca_2LaRuO_6 (right).

NEODYMIUM OSMATE
$Nd_4Os_6O_{19}$

F. ABRAHAM, J. TREHOUX and D. THOMAS, 1981.   J. Less-Common Metals, <u>77</u>, P23-P30.

Cubic, I23, a = 8.957 Å, Z = 2. Mo radiation, R = 0.027 for 410 reflexions. Nd in 8(c): x = 0.1574; Os in 12(e): x = 0.3615; O(1) in 12(d): x = 0.3228; O(2) in 2(a); O(3) in 24(f): 0.3408,0.2838,0.9785.

Isostructural with $La_4M_6O_{19}$ compounds (<u>1</u>). The structure is derived from that of cubic $KSbO_3$. It contains pairs of edge-sharing OsO_6 octahedra, linked by sharing corners to form an $Os_{12}O_{36}$ framework; holes in the framework are occupied by tetrahedral Nd_4O groups. Os-O = 1.95-2.01(1), Os-Os = 2.481(1) (a metal-metal bond), Nd-O = 2.44-2.56(1) (7 distances), 2.93(1) (3 distances) Å.

<u>1</u>. Structure Reports, <u>33</u>A, 353; <u>43</u>A, 214.

BARIUM RHODATE(IV) (4H-POLYTYPE)
$BaRhO_3$

B.L. CHAMBERLAND and J.B. ANDERSON, 1981. J. Solid State Chem., <u>39</u>, 114-119.

Hexagonal, $P6_3/mmc$, a = 5.744, c = 9.642 Å, Z = 4. Mo radiation, R = 0.044 for 132 reflexions.

Atomic positions

			x	y	z
Ba(1)	in	2(a)	0	0	0
Ba(2)		2(c)	1/3	2/3	1/4
Rh		4(f)	1/3	2/3	0.6137
O(1)		6(g)	1/2	0	0
O(2)		6(h)	-0.1799	-0.3598	1/4

The structure (Fig. 1) contains a four-layer stacking sequence (ABAC) of BaO_3 layers, with Rh in octahedral holes giving rise to face-sharing RhO_6 octahedra. Rh-O = 1.99, 2.01 (each x 3), Ba-O (12-coordination) = 2.87-3.00 Å.

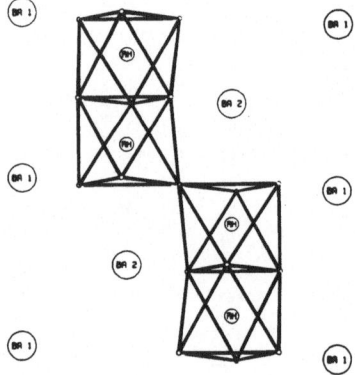

Fig. 1.   Structure of $BaRhO_3$ (4H).

LANTHANUM STRONTIUM COPPER OXIDE
$La_{0.8}Sr_{1.2}CuO_{3.4}$

N. NGUYEN, J. CHOISNET, M. HERVIEU and B. RAVEAU, 1981. J. Solid State Chem., <u>39</u>, 120-127.

Tetragonal, I4/mmm, a = 18.80, c = 12.94 Å, Z = 50 (subcell, I4/mmm, a' = a/5, c' = c, Z = 2). X-ray powder data and electron diffraction patterns.

The subcell is of K_2NiF_4 type (La/Sr in 4(c): z = 0.357; Cu in 2(a); O(1) in 4(e): z = 0.168; O(2) in 4(c); vacancies are in the O(2) site). The supercell involves slight displacements of La, Sr, and O. Phases with smaller Sr content have the tetra-gonal $LaSrCuO_4$ (K_2NiF_4) type and orthorhombic La_2CuO_4 type structures (<u>1</u>).

<u>1</u>. Structure Reports, <u>39A</u>, 252.

LANTHANUM BARIUM CUPRATE
$La_{3.6}Ba_{2.4}Cu_{1.8}O_{9.6}$

I.   C. MICHEL, L. ER-RAKHO and B. RAVEAU, 1981. J. Solid State Chem., <u>39</u>, 161-167.

Tetragonal, P4/mbm, a = 6.862, c = 5.871 Å, D_m = 6.67, Z = 1. Cu radiation, powder data. 2 La,Ba(1) in 2(b): 1/2,1/2,1/2; 4 La,Ba(2) in 4(g): x,1/2+x,0, x = 0.1730; 1.8 Cu in 2(c): 1/2,0,1/2; 2 O(1) in 2(a): 0,0,0; 7.6 O(2) in 8(k): x,1/2+x,z, x = 0.365, z = 0.276.

The structure (Fig. 1) contains square planar CuO_4 groups, linked by edge- and face-sharing MO_8 and MO_{10} polyhedra (M = La,Ba). Cu-O = 1.86, La,Ba-O = 2.47-2.98(2) Å.

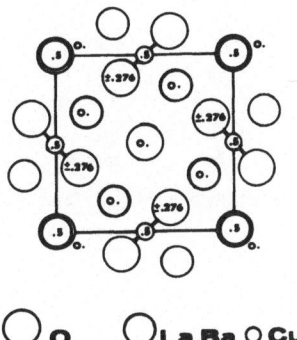

Fig. 1.   Structure of $La_{3.6}Ba_{2.4}Cu_{1.8}O_{9.6}$.

$La_3Ba_3Cu_6O_{14.1}$

II. L. ER-RAKHO, C. MICHEL, J. PROVOST and B. RAVEAU, 1981. J. Solid State Chem., <u>37</u>, 151-156.

Tetragonal, P4/mmm, a = 5.53 = √2 x a(perovskite), c = 11.72 Å = 3 x a(perovskite), Z = 1. Cu radiation, powder data.

The structure (Fig. 2) is an oxygen-defect perovskite, with ordered oxygen vacancies. Cu ions have square-planar, square-pyramidal 4+1, and distorted-octahedral 4+2 coordinations (the last is probably Cu(III)). Ba is probably mainly in site A_1.

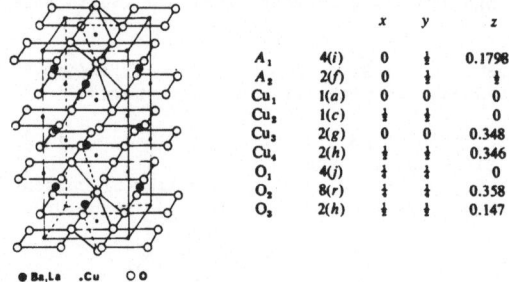

		x	y	z
A_1	4(i)	0	$\frac{1}{2}$	0.1798
A_2	2(f)	0	$\frac{1}{2}$	$\frac{1}{2}$
Cu_1	1(a)	0	0	0
Cu_2	1(c)	$\frac{1}{2}$	$\frac{1}{2}$	0
Cu_3	2(g)	0	0	0.348
Cu_4	2(h)	$\frac{1}{2}$	$\frac{1}{2}$	0.346
O_1	4(j)	$\frac{1}{2}$	$\frac{1}{2}$	0
O_2	8(r)	$\frac{1}{2}$	$\frac{1}{2}$	0.358
O_3	2(h)	$\frac{1}{2}$	$\frac{1}{2}$	0.147

● Ba,La   .Cu   O O

Fig. 2.   Structure of $La_3Ba_3Cu_6O_{14.1}$.

SODIUM DIAURATE(III)
$Na_6Au_2O_6$

H. KLASSEN and R. HOPPE, 1981. Z. Naturforsch., <u>36B</u>, 1395-1399.

Tetragonal, $P4_2/mnm$, a = 9.465, c = 4.506 Å, D_m = 5.20, Z = 2. Mo radiation, R = 0.055 for 230 reflexions.

The structure (Fig. 1) contains planar $Au_2O_6{}^{6-}$ groups linked by 4-coordinate Na ions (far from tetrahedral, with O-Na-O angles as large as 155°). Au-O = 1.995(8), Na-O = 2.29-2.67 Å (further O at 2.83, 2.86 Å).

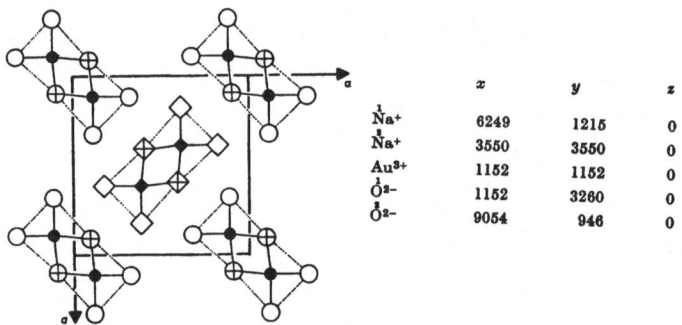

	x	y	z
$\overset{1}{Na}{}^+$	6249	1215	0
$\overset{2}{Na}{}^+$	3550	3550	0
Au^{3+}	1152	1152	0
$\overset{1}{O}{}^{2-}$	1152	3260	0
$\overset{2}{O}{}^{2-}$	9054	946	0

Fig. 1.   Structure of $Na_6Au_2O_6$, and atomic positional parameters (x 10^4).

BARIUM STRONTIUM LANTHANON OXIDES
$BaSr_2Ln_6O_{12}$ (Ln = Y, Er, Tm)

A.-R. SCHULZE and H. MÜLLER-BUSCHBAUM, 1981. Z. Naturforsch., <u>36B</u>, 837-839.

Hexagonal, $P6_3/m$, a = 10.299, 10.277, 10.215, c = 3.409, 3.385, 3.376 Å, Z = 1. R = 0.08-0.09 for 514-528 reflexions.

Atomic positions

		BaSr$_2$Y$_6$O$_{12}$			BaSr$_2$Er$_6$O$_{12}$			BaSr$_2$Tm$_6$O$_{12}$		
		x	y	z	x	y	z	x	y	z
1/2 Ba$_I$	2b	0,0	0,0	0,0	0,0	0,0	0,0	0,0	0,0	0,0
1/2 Ba$_{II}$	2a	0,0	0,0	0,25	0,0	0,0	0,25	0,0	0,0	0,25
Sr	2d	0,3333	0,6667	0,75	0,3333	0,6667	0,75	0,3333	0,6667	0,75
Ln	6h	0,3462	0,9984	0,25	0,3470	0,9991	0,25	0,3467	0,9997	0,25
O$_I$	6h	0,191	0,890	0,25	0,195	0,896	0,25	0,195	0,892	0,25
O$_{II}$	6h	0,529	0,132	0,25	0,526	0,129	0,25	0,523	0,128	0,25

The structures (Fig. 1) are similar to that of SrCa$_2$Sc$_6$O$_{12}$ (1); Ln, Sr, Ba(1) have octahedral coordinations, and Ba(2) has 9-coordination, the $\overline{\text{Ba}}$ positions being half-occupied. Ln-O = 2.18-2.39, Sr-O = 2.51-2.55, Ba-O = 2.70-3.21 Å.

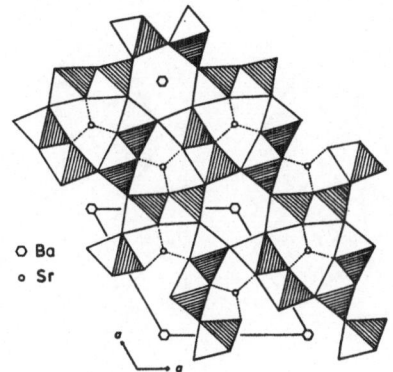

O Ba
o Sr

Fig. 1.   Structure of BaSr$_2$Ln$_6$O$_{12}$.

1. Structure Reports, 41A, 230.

BERYLLIUM PRASEODYMIUM OXIDE
Be$_4$Pr$_9$O$_{20}$

M. SCHWEIZER and H. MÜLLER-BUSCHBAUM, 1981. Z. anorg. Chem., 482, 173-178.

Orthorhombic, Pna2$_1$, a = 9.541, b = 6.557, c = 7.227 Å, Z = 1. R = 0.052 for 420 reflexions.

Atomic positions

	x	y	z
Be	0.1059	0.8061	0.5080
0.625 Pr(1)$^{4+}$	0.1770	0.0852	0.2500
0.625 Pr(2)$^{4+}$	0.3328	0.5724	0.2992
Pr(3)$^{3+}$	0.0255	0.5986	0.0165
O(1)	0.495	0.522	0.005
O(2)	0.275	0.833	0.014
O(3)	0.431	0.187	0.197
O(4)	0.067	0.701	0.319
O(5)	0.226	0.365	0.039

The structure contains BeO$_4$ tetrahedra, Be-O = 1.58-1.71 Å, and 8- and 9-coordinated Pr^{3+} and Pr^{4+} ions, the latter sites being only partially occupied; Pr-O = 2.31-2.83 Å.

CURITE

$Pb_3[U_8O_{24}(OH)_6] \cdot 3H_2O$

J.C. TAYLOR, W.I. STUART and I.A. MUMME, 1981.   J. Inorg. Nucl. Chem., **43**, 2419-2423.

Orthorhombic, Pnam, a = 12.551, b = 13.003, c = 8.390 Å, D_m = 7.37, z = 2.   Mo radiation, R = 0.09 for 1114 reflexions.   Previous study in **1**.

  Two U have pentagonal bipyramidal coordinations and one U has distorted octa-hedral coordination with two more-distant oxygen neighbours; these polyhedra share edges and corners to give U_8O_{30} layers parallel to (100) (Fig. 1).   Two Pb ions (ten coordination, 94% occupancy, and twelve coordination, 57% occupancy) and another O are located between the layers.

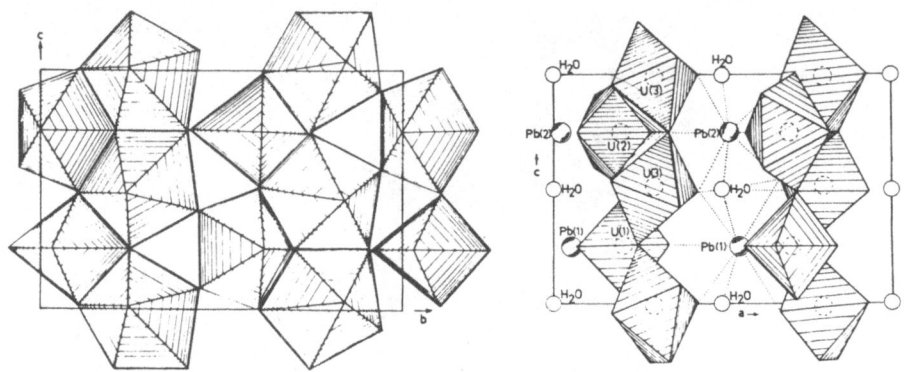

Fig. 1.   Structure of curite.

1. Structure Reports, **28**, 150.

CADMIUM URANATE(VI)

$CdUO_4$

T. YAMASHITA, T. FUJINO, N. MASAKI and H. TAGAWA, 1981.   J. Solid State Chem., **37**, 133-139.

α-Form, rhombohedral, R$\overline{3}$m, a = 6.233 Å, α = 36.12°, Z = 1 (hexagonal cell, a = 3.864, c = 17.46 Å, Z = 3).   Powder data.   U in 1(a): 0,0,0; Cd in 1(b): 1/2,1/2,1/2; O(1) and O(2) in 2(c): x,x,x, x = 0.113 and 0.350.

β-Form, orthorhombic, Cmmm, a = 7.023, b = 6.849, c = 3.514 Å, Z = 2.   Powder data. U in 2(a); Cd in 2(c); O(1) in 4(i): y = 0.278; O(2) in 4(h): x = 0.159.

  The α-form is isostructural with rhombohedral $SrUO_4$ (**1**), as previously described (**2**); U and Cd each have 8-coordination, U-O = 1.98, 2.25, $\overline{C}$d-O = 2.42, 2.61 Å.   The β-$CdUO_4$ structure is quite similar to that previously described (**3**), but in space group Cmmm (rather than Pbam); the structure contains chains of edge-sharing UO_6 octahedra, linked by octahedrally-coordinated Cd ions, U-O = 1.91, 2.08; Cd-O = 2.32, 2.40 Å.

1. Structure Reports, 43A, 353.
2. K.V. REŠETOV and L.M. KOVBA, 1966. J. Struct. Chem., 7, 589.
3. Structure Reports, 27, 405.

CAESIUM HYDROXIDE
CsOH

H. JACOBS and B. HARBRECHT, 1981. Z. Naturforsch., 36B, 270-271.

Below 497.5K, orthorhombic, Cmcm, a = 4.350, b = 11.99, c = 4.516 Å, at 298K, Z = 4. Powder data. Cs and OH in 4(c): 0,y,1/4, y = 0.14, 0.385, respectively. Cs-O = 2.94 (x 1), 3.15 (x 4), 3.75 (x 2) Å.

Above 497.5K, cubic, [Fm3m], a = 6.427 Å, at 503K, Z = 4. Powder data. NaCl-type structure.

TIVANITE
VTiO$_3$OH

I.E. GREY and E.H. NICKEL, 1981. Amer. Min., 66, 866-871.

Monoclinic, P2$_1$/c, a = 7.494, b = 4.552, c = 10.005 Å, β = 129.79°, Z = 4. Cu radiation, R = 0.12 for 97 reflexions (films, visual intensities, for multiply-twinned crystal).

The structure (Fig. 1) contains a hexagonal close-packed anion array, with metal ions in half the octahedral sites. It can be described as an ordered 1:1 intergrowth of rutile-type TiO$_2$ and diaspore-type VOOH (with M(1) = Ti, M(2) = V). M(1)-O = 1.87-2.09, mean 2.00(4), M(2)-O = 1.84-2.22, mean 2.03(4) Å.

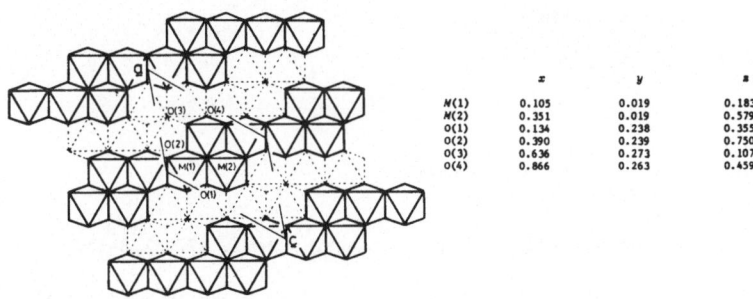

	x	y	z
M(1)	0.105	0.019	0.183
M(2)	0.351	0.019	0.579
O(1)	0.134	0.238	0.355
O(2)	0.390	0.239	0.750
O(3)	0.636	0.273	0.107
O(4)	0.866	0.263	0.459

Fig. 1.   Structure of tivanite.

PLATINIC ACID
H$_8$PtO$_6$ H$_2$Pt(OH)$_6$

G. BANDEL, C. PLATTE and M. TRÖMEL, 1981. Z. anorg. Chem., 472, 95-101.

Monoclinic, C2/c, a = 8.470, b = 7.195, c = 7.451 Å, β = 93.54°, D$_m$ = 4.2, Z = 4. Mo radiation, R = 0.067 for 325 reflexions.

The structure (Fig. 1) contains PtO_6 octahedra linked by strong O-H...O hydrogen bonds; Pt-O = 1.99-2.00, O...O = 2.59-2.66 Å. Ammonium hexahydroxoplatinate(IV), $(NH_4)_2Pt(OH)_6$, is isostructural with the K salt (1).

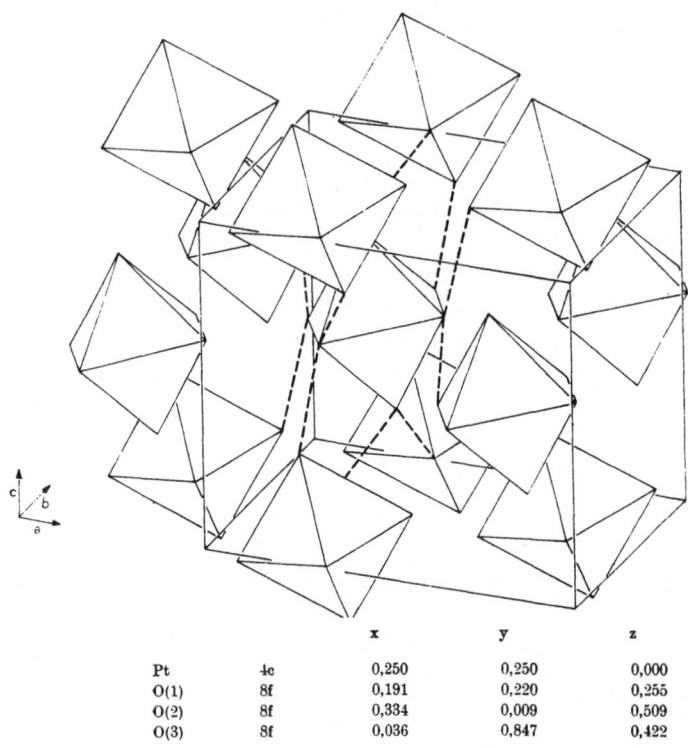

		x	y	z
Pt	4c	0,250	0,250	0,000
O(1)	8f	0,191	0,220	0,255
O(2)	8f	0,334	0,009	0,509
O(3)	8f	0,036	0,847	0,422

Fig. 1.   Structure of platinic acid.

1. Structure Reports, 9, 213.

BARIUM HEXAHYDROXOPLATINATE(IV) MONOHYDRATE
$BaPt(OH)_6 \cdot H_2O$

G. BANDEL, C. PLATTE and M. TRÖMEL, 1981. Z. anorg. Chem., 477, 178-182.

Monoclinic, $P2_1/m$, a = 6.284, b = 6.246, c = 8.574 Å, β = 108.19°, D_m = 4.66, Z = 2. Mo radiation, R = 0.073 for 765 reflexions.

The structure (Fig. 1) contains octahedral $Pt(OH)_6^{2-}$ ions linked by nine-coordinate Ba^{2+} ions; Pt-O = 1.98, Ba-O = 2.75-3.12(2) Å. The water molecule, which is easily removed, is held in the structure only by weak hydrogen bonds.

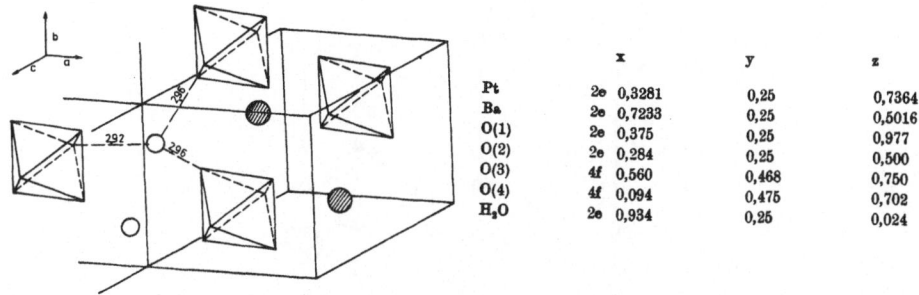

		x	y	z
Pt	2e	0,3281	0,25	0,7364
Ba	2e	0,7233	0,25	0,5016
O(1)	2e	0,375	0,25	0,977
O(2)	2e	0,284	0,25	0,500
O(3)	4f	0,560	0,468	0,750
O(4)	4f	0,094	0,475	0,702
H₂O	2e	0,934	0,25	0,024

Fig. 1.   Structure of barium hexahydroxoplatinate(IV) monohydrate.

YTTERBIUM HYDROXIDE OXIDE (MONOCLINIC)
YbO(OH)

A.N. CHRISTENSEN and B. LEBECH, 1981. Acta Cryst., B37, 425-427.

Monoclinic, $P2_1/m$, a = 5.87, b = 3.58, c = 4.27 Å, β = 109.3°, Z = 2. Neutron powder data. Atoms in 2(e): (x,3/4,z), x = 0.1918, 0.058, 0.571,0.582, z = 0.331, 0.771, 0.755, 0.988, for Yb, O(1), O(2), H.

    Structure as previously described (1). TbOOH orders magnetically at 7.6K, but YbOOH shows no magnetic transition down to 4.2K. The compounds also exist in a denser tetragonal modification (2).

1. Structure Reports, 40A, 215.
2. Ibid., 38A, 289.

POTASSIUM HYDROGEN SULPHIDE
KHS

H. JACOBS and C. ERTEN, 1981. Z. anorg. Chem., 473, 125-132.

Rhombohedral, $R\bar{3}m$, a = 4.957, c = 9.916 Å, D_m = 1.69, Z = 3. Mo radiation, R = 0.028 for 79 reflexions. 3 K in 3(a): 0,0,0; 3 S in 3(b): 0,0,1/2; 3 H in 6(c): 0,0,0.349. Previous studies in 1.

    The structure (Fig. 1) contains disordered HS⁻ ions. K-12K = 4.37, 4.96, K-6S = 3.31, S-12S = 4.37, 4.96, S-H = 1.5 Å.

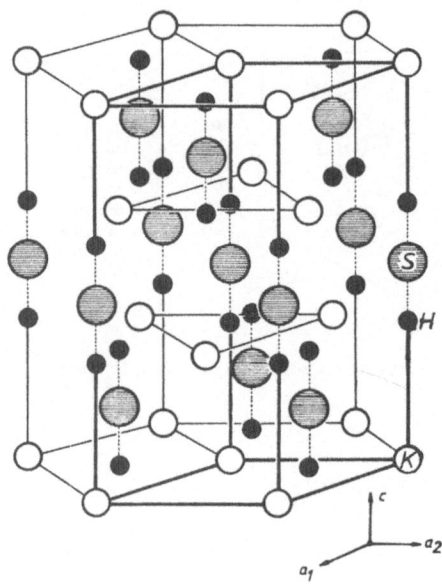

Fig. 1.   Structure of KHS.

1. Strukturbericht, 3, 8; Structure Reports, 37A, 270.

SODIUM DITHIODIHYDROXOGERMANATE PENTAHYDRATE
$Na_2GeS_2(OH)_2 \cdot 5H_2O$

B. KREBS and H.-J. WALLSTAB, 1981. Z. Naturforsch., 36B, 1400-1406.

Orthorhombic, Pbcn, a = 10.752, b = 13.787, c = 14.150 Å, D_m = 1.95, Z = 8.   R = 0.032.

The structure (Fig. 1) contains isolated tetrahedral $GeS_2(OH)_2^{2-}$ anions, linked by octahedrally-coordinated Na^+ ions and by an extensive system of O-H...O and O-H...S hydrogen bonds.   Na-O = 2.32-2.89 Å.

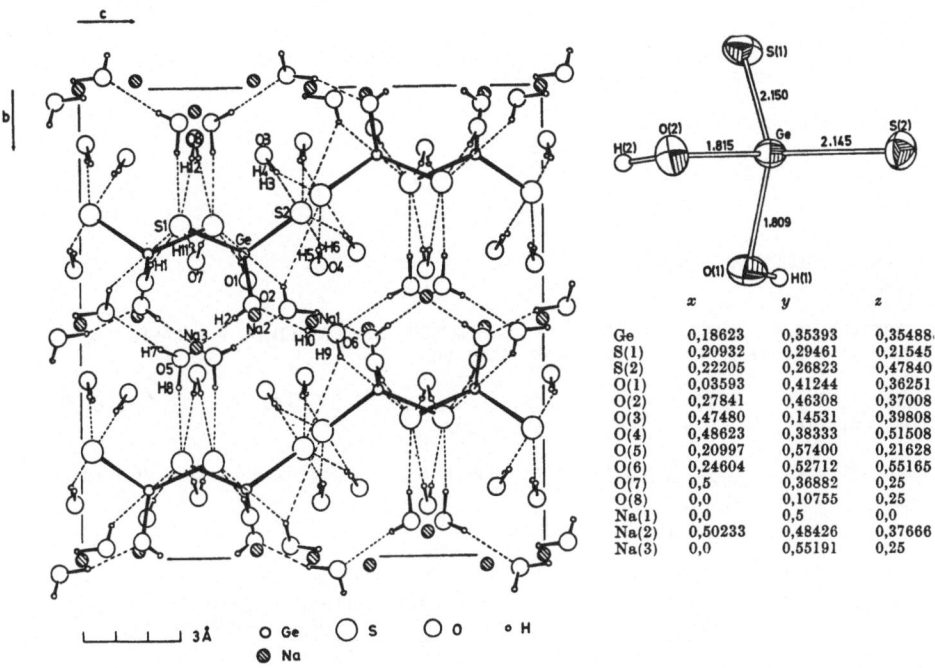

	x	y	z
Ge	0,18623	0,35393	0,35488
S(1)	0,20932	0,29461	0,21545
S(2)	0,22205	0,26823	0,47840
O(1)	0,03593	0,41244	0,36251
O(2)	0,27841	0,46308	0,37008
O(3)	0,47480	0,14531	0,39808
O(4)	0,48623	0,38333	0,51508
O(5)	0,20997	0,57400	0,21628
O(6)	0,24604	0,52712	0,55165
O(7)	0,5	0,36882	0,25
O(8)	0,0	0,10755	0,25
Na(1)	0,0	0,5	0,0
Na(2)	0,50233	0,48426	0,37666
Na(3)	0,0	0,55191	0,25

Fig. 1.   Structure of $Na_2GeS_2(OH)_2.5H_2O$ (S-Ge-S = 121, O-Ge-O = 96°).

PHOSPHORUS SULPHOXIDE
$P_4O_3S_6$

V.K. PALKINA, S.I. MAKSIMOVA and G.U. WOLF, 1980. Izv. Akad. Nauk SSSR, Neorg.
Mater., 16, 1466-1468.

Monoclinic, $P2_1/m$, a = 6.596, b = 12.289, c = 7.173 Å, β = 104.02°, D_m = 2.13, Z = 2. R = 0.048.

     The molecular structure is similar to that of P_4O_9 (1), with three bridging O, and three bridging and three terminal S. One P is bonded trigonally to 3 S, and the other P atoms are bonded tetrahedrally to 2 O and 2 S. P-S (bridging) = 2.13, 2.07 Å.

1. Structure Reports, 29, 300; 32A, 252.

SELENIUM SULPHUR
$Se_{1.1}S_{6.9}$ $Se_{3.7}S_{4.3}$

R.A. BOUDREAU and H.M. HAENDLER, 1981. J. Solid State Chem., 36, 289-296.

Monoclinic, P2/c, a = 8.34, 8.40, b = 13.11, 13.26, c = 9.30, 9.37 Å, β = 123.9, 124.5°, D_m = 2.10, 3.32, Z = 4. Cu radiation, R = 0.080, 0.097 for 309, 568 reflexions (film data).

γ-Sulphur structure (1, 2), with partially disordered S/Se distributions. Bond lengths = 2.02-2.19 Å, angles = 104-108°.

1. Structure Reports, 40A, 119.
2. Ibid., 43A, 354; 44A, 226; 45A, 275.

AMMONIUM DITHIOMOLYBDATE
$(NH_4)_2MoO_2S_2$

F.W. KUTZLER, R.A. SCOTT, J.M. BERG, K.O. HODGSON, S. DONIACH, S.P. CRAMER and C.H. CHANG, 1981. J. Amer. Chem. Soc., 103, 6083-6088.

Monoclinic, C2/c, a = 11.376, b = 7.300, c = 10.923 Å, β = 130.16°, D_m = 2.16, Z = 4. Mo radiation, R = 0.032 for 797 reflexions.

The structure (Fig. 1) contains tetrahedral anions, linked by the ammonium cations.

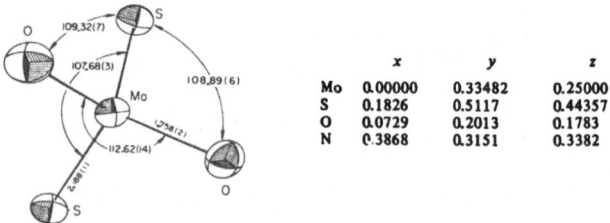

	x	y	z
Mo	0.00000	0.33482	0.25000
S	0.1826	0.5117	0.44357
O	0.0729	0.2013	0.1783
N	0.3868	0.3151	0.3382

Fig. 1.   The $MoO_2S_2^{2-}$ anion and atomic positional parameters in ammonium dithiomolybdate.

LANTHANUM ARSENIC OXYSULPHIDE
$La_4O_3(AsS_3)_2$

M. PALAZZI and S. JAULMES, 1981. Acta Cryst., B37, 1340-1342.

Orthorhombic, Ibam, a = 19.032, b = 12.051, c = 5.852 Å, Z = 4. Mo radiation, R = 0.048 for 690 reflexions.

The structure (Fig. 1) is derived from the LaO sheet structure, and contains ribbons of La_4O tetrahedra parallel to c and AsS_3 trigonal pyramids. La ions have 8-coordination, 4 O + 4 S for La(1) and 2 O + 6 S for La(2). La-O = 2.33-2.47, La-S = 3.01-3.17, As-S = 2.25-2.35(1) Å. As has 50% occupancy, an As...As distance of 1.29 Å precluding simultaneous occupation of two neighbouring sites.

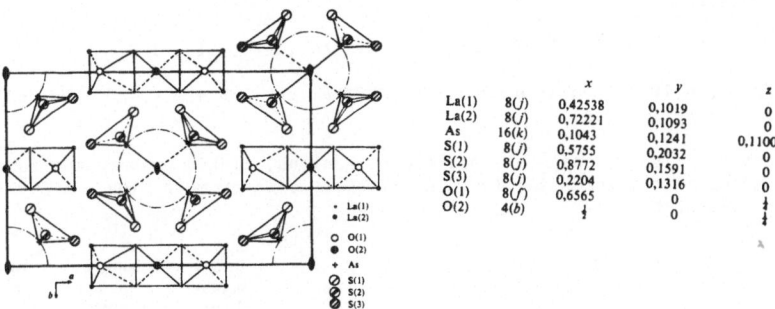

		x	y	z
La(1)	8(j)	0,42538	0,1019	0
La(2)	8(j)	0,72221	0,1093	0
As	16(k)	0,1043	0,1241	0,1100
S(1)	8(j)	0,5755	0,2032	0
S(2)	8(j)	0,8772	0,1591	0
S(3)	8(j)	0,2204	0,1316	0
O(1)	8(f)	0,6565	0	$\frac{1}{4}$
O(2)	4(b)	$\frac{1}{4}$	0	$\frac{1}{4}$

Fig. 1.   Structure of $La_4O_3(AsS_3)_2$ and atomic positional parameters
(As has 50% occupancy).

LANTHANUM COPPER OXYSULPHIDE
(LaO)CuS

M. PALAZZI, 1981. C.R. Acad. Sci. Paris, 292, 789-791.

Tetragonal, P4/nmm, a = 3.999, c = 8.53 Å, D_m = 6.14, Z = 2. Mo radiation, R =
0.029 for 182 reflexions. La in 2(c): 1/4,1/4,0.1484; Cu in 2(b): 1/4,3/4,1/2; S
in 2(c): 1/4,1/4,0.6633; O in 2(a): 1/4,3/4,0 [origin at centre].

The structure contains alternating LaO and CuS sheets, linked by La-S bonding.
Cu has tetrahedral coordination, Cu-4S = 2.437 Å. La-S = 3.253, La-O = 2.367 Å.

LANTHANUM SILVER OXYSULPHIDE
(LaO)AgS

M. PALAZZI and S. JAULMES, 1981. Acta Cryst., B37, 1337-1339.

Tetragonal, P4/nmm, a = 4.050, c = 9.039 Å, D_m = 6.3, Z = 2. Mo radiation, R = 0.055
for 154 reflexions. La in 2(c): 1/4,1/4,0.1356; Ag in 2(b): 3/4,1/4,1/2; S in 2(c):
1/4,1/4,0.6929; O in 2(a): 3/4,1/4,0.

The structure contains alternating LaO and AgS sheets. La is coordinated to
4 O at 2.367(1) and 4 S at 3.255(4) Å; Ag has tetrahedral coordination to 4 S at
2.673(6) Å (this long distance explains the easy removal of Ag, and the electrical
conductivity).

LANTHANUM NEODYMIUM OXYSULPHIDE
$LaNdO_{1.75}S$

G.M. KUZMIČEVA, I.V. PEREPELKIN and A.A. ELISEEV, 1980. Ž. Neorg. Khim., 25, 3212-
3214.

Trigonal, P3m1, a = 4.052, c = 6.958 Å, Z = 1. R = 0.065.

La and Nd are each coordinated to 4 O and 3 S atoms.

AMMONIUM TETRABORATE DIHYDRATE
$(NH_4)_2[B_4O_5(OH)_4] \cdot 2H_2O$

R. JANDA, G. HELLER and J. PICKARDT, 1981. Z. Kristallogr., <u>154</u>, 1-9.

Monoclinic, $P2_1$, a = 10.691, b = 10.646, c = 7.223 Å, β = 139.4°, Z = 2. Mo radiation, R = 0.092 for 1190 reflexions.

The structure (Fig. 1) contains isolated polyanions, linked by ammonium ions via a hydrogen bonding system which involves one of the two water molecules. B-O = 1.33-1.51, N...O = 2.83-3.40 Å.

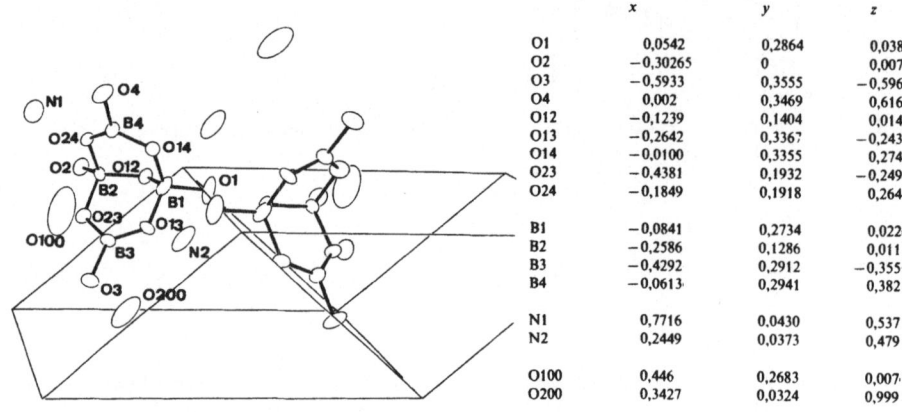

	x	y	z
O1	0,0542	0,2864	0,038
O2	−0,30265	0	0,007
O3	−0,5933	0,3555	−0,596
O4	0,002	0,3469	0,616
O12	−0,1239	0,1404	0,014
O13	−0,2642	0,3367	−0,243
O14	−0,0100	0,3355	0,274
O23	−0,4381	0,1932	−0,249
O24	−0,1849	0,1918	0,264
B1	−0,0841	0,2734	0,022
B2	−0,2586	0,1286	0,011
B3	−0,4292	0,2912	−0,355
B4	−0,0613	0,2941	0,382
N1	0,7716	0,0430	0,537
N2	0,2449	0,0373	0,479
O100	0,446	0,2683	0,007
O200	0,3427	0,0324	0,999

Fig. 1.   Structure of ammonium tetraborate dihydrate.

AMMONIUM PENTABORATE TETRAHYDRATE
$\alpha\text{-}NH_4[B_5O_6(OH)_4] \cdot 2H_2O$ $NH_4B_5O_8 \cdot 4H_2O$

V. DOMENECH, J. SOLANS and X. SOLANS, 1981. Acta Cryst., B<u>37</u>, 643-645.

Monoclinic, Pn, a = 7.115, b = 11.301, c = 7.183 Å, β = 99.92°, Z = 2. Mo radiation, R = 0.058 for 769 reflexions.

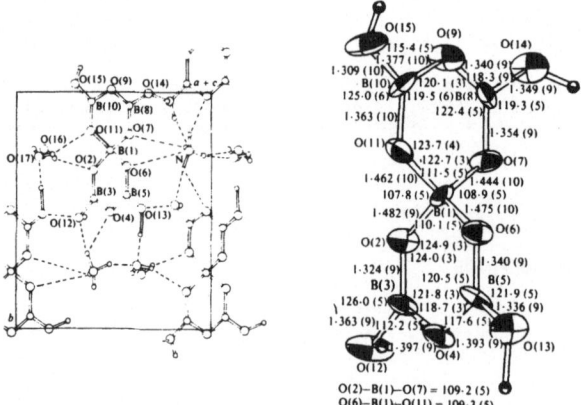

Fig. 1.   Structure of $\alpha\text{-}NH_4[B_5O_6(OH)_4] \cdot 2H_2O$.

The structure is related to that of K salt (1), and contains pentaborate anions linked by ammonium ions and water molecules (Fig. 1). The β-phase structure is quite similar (2).

1. Strukturbericht, 5, 23, 108; Structure Reports, 23, 409; 28, 168; 40A, 305.
2. Structure Reports, 34A, 353.

LITHIUM PEROXOBORATE
$Li_2[B_2(O_2)_2(OH)_4]$

A. PAWEL, G. HELLER and J. PICKARDT, 1981. Z. Kristallogr., 157, 251-257.

Monoclinic, $P2_1/c$, a = 7.040, b = 7.880, c = 5.085 Å, β = 97.98°, D_m = 1.97, Z = 2. Mo radiation, R = 0.045 for 533 reflexions.

The structure (Fig. 1) contains tetrahydroxo-di-μ-peroxo-diborate anions (1), linked by six-coordinate Li^+ ions. B-OH = 1.43, 1.45, B-O = 1.50, 1.51, O-O = $\overline{1}$.49, Li-O = 1.97-2.25 Å.

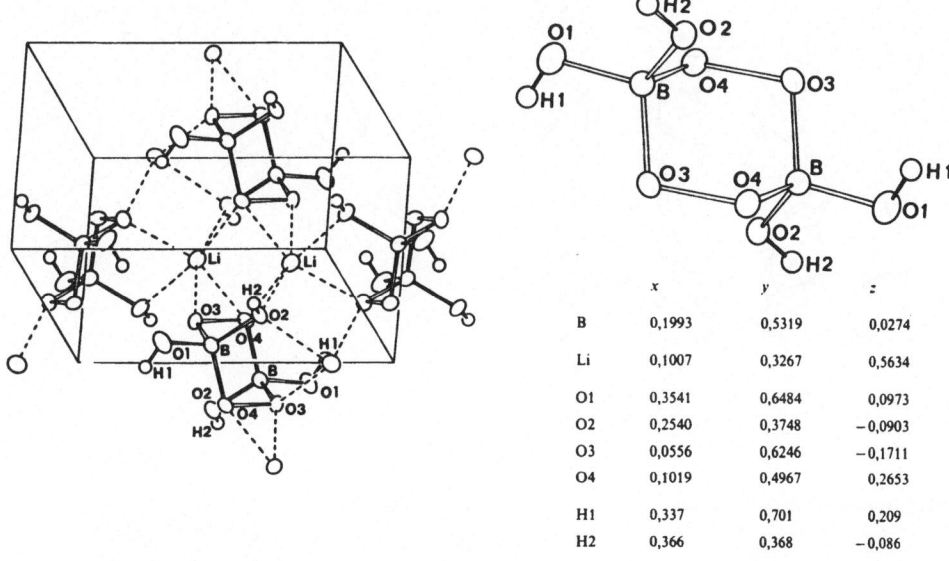

	x	y	z
B	0,1993	0,5319	0,0274
Li	0,1007	0,3267	0,5634
O1	0,3541	0,6484	0,0973
O2	0,2540	0,3748	−0,0903
O3	0,0556	0,6246	−0,1711
O4	0,1019	0,4967	0,2653
H1	0,337	0,701	0,209
H2	0,366	0,368	−0,086

Fig. 1.   Structure of lithium peroxoborate.

1. Structure Reports, 26, 487; 44A, 229.

LITHIUM BORACITE
$Li_5B_7O_{12.5}Cl$

M. VLASSE, A. LEVASSEUR and P. HAGENMULLER, 1981. Solid State Ionics, 2, 33-37.

Cubic, F23, a = 12.136 Å. Boracite-type structure (1), with modification of some Li and B coordinations.

1. Structure Reports, 39A, 265.

CARBOBORITE
$MgCa_2(CO_3)_2[B(OH)_4]_2 \cdot 4H_2O$

Z. MA, N. SHI, J. SHEN and Z. PENG, 1981. Bull. Minéral., 104, 578-581.

Monoclinic, $P2_1/n$, a = 11.011, b = 6.674, c = 10.692 Å, β = 116.64°, Z = 2. Mo radiation, R = 0.043 for 1964 reflexions.

The structure (Fig. 1) contains sheets of $B(OH)_4^-$ tetrahedra and CO_3^{2-} triangles linked by MgO_6 octahedra and CaO_8 polyhedra; the sheets are joined by hydrogen bonding. B-O = 1.45-1.48, C-O = 1.27-1.28, Mg-O = 2.01-2.09, Ca-O = 2.32-2.52 Å.

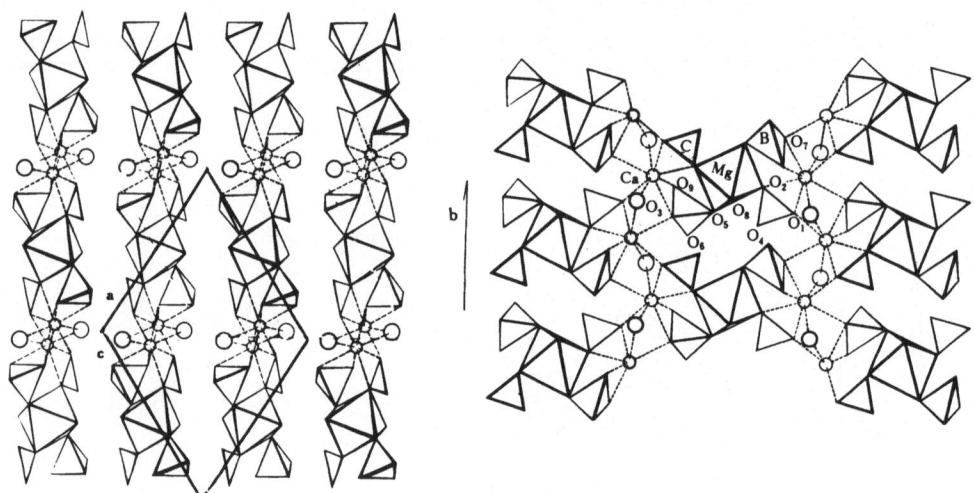

Fig. 1.   Structure of carboborite.

THALLIUM DIBORATE HYDRATE
$Tl_2[B_4O_6(OH)_2] \cdot 2H_2O$ $Tl_2O \cdot 2B_2O_3 \cdot 3H_2O$

THALLIUM PENTABORATE HYDRATE
$Tl[B_5O_6(OH)_4] \cdot 2H_2O$ $0.5 \times [Tl_2O \cdot 5B_2O_3 \cdot 8H_2O]$

I.   K.-H. WOLLER and G. HELLER, 1981. Z. Kristallogr., 156, 151-157.
II. Idem, 1981. Ibid., 156, 159-166.

Diborate
Orthorhombic, $P2_12_12_1$, a = 10.909, b = 9.762, c = 9.450 Å, Z = 4. Mo radiation, R = 0.037 for 1289 reflexions.

Pentaborate
Monoclinic, $P2_1/c$, a = 11.275, b = 7.155, c = 13.928 Å, β = 94.16°, Z = 4. Mo radiation, R = 0.055 for 2440 reflexions.

The diborate structure (Fig. 1) contains infinite chain polyanions, linked by 7-coordinate Tl ions and by hydrogen bonding via the water molecules. B-O = 1.43-1.54 (tetrahedral), 1.32-1.42 (trigonal), Tl-O = 2.59-3.46(3) Å.

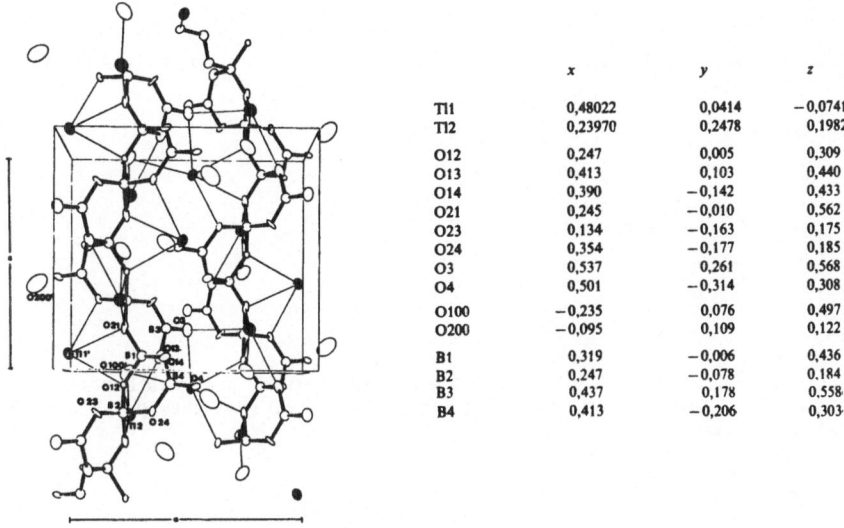

	x	y	z
Tl1	0,48022	0,0414	−0,0741
Tl2	0,23970	0,2478	0,1982
O12	0,247	0,005	0,309
O13	0,413	0,103	0,440
O14	0,390	−0,142	0,433
O21	0,245	−0,010	0,562
O23	0,134	−0,163	0,175
O24	0,354	−0,177	0,185
O3	0,537	0,261	0,568
O4	0,501	−0,314	0,308
O100	−0,235	0,076	0,497
O200	−0,095	0,109	0,122
B1	0,319	−0,006	0,436
B2	0,247	−0,078	0,184
B3	0,437	0,178	0,558
B4	0,413	−0,206	0,303

Fig. 1.   Structure of thallium diborate hydrate.

The pentaborate structure (Fig. 2) contains isolated polyanions, linked by Tl ions which have distorted octahedral coordination, and by hydrogen bonding via the water molecules. B-O = 1.46-1.49 (tetrahedral), 1.33-1.40 (trigonal), Tl-O = 2.84-3.02(1) Å.

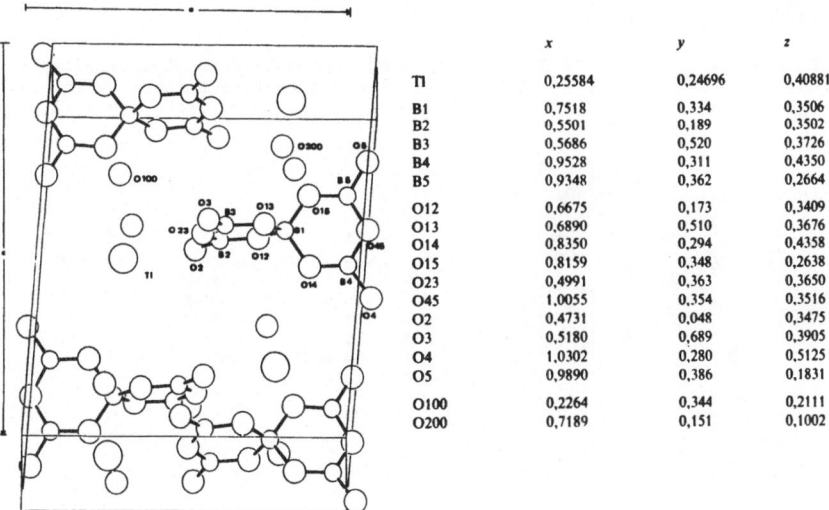

	x	y	z
Tl	0,25584	0,24696	0,40881
B1	0,7518	0,334	0,3506
B2	0,5501	0,189	0,3502
B3	0,5686	0,520	0,3726
B4	0,9528	0,311	0,4350
B5	0,9348	0,362	0,2664
O12	0,6675	0,173	0,3409
O13	0,6890	0,510	0,3676
O14	0,8350	0,294	0,4358
O15	0,8159	0,348	0,2638
O23	0,4991	0,363	0,3650
O45	1,0055	0,354	0,3516
O2	0,4731	0,048	0,3475
O3	0,5180	0,689	0,3905
O4	1,0302	0,280	0,5125
O5	0,9890	0,386	0,1831
O100	0,2264	0,344	0,2111
O200	0,7189	0,151	0,1002

Fig. 2.   Structure of thallium pentaborate hydrate.

YTTRIUM ALUMINUM BORATE
$YAl_3(BO_3)_4$

E.L. BELOKONEVA, A.V. AZIZOV, N.I. LEONJUK, M.A. SIMONOV and N.V. BELOV, 1981. Ž.
Strukt. Khim., 22, No. 3, 196-199 [J. Struct. Chem., 22, 476-477].

Rhombohedral, R32, a = 9.295, c = 7.243 Å, D_m = 3.72, Z = 3. Mo radiation, R = 0.037
for 450 reflexions.

Atomic positions

	x	y	z
Y	0	0	0
Al	0.5554	0	0
B(1)	0	0	1/2
B(2)	0.4436	0	1/2
O(1)	0.8512	0	1/2
O(2)	0.5907	0	1/2
O(3)	0.4499	0.1498	0.4786

Isostructural with the $NdAl_3$ (1) and $NdGa_3$ (2) borates and with huntite,
$CaMg_3(CO_3)_4$ (3). B-O = 1.37-1.38 (triangles), Y-O = 2.32 (trigonal prism), Al-O =
1.84-1.94 Å (octahedral).

1. Structure Reports, 40A, 226.
2. Ibid., 44A, 234.
3. Ibid., 27, 561.

POTASSIUM TANTALUM BORATE
$K_3Ta_3B_2O_{12}$

S.C. ABRAHAMS, L.E. ZYONTZ, J.L. BERNSTEIN, J.P. REMEIKA and A.S. COOPER, 1981.
J. Chem. Phys., 75, 5456-5460.

Hexagonal, P$\bar{6}$2m, a = 8.78158, c = 3.89902 Å, at 298K, D_m = 5.61, Z = 1. Mo radiation,
R = 0.013 for 1050 reflexions.

The structure contains triads of distorted TaO_6 octahedra stacked along c and
connected by planar BO_3 groups (Fig. 1); interstices contain K ions, which have 13
oxygen neighbours. Ta-O = 1.942-2.022, B-O = 1.364, K-O = 2.762-3.139(4) Å.

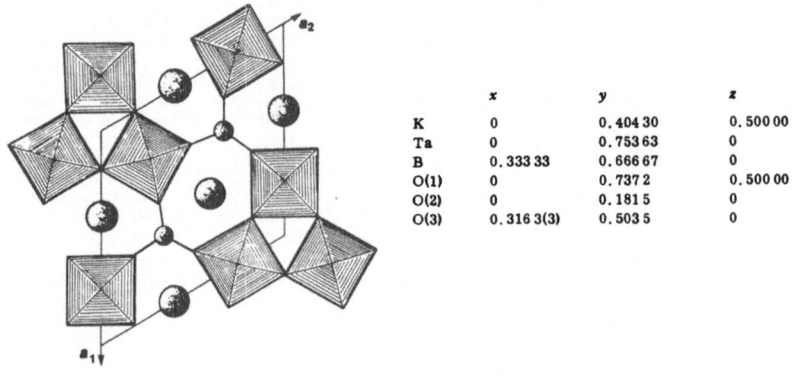

	x	y	z
K	0	0.40430	0.50000
Ta	0	0.75363	0
B	0.33333	0.66667	0
O(1)	0	0.7372	0.50000
O(2)	0	0.1815	0
O(3)	0.3163(3)	0.5035	0

Fig. 1.   Structure of $K_3Ta_3B_2O_{12}$.

NICKEL BROMINE BORACITE (ORTHORHOMBIC)
$Ni_3B_7O_{13}Br$

S.C. ABRAHAMS, J.L. BERNSTEIN and C. SVENSSON, 1981. J. Chem. Phys., 75, 1912-1918.

Orthorhombic, $Pca2_1$, a = 8.5218, b = 8.5127, c = 12.0408 Å, at 298K, D_m = 4.10, Z = 4. Mo radiation, R = 0.049 for 1531 reflexions.

Boracite structure (1), with six BO_4 tetrahedra and one BO_3 triangle (B is 0.14 Å out of plane) (Fig. 1); mean B-O = 1.478 (tetrahedra), 1.378 Å (triangle). Ni has 4 oxygen neighbours in a distorted square plane, Ni-O = 2.00-2.06 Å, with a Br at 2.66 Å completing 5-coordination, and a further Br at 3.37 Å.

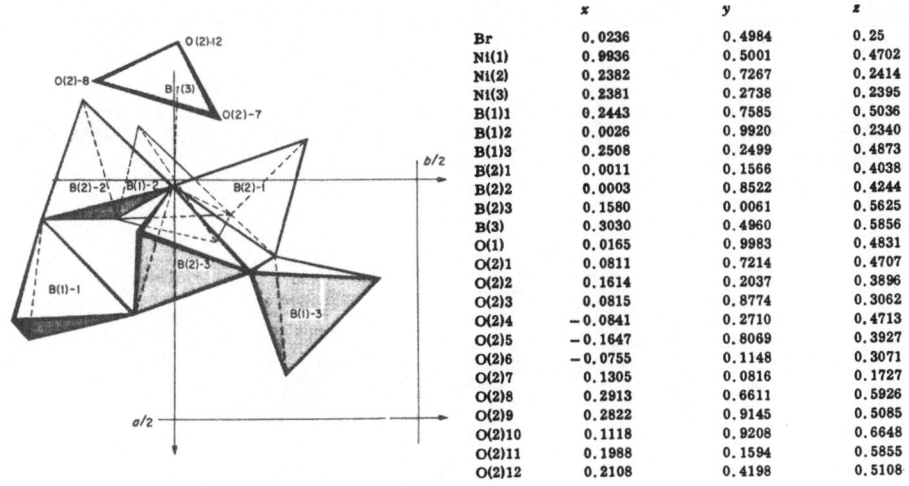

	x	y	z
Br	0.0236	0.4984	0.25
Ni(1)	0.9936	0.5001	0.4702
Ni(2)	0.2382	0.7267	0.2414
Ni(3)	0.2381	0.2738	0.2395
B(1)1	0.2443	0.7585	0.5036
B(1)2	0.0026	0.9920	0.2340
B(1)3	0.2508	0.2499	0.4873
B(2)1	0.0011	0.1566	0.4038
B(2)2	0.0003	0.8522	0.4244
B(2)3	0.1580	0.0061	0.5625
B(3)	0.3030	0.4960	0.5856
O(1)	0.0165	0.9983	0.4831
O(2)1	0.0811	0.7214	0.4707
O(2)2	0.1614	0.2037	0.3896
O(2)3	0.0815	0.8774	0.3062
O(2)4	−0.0841	0.2710	0.4713
O(2)5	−0.1647	0.8069	0.3927
O(2)6	−0.0755	0.1148	0.3071
O(2)7	0.1305	0.0816	0.1727
O(2)8	0.2913	0.6611	0.5926
O(2)9	0.2822	0.9145	0.5085
O(2)10	0.1118	0.9208	0.6648
O(2)11	0.1988	0.1594	0.5855
O(2)12	0.2108	0.4198	0.5108

Fig. 1.   Structure of nickel bromine boracite.

1. Structure Reports, 15, 284; 39A, 264.

COPPER(II) METABORATE
CuB_2O_4

G.K. ABDULLAEV and K.S. MAMEDOV, 1981. Ž. Strukt. Khim., 22, No. 4, 184-187.

Tetragonal, $I\bar{4}2d$, a = 11.506, c = 5.644 Å, Z = 12. Mo radiation, R = 0.047 for 518 reflexions.

Atomic positions

	x	y	z
Cu(1)	0	0	1/2
Cu(2)	0.0815	1/4	1/8
O(1)	0.1583	0.0707	0.4940
O(2)	0.2523	1/4	5/8
O(3)	1/4	0.0833	7/8
O(4)	0.0738	0.1891	0.7983
B(1)	0.1828	0.1469	0.6964
B(2)	0.0004	1/4	5/8

The structure is as previously determined (1), containing polyanions built from B_3O_6 groups of corner-sharing BO_4 tetrahedra, B-O = 1.46-1.49 Å. Cu ions have square-planar coordinations, Cu-O = 1.90-1.97 Å.

1. Structure Reports, 37A, 277.

SILVER(I) METABORATE
$AgBO_2$

G. BRACHTEL and M. JANSEN, 1981. Z. anorg. Chem., 478, 13-19.

Orthorhombic, Pbcn, a = 8.441, b = 8.680, c = 19.741 Å, Z = 16. Mo radiation, R = 0.054 for 1776 reflexions.

The structure (Fig. 1) contains an infinite polyanion with tetrahedral and trigonal planar B coordinations, B-O = 1.43-1.51 and 1.32-1.42 Å, O-B-O = 105-113 and 116-126°, respectively. The polyanions form sheets parallel to (001), linked by five independent Ag ions, with various coordinations.

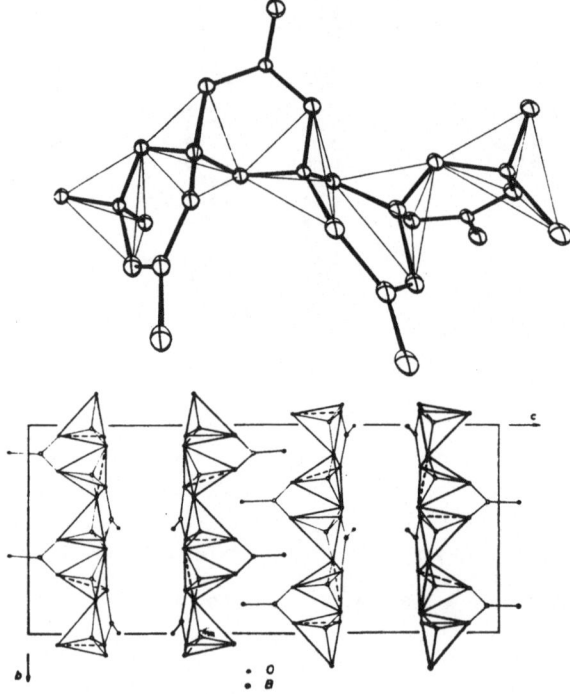

Fig. 1.   The polyanion in $AgBO_2$.

SILVER(I) ORTHOBORATE
Ag_3BO_3

M. JANSEN and W. SCHELD, 1981. Z. anorg. Chem., 477, 85-89.

Rhombohedral, R32, a = 9.873, c = 3.381 Å, Z = 3.   Mo radiation, R = 0.049 for 179 reflexions.   B in 3(a): 0,0,0; Ag and O in 9(d): x,0,0, x = 0.5005 and 0.8620, respectively.

The structure (Fig. 1) contains trigonal planar BO_3^{3-} ions, B-O = 1.363(6) Å, linked by Ag^+ ions which have non-linear two-coordination to form ...O-Ag-O-Ag... chains, Ag-O = 2.125(7) Å, O-Ag-O = 143, Ag-O-Ag = 117.2°.

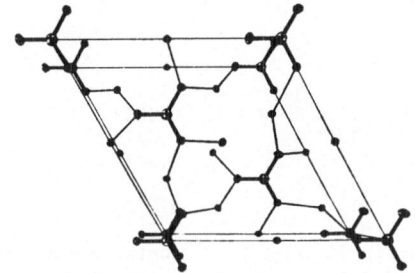

Fig. 1.   Structure of silver(I) orthoborate.

LITHIUM CADMIUM BORATE (HEXAGONAL)
$LiCdBO_3$

E.V. SOKOLOVA, M.A. SIMONOV and N.V. BELOV, 1980.   Kristallografija, 25, 1285-1286 [Soviet Physics - Crystallography, 25, 733-734].

Hexagonal, $P\overline{6}$, a = 8.324, c = 3.264 Å, D_m = 4.5, Z = 3.   Mo radiation, R = 0.037 for 453 reflexions.   Preliminary study in 1, and preliminary study of a triclinic form in 2.

The structure (Fig. 1) contains columns of BO_3 triangles, LiO_4 tetrahedra, and CdO_5 polyhedra.   B-O = 1.39(1), Li-O = 1.90-2.02(4), Cd-O = 2.21-2.28(1) Å.

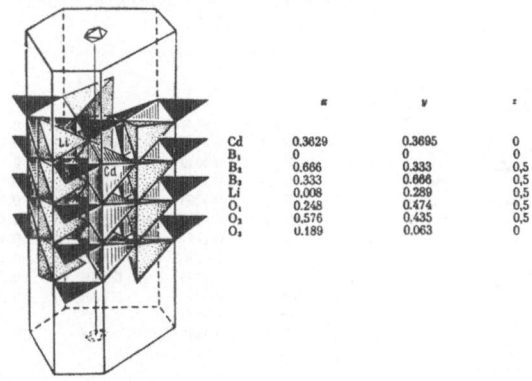

	x	y	z
Cd	0.3629	0.3695	0
B_1	0	0	0
B_2	0.666	0.333	0,5
B_3	0.333	0.666	0,5
Li	0.008	0.289	0,5
O_1	0.248	0.474	0,5
O_2	0.576	0.435	0,5
O_3	0.189	0.063	0

Fig. 1.   Structure of hexagonal $LiCdBO_3$.

1. Structure Reports, 44A, 329.
2. Ibid., 45A, 391.

LANTHANUM METABORATE
La(BO$_2$)$_3$

G.K. ABDULLAEV, K.S. MAMEDOV and G.G. DŽAFAROV, 1981. Kristallografija, $\underline{26}$, 837-840
[Soviet Physics - Crystallography, $\underline{26}$, 473-474].

Monoclinic, I2/c, a = 6.509, b = 8.172, c = 7.983 Å, β = 93.43°, D$_m$ = 4.18, Z = 4.
Mo radiation, R = 0.066 for 949 reflexions.

Atomic positions

	x	y	z
La	0	0.0509	1/4
O(1)	0.1445	0.3648	0.3568
O(2)	0.1411	0.4373	0.6418
O(3)	0.0455	0.1667	0.5482
B(1)	0	0.4695	1/4
B(2)	0.1088	0.3196	0.5214

The structure is as previously described ($\underline{1}$), containing a $[B_6O_{12}{}^{6-}]_n$ chain along $\underline{c}$ of two BO$_4$ tetrahedra and four BO$_3$ triangles sharing corners. The chains are linked by LaO$_{10}$ polyhedra which share edges to form layers parallel to (010). B-O = 1.473, 1.530 (tetrahedron), 1.332, 1.376, 1.384 (triangle), La-O = 2.418-2.863 Å.

$\underline{1}$. Structure Reports, $\underline{35A}$, 282; $\underline{38A}$, 298; $\underline{41A}$, 290.

SODIUM NEODYMIUM BORATE
Na$_3$Nd(BO$_3$)$_2$

J. MASCETTI, M. VLASSE and C. FOUASSIER, 1981. J. Solid State Chem., $\underline{39}$, 288-293.

Monoclinic, P2$_1$/c, a = 6.618, b = 8.810, c = 12.113 Å, β = 122.27°, D$_m$ = 3.70, Z = 4.
Mo radiation, R = 0.040 for 5998 reflexions.

The structure (Fig. 1) contains isolated BO$_3$ triangles, linked by eight-coordinated Nd, and by Na ions which have 7-, 6-, and unusual (3+1)-coordinations. B-O = 1.36-1.40(1), Nd-O = 2.36-2.53, Na-O = 2.24-2.68 and 2.96 Å.

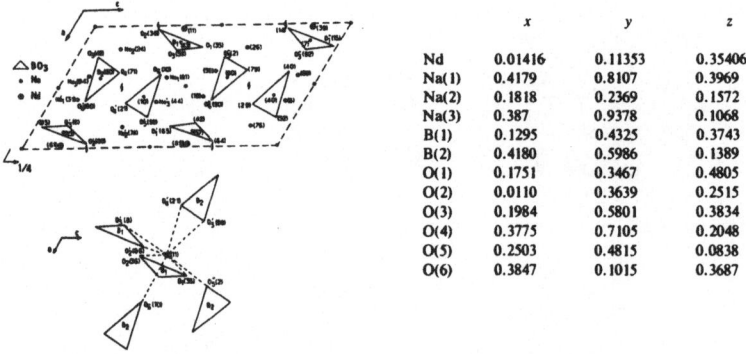

	x	y	z
Nd	0.01416	0.11353	0.35406
Na(1)	0.4179	0.8107	0.3969
Na(2)	0.1818	0.2369	0.1572
Na(3)	0.387	0.9378	0.1068
B(1)	0.1295	0.4325	0.3743
B(2)	0.4180	0.5986	0.1389
O(1)	0.1751	0.3467	0.4805
O(2)	0.0110	0.3639	0.2515
O(3)	0.1984	0.5801	0.3834
O(4)	0.3775	0.7105	0.2048
O(5)	0.2503	0.4815	0.0838
O(6)	0.3847	0.1015	0.3687

Fig. 1.   Structure of Na$_3$Nd(BO$_3$)$_2$, neodymium coordination and atomic positional parameters.

EUROPIUM(II) ORTHOBORATE
$Eu_3(BO_3)_2$

K.-I. MACHIDA, G.-Y. ADACHI, H. HATA and J. SHIOKAWA, 1981. Bull. Chem. Soc. Japan,
54, 1052-1055.

Rhombohedral, $R\bar{3}c$, a = 9.069, c = 12.542 Å, D_m = 6.31, Z = 6. Mo radiation, R = 0.082
for 259 reflexions. Eu in 18(e): (0.3075,0,1/4); B in 12(c): (0,0,0.39), O in 36(d):
(0.164,0.033,0.396).

Isostructural with the Ca salt (1). The structure (Fig. 1) contains isolated
BO_3 triangles linked by 8-coordinate Eu ions. B-O = 1.36, Eu-O = 2.36-2.95 Å.

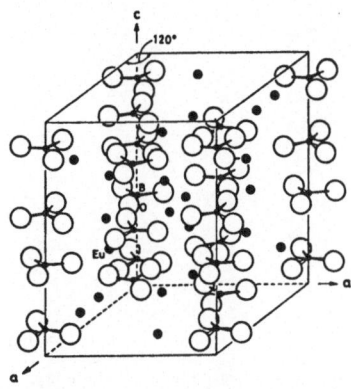

Fig. 1.   Structure of europium(II) orthoborate.

1. Structure Reports, 34A, 356; 41A, 285.

EUROPIUM(II) BORATE CHLORIDE
EUROPIUM(II) BORATE BROMIDE
$Eu_2B_5O_9X$ (X = Cl, Br)

K.-I. MACHIDA, G.-Y. ADACHI, Y. MORIWAKI and J. SHIOKAWA, 1981. Bull. Chem. Soc.
Japan, 54, 1048-1051.

Orthorhombic, Pnn2, a = 11.364, 11.503, b = 11.301, 11.382, c = 6.504, 6.484 Å, D_m =
4.30, 4.53, for X = Cl, Br, Z = 4. Mo radiation, R = 0.054, 0.047 for 1967, 1979
reflexions.

The structures (Fig. 1) contain infinite three-dimensional $(B_5O_9)_n$ frameworks
of three BO_4 tetrahedra and two BO_3 triangles, with Eu^{2+} and X^- in tunnels along c;
Eu ions have 2 X and 7 O neighbours. B-O = 1.44-1.49 (tetrahedra), 1.30-1.41(3)
(triangles), Eu-O = 2.47-3.00, Eu-Cl = 2.96-3.05 Å.

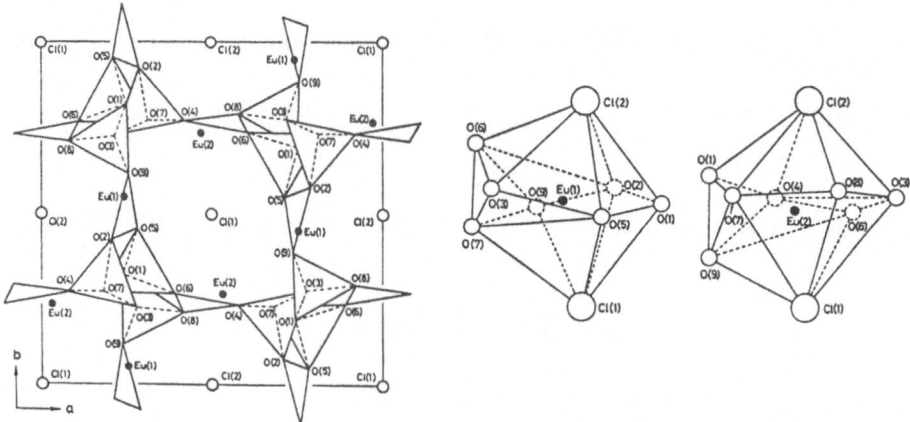

Fig. 1.   Structure of $Eu_2B_5O_9Cl$.

CALCIUM ERBIUM GERMANIUM BORATE
$Ca_3Er_3Ge_2BO_{13}$

J. CHENAVAS, I.E. GREY, J.C. GUITEL, J.C. JOUBERT, M. MAREZIO, J.P. REMEIKA and A.S. COOPER, 1981. Acta Cryst., B37, 1343-1346.

Cubic, $F\bar{4}3m$, a = 10.452 Å, Z = 4.   Ag radiation, R = 0.026 for 491 reflexions.

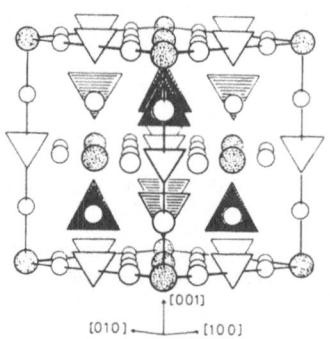

The structure of $(Ca,Er)_6Ge_2BO_{13}$ viewed approximately along [110]. Shaded and unshaded tetrahedra correspond to $[GeO_4]^{4-}$ and $[BO_4]^{5-}$ respectively. Shaded and unshaded circles represent interstitial oxygen, O(4), and (Ca,Er) respectively. The unit-cell outline is marked.

		x	y	z
$Ca_{3+x}Er_{3+x}Pb_{x+x}$	24(f)	0·22807 (3)	0	0
Ge(1)	4(c)	¼	¼	¼
Ge(2)	4(d)	¾	¾	¾
B	4(b)	½	½	½
O(1)	16(e)	0·1512	0·1512	0·1512
O(2)	16(e)	0·4180	0·4180	0·4180
O(3)	16(e)	0·8461	0·8461	0·8461
O(4)	4(a)	0	0	0

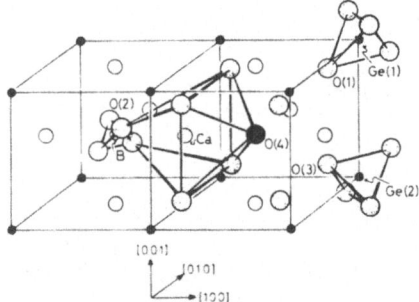

Partial representation of the structure for $(Ca,Er)_6Ge_2BO_{13}$, showing its relationship to fluorite. Two fluorite-like face-centred cubic subcells are outlined. Small filled and open circles represent Ge and B respectively; medium circles correspond to (Ca,Er) and large stippled circles to oxygen. The large filled circle represents the interstitial oxygen, O(4). Oxygen polyhedra around independent cations are outlined. Two (Ca,Er) atoms have been omitted for clarity. Note that the origin of the cubic cell of $(Ca,Er)_6Ge_2BO_{13}$ is on the O(4) oxygen atom.

Fig. 1.   Structure of $Ca_3Er_3Ge_2BO_{13}$.

The structure (Fig. 1) contains a fluorite-like cation sublattice, with Ge at cube corners and Ca/Er at face centres. Displacement of oxygen atoms along the sub-cell <111> to give tetrahedral coordination around Ge creates large interstitial sites at the subcell body centres, half of which are occupied by BO_4 tetrahedra and the other half by O atoms. Ca/Er-O = 2.33-2.40 (7-coordination), Ge-O = 1.74, 1.79, B-O = 1.49 Å. An alternative description is as a perovskite-derivative structure, with a relationship to sulphohalite.

TESCHEMACHERITE (AMMONIUM BICARBONATE)
NH_4HCO_3

F. PERTLIK, 1981. Tschermaks Miner. Petr. Mitt., _29_, 67-74.

Orthorhombic, Pccn, a = 7.255, b = 10.709, c = 8.746 Å, Z = 8. Mo radiation, R = 0.056 for 844 reflexions. Previous study in _1_.

The structure (Fig. 1) contains chains along _c_ of hydrogen-bonded bicarbonate anions, linked by hydrogen bonds via ammonium ions. C-O = 1.255, C-OH = 1.256(2) Å.

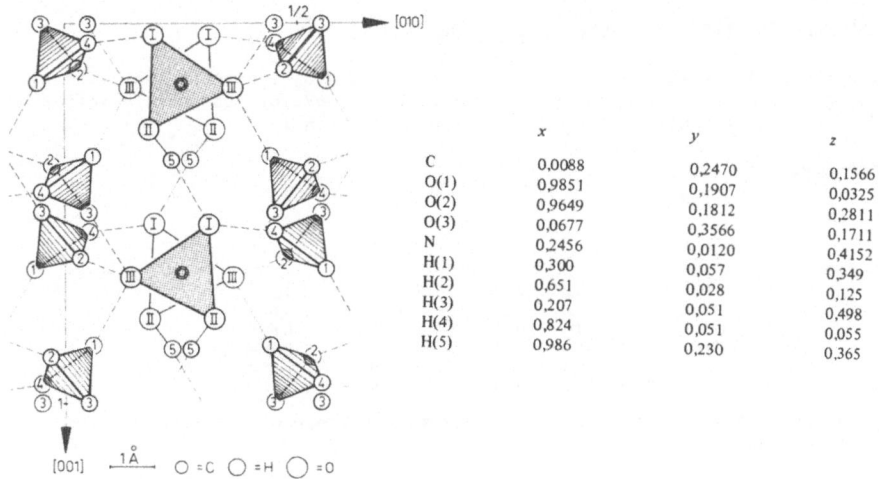

	x	y	z
C	0,0088	0,2470	0,1566
O(1)	0,9851	0,1907	0,0325
O(2)	0,9649	0,1812	0,2811
O(3)	0,0677	0,3566	0,1711
N	0,2456	0,0120	0,4152
H(1)	0,300	0,057	0,349
H(2)	0,651	0,028	0,125
H(3)	0,207	0,051	0,498
H(4)	0,824	0,051	0,055
H(5)	0,986	0,230	0,365

Fig. 1.   Structure of NH_4HCO_3.

_1_. Structure Reports, _13_, 294.

MAGNESITE, CALCITE, RHODOCHROSITE, SIDERITE, SMITHONITE, DOLOMITE
$MgCO_3$ $CaCO_3$ $MnCO_3$ $FeCO_3$ $ZnCO_3$ $MgCa(CO_3)_2$

H. EFFENBERGER, K. MEREITER and J. ZEMANN, 1981. Z. Kristallogr., _156_, 233-243.

Dolomite, rhombohedral, $R\overline{3}$, a = 4.812, c = 16.020 Å, Z = 3. Mo radiation, R = 0.021 for 596 reflexions. Ca in 3(a): 0,0,0; Mg in 3(b): 0,0,1/2; C in 6(c): 0,0,0.2429; O in 18(f): 0.2480,-0.0347,0.2440.

Others, rhombohedral, R$\bar{3}$c, a = 4.6328, 4.9896, 4.7682, 4.6916, 4.6526, c = 15.0129, 17.0610, 15.6354, 15.3796, 15.0257 Å, Z = 6. Mo radiation, R = 0.020, 0.022, 0.015, 0.013, 0.013 for 253, 340, 288, 274, 264 reflexions. M in 6(b): 0,0,0; C in 6(a): 0,0,1/4; O in 18(e): x,0,1/4, x = 0.2774, 0.2568, 0.2699, 0.2743, 0.2764.

Structures are as previously described for magnesite (1), calcite (2), rhodochrosite (3), siderite (4), smithonite (5), and dolomite (6). All $\overline{M}O_6$ octahedra are elongated parallel to c. Some short O...O contacts (2.85 Å) are noted.

1. Strukturbericht, 1, 295, 317; Structure Reports, 39A, 271; 42A, 422.
2. Strukturbericht, 1, 292, 317, 318; Structure Reports, 21, 354; 30A, 408; 42A, 422; 43A, 235.
3. Strukturbericht, 1, 295, 326; 2, 391; Structure Reports, 32A, 415.
4. Strukturbericht, 1, 295, 326.
5. Strukturbericht, 1, 295, 317, 326; 2, 391.
6. Strukturbericht, 1, 303, 317, 324; Structure Reports, 23, 419; 43A, 235.

MONOHYDROCALCITE
$CaCO_3.H_2O$

H. EFFENBERGER, 1981. Mh. Chem., 112, 899-909.

Trigonal subcell, P3$_1$21, a = 6.093, c = 7.545 Å, Z = 3; supercell, P3$_1$, a = $\sqrt{3}$.a(subcell), c = c(subcell), Z = 9. Mo radiation, R = 0.067 for 408 subcell reflexions, 0.139 for 634 supercell reflexions. Previous study in 1.

Atomic positions

Subcell

			x	y	z
Ca	3(a)	1.0	0.2769	0	1/3
O w	3(b)	1.0	0.6027	0	5/6
O(1)	6(c)	1.0	0.4560	0.2744	0.0696
O(2)	6(c)	0.5	0.0691	−0.0504	0.7856
C	6(c)	0.5	0.7188	−0.0261	0.3636
H	6(c)	1.0	0.552	0.095	0.916

Supercell (parameters in square brackets are those of the substructure in the supercell setting)

	x		y		z	
Ca(1)	0.1853	[0.1846]	0.0879	[0.0923]	1/3	[1/3]
Ca(2)	0.8413	[0.8513]	0.4235	[0.4256]	0.3271	[1/3]
Ca(3)	0.5259	[0.5179]	0.7641	[0.7590]	0.3398	[1/3]
O w(1)	0.392	[0.4018]	0.191	[0.2009]	0.835	[5/6]
O w(2)	0.066	[0.0685]	0.541	[0.5342]	0.824	[5/6]
O w(3)	0.748	[0.7351]	0.877	[0.8676]	0.849	[5/6]
O(1—1)	0.206	[0.2125]	0.240	[0.2435]	0.064	[0.0696]
O(1—2)	0.020	[0.0310]	0.245	[0.2435]	0.936	[0.9304]
O(1—3)	0.885	[0.8792]	0.587	[0.5768]	0.088	[0.0696]
O(1—4)	0.701	[0.6976]	0.579	[0.5768]	0.936	[0.9304]
O(1—5)	0.547	[0.5459]	0.899	[0.9102]	0.066	[0.0696]
O(1—6)	0.370	[0.3643]	0.912	[0.9102]	0.927	[0.9304]
O(2—1)	0.060	[0.0629]	0.062	[0.0567]	0.879	[0.8811]
O(2—2)	0.660	[0.6604]	0.400	[0.3899]	0.123	[0.1189]
O(2—3)	0.401	[0.3962]	0.723	[0.7233]	0.879	[0.8811]
C(1)	0.092	[0.1024]	0.179	[0.1788]	0.965	[0.9698]
C(2)	0.750	[0.7430]	0.516	[0.5121]	0.038	[0.0303]
C(3)	0.429	[0.4358]	0.836	[0.8454]	0.973	[0.9698]
H(1)		[0.336]		[0.216]		[0.916]
H(2)		[0.003]		[0.549]		[0.916]
H(3)		[0.670]		[0.882]		[0.916]

The structure (Fig. 1) contains planar carbonate groups, eight-coordinate Ca ions, and water molecules. The carbonate groups are disordered in the sub-structure, and the superstructure results from ordering of these groups. C-O = 1.26-1.32, Ca-O = 2.34-2.64, O-H...O = 2.69-3.09 Å.

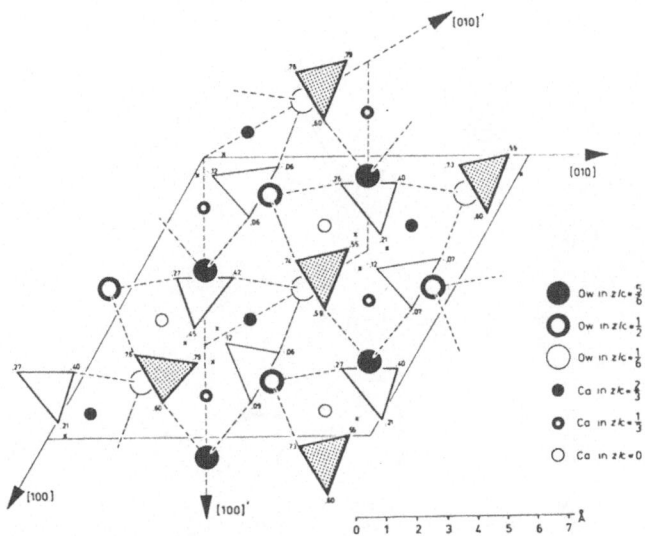

Fig. 1.   Structure of $CaCO_3 \cdot H_2O$.

1. I. KOHATSU and J.W. McCAULEY, 1973. Amer. Min., 58, 1102.

FAIRCHILDITE
$K_2Ca(CO_3)_2$

F. PERTLIK, 1981. Z. Kristallogr., 157, 199-205.

Hexagonal, $P6_3/mmc$, a = 5.294, c = 13.355 Å, Z = 2. Mo radiation, R = 0.10 for 197 reflexions, for synthetic specimen.

The structure (Fig. 1) contains two independent carbonate groups, one of which is disordered over six positions. K and Ca are statistically distributed in two sites, with 7- and 8-coordinations.

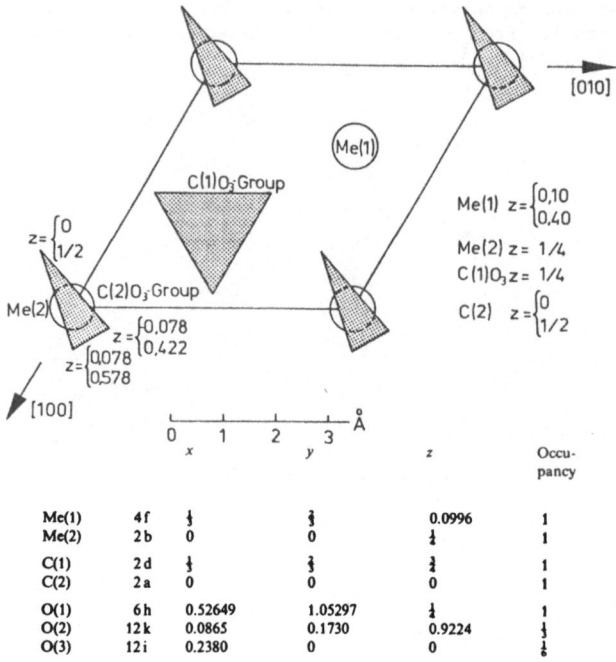

Fig. 1.   Structure of fairchildite; only one of the C(2)O₃ orientations is shown.

TUNISITE

$NaCa_2Al_4(CO_3)_4(OH)_8Cl$

H. EFFENBERGER, F. KLUGER, F. PERTLIK and J. ZEMANN, 1981. Tschermaks Min. Petr. Mitt., 28, 65-77.

Tetragonal, P4/nmm, a = 11.198, c = 6.564 Å, Z = 2. Mo radiation, R = 0.059 for 692 reflexions.

The structure (Fig. 1) contains $Al(OH)_2CO_3^-$ sheets of edge- and corner-sharing octahedra; Al-O = 1.852-1.926, C-O = 1.269, 1.284 Å. Na is coordinated to 4 O at 2.429 Å and 1 Cl at 3.075 Å; Ca has 10 O neighbours at 2.346, 2.755, 2.829 Å; a high thermal parameter for Ca suggests disorder over two positions 0.4 Å apart.

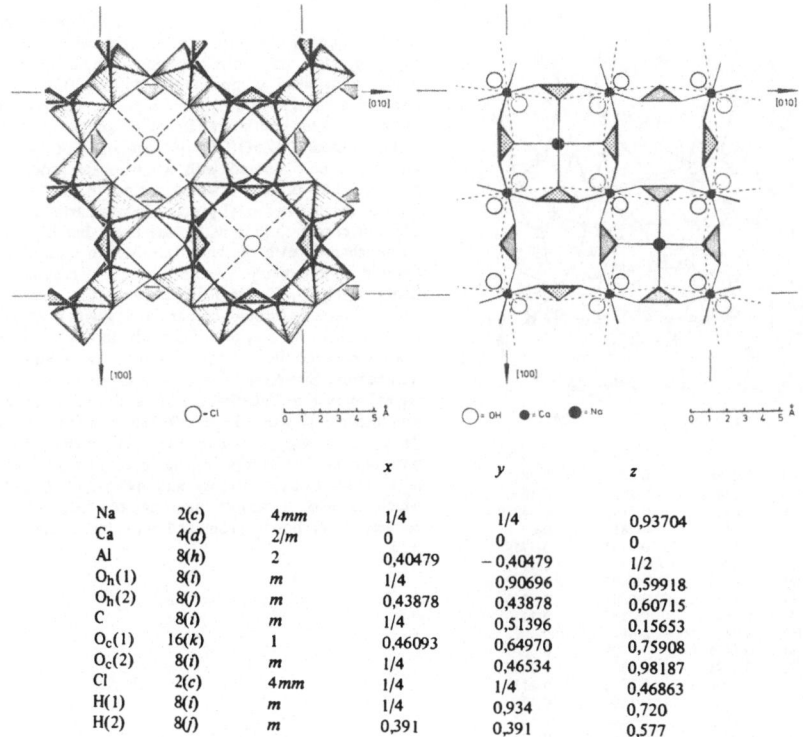

			x	y	z
Na	2(c)	4mm	1/4	1/4	0,93704
Ca	4(d)	2/m	0	0	0
Al	8(h)	2	0,40479	−0,40479	1/2
$O_h(1)$	8(i)	m	1/4	0,90696	0,59918
$O_h(2)$	8(j)	m	0,43878	0,43878	0,60715
C	8(i)	m	1/4	0,51396	0,15653
$O_c(1)$	16(k)	1	0,46093	0,64970	0,75908
$O_c(2)$	8(i)	m	1/4	0,46534	0,98187
Cl	2(c)	4mm	1/4	1/4	0,46863
H(1)	8(i)	m	1/4	0,934	0,720
H(2)	8(j)	m	0,391	0,391	0,577

Fig. 1.   Structure of tunisite.

LOSEYITE

$Mn_{3.48}Zn_{2.99}Mg_{0.53}(CO_3)_2(OH)_{10}$

R.J. HILL, 1981. Acta Cryst., B37, 1323-1328.

Monoclinic, A2/a, a = 16.408, b = 5.540, c = 15.150 Å, β = 95.48°, Z = 4. Cu
radiation, R = 0.069 for 816 reflexions.

    The structure is illustrated and described in Fig. 1. Mn-O = 2.12-2.20, Zn-O =
1.86-2.01, C-O = 1.27-1.32, O-H...O = 2.68-3.08(1) Å.

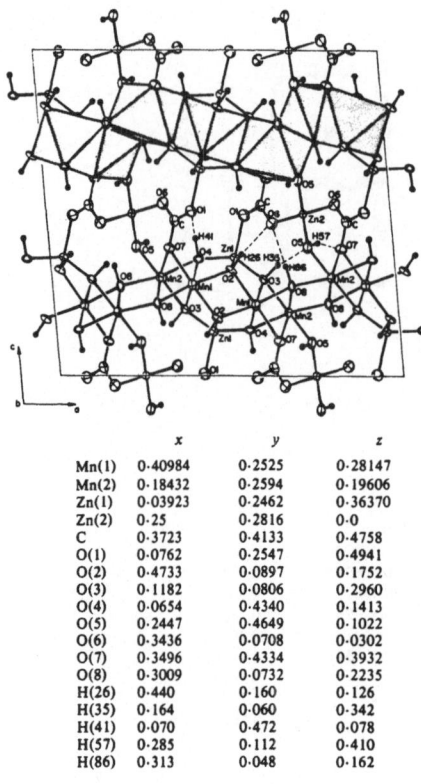

The fundamental building unit of the structure is a regularly stepped sheet of edge- and corner-sharing manganese octahedra, with ideal composition $[Mn_4\square_1O_2(OH)_{10}]^{6-}$, oriented parallel to (001). The vacant octahedral site, $\square$, in this sheet unit shares its upper and lower faces with tetrahedral $[ZnO(OH)_3]^{3-}$ groups. The two O atoms are each shared with a carbonate group on opposite sides of, and oriented normal to, the layers. Two of the ten hydroxyl groups do not participate in edge- or corner-sharing within the layer, but both are corner-shared with a zinc tetrahedron possessing twofold symmetry with the composition $[ZnO_2(OH)_2]^{4-}$. Connectivity between the sheets is achieved directly through corner-sharing with this latter zinc tetrahedron, and also indirectly through corner-sharing between the carbonate group and both kinds of tetrahedron. Site-population analyses carried out during refinement indicate that the Mn and asymmetric Zn sites contain between 6 and 9% Mg, and the twofold Zn site 33% Mg (all within ±1%). A set of H atom positions has been proposed, based on considerations of O atom electroneutrality and hybridization state, which suggest a simple hydrogen-bonding scheme between the carbonate group and both zinc tetrahedra.

	x	y	z
Mn(1)	0·40984	0·2525	0·28147
Mn(2)	0·18432	0·2594	0·19606
Zn(1)	0·03923	0·2462	0·36370
Zn(2)	0·25	0·2816	0·0
C	0·3723	0·4133	0·4758
O(1)	0·0762	0·2547	0·4941
O(2)	0·4733	0·0897	0·1752
O(3)	0·1182	0·0806	0·2960
O(4)	0·0654	0·4340	0·1413
O(5)	0·2447	0·4649	0·1022
O(6)	0·3436	0·0708	0·0302
O(7)	0·3496	0·4334	0·3932
O(8)	0·3009	0·0732	0·2235
H(26)	0·440	0·160	0·126
H(35)	0·164	0·060	0·342
H(41)	0·070	0·472	0·078
H(57)	0·285	0·112	0·410
H(86)	0·313	0·048	0·162

Fig. 1.   Structure of loseyite.

μ-CARBONATO-DI-μ-HYDROXO-BIS(TRIAMMINECOBALT(III)) SULPHATE PENTAHYDRATE
$[(NH_3)_3Co(OH)_2(CO_3)Co(NH_3)_3]SO_4.5H_2O$

M.R. CHURCHILL, R.A. LASHEWYCZ, K. KOSHY and T.P. DASGUPTA, 1981.   Inorg. Chem., 20, 376-381.

Triclinic, P$\bar{1}$, a = 6.691, b = 11.285, c = 11.825 Å, α = 92.78, β = 99.10, γ = 101.50°, Z = 2. Mo radiation, R = 0.029 for 2258 reflexions.

The cation (Fig. 1) has approximate C_{2v} symmetry, with octahedral coordination at each Co; Co-OH = 1.896, 1.922; 1.897, 1.918(2), Co-OCO$_2$ = 1.899, 1.903, Co-N = 1.929-1.947, Co...Co = 2.817(1) Å. In the tetrahedral sulphate anion, S-O = 1.461-1.487 Å, and there is an extensive hydrogen-bonding system.

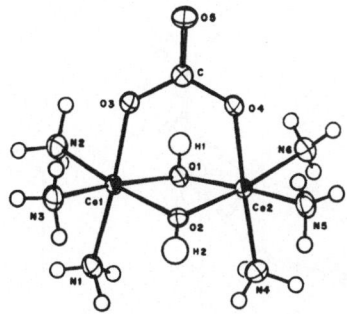

Fig. 1.   The μ-carbonato-di-μ-hydroxo-bis(triamminecobalt(III)) cation.

POTASSIUM BIS(CARBONATO)CUPRATE(II)
$K_2Cu(CO_3)_2$

B.N. FIGGIS, P.A. REYNOLDS, A.H. WHITE and G.A. WILLIAMS, 1981. J. Chem. Soc.,
Dalton, 371-376.

Orthorhombic, Fdd2, a = 11.425, b = 17.658, c = 6.154 Å, D_m = 2.79, Z = 8. Mo
radiation, R = 0.029 for 1693 reflexions.

     Structure as previously determined (1); the electron density distribution is
now studied.

1. Structure Reports, 46A, 307.

THALLIUM(I) COPPER(II) CARBONATE
$Tl_2Cu(CO_3)_2$

H. EHRHARDT, R. LEMOR and H. SEIDEL, 1981. Z. anorg. Chem., 477, 183-195.

Monoclinic, $P2_1/c$, a = 7.583, b = 9.799, c = 9.119 Å, β = 111.51°, D_m = 6.21, Z = 4.
R = 0.12 for 963 reflexions.

     The structure (Fig. 1) contains Cu_2 pairs bridged by four bidentate carbonate
groups, these units being linked by Cu-O bonds; each Cu thus has octahedral coordination, Cu-Cu = 2.583, Cu-O = 1.94-1.98 (equatorial), 2.17 Å (apical). Each Tl
ion has seven oxygen neighbours, Tl-O = 2.57-3.28 Å.

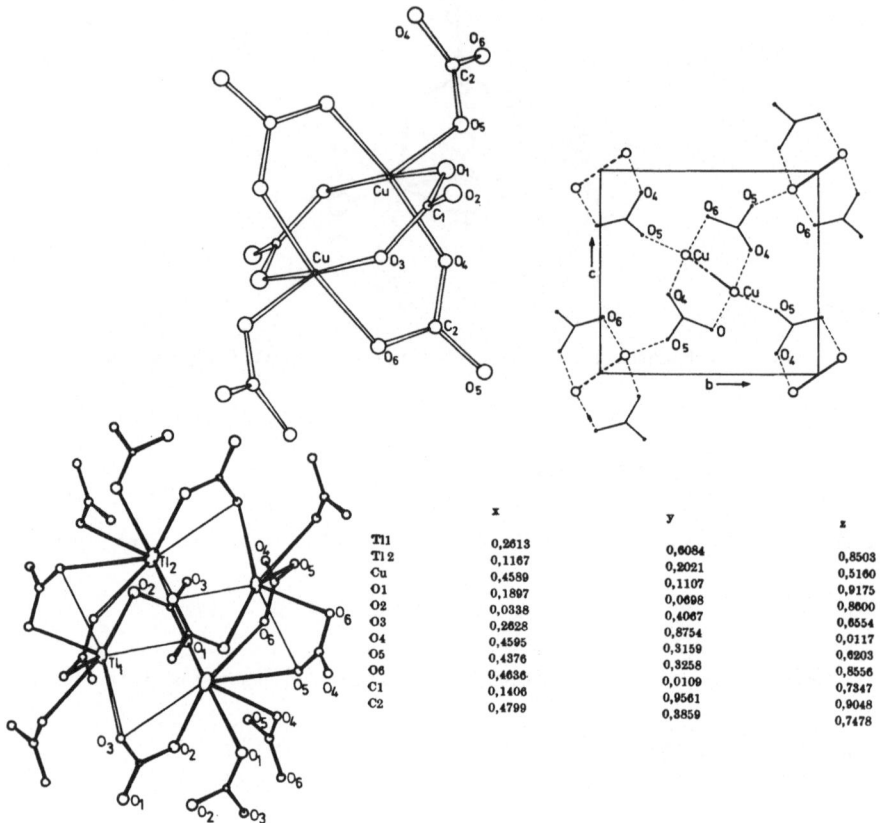

	x	y	z
Tl1	0,2613	0,6084	0,8503
Tl2	0,1167	0,2021	0,5160
Cu	0,4589	0,1107	0,9175
O1	0,1897	0,0698	0,8600
O2	0,0338	0,4067	0,6554
O3	0,2628	0,8754	0,0117
O4	0,4595	0,3159	0,6203
O5	0,4376	0,3258	0,8556
O6	0,4636	0,0109	0,7347
C1	0,1406	0,9561	0,9048
C2	0,4799	0,3859	0,7478

Fig. 1.   Structure of $Tl_2Cu(CO_3)_2$.

GADOLINIUM BICARBONATE PENTAHYDRATE
$Gd(HCO_3)_3 \cdot 5H_2O$ $[Gd(HCO_3)_3(H_2O)_4] \cdot H_2O$

N.G. FURMANOVA, L.V. SOBOLEVA and L.M. BELJAEV, 1981. Kristallografija, 26, 312-315 [Soviet Physics - Crystallography, 26, 176-178].

Monoclinic, $P2_1/a$, a = 6.877, b = 9.575, c = 18.871 Å, γ = 102.61°, Z = 4. Mo radiation, R = 0.117 for 1906 reflexions.

The structure (Fig. 1) contains $Gd(HCO_3)_3(H_2O)_4$ complexes, linked together by hydrogen bonding which incorporates the fifth water molecule. Gd has 10-coordination.

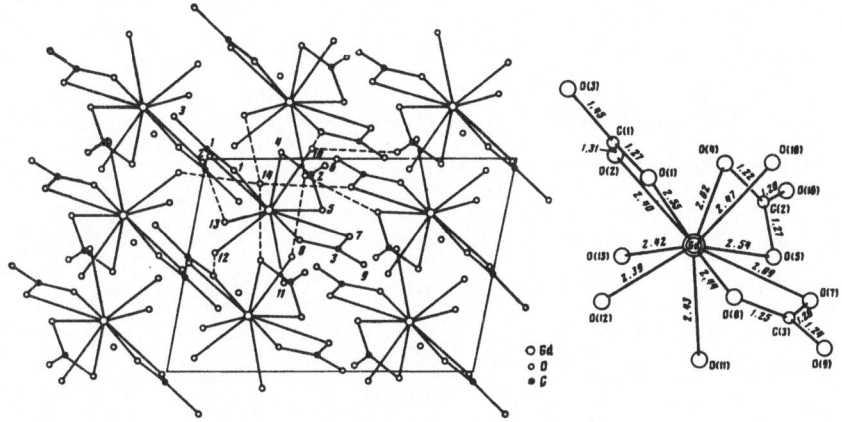

Fig. 1.   Structure of gadolinium bicarbonate pentahydrate.

ANDERSONITE
$Na_2Ca[UO_2(CO_3)_3].5\cdot6H_2O$

A. CODA, A. DELLA GIUSTA and V. TAZZOLI, 1981. Acta Cryst., B<u>37</u>, 1496-1500.

Rhombohedral, R$\overline{3}$m, a = 17.902, c = 23.734 Å, D$_m$ = 2.8, Z = 18. Mo radiation, R$_W$ = 0.049 for 794 reflexions.

The structure (Fig. 1) contains $(UO_2)O_6$ hexagonal bipyramids, in which the six equatorial O atoms belong to three bidentate carbonate groups; these anions are linked by 2 Na and 1 Ca with 6-, 6-, and 7-coordinations. Only five water molecules were located, the remainder possibly being statistically distributed in a channel in the structure. U-O = 1.80 (uranyl), 2.44 (carbonate), C-O = 1.26-1.31, Na-O = 2.30-2.53, Ca-O = 2.31-2.55 Å.

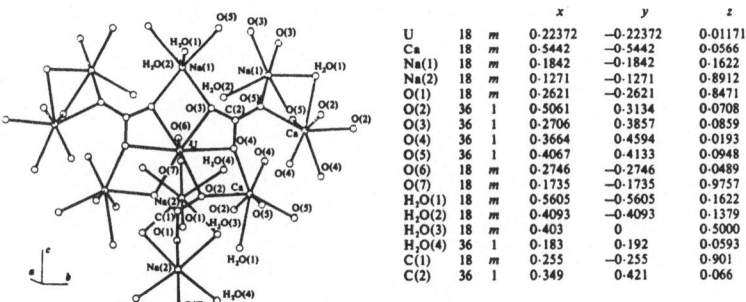

			x	y	z
U	18	m	0.22372	−0.22372	0.01171
Ca	18	m	0.5442	−0.5442	0.0566
Na(1)	18	m	0.1842	−0.1842	0.1622
Na(2)	18	m	0.1271	−0.1271	0.8912
O(1)	18	m	0.2621	−0.2621	0.8471
O(2)	36	1	0.5061	0.3134	0.0708
O(3)	36	1	0.2706	0.3857	0.0859
O(4)	36	1	0.3664	0.4594	0.0193
O(5)	36	1	0.4067	0.4133	0.0948
O(6)	18	m	0.2746	−0.2746	0.0489
O(7)	18	m	0.1735	−0.1735	0.9757
H$_2$O(1)	18	m	0.5605	−0.5605	0.1622
H$_2$O(2)	18	m	0.4093	−0.4093	0.1379
H$_2$O(3)	18	m	0.403	0	0.5000
H$_2$O(4)	36	1	0.183	0.192	0.0593
C(1)	18	m	0.255	−0.255	0.901
C(2)	36	1	0.349	0.421	0.066

Fig. 1.   Structure of andersonite.

BARIUM NITRITE MONOHYDRATE
Ba(NO$_2$)$_2$.H$_2$O

G. SCHÄFER and K.F. FISCHER, 1981. Z. Kristallogr., 155, 75-79.

Hexagonal, P6$_1$, a = 7.083, c = 17.967 Å, Z = 6. Mo radiation, R = 0.021, 0.020 for 780
reflexions, at room temperature and 133K. Previous study in 1.

Atomic positions (room temperature and 133K)

	x	y	z
Ba	0,42662	0,85120	0,25
	0,42853	0,84970	
N(1)	0,4497	0,5659	0,0914
	0,4475	0,5610	0,0910
O(1A)	0,2975	0,5871	0,1196
	0,2939	0,5858	0,1173
O(1B)	0,6026	0,6172	0,1349
	0,6016	0,6115	0,1355
N(2)	0,2469	0,0297	0,1052
	0,2408	0,0263	0,1045
O(2A)	0,1095	−0,1354	0,1386
	0,1052	−0,1405	0,1386
O(2B)	0,4388	0,1038	0,1279
	0,4352	0,1009	0,1262
H$_2$O	0,8459	0,0766	0,1673
	0,8439	0,0726	0,1667

N-O = 1.24-1.27(1) Å, O-N-O = 113°, Ba-7 O and 2 H$_2$O = 2.77-3.04, Ba-1 N = 3.10 Å.

1. Structure Reports, 11, 355.

trans-TETRAAMMINEDINITRONICKEL(II)
Ni(NH$_3$)$_4$(NO$_2$)$_2$

B.N. FIGGIS, P.A. REYNOLDS, A.H. WHITE, G.A. WILLIAMS and S. WRIGHT, 1981. J. Chem.
Soc., Dalton, 997-1003.

Monoclinic, C2/m, a = 10.621, 10.628, b = 6.812, 6.887, c = 5.912, 5.974 Å, β =
114.83, 114.80°, at 130, 295K, D$_m$ (295K) = 1.82, Z = 2. Mo radiation, R = 0.028,
0.051 for 900, 1264 reflexions at 130, 295K. Previous study in 1.

Atomic positions (130K)

	x	y	z
Ni(1)	0	0	0
N(1)	0.2000	0	-0.0066
N(2)	0.0655	0.2235	0.2714
O(1)	0.3086	0	0.1911
O(2)	0.2134	0	-0.2091
H(1)	0.1004	0.1792	0.4370
H(2)	0.1186	0.3014	0.2479
H(3)	-0.0007	0.3023	0.2424

   Ni has trans-octahedral coordination, Ni-NO$_2$ = 2.142, Ni-NH$_3$ = 2.109, N-O =
1.258 and 1.270(2) Å; the difference between the two N-O lengths results from intermolecular hydrogen bonding, N-H...O = 3.151 and 3.189 (to O(2)) and 3.183 Å (to O(1)).

1. Structure Reports, 21, 407; 23, 426.

SILVER(I) NITRITE
$AgNO_2$

S. OHBA and Y. SAITO, 1981. Acta Cryst., B**37**, 1911-1913.

Orthorhombic, Imm2, a = 3.528, b = 6.172, c = 5.181 Å, Z = 2. Ag radiation, R = 0.024 for 1128 reflexions. Ag in 2(a): 0,0,0; N in 2(a): 0,0,0.4446; O in 4(d): 0,0.1701, 0.5747.

The structure is as previously described (**1**). N-O = 1.248(3) Å, O-N-O = 114.6°, Ag-O = 2.441(3), Ag-N = 2.304(2) Å.

1. Strukturbericht, **4**, 41, 153; Structure Reports, **27**, 573.

SODIUM ANTIMONY(III) TRIFLUORIDE NITRATE MONOHYDRATE
$NaSbF_3NO_3.H_2O$

M. BOURGAULT, B. DUCOURANT, D. MASCHERPA-CORRAL and R. FOURCADE, 1981. J. Fluor. Chem., **17**, 305-315.

Orthorhombic, Pbca, a = 18.18, b = 11.505, c = 5.660 Å, D_m = 3.12, Z = 8. Mo radiation, R = 0.047 for 1301 reflexions.

The structure (Fig. 1) contains double anionic layers normal to **a**, in which Sb has distorted dodecahedral 8-coordination, including a lone electron-pair, Sb-F = 1.92, 1.95, 1.97, Sb-O = 2.45, 2.83, 3.01, 3.09(1) Å. These layers are linked by the Na ions and by O-H...O and O-H...F hydrogen bonds via the water molecules.

	x	y	z
Sb^I	0,36646	0,60123	0,1620
Na^I	0,4841	0,1433	0,8183
$F(1^I)$	0,4460	0,6931	0,028
$F(2^I)$	0,4074	0,4699	0,998
$F(3^I)$	0,4387	0,5519	0,402
$O(1^I)$	0,0886	0,6999	0,145
$O(2^I)$	0,2266	0,5698	0,860
$O(3^I)$	0,3284	0,6484	0,757
$O(4^I)$	0,2321	0,6853	0,558
N^I	0,2614	0,6337	0,723

Fig. 1.   Structure of $NaSbF_3NO_3.H_2O$.

POTASSIUM ANTIMONY(III) TRIFLUORIDE NITRATE
$KSbF_3NO_3$

M. BOURGAULT, B. DUCOURANT, D. MASCHERPA-CORRAL and R. FOURCADE, 1981. J. Fluor. Chem., **17**, 215-224.

Orthorhombic, Pbca, a = 7.911, b = 7.861, c = 18.092 Å, D_m = 3.29, Z = 8. Mo radiation, R = 0.044 for 1179 reflexions.

The structure (Fig. 1) contains double layers normal to **c** of $F_3Sb-ONO_2$ groups linked by longer Sb...O bonds; Sb has monocapped-octahedral coordination, including the lone electron-pair, Sb-F = 1.93, 1.93, 1.97, Sb-O = 2.52, 2.84, 2.87(1) Å. The anion layers are linked by 8-coordinate K ions, K-O/F = 2.70-2.99 Å.

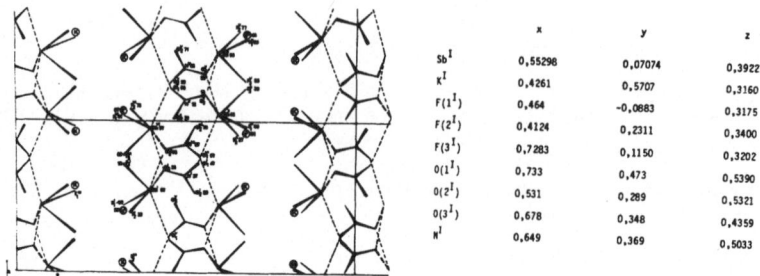

	x	y	z
Sb^I	0,55298	0,07074	0,39221
K^I	0,4261	0,5707	0,3160
F(1^I)	0,464	−0,0883	0,3175
F(2^I)	0,4124	0,2311	0,3400
F(3^I)	0,7283	0,1150	0,3202
O(1^I)	0,733	0,473	0,5390
O(2^I)	0,531	0,289	0,5321
O(3^I)	0,678	0,348	0,4359
N^I	0,649	0,369	0,5033

Fig. 1.   Structure of KSbF$_3$NO$_3$.

POTASSIUM HYDROGENPHOSPHITE - PHOSPHOROUS ACID (2:1)
$2KH_2PO_3 \cdot H_3PO_3$

J. LOUB and H. PAULUS, 1981. Acta Cryst., B$\underline{37}$, 2058-2059.

Triclinic, P$\bar{1}$, a = 8.590, b = 9.010, c = 7.576 Å, α = 112.58, β = 87.88, γ = 101.21°, D_m = 2.00, Z = 2. Mo radiation, R = 0.024 for 2330 reflexions.

The structure (Fig. 1) contains a three-dimensional hydrogen-bonded network of $HPO_2(OH)^-$ ions and $HPO(OH)_2$ molecules, and columns of 7-coordinated K$^+$ ions. Mean P-O = 1.49(2), P-OH = 1.56(1), P-H = 1.27(5) Å, angles at P = 101-118°, O-H...O = 2.53-2.65 Å.

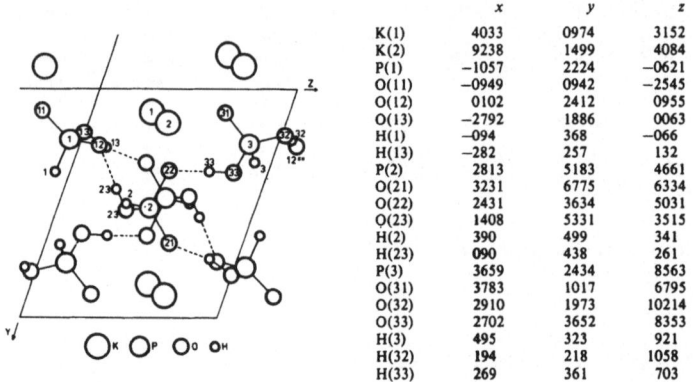

	x	y	z
K(1)	4033	0974	3152
K(2)	9238	1499	4084
P(1)	−1057	2224	−0621
O(11)	−0949	0942	−2545
O(12)	0102	2412	0955
O(13)	−2792	1886	0063
H(1)	−094	368	−066
H(13)	−282	257	132
P(2)	2813	5183	4661
O(21)	3231	6775	6334
O(22)	2431	3634	5031
O(23)	1408	5331	3515
H(2)	390	499	341
H(23)	090	438	261
P(3)	3659	2434	8563
O(31)	3783	1017	6795
O(32)	2910	1973	10214
O(33)	2702	3652	8353
H(3)	495	323	921
H(32)	194	218	1058
H(33)	269	361	703

Fig. 1.   Structure of $2KH_2PO_3 \cdot H_3PO_3$ and atomic positional parameters
(x 10^4 for K, P, O; x 10^3 for H).

BERYLLIUM POLYPHOSPHATE (POLYMORPH III)
$Be(PO_3)_2$

E. SCHULTZ and F. LIEBAU, 1981. Z. Kristallogr., $\underline{154}$, 115-126.

At 119°C, orthorhombic, $C222_1$, a = 9.968, b = 10.080, c = 8.692 Å, Z = 8. Mo radiation, film data, R = 0.113 for 599 reflexions. Below 96°C, monoclinic, $P2_1$, a = 14.062, b = 8.629, c = 7.091 Å, β = 90.80°, at 18°C, Z = 8. The structure of polymorph II has been described previously (1).

The structure (Fig. 1) contains strongly-folded polyphosphate chains with a repeat of eight tetrahedra; the chains are joined by BeO_4 tetrahedra to give a three-dimensional framework with the same topology as silica K (keatite). P-O = 1.55-1.60 (bridging), 1.45-1.48 (terminal), Be-O = 1.55-1.66 Å, P-O-P = 139°.

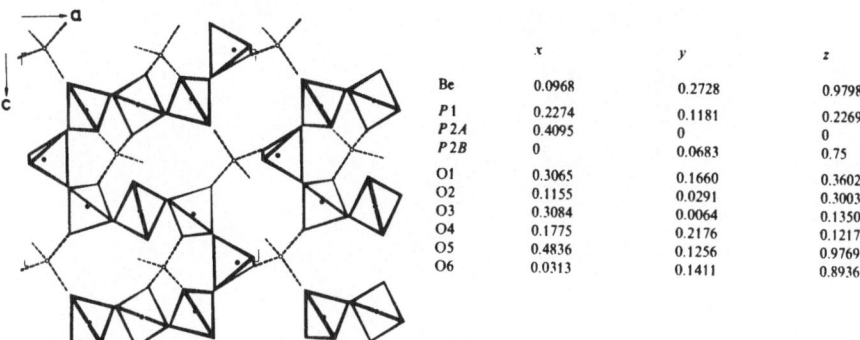

	x	y	z
Be	0.0968	0.2728	0.9798
P1	0.2274	0.1181	0.2269
P2A	0.4095	0	0
P2B	0	0.0683	0.75
O1	0.3065	0.1660	0.3602
O2	0.1155	0.0291	0.3003
O3	0.3084	0.0064	0.1350
O4	0.1775	0.2176	0.1217
O5	0.4836	0.1256	0.9769
O6	0.0313	0.1411	0.8936

Fig. 1.   Structure of $Be(PO_3)_2$-III at 119°C.

1. Structure Reports, 43A, 244.

LITHIUM THALLIUM PHOSPHITE
$LiTlHPO_3$

M. RAFIQ, J. DURAND and L. COT, 1981. Rev. Chim. Minér., 18, 1-8.

Monoclinic, C2, a = 14.284, b = 5.091, c = 5.329 Å, β = 89.19°, D_m = 4.91, Z = 4. Mo radiation, R = 0.028 for 280 reflexions.

Atomic positions

	x	y	z
Tl	0.0914	0	0.2501
P	0.3582	0.041	0.302
O(1)	0.284	-0.086	0.138
O(2)	0.353	-0.007	0.577
O(3)	0.359	0.329	0.278
Li	0.754	0.049	0.191
H	0.452	0.024	0.248

The structure (Fig. 1) contains sheets of HPO_3 and LiO_4 tetrahedra, linked by Tl ions which have stereochemically-active lone pairs. P-O = 1.43-1.53, P-H = 1.37, Li-O = 1.92-1.97 Å.

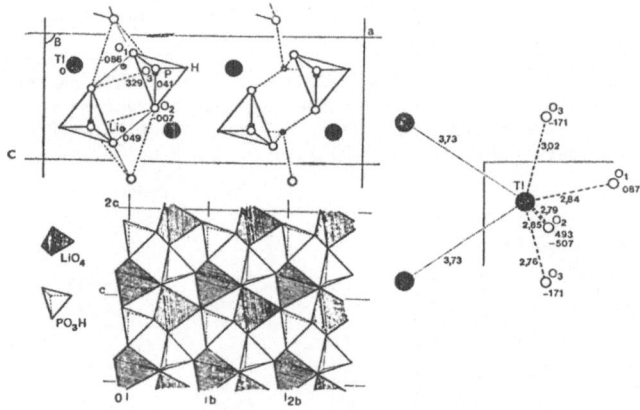

Fig. 1.   Structure of LiTlHPO₃.

ANTIMONY(III) PHOSPHITE
Sb₂(HPO₃)₃

J. LOUB and H. PAULUS, 1981. Acta Cryst., B$\underline{37}$, 1106-1107.

Triclinic, P$\bar{1}$, a = 9.182, b = 8.353, c = 7.220 Å, α = 68.21, β = 79.52, γ = 66.47°,
D_m = 3.33, Z = 2. Mo radiation, R = 0.020 for 2150 reflexions.

    The structure (Fig. 1) contains HPO_3^{2-} tetrahedra, linked by Sb ions which
have trigonal bipyramidal SbO_4E coordination (E = equatorial lone-pair). P-O =
1.506-1.550(3), Sb-O = 1.980-2.091 (equatorial), 2.143-2.233 Å (axial), O-P-O =
106.8-116.0°.

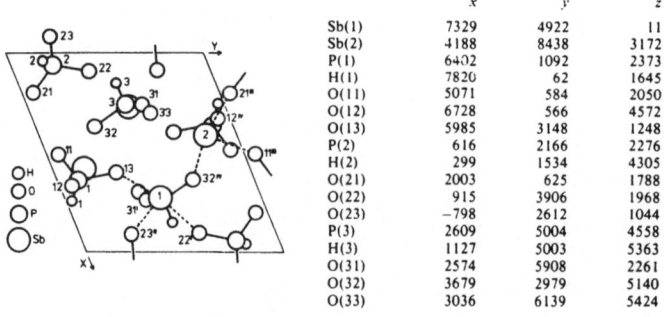

	x	y	z
Sb(1)	7329	4922	11
Sb(2)	4188	8438	3172
P(1)	6402	1092	2373
H(1)	7820	62	1645
O(11)	5071	584	2050
O(12)	6728	566	4572
O(13)	5985	3148	1248
P(2)	616	2166	2276
H(2)	299	1534	4305
O(21)	2003	625	1788
O(22)	915	3906	1968
O(23)	−798	2612	1044
P(3)	2609	5004	4558
H(3)	1127	5003	5363
O(31)	2574	5908	2261
O(32)	3679	2979	5140
O(33)	3036	6139	5424

Fig. 1.   Structure of antimony(III) phosphite, and atomic
            positional parameters (x 10⁴).

AMMONIUM HYDROGENSULPHATE DIHYDROGENPHOSPHATE
NH₄HSO₄.NH₄H₂PO₄

M.T. AVERBUCH-POUCHOT, 1981. Mater. Res. Bull., $\underline{16}$, 407-411.

Monoclinic, P2$_1$/n, a = 7.723, b = 7.540, c = 7.482 Å, β = 101.32°, Z = 2.   Ag
radiation, R = 0.028 for 813 reflexions.

Isostructural with the K salt (1).   The structure (Fig. 1) contains randomly
distributed PO$_4$ and SO$_4$ tetrahedra and ammonium ions, linked by hydrogen bonding.
P,S-O = 1.48-1.52, N-H...O = 2.87-3.01, O-H...O = 2.55 Å (including one hydrogen bond
disordered across a symmetry centre).

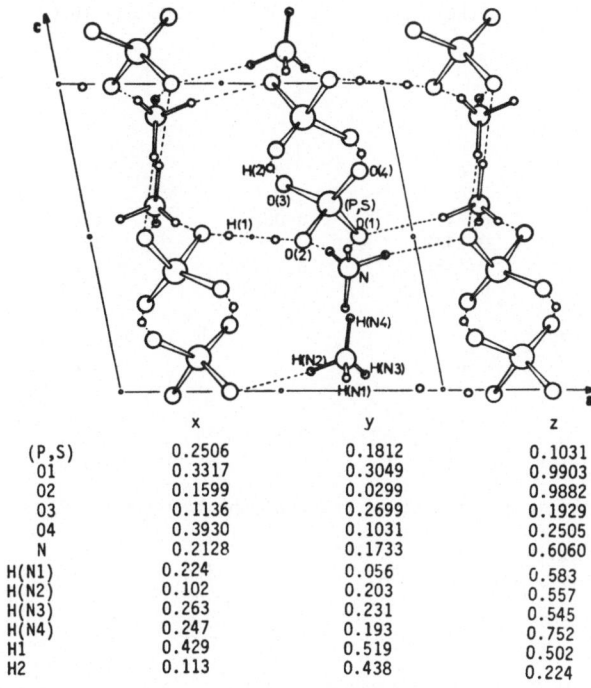

	x	y	z
(P,S)	0.2506	0.1812	0.1031
O1	0.3317	0.3049	0.9903
O2	0.1599	0.0299	0.9882
O3	0.1136	0.2699	0.1929
O4	0.3930	0.1031	0.2505
N	0.2128	0.1733	0.6060
H(N1)	0.224	0.056	0.583
H(N2)	0.102	0.203	0.557
H(N3)	0.263	0.231	0.545
H(N4)	0.247	0.193	0.752
H1	0.429	0.519	0.502
H2	0.113	0.438	0.224

Fig. 1.   Structure of NH$_4$HSO$_4$.NH$_4$H$_2$PO$_4$.

1.   Structure Reports, 46A, 318.

SODIUM DIHYDROGEN PHOSPHATE
NaH$_2$PO$_4$

R.N.P. CHOUDHARY, R.J. NELMES and K.D. ROUSE, 1981.   Chem. Phys. Letters, 78, 102-105.

Monoclinic, P2$_1$/c, a = 6.808, b = 13.491, c = 7.331 Å, β = 92.88°, Z = 8.   Neutron
radiation, R = 0.052 for 679 reflexions.

The structure is as previously determined by X-ray methods (1).   All hydrogen
bonds are asymmetric, O...O = 2.49-2.64, H...O = 1.46-1.64 Å, O-H...O = 168-178°.

1.   Structure Reports, 40A, 235.

RUBIDIUM DIHYDROGEN PHOSPHATE
RbH_2PO_4

N.S.J. KENNEDY and R.J. NELMES, 1980.   J. Phys. C: Solid State Phys., 13, 4841-4853.

Paraelectric phase, room-temperature and T_C+5K, tetragonal, $I\bar{4}2d$, a = 7.607, 7.586, c = 7.299, 7.254 Å, Z = 4.   Neutron radiation, R = 0.033, 0.046 for 135, 116 reflexions.

Ferroelectric phase, 77K, orthorhombic, Fdd2, a = 10.800, b = 10.672, c = 7.242 Å, Z = 8.   Neutron powder data.

Atomic positions

Paraelectric phase

	x	y	z
Rb	0	0	1/2
P	0	0	0
O room-temp.	0.1419	0.0859	0.1201
∿150K	0.1439	0.0862	0.1207
H room-temp.	0.1385	0.2225	0.1222
∿150K	0.1421	0.2247	0.1214

Ferroelectric phase

	x	y	z
Rb	0	0	0.5165
P	0	0	0
O(1)	0.1146	-0.0282	-0.1116
O(2)	0.0283	0.1141	0.1334
H	-0.0436	0.1817	0.1403

The structures (1) are similar to those of the K compound (2).

1. Structure Reports, 44A, 245; 45A, 300.
2. Strukturbericht, 1, 362, 393, 394; 2, 454; 17, 478; 46A, 416.

CAESIUM DIHYDROGEN PHOSPHATE (PARAELECTRIC PHASE)
CsH_2PO_4

H. MATSUNAGA, K. ITOH and E. NAKAMURA, 1980.   J. Phys. Soc. Japan, 48, 2011-2014.

Monoclinic, $P2_1/m$, a = 7.912, b = 6.383, c = 4.881 Å, β = 107.73°, Z = 2.   Mo radiation, R = 0.029 for 780 reflexions.

Atomic positions

	x	y	z
Cs	0.2657	1/4	0.0354
P	0.2370	3/4	0.5293
O(1)	0.3898	3/4	0.3874
O(2)	0.3222	3/4	0.8447
O(3)	0.1266	0.5540	0.4178
H(1)	0.348	3/4	0.195
H(2)	0	1/2	1/2

The structure is as previously described (1), containing tetrahedral $PO_2(OH)_2^-$ anions, linked by Cs ions and by two hydrogen bonds, one of which is disordered (H(2) distributed in two position displaced from 0,1/2,1/2).

1. Structure Reports, 42A, 337.

MAGNESIUM PHOSPHATE 22-HYDRATE (POLYMORPH II)
$Mg_3(PO_4)_2 \cdot 22H_2O$

M. CATTI, M. FRANCHINI-ANGELA and G. IVALDI, 1981. Z. Kristallogr., 155, 53-64.

Triclinic, $P\bar{1}$, a = 6.937, b = 6.932, c = 16.132 Å, α = 82.15, β = 89.72, γ = 119.49°, D_m = 1.64, Z = 1. Mo radiation, R = 0.042 for 1826 reflexions.

   The structure (Fig. 1) contains (001) layers of $Mg(H_2O)_6$ octahedra, PO_4 tetra-hedra, and lattice water molecules, linked by hydrogen bonds. Mg-O = 2.039-2.112, P-O = 1.534-1.541(3), O-H...O = 2.61-3.26 Å. Another polymorph with different stacking of layers has been described previously (1).

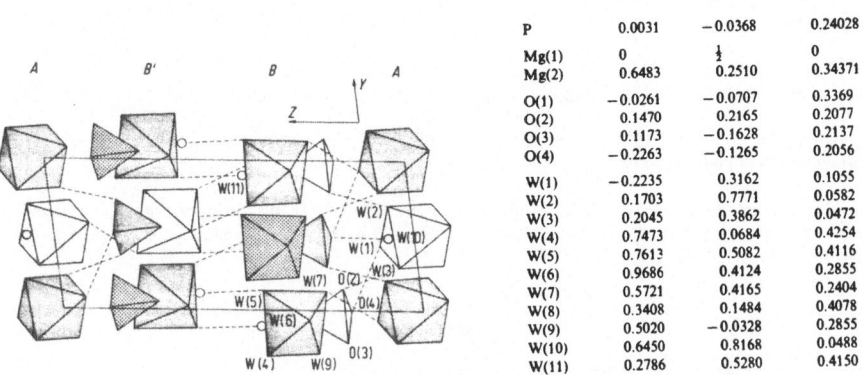

	x	y	z
P	0.0031	−0.0368	0.24028
Mg(1)	0	$\frac{1}{2}$	0
Mg(2)	0.6483	0.2510	0.34371
O(1)	−0.0261	−0.0707	0.3369
O(2)	0.1470	0.2165	0.2077
O(3)	0.1173	−0.1628	0.2137
O(4)	−0.2263	−0.1265	0.2056
W(1)	−0.2235	0.3162	0.1055
W(2)	0.1703	0.7771	0.0582
W(3)	0.2045	0.3862	0.0472
W(4)	0.7473	0.0684	0.4254
W(5)	0.7613	0.5082	0.4116
W(6)	0.9686	0.4124	0.2855
W(7)	0.5721	0.4165	0.2404
W(8)	0.3408	0.1484	0.4078
W(9)	0.5020	−0.0328	0.2855
W(10)	0.6450	0.8168	0.0488
W(11)	0.2786	0.5280	0.4150

Fig. 1.   Structure of magnesium phosphate 22-hydrate (polymorph II).

1. Structure Reports, 45A, 301.

BROMAPATITE
$Ca_5(PO_4)_3Br$

J.C. ELLIOTT, E. DYKES and P.E. MACKIE, 1981. Acta Cryst., B37, 435-438.

Hexagonal, $P6_3/m$, a = 9.761, c = 6.739 Å, Z = 2. Mo radiation, R_w = 0.041 for 673 reflexions.

Atomic positions (x 10^4 for decimal fractions)

	Occupancy	x	y	z
O(1)	9954	3533	4972	$\frac{1}{4}$
O(2)	9928	5954	4642	$\frac{1}{4}$
O(3)	10052	3572	2713	662
P	9654	4124	3785	$\frac{1}{4}$
Ca(1)	9861	$\frac{1}{3}$	$\frac{2}{3}$	45
Ca(2)	9834	2672	121	$\frac{1}{4}$
Br(1)	9522	0	0	0
Br(2)	300	0	0	1032

   Typical hexagonal apatite structure, with most of the Br at (0,0,0) and about 2% at (0,0,0.103).

BARIUM DIHYDROGENPHOSPHATE
$Ba(H_2PO_4)_2$

P.G. LENHERT, 1981. Acta Cryst., B$\underline{37}$, 1321-1322.

Comparison of two independent analyses for the triclinic and orthorhombic forms ($\underline{1}$), and derivation of weighted-mean atomic coordinates [not listed].

$\underline{1}$. Structure Reports, $\underline{43}$A, 250; $\underline{44}$A, 246, 247.

SODIUM BARIUM PHOSPHATE
$NaBaPO_4$

A.W. KOLSI, M. QUARTON and W. FREUNDLICH, 1981. J. Solid State Chem., $\underline{36}$, 107-111.

Monoclinic, C2/m, a = 9.743, b = 5.622, c = 7.260 Å, β = 90.10°, Z = 4. Mo radiation, R = 0.058 for 761 reflexions.

The structure (Fig. 1) contains PO_4 tetrahedra, NaO_6 octahedra, BaO_{12} and MO_{10} polyhedra (M = disordered Na + Ba); P-O = 1.48-1.54, Na-O = 2.34, 2.38, Ba-O = 2.77-3.26, M-O = 2.59-3.03 Å.

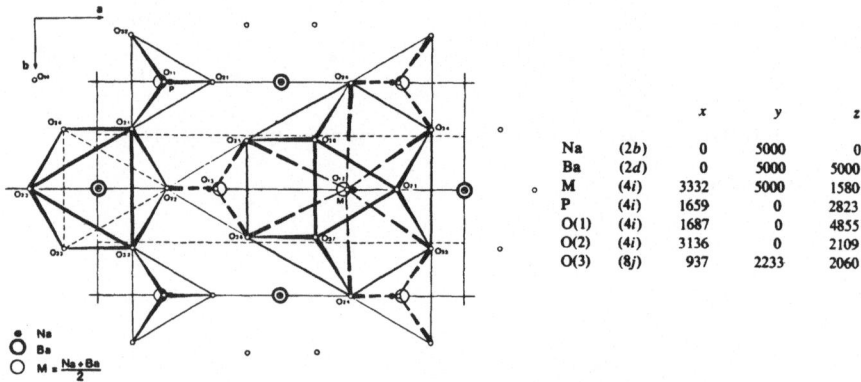

		x	y	z
Na	(2b)	0	5000	0
Ba	(2d)	0	5000	5000
M	(4i)	3332	5000	1580
P	(4i)	1659	0	2823
O(1)	(4i)	1687	0	4855
O(2)	(4i)	3136	0	2109
O(3)	(8j)	937	2233	2060

Fig. 1.   Structure of $NaBaPO_4$, and atomic positional parameters (x 10^4).

ALUMINUM TRIS(DIHYDROGENPHOSPHATE) (RHOMBOHEDRAL FORM)
$Al(H_2PO_4)_3$

D. BRODALLA, R. KNIEP and D. MOOTZ, 1981. Z. Naturforsch., $\underline{36}$B, 907-909.

Rhombohedral, R$\overline{3}$c, a = 7.858, c = 24.956 Å, D_m = 2.37, Z = 6. Mo radiation, R = 0.032 for 373 reflexions.

The structure (Fig. 1) contains a three-dimensional arrangement of corner-sharing AlO_6 octahedra and $O_2P(OH)_2$ tetrahedra; each OH group is involved in two hydrogen bonds, with disordered hydrogen positions. Al-O = 1.887(1), P-O = 1.489, P-OH = 1.569, O-H...O = 2.654, 2.754 Å. Another form with chains of octahedra and tetrahedra is described in $\underline{1}$.

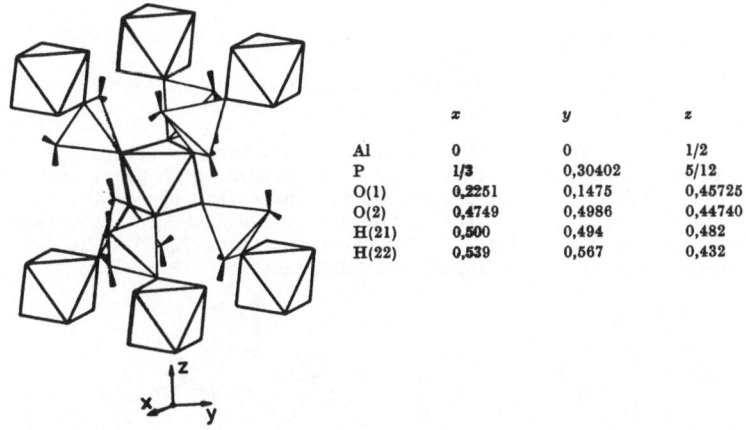

	x	y	z
Al	0	0	1/2
P	1/3	0,30402	5/12
O(1)	0,2251	0,1475	0,45725
O(2)	0,4749	0,4986	0,44740
H(21)	0,500	0,494	0,482
H(22)	0,539	0,567	0,432

Fig. 1.   Structure of Al(H$_2$PO$_4$)$_3$; H atoms have occupancy = 0.5.

<u>1</u>. Structure Reports, <u>44A</u>, 328.

LEAD(II) DIHYDROGENPHOSPHATE
Pb(H$_2$PO$_4$)$_2$

P. VASIĆ, B. PRELESNIK, R. HERAK and M. ČURIĆ, 1981. Acta Cryst., B<u>37</u>, 660-662.

Triclinic, P$\overline{1}$, a = 9.029, b = 5.863, c = 7.815 Å, α = 96.92, β = 119.56, γ = 104.92°,
Z = 2. Mo radiation, R = 0.099 for 1730 reflexions.

The structure (Fig. 1) contains phosphate tetrahedra and 7-coordinate Pb^{2+} ions,
the PbO$_7$ polyhedra sharing edges to form chains along $\underline{b}$; P-O = 1.51-1.60(2), Pb-O =
2.44-2.88(1) Å. The chains are linked by hydrogen bonds, two of which are across
centres of symmetry (O-H...O = 2.44, 2.46 Å); other O-H...O = 2.64-2.74 Å.

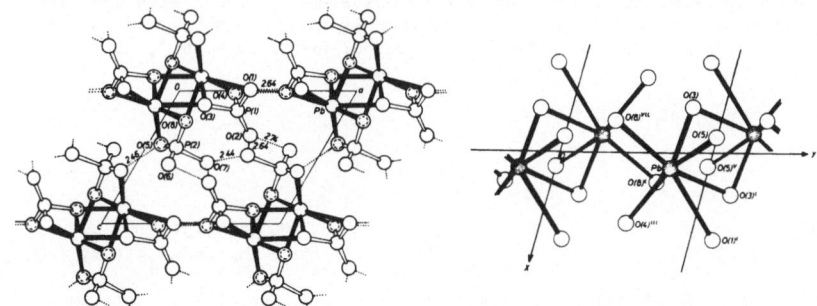

Fig. 1.   Structure of lead(II) dihydrogenphosphate.

VANADYL PHOSPHATE
VOPO$_4$

M. TACHEZ, F. THEOBALD and E. BORDES, 1981. J. Solid State Chem., $\underline{40}$, 280-283.

α_I- and α_{II}-phases, tetragonal, P4/n, a = 6.20, 6.014, c = 4.11, 4.434 Å, Z = 2.
Poweder data.

The α_I-VOPO$_4$ structure ($\underline{1}$) is isostructural with, and with positional parameters
very close to those of, α-VOSO$_4$ ($\underline{2}$). The α_{II}-VOPO$_4$ structure is as previously
described ($\underline{3}$); MoOPO$_4$, NbOPO$_4$, VOMoO$_4$, and TaOPO$_4$ also have this structure. The
α_{II}-types are prepared by heating a mixture of oxides at 800-1200°C; the α_I-types
are obtained from dehydration of hydrates at temperatures below 220°C. The difference between the two structures is mainly in the V positions (Fig. 1).

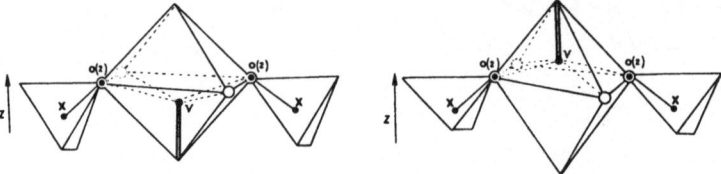

Fig. 1.   Part of the structures of α_I- (left) and α_{II}-VOPO$_4$ (right).

$\underline{1}$. E. BORDES, P. COURTINE and G. PANNETIER, 1973. Ann. Chim., $\underline{8}$, 105.
$\underline{2}$. Structure Reports, $\underline{35A}$, 373.
$\underline{3}$. Ibid., $\underline{39A}$, 286; $\underline{42A}$, 342.

ARROJADITE
KNa$_4$CaMn$_4$Fe$_{10}$Al(PO$_4$)$_{12}$(OH,F)$_2$

S. MERLINO, M. MELLINI and P.F. ZANAZZI, 1981. Acta Cryst., B$\underline{37}$, 1733-1736.

Monoclinic, C2/c, a = 16.526, b = 10.057, c = 24.730 Å, β = 105.78°, Z = 4. Mo
radiation, R = 0.084 for 6910 reflexions.

The results are similar to those of a previous study ($\underline{1}$), but with additional,
partially-occupied Na and Ca sites. The structure is rather complex and contains
two structural layers of PO$_4$, AlO$_6$, MO$_4$, MO$_5$, MO$_6$ polyhedra (M = Mn^{2+}, Fe^{2+}), with
K, Na, Ca in cavities.

$\underline{1}$. Structure Reports, $\underline{45A}$, 403.

DICKINSONITE   ARROJADITE
KNa$_4$CaMn$_{14}$Al(OH)$_2$(PO$_4$)$_{12}$ KNa$_4$CaFe$_{14}$Al(OH)$_2$(PO$_4$)$_{12}$

P.B. MOORE, T. ARAKI, S. MERLINO, M. MELLINI and P.F. ZANAZZI, 1981. Amer. Min.,
$\underline{66}$, 1034-1049.

Monoclinic, A2/a, a = 24.940, 24.692, b = 10.131, 10.031, c = 16.722, 16.453 Å,
β = 105.60, 105.72°, D$_m$ = 3.41, 3.59, Z = 4. Mo radiation, R = 0.078, 0.075 for
7740, 7116 reflexions. Previous studies in $\underline{1}$ and $\underline{2}$.

The arrojadite-dickinsonite structure is a complex one, related to that of wyllieite, $Na_2Fe_2Al(PO_4)_3$ (3). The coordination polyhedra of the larger cations include six octahedra, one tetrahedron, one square pyramid, one seven-coordinate polyhedron, two distorted cubes, one non-cubic polyhedron of order eight, two of ten, and one of twelve coordination. Some cation sites are disordered and one PO_4 tetrahedron is also disordered.

1. Structure Reports, 45A, 403.
2. Preceding report.
3. Structure Reports, 40A, 247.

FILLOWITE
$Na_2Ca(Mn,Fe)_7(PO_4)_6$

T. ARAKI and P.B. MOORE, 1981. Amer. Min., 66, 827-842.

Rhombohedral, $R\overline{3}$, a = 15.282, c = 43.507 Å, Z = 18. Mo radiation, R = 0.069 for 6891 reflexions.

The structure can be derived from that of glaserite. It contains six distinct PO_4 tetrahedra, nine MO_6 octahedra, four 5-coordinate, one 7-coordinate, and one 8-coordinate (Ca, Fig. 1) polyhedra.

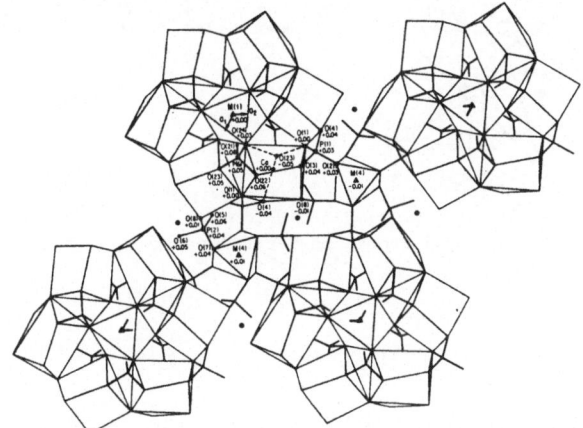

Fig. 1.   Part of the structure of fillowite, showing the CaO_8 polyhedra.

SIDORENKITE
$Na_3Mn(PO_4)(CO_3)$

T.A. KUROVA, N.G. ŠUMJATSKAJA, A.A. VORONKOV and Ju.A. PJATENKO, 1980. Mineral. Ž., 2, 65-70.

Monoclinic, $P2_1/m$, a = 8.997, b = 5.163, c = 6.741 Å, γ = 90.16°, D_m = 2.90, Z = 2. R = 0.035.

The structure contains chains of MnO_6 octahedra, PO_4 tetrahedra, and CO_3 triangles, linked by 6- and 7-coordinate Na^+ ions.

IRON PHOSPHATE OXIDE

Fe_2PO_5 $\qquad\qquad\qquad\qquad\qquad\qquad\qquad$ $Fe(II)Fe(III)O(PO_4)$

A. MODARESSI, A. COURTOIS, R. GERARDIN, B. MALAMAN and C. GLEITZER, 1981. J. Solid State Chem., 40, 301-311.

Orthorhombic, Pnma, a = 7.378, b = 6.445, c = 7.471 Å, D_m = 4.18, Z = 4. Mo radiation, R = 0.027 for 960 reflexions.

The structure (Fig. 1) contains chains along b of FeO_6 octahedra and PO_4 tetrahedra; Fe(II)-O = 2.06-2.24, Fe(III)-O = 1.86-2.22, P-O = 1.51-1.55 Å.

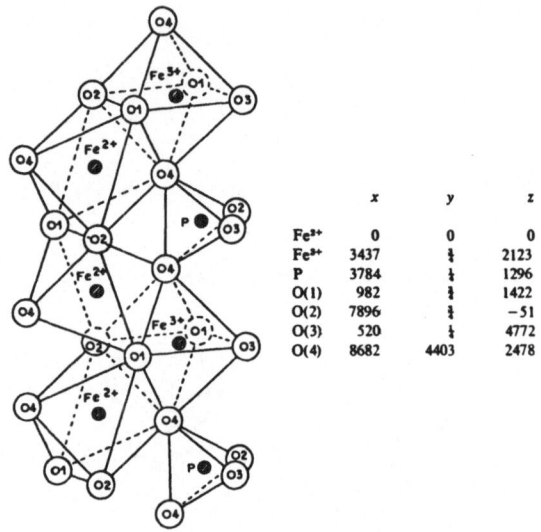

	x	y	z
Fe^{3+}	0	0	0
Fe^{3+}	3437	¼	2123
P	3784	¼	1296
O(1)	982	¼	1422
O(2)	7896	¼	−51
O(3)	520	¼	4772
O(4)	8682	4403	2478

Fig. 1.   Structure of Fe_2PO_5 and atomic positional parameters (x 10^4).

LIPSCOMBITE

$Fe_{25}(PO_4)_{14}(OH)_{24}$

E.N. MATVIENKO, O.V. JAKUBOVIČ, M.A. SIMONOV and N.V. BELOV, 1981. Ž. Strukt. Khim., 22, No. 1, 121-125 [J. Struct. Chem., 22, 91-95].

Tetragonal, $P4_32_12$, a = 14.961, c = 12.740 Å, D_m = 3.7, Z = 2. Mo radiation, R = 0.089 for 793 reflexions for a synthetic specimen. Previous studies in 1.

The structure contains two types of linear cluster of three face-sharing FeO_6 octahedra; one type is isolated, the other is linked by two additional Fe octahedra to form chains. Clusters and chains are linked by PO_4 tetrahedra. There is partial ordering of Fe^{2+} and Fe^{3+}, $Fe(III)_7[Fe(III)_{0.5}Fe(II)_{0.5}]_{18}(PO_4)_{14}(OH)_{24}$.

1. Structure Reports, 15, 263; 27, 603.

MAGNIOTRIPLITE

$(Mg,Fe,Mn)_2(PO_4)(F,OH)$

C. TADINI, 1981. Bull. Minéral., 104, 677-680.

Monoclinic, Ia or I2/a, a = 12.035, b = 6.432, c = 9.799 Å, β = 108.12°, Z = 8. Mo
radiation, R = 0.034 and 0.033 for 1566 reflexions in Ia and I2/a.

Atomic positions (I2/a)

	x	y	z
M(1)	0.1906	-0.0227	0.1964
M(2)	0.0917	0.1436	0.4553
P	0.0753	0.6517	0.3810
O(1)	0.0557	0.8294	0.4758
O(2)	0.9612	0.6029	0.2629
O(3)	0.1699	0.7099	0.3130
O(4)	0.1174	0.4615	0.4787
F(1)	0.2588	0.0992	0.4134
F(2)	0.2832	0.1698	0.3343

M(1) = 0.29 Mg + 0.71 Fe
M(2) = 0.60 Mg + 0.40 Fe
F(1) = F(2) = 0.67 F + 0.33 OH (occupancy 0.5)

Isostructural with triplite (1), probably with disordered F in the higher symmetry
space group.

1. Structure Reports, 41A, 424.

TETRAAMMINEPHOSPHATOCOBALT(III) TRIHYDRATE
$Co(NH_3)_4(PO_4).3H_2O$

X. SOLANS, J. RIUS and C. MIRAVITLLES, 1981. Z. Kristallogr., 157, 207-214.

Monoclinic, $P2_1$, a = 9.730, b = 7.301, c = 6.880 Å, β = 92.15°, D_m = 1.87, Z = 2. Mo
radiation, R = 0.044 for 729 reflexions.

The structure (Fig. 1) contains molecules with a four-membered chelate ring,
linked into layers by hydrogen bonding; the layers are joined by further hydrogen
bonding via the water molecules.

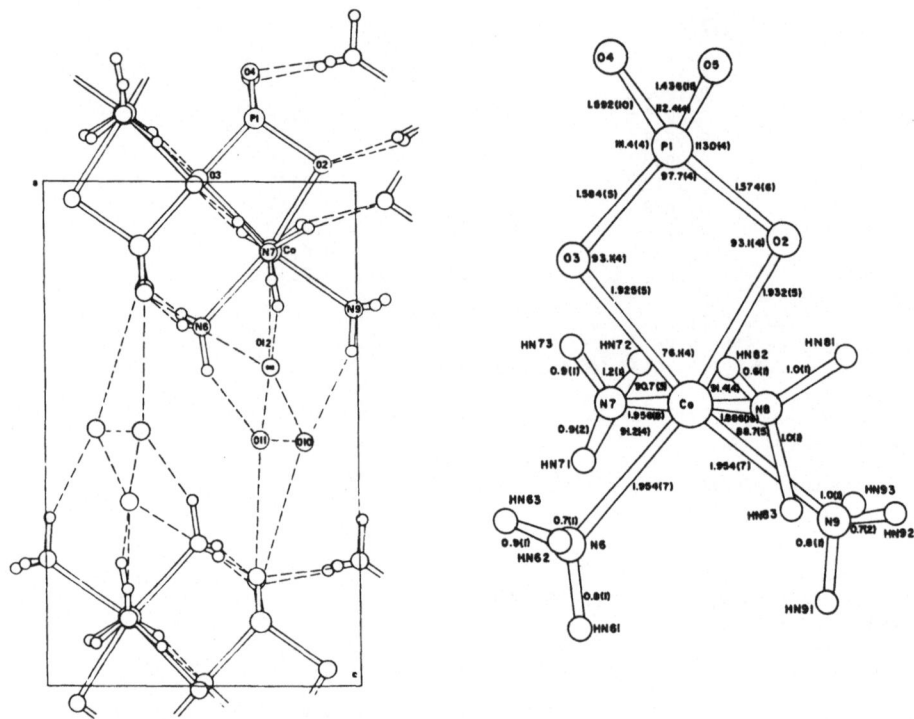

Fig. 1.   Structure of Co(NH$_3$)$_4$(PO$_4$).3H$_2$O.

RHODIUM(III) PHOSPHATE   RHODIUM(III) ARSENIC(V) OXIDE
RhPO$_4$ RhAsO$_4$

G. ENGEL, 1981.   J. Less-Common Metals, 77, P41-P46.

RhPO$_4$, orthorhombic, I222, I2$_1$2$_1$2$_1$, Imm2, or Immm, a = 10.391, b = 13.091, c = 6.391
Å, Z = 16. Powder data; positional parameters not determined. Assumed isostructural
with high-temperature CrPO$_4$ (1).

RhAsO$_4$, tetragonal, P4$_2$/mnm, a = 4.460, c = 2.973 Å, Z = 1. Powder data. Rh/As in
2(a); O in 4(f): x = 0.306. Rutile structure (2), with statistical distribution of
Rh and As (octahedral coordination).

1.   Crystal Data, volume 2, O-167.
2.   Strukturbericht, 1, 155.

COPPER(II) PHOSPHATE HYDROXIDE (TRICLINIC PSEUDOMALACHITE)
Cu$_5$(PO$_4$)$_2$(OH)$_4$

I.   G.L. SHOEMAKER, J.B. ANDERSON and E. KOSTINER, 1981. Amer. Min., 66, 169-175.
II. G.L. SHOEMAKER and E. KOSTINER, 1981. Ibid., 66, 176-181.

Triclinic, P$\bar{1}$, a = 4.445, b = 5.873, c = 8.668 Å, α = 103.62, β = 90.35, γ = 93.02°,
Z = 1. Mo radiation, R = 0.021 for 1350 reflexions.

As in the other two polymorphs (1, 2), the structure (Fig. 1) contains sheets of
edge-sharing CuO_6 distorted octahedra (4 short and 2 long bonds); the sheets are
linked by phosphate tetrahedra and hydrogen bonds. Cu-O = 1.92-2.04, 2.29-2.83,
P-O = 1.51-1.58 Å.

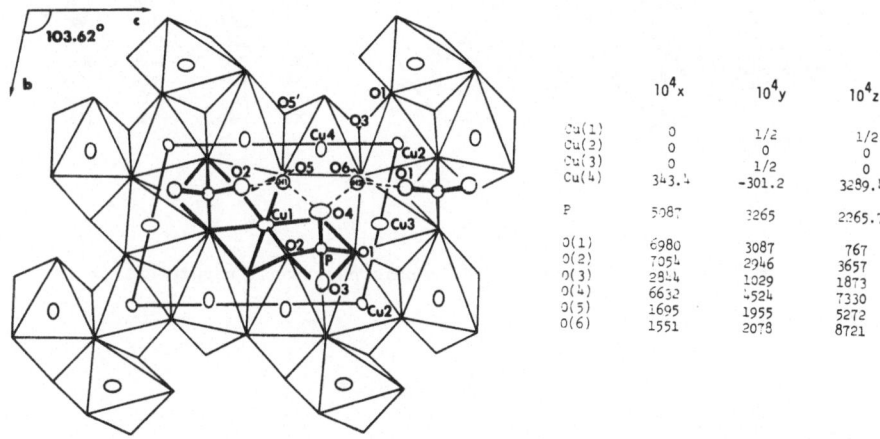

	$10^4 x$	$10^4 y$	$10^4 z$
Cu(1)	0	1/2	1/2
Cu(2)	0	0	0
Cu(3)	0	1/2	0
Cu(4)	343.4	-301.2	3289.4
P	5087	3265	2265.7
O(1)	6980	3087	767
O(2)	7054	2946	3657
O(3)	2844	1029	1873
O(4)	6632	4524	7330
O(5)	1695	1955	5272
O(6)	1551	2078	8721

Fig. 1.   Structure of triclinic $Cu_5(PO_4)_2(OH)_4$.

1. Structure Reports, 28, 192.
2. Ibid., 43A, 259.

CALCIUM COPPER(II) PHOSPHATE
$Ca_3Cu_3(PO_4)_4$

J.B. ANDERSON, E. KOSTINER and F.A. RUSZALA, 1981. J. Solid State Chem., 39, 29-34.

Monoclinic, P2$_1$/a, a = 17.619, b = 4.8995, c = 8.917 Å, β = 124.08°, Z = 2.   Mo
radiation, R = 0.037 for 1388 reflexions.

The structure (Fig. 1) contains PO_4 tetrahedra linked by Cu and Ca ions. The
two Cu coordination polyhedra are similar to those in $Cu_3(PO_4)_2$ (1), square-planar
(Cu-O = 1.93, 1.96 (each x 2) Å) and irregular five coordination (Cu-O = 1.91-2.17 Å).
Ca ions have 6- and 9-coordinations, Ca-O = 2.29-3.04 Å.

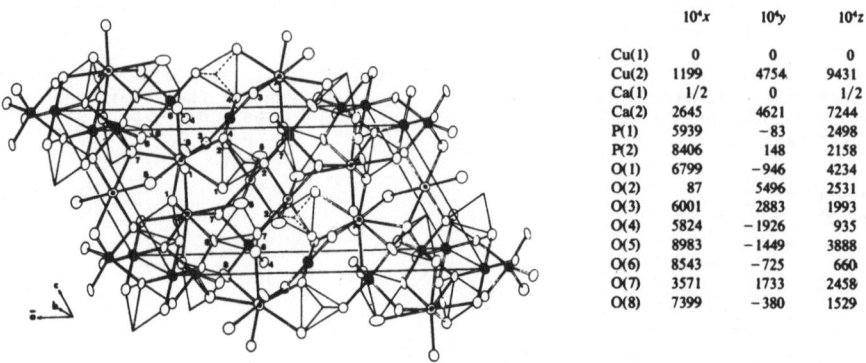

	10^4x	10^4y	10^4z
Cu(1)	0	0	0
Cu(2)	1199	4754	9431
Ca(1)	1/2	0	1/2
Ca(2)	2645	4621	7244
P(1)	5939	−83	2498
P(2)	8406	148	2158
O(1)	6799	−946	4234
O(2)	87	5496	2531
O(3)	6001	2883	1993
O(4)	5824	−1926	935
O(5)	8983	−1449	3888
O(6)	8543	−725	660
O(7)	3571	1733	2458
O(8)	7399	−380	1529

Fig. 1.   Structure of $Ca_3Cu_3(PO_4)_4$, and atomic positional parameters (x 10^4).

1. Structure Reports, 43A, 257.

ZINC NICKEL PHOSPHATE
γ-$(Zn_{0.70}Ni_{0.30})_3(PO_4)_2$

A.G. NORD and T. STEFANIDIS, 1981. Acta Cryst., B37, 1509-1511.

Monoclinic, $P2_1/n$, a = 7.505, b = 8.316, c = 5.056 Å, β = 94.48°, Z = 2. Neutron powder data.

Atomic positions (x 10^3)

	x	y	z
M(1)	617	144	96
M(2)	0	0	500
P	191	198	28
O(1)	42	138	829
O(2)	122	202	302
O(3)	258	363	947
O(4)	364	76	48

The structure is as previously described (1), with Zn mainly in the five-coordinate distorted trigonal bipyramidal M(1) site, and Ni mainly in the octahedral M(2) site; M(1)-O = 1.97-2.26, M(2)-O = 2.03-2.19 Å.

1. Structure Reports, 28, 189; 43A, 259.

MERCURY(II) HYDROGENPHOSPHATE
$HgHPO_4$

E. DUBLER, L. BECK, L. LINOWSKY and G.B. JAMESON, 1981. Acta Cryst., B37, 2214-2217.

Triclinic, P$\bar{1}$, a = 6.288, b = 7.309, c = 7.276 Å, α = 79.37, β = 85.27, γ = 82.85°, Z = 4. Mo radiation, R = 0.041 for 1983 reflexions.

The structure contains phosphate tetrahedra, linked into chains by three strong hydrogen bonds, two of which are across centres of symmetry (Fig. 1). The chains are linked by Hg ions which have approximately linear coordinations, Hg-O = 2.07(1) Å, with 4 or 5 further oxygens at 2.47-2.92 Å.

	x	y	z
Hg(1)	0·12495	0·16482	0·29934
Hg(2)	0·44481	0·20692	0·82592
P(1)	0·6124	0·2606	0·3129
P(2)	0·9433	0·2940	0·7903
O(11)	0·4521	0·1495	0·2340
O(12)	0·8086	0·1144	0·3617
O(13)	0·5034	0·3298	0·4887
O(14)	0·6690	0·4343	0·1747
O(21)	1·1261	0·2048	0·9172
O(22)	0·7772	0·1518	0·8215
O(23)	1·0320	0·3397	0·5874
O(24)	0·8384	0·4804	0·8492

Fig. 1.   Chain of phosphate tetrahedra and atomic positonal parameters in HgHPO₄.

CERIUM(III) PHOSPHATE
CePO$_4$

G.W. BEALL, L.A. BOATNER, D.F. MULLICA and W.O. MILLIGAN, 1981.   J. Inorg. Nucl. Chem., 43, 101-105.

Monoclinic, P2$_1$/n, a = 6.777, b = 6.993, c = 6.445 Å, β = 103.54°, Z = 4.   Mo radiation, R = 0.028 for 593 reflexions; synthetic material.

Atomic positions

	x	y	z
Ce	0.2818	0.1591	0.1000
P	0.3050	0.1663	0.6124
O(1)	0.2494	0.0059	0.4439
O(2)	0.3813	0.3314	0.4995
O(3)	0.4734	0.1061	0.8040
O(4)	0.1282	0.2163	0.7086

Monazite-type structure (e.g. cheralite, 1), with PO₄ tetrahedra linked by 9-coordinate Ce ions. P-O = 1.52-1.53, Ce-O = $\bar{2}$.45-2.78 Å.

1.   Structure Reports, 32A, 359.

VITUSITE
Na$_3$Ce(PO$_4$)$_2$

O.G. KARPOV, D.Ju. PUŠČAROVSKIJ, A.P. KHOMJAKOV, E.A. POBEDIMSKAJA and N.V. BELOV, 1980.   Kristallografija, 25, 1135-1141 [Soviet Physics - Crystallography, 25, 650-653].

Orthorhombic, Pca2$_1$(pseudo-cell), a = 14.091, b = 5.357, c = 18.740 Å, D$_m$ = 3.60, Z = 8.   Mo radiation, R = 0.13 for 1833 reflexions.

Two sets of atomic coordinates are derived, corresponding to different structural blocks, alternation of which gives rise to superlattice reflexions, increase in period along b, and a decrease in symmetry. The first block has a structure (Fig. 1) similar to that of Na₃La(VO₄)₂ (1), with PO₄ tetrahedra linked by 8-coordinate Ce and 6-coordinate Na; the second block has a displacement of 0.25 in x coordinates, which gives rise to 4-coordination for one Na.

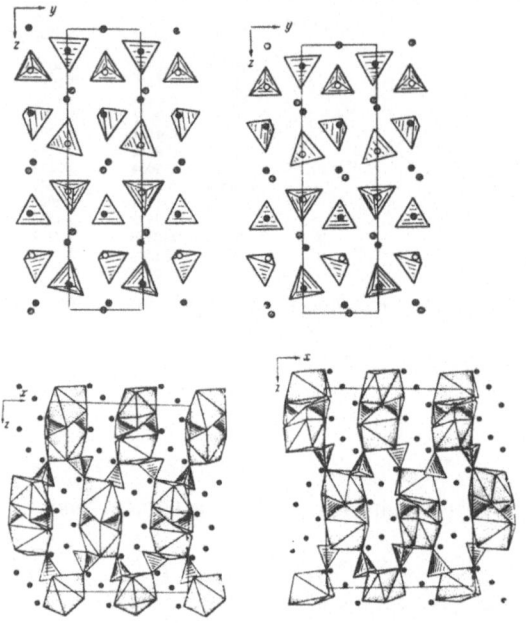

Fig. 1.   Structure of vitusite, showing the two structural blocks
            (left and right).

1. Structure Reports, 42A, 259.

POTASSIUM HYDROGEN PHOSPHATE PYROPHOSPHATE
$K_2(H_2PO_4)(H_3P_2O_7)$

A. LARBOT, J. DURAND, S. VILMINOT and A. NORBERT, 1981. Acta Cryst., B37, 1023-1027.

Monoclinic, C2/c, a = 31.272, b = 7.428, c = 9.253 Å, β = 99.85°, D_m = 2.21, Z = 8.
Cu radiation, R = 0.075 for 526 reflexions.

The structure (Fig. 1) contains $O_2P(OH)_2^-$ and $(HO)O_2P\text{-}O\text{-}PO(OH)_2^-$ ions, linked by hydrogen bonds and by the K ions. P-O = 1.46-1.57 Å, P-O-P = 139°.

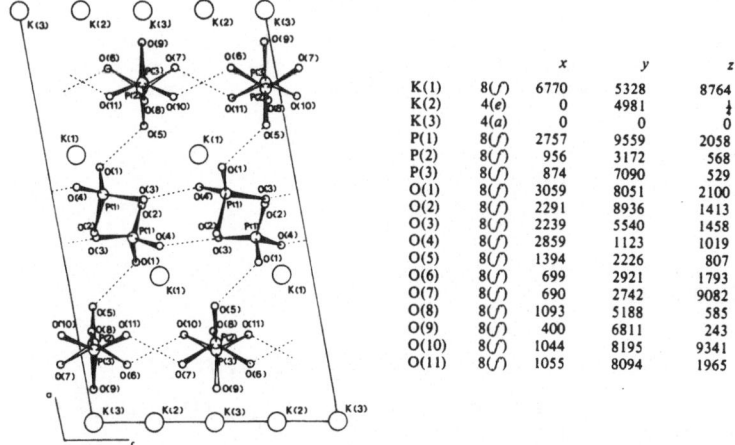

		x	y	z
K(1)	8(_f_)	6770	5328	8764
K(2)	4(_e_)	0	4981	¼
K(3)	4(_a_)	0	0	0
P(1)	8(_f_)	2757	9559	2058
P(2)	8(_f_)	956	3172	568
P(3)	8(_f_)	874	7090	529
O(1)	8(_f_)	3059	8051	2100
O(2)	8(_f_)	2291	8936	1413
O(3)	8(_f_)	2239	5540	1458
O(4)	8(_f_)	2859	1123	1019
O(5)	8(_f_)	1394	2226	807
O(6)	8(_f_)	699	2921	1793
O(7)	8(_f_)	690	2742	9082
O(8)	8(_f_)	1093	5188	585
O(9)	8(_f_)	400	6811	243
O(10)	8(_f_)	1044	8195	9341
O(11)	8(_f_)	1055	8094	1965

Fig. 1.   Structure of $K_2(H_2PO_4)(H_3P_2O_7)$, and atomic positional parameters ($\times 10^4$).

(PYROPHOSPHATO)TETRAAQUOCHROMIUM(III) TRIHYDRATE(α,β-BIDENTATE)   (CHROMIUM(III) HYDROGEN PYROPHOSPHATE HEPTAHYDRATE)

$Cr(H_2O)_4HP_2O_7 \cdot 3H_2O$ $CrHP_2O_7 \cdot 7H_2O$

E.A. MERRITT, M. SUNDARALINGAM and D. DUNAWAY-MARIANO, 1981.   J. Amer. Chem. Soc., 103, 3565-3567.

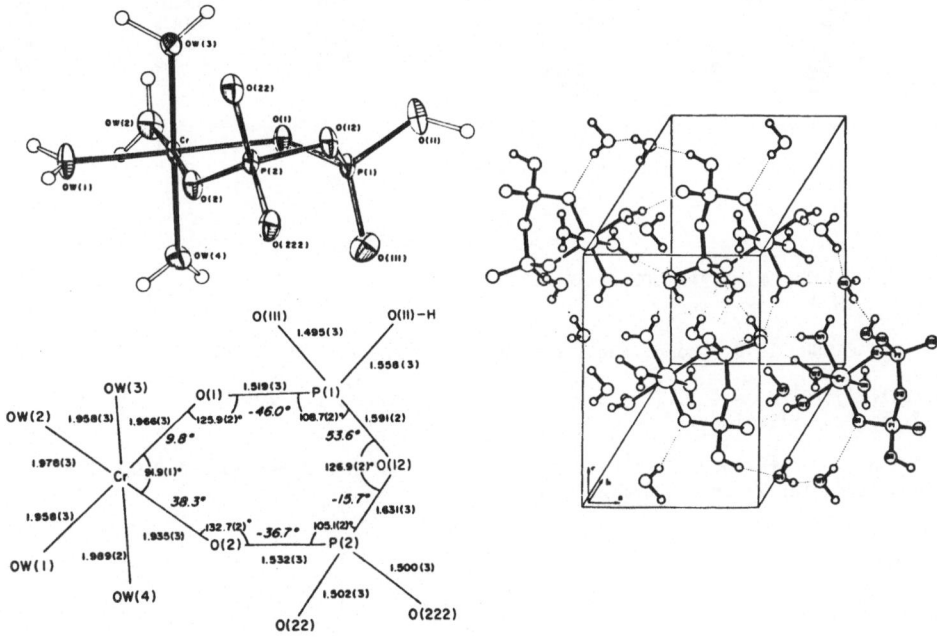

Fig. 1.   Structure of $Cr(H_2O)_4HP_2O_7 \cdot 3H_2O$.

Triclinic, $P\bar{1}$, a = 7.038, b = 8.858, c = 10.636 Å, α = 69.19, β = 89.54, γ = 68.37°, Z = 2. Cu radiation, R = 0.063 for 2401 reflexions.

The structure (Fig. 1) contains chelate molecules, linked by hydrogen bonding, directly and via the non-coordinated water molecules. The boat conformation of the chelate ring precludes intramolecular hydrogen bonding, in contrast to related cobalt compounds (1, 2).

1. Structure Reports, 46A, 333.
2. This volume, p. 295.

PLATINUM(IV) DIPHOSPHATE
PtP_2O_7

B. WELLMANN and F. LIEBAU, 1981. J. Less-Common Metals, 77, P31-P39.

Monoclinic, $P2_1/n$, a = 7.095, b = 7.883, c = 9.302 Å, β = 111.37°, Z = 4. Cu radiation, R = 0.101 for 646 reflexions (films, densitometer intensities).

The structure (Fig. 1) contains almost-eclipsed $O_3P-O-PO_3^{4-}$ anions, linked by octahedrally-coordinated Pt^{4+} ions. P-O = 1.47-1.53 (terminal), 1.57, 1.59(3) (bridging) Å, O-P-O = 103-118, P-O-P = 132°, Pt-O = 1.95-2.02(3) Å.

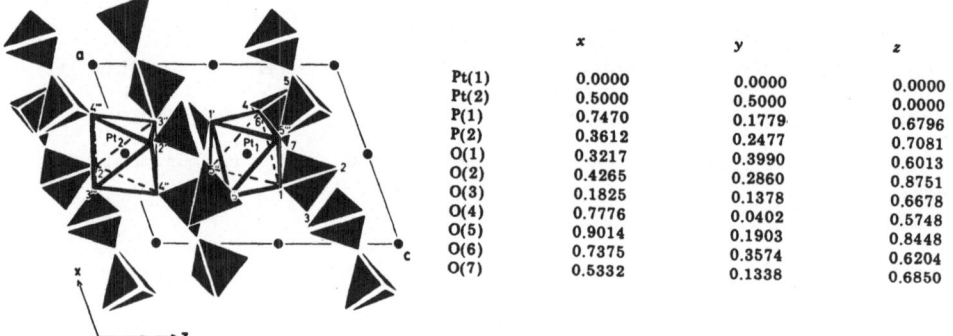

	x	y	z
Pt(1)	0.0000	0.0000	0.0000
Pt(2)	0.5000	0.5000	0.0000
P(1)	0.7470	0.1779	0.6796
P(2)	0.3612	0.2477	0.7081
O(1)	0.3217	0.3990	0.6013
O(2)	0.4265	0.2860	0.8751
O(3)	0.1825	0.1378	0.6678
O(4)	0.7776	0.0402	0.5748
O(5)	0.9014	0.1903	0.8448
O(6)	0.7375	0.3574	0.6204
O(7)	0.5332	0.1338	0.6850

Fig. 1.   Structure of platinum(IV) diphosphate.

AMMONIUM BARIUM TRIMETAPHOSPHATE MONOHYDRATE
$NH_4BaP_3O_9 \cdot H_2O$

ALI FUAT CESUR, 1979. Comm. Fac. Sci. Univ. Ankara, A2, 28, 1-10.

Monoclinic, $P2_1/n$, a = 11.705, b = 12.120, c = 7.558 Å, β = 101.05°, Z = 4. R = 0.12 for 863 reflexions.

The structure contains cyclic $P_3O_9^{3-}$ anions, with a six-membered P_3O_3 ring. Ba has irregular 7- or 8-coordination.

TETRAAMMINE(DIHYDROGENTRIPHOSPHATO)COBALT(III) MONOHYDRATE (α,γ-BIDENTATE)
$Co(NH_3)_4(H_2P_3O_{10})\cdot H_2O$

I.   E.A. MERRITT, M. SUNDARALINGAM and R.D. CORNELIUS, 1980. J. Amer. Chem. Soc.,
     102, 6151-6153.
II. Idem, 1981. Acta Cryst., B37, 657-659.

Monoclinic, $P2_1/n$, a = 7.234, b = 14.106, c = 12.113 Å, β = 92.91°, Z = 4. Cu radiation, R = 0.046 for 2369 reflexions.

    This coordination isomer (Fig. 1) contains an 8-membered chelate ring in a boat
conformation stabilized by two N-H...O hydrogen bonds. The β,γ-bidentate isomer has
been described previously (1).

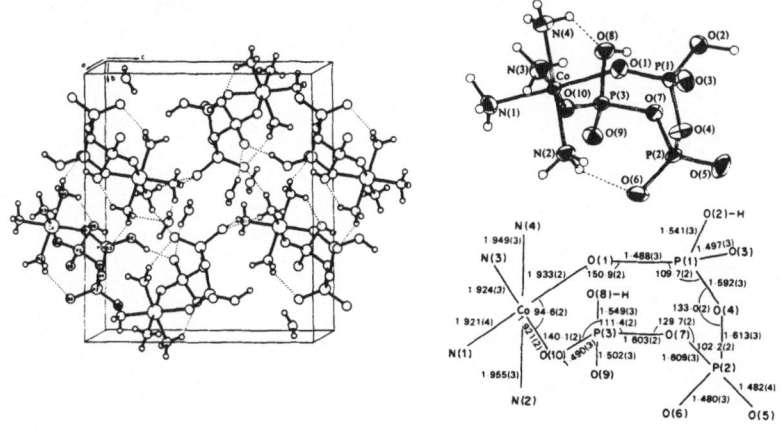

Fig. 1.   Structure of $Co(NH_3)_4(H_2P_3O_{10})\cdot H_2O$ (α,γ-bidentate isomer).

1. Structure Reports, 46A, 333.

TRIAMMINE(DIHYDROGENTRIPHOSPHATO)COBALT(III) (α,β,γ-TRIDENTATE)
$Co(NH_3)_3(H_2P_3O_{10})$

E.A. MERRITT and M. SUNDARALINGAM, 1981. Acta Cryst., B37, 1505-1509.

Tetragonal, $P4_2/mbc$, a = 12.892, c = 13.519 Å, Z = 8. Cu radiation, R = 0.058 for
1069 reflexions.

    The molecule (Fig. 1) contains two fused six-membered rings formed by facial
coordination of one O from each of the three phosphate residues; the molecules are
linked by N-H...O (2.89, 3.02 Å) and O-H...O (2.57 Å) hydrogen bonds. Related
β,γ- and α,γ-bidentate compounds have been described (1, 2).

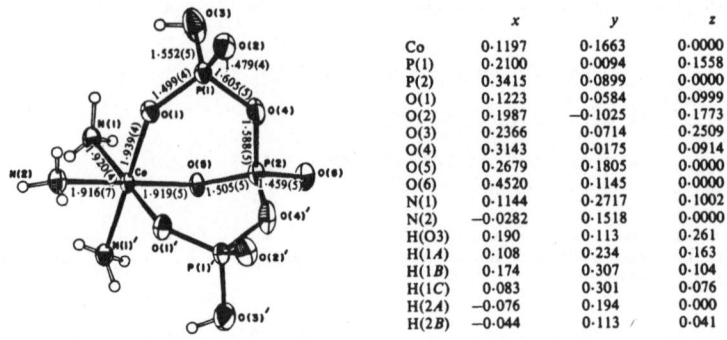

	x	y	z
Co	0·1197	0·1663	0·0000
P(1)	0·2100	0·0094	0·1558
P(2)	0·3415	0·0899	0·0000
O(1)	0·1223	0·0584	0·0999
O(2)	0·1987	−0·1025	0·1773
O(3)	0·2366	0·0714	0·2509
O(4)	0·3143	0·0175	0·0914
O(5)	0·2679	0·1805	0·0000
O(6)	0·4520	0·1145	0·0000
N(1)	0·1144	0·2717	0·1002
N(2)	−0·0282	0·1518	0·0000
H(O3)	0·190	0·113	0·261
H(1A)	0·108	0·234	0·163
H(1B)	0·174	0·307	0·104
H(1C)	0·083	0·301	0·076
H(2A)	−0·076	0·194	0·000
H(2B)	−0·044	0·113	0·041

Fig. 1.   Structure of α,β,γ-tridentate $Co(NH_3)_3(H_2P_3O_{10})$.

1. Structure Reports, 46A. 333.
2. Preceding report.

URANYL ULTRAPHOSPHATE
$(UO_2)_2P_6O_{17}$

Ju.E. GORBUNOVA, S.A. LINDE and A.V. LAVROV, 1981. Ž. Neorg. Khim., 26, 713-717.

Monoclinic, Cc, a = 8.653, b = 11.092, c = 17.453 Å, β = 106.12°, Z = 4. R = 0.087.

The structure contains polymeric $[P_6O_{17}^{4-}]_n$ anions of corner-sharing PO_4 tetrahedra; U ions have pentagonal bipyramidal coordination.

VLADIMIRITE (TRICLINIC)
$Ca_5H_2(AsO_4)_4 \cdot 5H_2O$

M. CATTI and G. IVALDI, 1981. Z. Kristallogr., 157, 119-130.

Triclinic, P1̄, a = 8.286, b = 6.673, c = 9.743 Å, α = 86.58, β = 111.10, γ = 99.74°, Z = 1. Mo radiation, R = 0.088 for 1200 reflexions.

The material is obtained by dehydration of the enneahydrate, ferrarisite (1), which involves loss of interlayer water, and disorder of one Ca ion and three water molecules (Fig. 1).

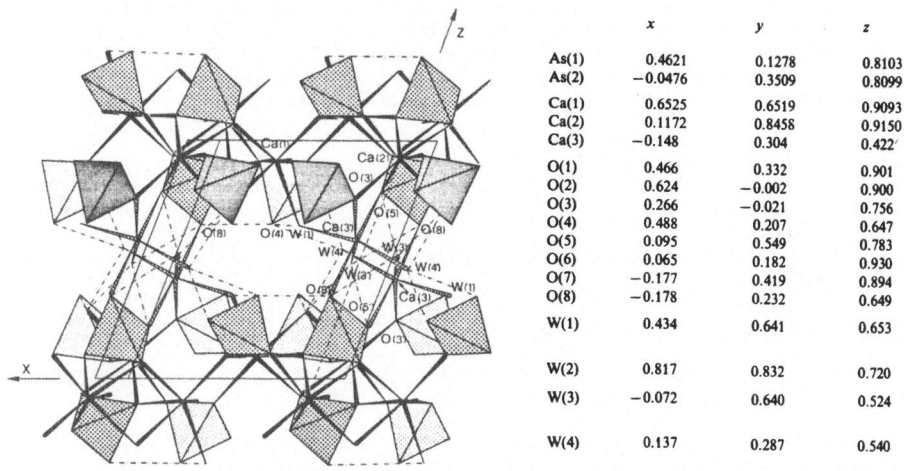

	x	y	z
As(1)	0.4621	0.1278	0.8103
As(2)	−0.0476	0.3509	0.8099
Ca(1)	0.6525	0.6519	0.9093
Ca(2)	0.1172	0.8458	0.9150
Ca(3)	−0.148	0.304	0.422
O(1)	0.466	0.332	0.901
O(2)	0.624	−0.002	0.900
O(3)	0.266	−0.021	0.756
O(4)	0.488	0.207	0.647
O(5)	0.095	0.549	0.783
O(6)	0.065	0.182	0.930
O(7)	−0.177	0.419	0.894
O(8)	−0.178	0.232	0.649
W(1)	0.434	0.641	0.653
W(2)	0.817	0.832	0.720
W(3)	−0.072	0.640	0.524
W(4)	0.137	0.287	0.540

Fig. 1.   Structure of triclinic vladimirite; occupancy = 0.5 for Ca(3),
            W(1), W(3), W(4).

<u>1</u>. Structure Reports, <u>46</u>A, 337.

PICROPHARMACOLITE
$Ca_4Mg(HAsO_4)_2(AsO_4)_2 \cdot 11H_2O$

M. CATTI, G. FERRARIS and G. IVALDI, 1981. Amer. Min., <u>66</u>, 385-391.

Triclinic, $P\bar{1}$, a = 13.547, b = 13.500, c = 6.710 Å, α = 99.85, β = 96.41, γ = 91.60°,
D_m = 2.62, Z = 2. Mo radiation, R = 0.087 for 1611 reflexions.

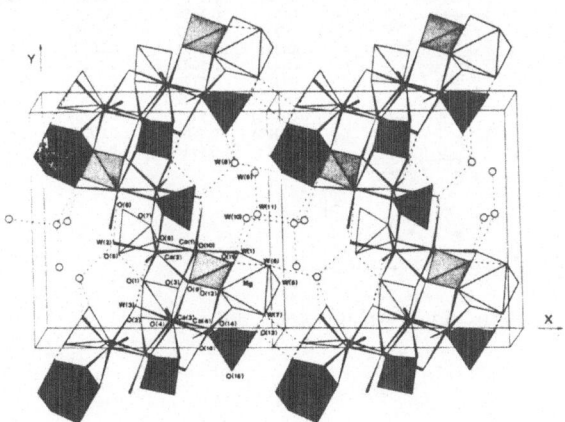

Fig. 1.   Structure of picropharmacolite.

The structure (Fig. 1) contains corrugated (100) layers of edge- and corner-sharing AsO_4 tetrahedra, MgO_6 octahedra, CaO_6 distorted octahedra, and CaO_7 polyhedra; the layers are linked only by hydrogen bonds, most of which are via four interlayer water molecules. As-O = 1.60-1.77, Mg-O = 2.03-2.19, Ca-O = 2.29-2.73 Å. The structure shows similarities with those of guerinite and ferrarisite (1), which contain polyhedral layers, but linked by Ca-O bonds in addition to hydrogen bonds.

1. Structure Reports, 40A, 260; 46A, 337.

IRON(III) TRIS(DIHYDROGENARSENATE) PENTAHYDRATE
$Fe(H_2AsO_4)_3.5H_2O$

A. BOUDJADA and J.C. GUITEL, 1981. Acta Cryst., B37, 1402-1405.

Monoclinic, $P2_1/n$, a = 15.25, b = 19.60, c = 4.72 Å, β = 91.8°, Z = 4. Mo radiation, R = 0.058 for 1910 reflexions.

The structure (Fig. 1) contains $[Fe(H_2AsO_4)_3]_n$ columns along c of corner-sharing FeO_6 octahedra and AsO_4 tetrahedra, with zeolitic water between the columns. As-O = 1.64-1.71, Fe-O = 1.96-2.02, O-H...O = 2.75-2.94(1) Å.

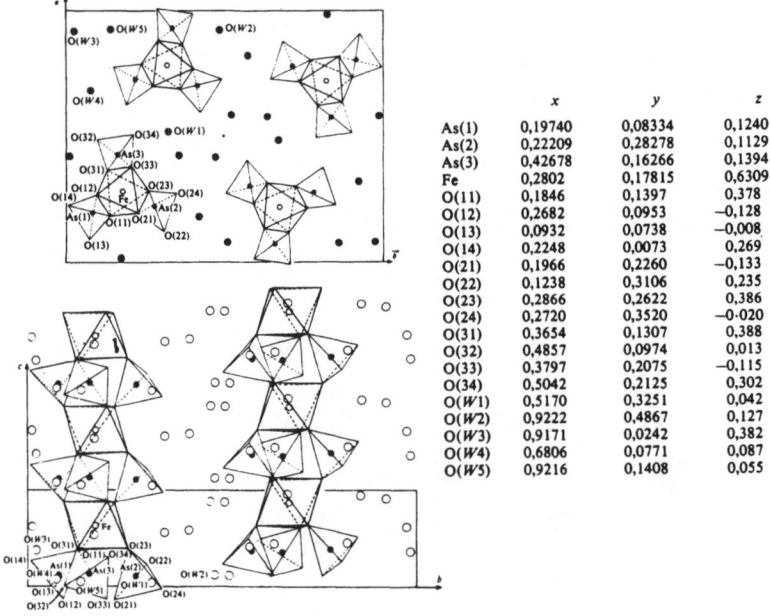

	x	y	z
As(1)	0,19740	0,08334	0,1240
As(2)	0,22209	0,28278	0,1129
As(3)	0,42678	0,16266	0,1394
Fe	0,2802	0,17815	0,6309
O(11)	0,1846	0,1397	0,378
O(12)	0,2682	0,0953	−0,128
O(13)	0,0932	0,0738	−0,008
O(14)	0,2248	0,0073	0,269
O(21)	0,1966	0,2260	−0,133
O(22)	0,1238	0,3106	0,235
O(23)	0,2866	0,2622	0,386
O(24)	0,2720	0,3520	−0,020
O(31)	0,3654	0,1307	0,388
O(32)	0,4857	0,0974	0,013
O(33)	0,3797	0,2075	−0,115
O(34)	0,5042	0,2125	0,302
O(W1)	0,5170	0,3251	0,042
O(W2)	0,9222	0,4867	0,127
O(W3)	0,9171	0,0242	0,382
O(W4)	0,6806	0,0771	0,087
O(W5)	0,9216	0,1408	0,055

Fig. 1.   Structure of $Fe(H_2AsO_4)_3.5H_2O$.

SILVER COBALT ARSENATE
$AgCo_3H_2(AsO_4)_3$

SILVER ZINC ARSENATE
$AgZn_3H_2(AsO_4)_3$

P. KELLER, H. RIFFEL, F. ZETTLER and H. HESS, 1981. Z. anorg. Chem., <u>474</u>, 123-134.

Monoclinic, C2/c, a = 12.159, 12.169, b = 12.438, 12.495, c = 6.782, 6.755 Å, β = 113.16, 112.77°, Z = 4. Mo radiation, R = 0.039, 0.049 for 1142, 1118 reflexions.

The structures (Fig. 1) contain chains of edge-sharing MO_6 octahedra (M = Co, Zn), linked by AsO_4 tetrahedra and square-planar coordinated Ag ions. Co-O = 2.13-2.14, Zn-O = 2.13-2.15, As-O = 1.68-1.72, Ag-O = 2.40-3.07 Å. Hydrogen bonds are probably present.

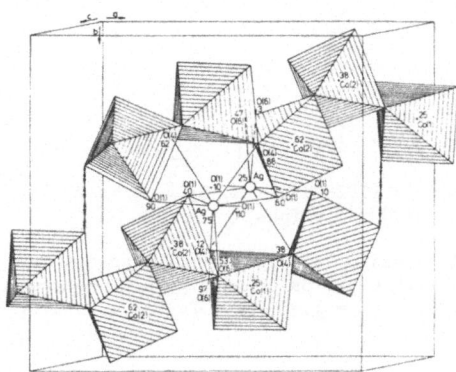

Fig. 1.   Co and Ag coordinations in $AgCo_3H_2(AsO_4)_3$.

KORITNIGITE
$ZnAsO_3(OH).H_2O$

P. KELLER, H. HESS and H. RIFFEL, 1980. Neues. Jb. Miner., Abh., <u>138</u>, 316-332.

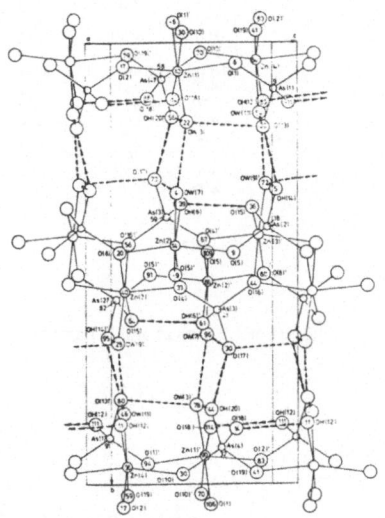

Fig. 1.   Structure of koritnigite.

Triclinic, $P\bar{1}$, a = 7.948, b = 15.829, c = 6.668 Å, α = 90.86, β = 96.56, γ = 90.05°, Z = 8. Mo radiation, R = 0.048 for 3421 reflexions.

Isostructural with the Co compound (1), except for differences in occupation of O and OH sites. The structure (Fig. 1) contains $AsO_3(OH)$ tetrahedra, linked by ZnO_6 octahedra and by hydrogen bonds. As-OH = 1.73-1.74, As-O = 1.65-1.69, Zn-O = 2.05-2.50(1) Å.

1. Structure Reports, 45A, 322.

MAPIMITE
$Zn_2Fe_3(AsO_4)_3(OH)_4 \cdot 10H_2O$

D. GINDEROW and F. CESBRON, 1981. Acta Cryst., B37, 1040-1043.

Monoclinic, Cm, a = 11.415, b = 11.259, c = 8.661 Å, β = 107.74°, D_m = 2.95, Z = 2. Mo radiation, R = 0.043 for 3140 reflexions.

The structure (Fig. 1) contains clusters of five edge-sharing MO_6 octahedra, linked by AsO_4 tetrahedra, with cavities containing some non-coordinated water molecules. As-O = 1.66-1.70, Zn-O = 1.97-2.24, Fe-O = 1.93-2.11 Å.

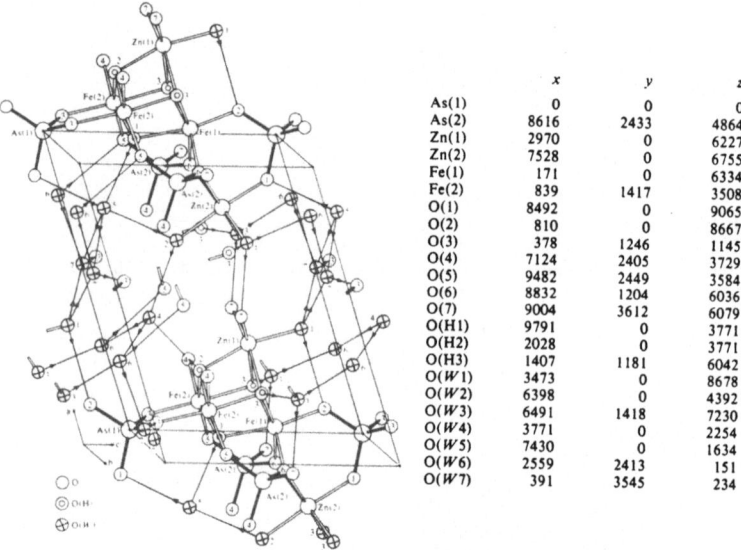

	x	y	z
As(1)	0	0	0
As(2)	8616	2433	4864
Zn(1)	2970	0	6227
Zn(2)	7528	0	6755
Fe(1)	171	0	6334
Fe(2)	839	1417	3508
O(1)	8492	0	9065
O(2)	810	0	8667
O(3)	378	1246	1145
O(4)	7124	2405	3729
O(5)	9482	2449	3584
O(6)	8832	1204	6036
O(7)	9004	3612	6079
O(H1)	9791	0	3771
O(H2)	2028	0	3771
O(H3)	1407	1181	6042
O(W1)	3473	0	8678
O(W2)	6398	0	4392
O(W3)	6491	1418	7230
O(W4)	3771	0	2254
O(W5)	7430	0	1634
O(W6)	2559	2413	151
O(W7)	391	3545	234

Fig. 1.   Structure of mapimite, and atomic positonal parameters (x 10^4).

MANGANESE(II) SULPHITE
α-MnSO$_3$

A. MAGNUSSON and L.-G. JOHANSSON, 1981. Acta Cryst., B37, 1400-1401.

Rhombohedral, R$\bar{3}$, a = 7.912 Å, α = 109.241°, Z = 6. Mo radiation, R = 0.029 for 1152 reflexions.

Atomic positions

	x	y	z
Mn	0·36530	0·18610	0·68143
S	0·35000	0·02485	0·21003
O(1)	0·5535	0·2301	0·3405
O(2)	0·3101	−0·0203	0·3725
O(3)	0·4127	−0·1322	0·1190

Isostructural with the Fe compound (1). The structure contains trigonal pyramidal SO_3^{2-} ions, linked by octahedrally-coordinated Mn^{2+} ions (6 O from six different sulphite anions). Mn-O = 2.151-2.242, S-O = 1.536-1.541(2) Å, O-S-O = 103.2-103.5(1)°.

β-MnSO$_3$

A. MAGNUSSON, L.-G. JOHANSSON and O. LINDQVIST, 1981. Acta Cryst., B37, 1108-1110.

Monoclinic, P2$_1$/a, a = 8.227, b = 12.199, c = 5.4193 Å, β = 79.741°, Z = 8. Mo radiation, R = 0.030 for 1523 reflexions.

The structure (Fig. 1) contains SO_3^{2-} trigonal pyramids, linked into a three-dimensional framework by MnO$_6$ distorted trigonal prisms. S-O = 1.529-1.546(3) Å, O-S-O = 97.7-105.9°, Mn-O = 2.147-2.381 Å.

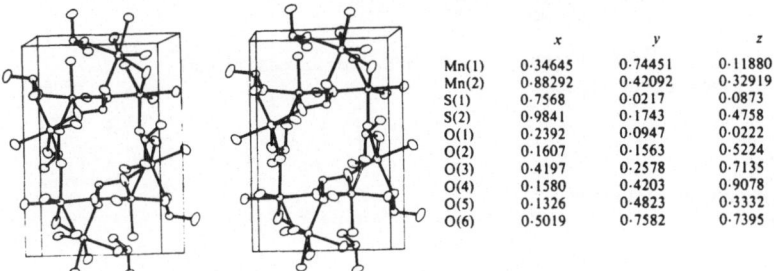

	x	y	z
Mn(1)	0·34645	0·74451	0·11880
Mn(2)	0·88292	0·42092	0·32919
S(1)	0·7568	0·0217	0·0873
S(2)	0·9841	0·1743	0·4758
O(1)	0·2392	0·0947	0·0222
O(2)	0·1607	0·1563	0·5224
O(3)	0·4197	0·2578	0·7135
O(4)	0·1580	0·4203	0·9078
O(5)	0·1326	0·4823	0·3332
O(6)	0·5019	0·7582	0·7395

Fig. 1.   Structure of β-manganese(II) sulphite.

1. Structure Reports, 46A, 345.

SODIUM TETRASULPHITOPALLADATE(II) DIHYDRATE
Na$_6$Pd(SO$_3$)$_4$·2H$_2$O

D. MESSER, D.K. BREITINGER and W. HAEGLER, 1981. Acta Cryst., B37, 19-23.

Tetragonal, I4$_1$/a, a = 16.488, c = 10.663 Å, D$_m$ = 2.69, Z = 8. Mo radiation, R = 0.041 for 1969 reflexions.

The structure (Fig. 1) contains two independent distorted-square-planar $Pd(SO_3)_4^{6-}$ anions, with long Pd-S bonds, 2.316 and 2.341(1) Å. The anions are stacked alternately to form rods along $\underline{c}$. Connection within and between the rods is via 6-coordinate Na ions, the arrangement producing channels filled with right- and left-handed helices of hydrogen-bonded water molecules, O-H...O = 2.68 Å. S-O = 1.479-1.494(3) Å, O-S-O = 107-114°, Na-O = 2.31-2.44 and 2.48-2.90 Å.

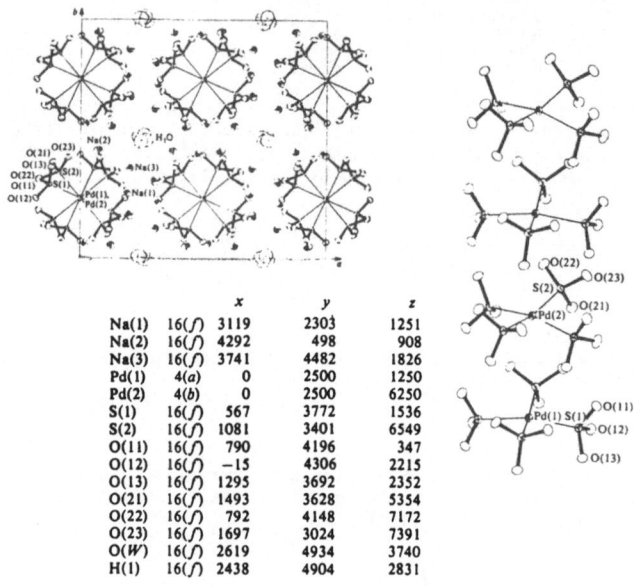

		x	y	z
Na(1)	16(f)	3119	2303	1251
Na(2)	16(f)	4292	498	908
Na(3)	16(f)	3741	4482	1826
Pd(1)	4(a)	0	2500	1250
Pd(2)	4(b)	0	2500	6250
S(1)	16(f)	567	3772	1536
S(2)	16(f)	1081	3401	6549
O(11)	16(f)	790	4196	347
O(12)	16(f)	−15	4306	2215
O(13)	16(f)	1295	3692	2352
O(21)	16(f)	1493	3628	5354
O(22)	16(f)	792	4148	7172
O(23)	16(f)	1697	3024	7391
O(W)	16(f)	2619	4934	3740
H(1)	16(f)	2438	4904	2831

Fig. 1.   Structure of $Na_6Pd(SO_3)_4.2H_2O$, and atomic positional parameters (x 10^4).

TRIAMMONIUM HYDROGEN DISULPHATE
$(ND_4)_3D(SO_4)_2$

M. TANAKA and Y. SHIOZAKI, 1981. Acta Cryst., B<u>37</u>, 1171-1174.

Monoclinic, A2/a, a = 10.158, b = 5.860, c = 15.401 Å, β = 101.88°, Z = 4. Mo radiation, R = 0.033 for 1753 reflexions.

Atomic positions (x 10^4 for non-deuterium, x 10^3 for deuterium atoms)

	x	y	z		x	y	z
N(1)	7500	2687	0	D(1)	304	665	25
N(2)	6527	7252	1989	D(2)	231	785	36
S	4613	2190	1142	D(3)	721	717	186
O(1)	3978	272	1491	D(4)	663	702	252
O(2)	4422	1863	142	D(5)	610	640	178
O(3)	6056	2237	1503	D(6)	621	840	191
O(4)	3982	4340	1289	D(7)	29	76	−7

The structure is as previously described for the H compound (<u>1</u>), except that the O-H...O hydrogen bond is 0.019 Å longer, with a double-minimum potential well.

<u>1</u>. Structure Reports, <u>44</u>A, 268.

AMMONIUM LITHIUM SULPHATE (HIGH-TEMPERATURE FERROELECTRIC PHASE)
NH$_4$LiSO$_4$

K. ITOH, H. ISHIKURA and E. NAKAMURA, 1981. Acta Cryst., B$\underline{37}$, 664-666.

Orthorhombic, Pmcn, a = 5.299, b = 9.199, c = 8.741 Å, at 478K, Z = 4. Mo radiation, R = 0.055 for 1060 reflexions.

The structure (Fig. 1) is similar to that at room temperature ($\underline{1}$), but with disorder giving rise to an additional mirror plane. S-O = 1.423-1.463, Li-O = 1.892-1.953, NH$_4$-O = 2.892-3.487 Å.

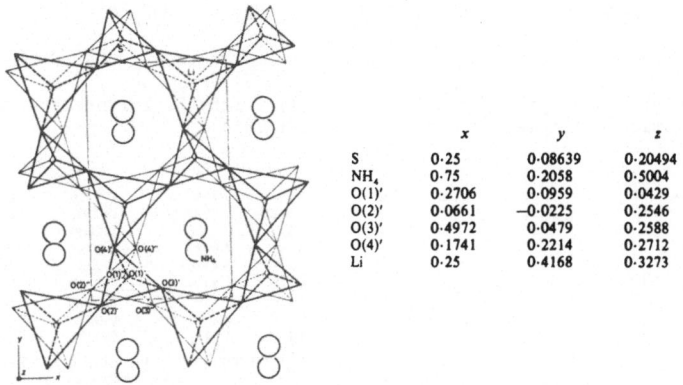

	x	y	z
S	0·25	0·08639	0·20494
NH$_4$	0·75	0·2058	0·5004
O(1)'	0·2706	0·0959	0·0429
O(2)'	0·0661	−0·0225	0·2546
O(3)'	0·4972	0·0479	0·2588
O(4)'	0·1741	0·2214	0·2712
Li	0·25	0·4168	0·3273

Fig. 1.   Structure of NH$_4$LiSO$_4$ (the two disordered arrangements are shown).

$\underline{1}$. Structure Reports, $\underline{34A}$, 305.

SODIUM SULPHATE (POLYMORPH III)
Na$_2$SO$_4$

B.N. MEHROTRA, 1981. Z. Kristallogr., $\underline{155}$, 159-163.

Orthorhombic, Cmcm, a = 5.607, b = 8.955, c = 6.967 Å, Z = 4. Mo radiation, R = 0.045 for 1032 reflexions. The material is the stable modification in the range 200-228°C, and can be quenched from the melt to room temperature. Previous studies in $\underline{1}$; several other polymorphs are known (e.g. thenardite, $\underline{2}$).

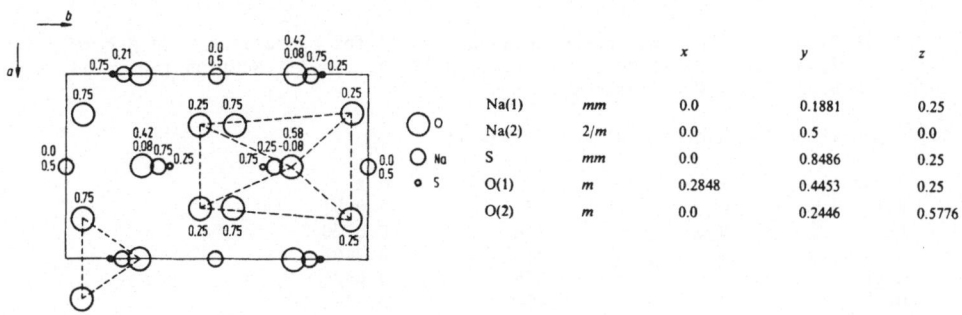

		x	y	z
Na(1)	mm	0.0	0.1881	0.25
Na(2)	2/m	0.0	0.5	0.0
S	mm	0.0	0.8486	0.25
O(1)	m	0.2848	0.4453	0.25
O(2)	m	0.0	0.2446	0.5776

Fig. 1.   Structure of Na$_2$SO$_4$-III.

The structure (Fig. 1) contains SO_4 tetrahedra linked by two types of NaO_6 octahedra. S-O = 1.463, 1.485(1) Å, Na-O = 2.338-2.803 Å.

1. Structure Reports, 8, 189; 18, 474.
2. Ibid., 41A, 343.

SODIUM SULPHATE CHLORIDE PERHYDRATE
$4Na_2SO_4.2D_2O_2.NaCl$

J.M. ADAMS, V. RAMDAS and A.W. HEWAT, 1981. Acta Cryst., B37, 915-917.

Tetragonal, P4/mnc, a = 10.527, c = 8.408 Å, Z = 2. Neutron powder data.

The structure is as previously described (1), with the arrangement of the disordered peroxide molecules now established (Fig. 1).

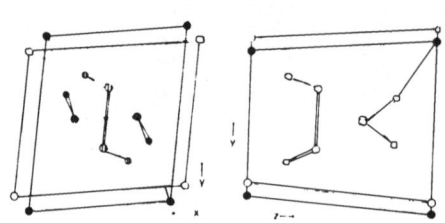

	x	y	z
Na(1)	0·7569	0·9796	0·1910
Na(2)	0·0000	0·0000	0·5000
Cl(1)	0·0000	0·0000	0·0000
S(1)	0·6987	0·2768	0·0000
O(1)	0·8346	0·2981	0·0000
O(2)	0·6369	0·3351	0·1432
O(3)	0·6691	0·1369	0·0000
O(4)*	0·4963	0·0651	0·1932
D(1)*	0·5493	0·0902	0·1055

* These atoms have a site occupancy of 0·5. The isotopic ratio of D to H in the D(1) site is 0·80 : 0·20.

Fig. 1.   Two views of the disordered D_2O_2 molecules (occupancy = 0.5) and atomic positional parameters in $4Na_2SO_4.2D_2O_2.NaCl$.

1. Structure Reports, 44A, 271.

POTASSIUM SULPHATE
K_2SO_4

H. ARNOLD, W. KURTZ, A. RICHTER-ZINNIUS, J. BETHKE and G. HEGER, 1981. Acta Cryst., B37, 1643-1651.

β-Form, at 832K, Pmcn, a = 5.927, b = 10.318, c = 7.882 Å, Z = 4. Neutron radiation, R = 0.071 for 469 reflexions.

α-Form, at 847, 913K, $P6_3/mmc$, ortho-hexagonal cell (for comparison with β-form), a = 5.886, 5.917, b = 10.209, 10.240, c = 8.118, 8.182 Å, Z = 4. Neutron radiation, R = 0.090, 0.055 for 70, 100 reflexions.

Atomic positions (orthohexagonal cell, Fig. 1)

β-form	x	y	z
K(1)	0	0.0364	-0.0081
K(2)	0	0.3362	0.6920
S	0	0.3322	0.2401
O(1)	0	0.3345	0.0596
O(2)	0	0.1988	0.2942
O(3)	0.1996	0.3967	0.2973

α-form (913K)   Apex model   Edge model

	x	y	z	x	y	z
K(1)	0	0	0	0	0	0
K(2)	0	1/3	3/4	0	1/3	3/4
S	0	1/3	1/4	0	1/3	1/4
O(1)	0	1/3	0.0718	0	0.1966	1/4
O(2)	0	0.2041	0.2983	0	0.2552	0.3900
O(3)	0.1938	0.3979	0.299			

The β-phase structure (Fig. 1) is as previously described (1). Two disordered models are possible for the α-phase (Fig. 2); the 'edge' model is more likely at 847 and 913K, but the 'apex' model seems more likely at 1073K (X-ray data of 2).

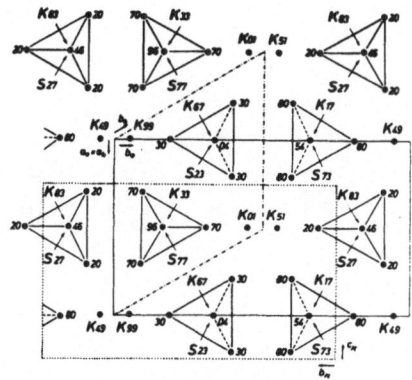

Fig. 1.   β-K$_2$SO$_4$ structure; full lines indicate the cell used in this report, dotted lines the conventional orthorhombic cell, and dot-dashed lines the hexagonal cell of the α-form.

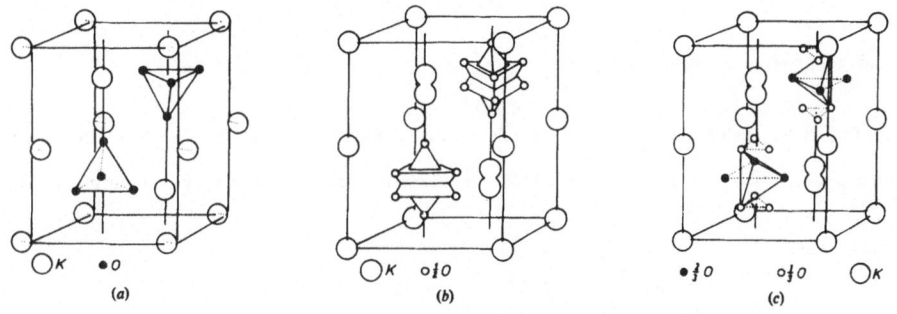

Fig. 2.   β-K$_2$SO$_4$ (a), and apex (b) and edge (c) models for α-K$_2$SO$_4$.

1. Strukturbericht, 2, 86, 423; Structure Reports, 22, 447; 33A, 367; 38A, 330.
2. Structure Reports, 46A, 347.

CAESIUM HYDROGENSULPHATE
CsHSO$_4$

K. ITOH, T. OZAKI and E. NAKAMURA, 1981. Acta Cryst., B$\underline{37}$, 1908-1909.

Monoclinic, P2$_1$/m, a = 7.3039, b = 5.8099, c = 5.4908 Å, β = 101.51°, Z = 2. Mo
radiation, R = 0.038 for 1442 reflexions.

The material is not isomorphous with other alkali-metal hydrogensulphates, but
the structure resembles that of CsH$_2$PO$_4$ ($\underline{1}$, $\underline{2}$). Tetrahedral HSO$_4^-$ ions are linked
by a disordered hydrogen bond and by Cs$^+$ ions (Fig. 1). S-O(H) = 1.507, S-O = 1.421,
1.444, Cs-O = 3.110-3.259(5), O...H...O = 2.437(7) Å.

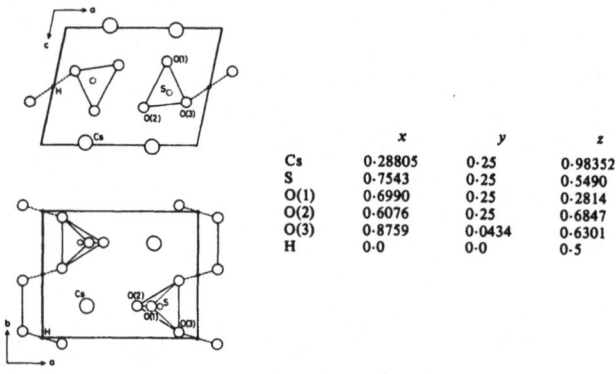

	x	y	z
Cs	0·28805	0·25	0·98352
S	0·7543	0·25	0·5490
O(1)	0·6990	0·25	0·2814
O(2)	0·6076	0·25	0·6847
O(3)	0·8759	0·0434	0·6301
H	0·0	0·0	0·5

Fig. 1.   Structure of caesium hydrogensulphate.

$\underline{1}$. Structure Reports, $\underline{42A}$, 337.
$\underline{2}$. This volume, p. 280.

MAGNESIUM HYDROXIDE SULPHATE HYDRATE
MgSO$_4$.1/3Mg(OH)$_2$.1/3H$_2$O

K.D. KEEFER, M.F. HOCHELLA and B.H.W.S. de JONG, 1981. Acta Cryst., B$\underline{37}$, 1003-1006.

Tetragonal, I4$_1$/amd, a = 5.242, c = 12.995 Å, Z = 4. Mo radiation, R = 0.041 for 160
reflexions.

Atomic positions

				x	y	z
4	S	in	4(a)	0	3/4	1/8
16/3	Mg		8(d)	0	0	1/2
16	O(1)		16(h)	0	0.5195	0.1895
4	O(2)		4(b)	0	1/4	3/8
16/3	H		16(h)	0	0.42	0.41

The structure (Fig. 1) contains chains of face-sharing oxygen octahedra, two-
thirds of which have Mg at their centres. The chains are arranged in layers perpendicular to $\underline{c}$, and are linked within and between the layers by SO$_4$ tetrahedra.

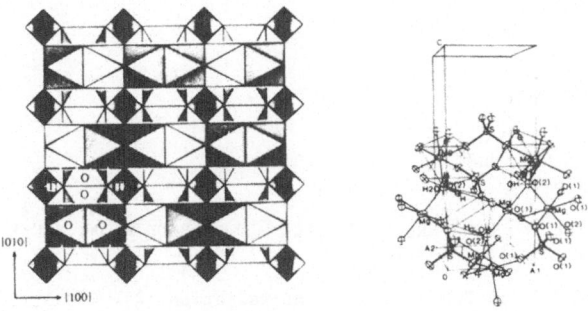

Fig. 1.   Structure of magnesium hydroxide sulphate hydrate.

KHADEMITE
Al(SO$_4$)F.5H$_2$O

B. BACHET, F. CESBRON and R. CHEVALIER, 1981. Bull. Minéral., 104, 19-22.

Orthorhombic, Pbca, a = 11.181, b = 13.048, c = 10.885 Å, D$_m$ = 1.925, Z = 8. Cu radiation, R = 0.052 for 1126 reflexions.

The structure contains sheets parallel to (100) of SO$_4$ tetrahedra, alternating with sheets of Al(H$_2$O)$_6$ and Al(H$_2$O)$_4$F$_2$ octahedra, connected by hydrogen bonds. S-O = 1.477, Al-O = 1.853-1.966, Al-F = 1.734 Å.

HYDROXYLAMMONIUM ALUMINUM ALUM
(NH$_3$OH)Al(SO$_4$)$_2$.12H$_2$O

A.M. ABDEEN, G. WILL and A. WEISS, 1981. Z. Kristallogr., 154, 45-57.

Cubic, Pa3, a = 12.328 Å, Z = 4. Neutron radiation, R = 0.054 for 2288 reflexions.

Atomic positions (x 10^4)

		Occu-pancy	x	y	z
Al	4(a)	1	0	0	0
S	8(c)	1	3022	= x	= x
O$_x$(1)	8(c)	1	2328	= x	= x
O$_x$(2)	24(d)	1	6913	7526	911
N	8(c)	⅓	5130	= x	= x
H(N)	24(d)	⅓	815	5287	236
O	8(c)	⅓	5433	= x	= x
H(O)	24(d)	⅙	4226	5169	3968
O$_w$(1)	24(d)	1	8509	9796	186
O$_w$(2)	24(d)	1	419	1373	2984
H(1)	24(d)	1	5417	2011	4732
H(2)	24(d)	1	4137	1837	4525
H(3)	24(d)	1	2138	4933	1958
H(4)	24(d)	1	2127	6092	1695

α-Alum structure (1), with disorder of the NH$_3$OH$^+$ about the (1/2,1/2,1/2) position.

<u>1</u>. Strukturbericht, <u>3</u>, 108; Structure Reports, <u>27</u>, 813.

AMMONIUM ALUMINUM ALUM
$NH_4Al(SO_4)_2 \cdot 12H_2O$

METHYLAMMONIUM ALUMINUM ALUM
$(CH_3NH_3)Al(SO_4)_2 \cdot 12H_2O$

A.M. ABDEEN, G. WILL, W. SCHÄFER, A. KIRFEL, M.O. BARGOUTH, K. RECKER and A. WEISS, 1981. Z. Kristallogr., <u>157</u>, 147-166.

Cubic, Pa3, a = 12.242, 12.322 Å, Z = 4. Mo radiation, R = 0.053, 0.068 for 905, 877 reflexions, and neutron radiation, R = 0.044, 0.063 for 190, 194 reflexions.

Alum structure (<u>1</u>, <u>2</u>), with NH_4^+ and $CH_3NH_3^+$ distributed on 8(c) and 24(d) sites in two orientations of equal probability. Sulphate anions are also disordered (17% and 4%, respectively).

<u>1</u>. Strukturbericht, <u>3</u>, 108; Structure Reports, <u>27</u>, 813.
<u>2</u>. This volume, preceding report.

CAESIUM ALUMS
$CsM(III)[SO_4]_2 \cdot 12H_2O$,   M = V, Cr, Mn, Fe, Co, Al, Ga, In

J.K. BEATTIE, S.P. BEST, B.W. SKELTON and A.H. WHITE, 1981. J. Chem. Soc., Dalton, 2105-2111.

Cubic, Pa3, a = 12.452, 12.413, 12.432, 12.449, 12.292, 12.357, 12.419, 12.540 Å, Z = 4. Mo radiation, R = 0.030-0.050 for 302-818 reflexions.

Atomic positions

β-Alums (V; similar values for Cr, Mn, Fe, Al, Ga, In)

			x	y	z
Cs	in	4(b)	1/2	1/2	1/2
V		4(a)	0	0	0
S		8(c)	0.3268	0.3268	0.3268
O(1)		8(c)	0.2584	0.2584	0.2584
O(2)		24(d)	0.2790	0.3370	0.4354
O(a)		24(d)	0.0521	0.2109	0.3426
O(b)		24(d)	0.1600	-0.0006	-0.0021

α-Alum (Co)

Cs	in	4(b)	1/2	1/2	1/2
Co		4(a)	0	0	0
S		8(c)	0.3178	0.3178	0.3178
O(1)		8(c)	0.2492	0.2492	0.2492
O(2)		24(d)	0.3296	0.2690	0.4255
O(a)		24(d)	0.0480	0.1508	0.2875
O(b)		24(d)	0.1523	0.0006	-0.0057

Alum structures (<u>1</u>).

<u>1</u>. Strukturbericht, <u>3</u>, 108.

AMMONIUM INDIUM(III) SULPHATE (HIGH-TEMPERATURE)
$(NH_4)_3In(SO_4)_3$

B. JOLIBOIS, G. LAPLACE, F. ABRAHAM and G. NOWOGROCKI, 1981. J. Solid State Chem., <u>40</u>,
69-74.

Rhombohedral, R3c, a = 15.531, c = 9.163 Å, at 120°C, Z = 6. Mo radiation, R = 0.023
for 570 reflexions, at 140°C. The material is obtained by heating the low-temperature
monoclinic form (<u>1</u>).

    The structure (Fig. 1) contains $In(SO_4)_3$ columns of InO_6 octahedra and SO_4 tetra-
hedra, as in the monoclinic form (<u>1</u>). In-O = 2.115, 2.149(8), S-O = 1.43-1.49(1) Å.

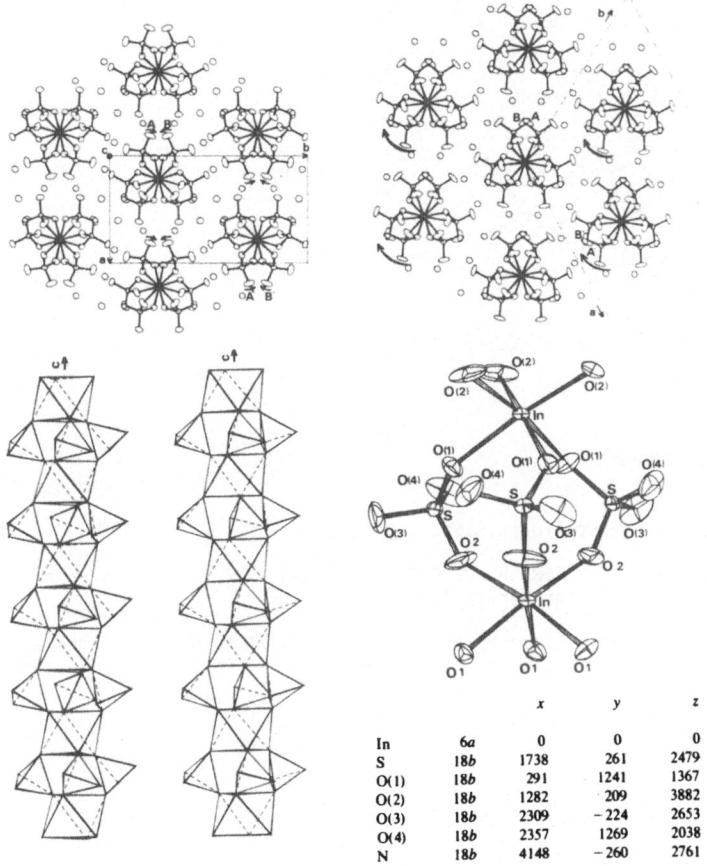

		x	y	z
In	6a	0	0	0
S	18b	1738	261	2479
O(1)	18b	291	1241	1367
O(2)	18b	1282	209	3882
O(3)	18b	2309	-224	2653
O(4)	18b	2357	1269	2038
N	18b	4148	-260	2761

Fig. 1.   Structure of low- (left) and high-temperature (right) $(NH_4)_3In(SO_4)_3$.

<u>1</u>. Structure Reports, <u>46A</u>, 349.

CHUKHROVITE
$Ca_4AlSi(SO_4)F_{13}.12H_2O$

M. MATHEW, S. TAKAGI, K.R. WAERSTAD and A.W. FRAZIER, 1981. Amer. Min., 66, 392-397.

Cubic, Fd3, a = 16.710 Å, D_m = 2.20, Z = 8. Mo radiation, R = 0.025 for 481 reflexions, for synthetic material.

The structure (Fig. 1) contains $(Al,Si)F_6$ octahedra, and a F^- ion surrounded tetrahedrally by Ca ions. Ca has monocapped octahedral coordination to 6 F and 1 H_2O, and the MF_6 octahedra and Ca polyhedra share corners. The sulphate ion is hydrogen bonded to 12 water molecules. (Al,Si)-F = 1.735, Ca-F = 2.361, 2.411, Ca-O = 2.414, S-O = 1.477, O-H...O = 2.807, O-H...F = 2.881 Å.

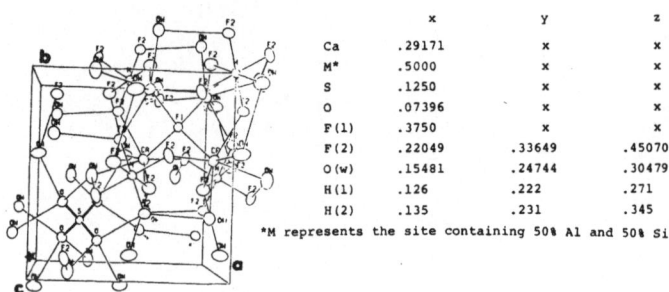

	x	y	z
Ca	.29171	x	x
M*	.5000	x	x
S	.1250	x	x
O	.07396	x	x
F(1)	.3750	x	x
F(2)	.22049	.33649	.45070
O(w)	.15481	.24744	.30479
H(1)	.126	.222	.271
H(2)	.135	.231	.345

*M represents the site containing 50% Al and 50% Si

Fig. 1.   Structure of synthetic chukhrovite.

LEAD(II) OXIDE SULPHATE
α-$Pb_3O_2(SO_4)$ 2PbO.$PbSO_4$

K. SAHL, 1981. Z. Kristallogr., 156, 209-217.

Monoclinic, $P2_1/m$, a = 7.168, b = 5.771, c = 8.036 Å, β = 102.40°, D_m = 7.7, Z = 2. Mo radiation, R = 0.066 for 978 reflexions.

The structure (Fig. 1) contains double chains along b of edge-sharing OPb_4 tetrahedra; the chains are linked by SO_4 tetrahedra. Pb ions have 7-, 5-, and 9-coordinations.

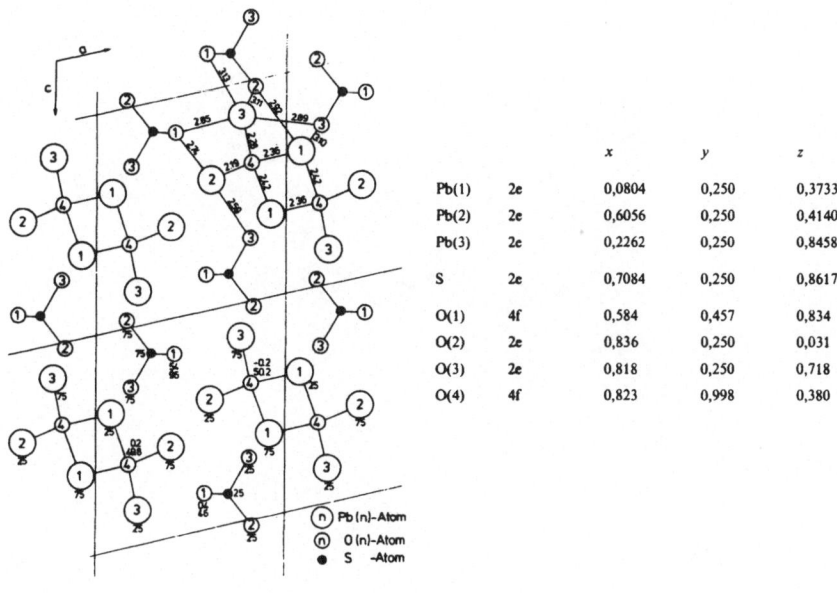

		x	y	z
Pb(1)	2e	0,0804	0,250	0,3733
Pb(2)	2e	0,6056	0,250	0,4140
Pb(3)	2e	0,2262	0,250	0,8458
S	2e	0,7084	0,250	0,8617
O(1)	4f	0,584	0,457	0,834
O(2)	2e	0,836	0,250	0,031
O(3)	2e	0,818	0,250	0,718
O(4)	4f	0,823	0,998	0,380

Fig. 1.   Structure of α-$Pb_3O_2(SO_4)$.

AMMONIUM ANTIMONY(III) FLUORIDE SULPHATE
$(NH_4)_6[Sb_4F_{12}(SO_4)_3]$

A.A. UDOVENKO, V.N. BUTENKO, R.L. DAVIDOVIČ, V.G. ANDRIANOV, M.Ju. ANTIPIN and A.I.
JANOVSKIJ, 1981. Kristallografija, <u>26</u>, 488-494 [Soviet Physics - Crystallography,
<u>26</u>, 277-280].

Trigonal, P3, a = 17.104, 16.993, c = 7.543, 7.514 Å, at room temperature, -120°C,
Z = 3. Mo radiation, R = 0.035, 0.026 for 2935, 3658 reflexions.

The structure (Fig. 1) contains three independent but structurally similar
infinite anion columns, linked by ammonium ions. The columns contain $Sb_3S_3O_6$ rings
with a central SbF_3 group; both types of Sb have capped-octahedral coordination,
SbF_4O_2E and SbF_3O_3E (E = lone-pair).

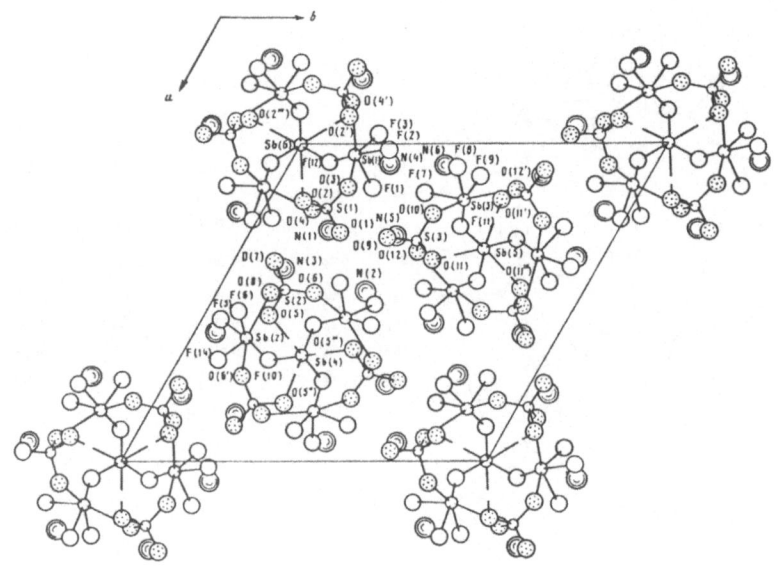

Fig. 1.   Structure of ammonium antimony(III) fluoride sulphate.

POTASSIUM ANTIMONY(III) SULPHATE FLUORIDE
$K_2SO_4(SbF_3)_2$

M. BOURGAULT, B. DUCOURANT, B. BONNET and R. FOURCADE, 1981.   J. Solid State Chem.,
36, 183-189.

Monoclinic, $P2_1/c$, a = 9.225, b = 5.632, c = 19.379 Å, β = 103.14°, D_m = 3.56, Z =
4. Mo radiation, R = 0.035 for 2264 reflexions.

The structure (Fig. 1) contains trigonal pyramidal SbF_3 units and tetrahedral
SO_4^{2-} ions, linked into sheets parallel to (102) by longer contacts which complete an
SbF_4O_2E monocapped octahedron and SbF_3O_2E octahedron (E = lone-pair). The sheets are
linked by the K^+ ions. Sb-F = 1.91-1.97, 2.80, Sb-O = 2.43-2.59, S-O = 1.44-1.47(1) Å.

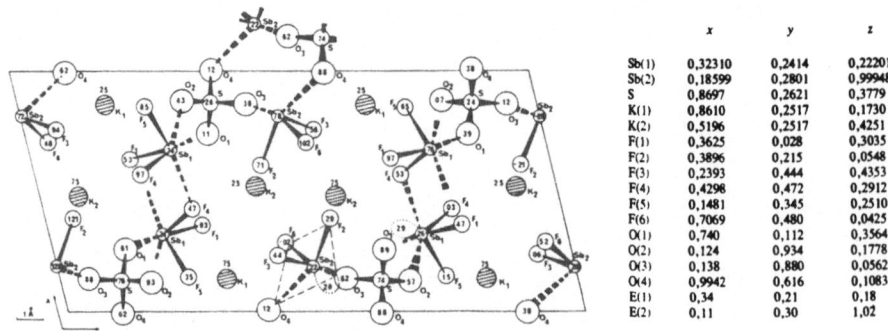

	x	y	z
Sb(1)	0,32310	0,2414	0,22201
Sb(2)	0,18599	0,2801	0,99948
S	0,8697	0,2621	0,3779
K(1)	0,8610	0,2517	0,1730
K(2)	0,5196	0,2517	0,4251
F(1)	0,3625	0,028	0,3035
F(2)	0,3896	0,215	0,0548
F(3)	0,2393	0,444	0,4353
F(4)	0,4298	0,472	0,2912
F(5)	0,1481	0,345	0,2510
F(6)	0,7069	0,480	0,0425
O(1)	0,740	0,112	0,3564
O(2)	0,124	0,934	0,1778
O(3)	0,138	0,880	0,0562
O(4)	0,9942	0,616	0,1083
E(1)	0,34	0,21	0,18
E(2)	0,11	0,30	1,02

Fig. 1.   Structure of $K_2SO_4(SbF_3)_2$.

SODIUM HAFNIUM SULPHATE HYDRATES

$Na_4Hf(SO_4)_4 \cdot 2H_2O$ $Hf(SO_4)_2 \cdot 2Na_2SO_4 \cdot 2H_2O$

$Na_2Hf(SO_4)_3 \cdot 3H_2O$ $Hf(SO_4)_2 \cdot Na_2SO_4 \cdot 3H_2O$

I.   D.L. ROGAČEV, L.M. DIKAREVA, V.Ja. KUZNECOV and G.G. SADIKOV, 1981. Ž. Strukt.
      Khim., 22, No. 3, 191-194 [J. Struct. Chem., 22, 470-473].
II.   D.L. ROGAČEV, L.M. DIKAREVA, V.P. NIKOLAEV and V.Ja. KUZNECOV, 1981. Ibid.,
      22, No. 3, 194-196 [Ibid., 22, 473-475].

$Hf(SO_4)_2 \cdot 2Na_2SO_4 \cdot 2H_2O$, monoclinic, Cc, a = 8.786, b = 9.907, c = 17.730 Å, β = 92.09°,
Z = 4. Mo radiation, R = 0.071 for 1980 reflexions.

$Hf(SO_4)_2 \cdot Na_2SO_4 \cdot 3H_2O$, orthorhombic, $P2_12_12_1$, a = 7.091, b = 7.791, c = 22.179 Å, Z =
4. Mo radiation, R = 0.037 for 1890 reflexions.

     The structure of the first compound (Fig. 1) contains isolated $Hf(SO_4)_4(H_2O)_2{}^{4-}$
anions, linked by Na^+ ions. Hf has dodecahedral 8-coordination, Hf-O = 2.09-2.33 Å,
and Na ions have 4-, 5-, and 7-coordinations, Na-O = 2.22-3.18 Å; S-O = 1.42-1.61 Å.

     The second compound is isostructural with $Na_2Zr(SO_4)_3 \cdot 3H_2O$ (1). The structure
contains infinite $[Hf(SO_4)_3(H_2O)_2{}^{2-}]_n$ spiral chains along a, linked by Na ions and
the additional water molecule. Hf has 8-coordination, Hf-O = 2.08-2.26 Å, and Na ions
have 6-coordinations, Na-O = 2.29-2.71 Å; S-O = 1.44-1.54 Å.

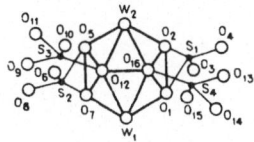

Fig. 1.   Hf coordination in $Hf(SO_4)_2 \cdot 2Na_2SO_4 \cdot 2H_2O$.

1. Structure Reports, 37A, 307.

VANADYL SULPHATE - SULPHURIC ACID (2:1)

$2VOSO_4 \cdot H_2SO_4$

M. TACHEZ and F. THÉOBALD, 1981. Acta Cryst., B37, 1978-1982.

Tetragonal, $P4_2/mnm$, a = 8.971, c = 15.594 Å, D_m = 2.18, Z = 4. Mo radiation, R =
0.077 for 340 reflexions.

     The structure (Fig. 1) contains SO_4 tetrahedra linked by corner-sharing with
VO_6 octahedra to form $VOSO_4$ layers, which are linked by H_2SO_4 molecules. S-O = 1.42-
1.57, S-OH = 1.71, V-O = 1.57, 2.00-2.17(4) Å. There appears to be no hydrogen
bonding.

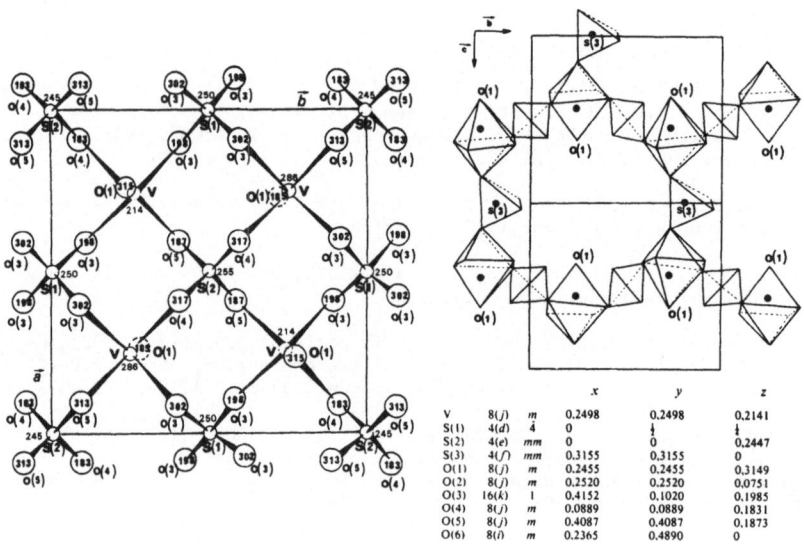

			x	y	z
V	8(j)	m	0.2498	0.2498	0.2141
S(1)	4(d)	4	0	¼	¼
S(2)	4(e)	mm	0	0	0.2447
S(3)	4(f)	mm	0.3155	0.3155	0
O(1)	8(j)	m	0.2455	0.2455	0.3149
O(2)	8(j)	m	0.2520	0.2520	0.0751
O(3)	16(k)	l	0.4152	0.1020	0.1985
O(4)	8(j)	m	0.0889	0.0889	0.1831
O(5)	8(j)	m	0.4087	0.4087	0.1873
O(6)	8(l)	m	0.2365	0.4890	0

Fig. 1.   Structure of $2VOSO_4 \cdot H_2SO_4$.

CHROMIUM HYDROXIDE SULPHATE MONOHYDRATE
$Cr(OH)SO_4 \cdot H_2O$

A. RIOU and A. BONNIN, 1981. Acta Cryst., B37, 1031-1035.

Monoclinic, Cc, a = 12.477, b = 7.259, c = 14.382 Å, β = 93.99°, D_m = 2.80, Z = 12.
Mo radiation, R = 0.085 for 2132 reflexions.

The structure (Fig. 1) contains $CrO_3(OH)_2(OH_2)$ octahedra, linked into infinite chains by single hydroxo bridges, and further linked into a three-dimensional framework by sulphate tetrahedra. Cr-O = 1.91-2.03, S-O = 1.45-1.50(1) Å.

Fig. 1.   Structure of $Cr(OH)SO_4 \cdot H_2O$.

FIBROFERRITE
Fe(OH)SO$_4$.5H$_2$O

F. SCORDARI, 1981. Tschermaks Min. Petr. Mitt., <u>28</u>, 17-29.

Rhombohedral, R$\overline{3}$, a = 24.176 [in abstract, 20.177 in text is incorrect], c = 7.656 Å,
D$_m$ = 1.95, Z = 18. Cu radiation, R = 0.076 for 818 reflexions.

The structure (Fig. 1) contains hydroxo-bridged chains of corner-sharing
Fe(OH)$_2$(H$_2$O)$_2$O$_2$ octahedra and SO$_4$ tetrahedra; the chains are related to those in
butlerite and parabutlerite (<u>1</u>). The chains are linked by hydrogen bonds, with
channels occupied by additional water molecules. Fe-O = 1.94-2.07, S-O = 1.45-1.50,
O-H...O = 2.61-3.18 Å.

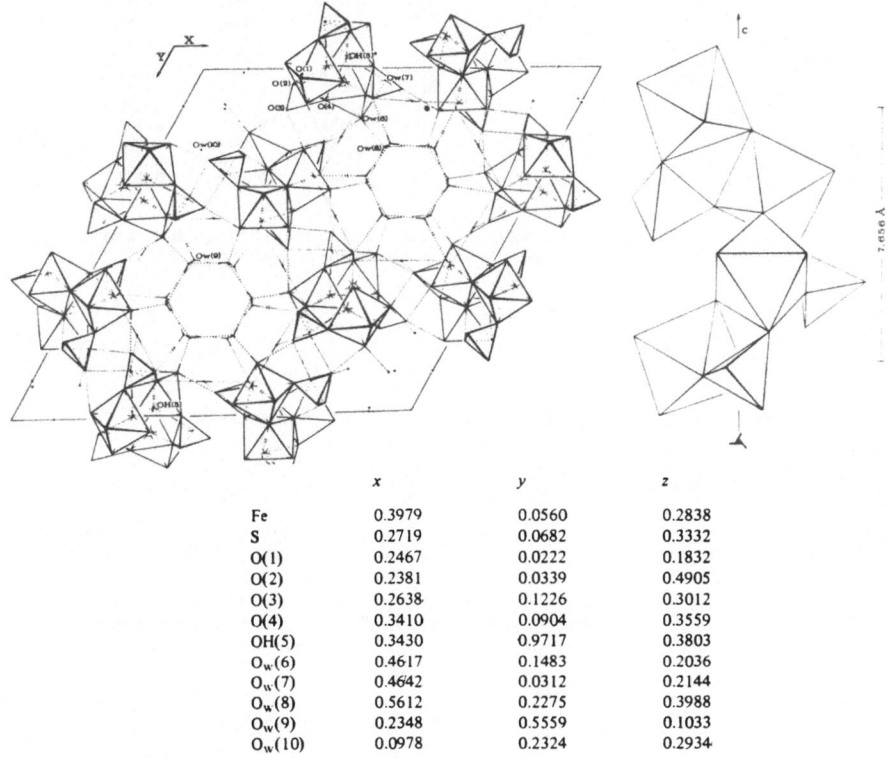

	x	y	z
Fe	0.3979	0.0560	0.2838
S	0.2719	0.0682	0.3332
O(1)	0.2467	0.0222	0.1832
O(2)	0.2381	0.0339	0.4905
O(3)	0.2638	0.1226	0.3012
O(4)	0.3410	0.0904	0.3559
OH(5)	0.3430	0.9717	0.3803
O$_w$(6)	0.4617	0.1483	0.2036
O$_w$(7)	0.4642	0.0312	0.2144
O$_w$(8)	0.5612	0.2275	0.3988
O$_w$(9)	0.2348	0.5559	0.1033
O$_w$(10)	0.0978	0.2324	0.2934

Fig. 1.   Structure of fibroferrite.

<u>1</u>. Structure Reports, <u>35</u>A, 375; <u>37</u>A, 309.

POTASSIUM SODIUM IRON(III) SULPHATE HYDRATE
K$_3$(K,H$_2$O)$_6$Na$_2$(Na,H$_3$O$^+$,H$_2$O)$_6$Fe$_6$O$_2$(SO$_4$)$_{12}$.6H$_2$O

F. SCORDARI, 1981. Acta Cryst., B<u>37</u>, 312-317.

Rhombohedral, $R\bar{3}$, a = 18.225 Å, α = 30.515° (hexagonal cell, a = 9.588, c = 51.959 Å), Z = 1. Mo radiation, R = 0.079 for 1607 reflexions.

The material is obtained by a solid-state transformation of a hydrated iron sulphate described previously (1). This transformation is a topotactic reaction involving a translation of adjacent sandwich sheets, $[Na_2K_2Fe_6O_2(SO_4)_{12}(H_2O_6)]^{6-}$ (Fig. 1), and a tripling of the hexagonal $\underline{c}$ axis.

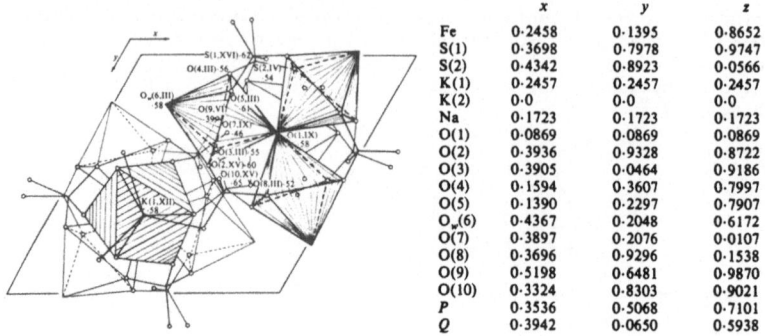

	x	y	z
Fe	0·2458	0·1395	0·8652
S(1)	0·3698	0·7978	0·9747
S(2)	0·4342	0·8923	0·0566
K(1)	0·2457	0·2457	0·2457
K(2)	0·0	0·0	0·0
Na	0·1723	0·1723	0·1723
O(1)	0·0869	0·0869	0·0869
O(2)	0·3936	0·9328	0·8722
O(3)	0·3905	0·0464	0·9186
O(4)	0·1594	0·3607	0·7997
O(5)	0·1390	0·2297	0·7907
O_w(6)	0·4367	0·2048	0·6172
O(7)	0·3897	0·2076	0·0107
O(8)	0·3696	0·9296	0·1538
O(9)	0·5198	0·6481	0·9870
O(10)	0·3324	0·8303	0·9021
P	0·3536	0·5068	0·7101
Q	0·3942	0·0650	0·5938

Fig. 1.   The sandwich sheet in the hydrated iron sulphate (hexagonal axes), and atomic positional parameters (rhombohedral axes).

1. Structure Reports, 46A, 358.

μ-HYPEROXO-BIS(PENTAAMMINECOBALT) TRIS(HYDROGENSULPHATE) MONOSULPHATE
$[Co_2(NH_3)_{10}(O_2)](HSO_4)_3(SO_4)$

W.P. SCHAEFER, S.E. EALICK and R.E. MARSH, 1981. Acta Cryst., B37, 34-38.

Orthorhombic, $P2_12_12_1$, a = 16.360, 16.198, b = 13.946, 13.842, c = 9.978, 9.963 Å, at 293, 17K, D_m = 2.060 (293K), Z = 4. Mo radiation, R = 0.066, 0.066 for 2377, 2257 reflexions.

The structure determined previously (1) is confirmed, and the hydrogen bonding pattern is now established. The sulphate group has a twofold disorder (72:28) at room temperature.

1. Structure Reports, 31A, 206.

CADMIUM SULPHATE HYDRATE
$3CdSO_4.8H_2O$

R. CAMINITI and G. JOHANSSON, 1981. Acta Chem. Scand., A35, 451-455.

Monoclinic, C2/c, a = 14.818, b = 11.903, c = 9.468 Å, β = 97.36°, D_m = 3.09, Z = 4. Mo radiation, R = 0.022 for 1340 reflexions.

The structure is essentially as previously described (1), with SO_4 tetrahedra and $CdO_4(OH_2)_2$ octahedra; one water molecule is not bonded to $C\bar{d}$, but is involved in four hydrogen bonds. S-O = 1.464-1.484, Cd-O = 2.268-2.319(3) Å.

1. Strukturbericht, 4, 52, 182.

SODIUM LANTHANUM SULPHATE
$NaLa(SO_4)_2$

S.M. ČIŽOV, A.N. POKROVSKIJ and L.M. KOVBA, 1981.   Kristallografija, 26, 834-836
[Soviet Physics - Crystallography, 26, 471-472].

Triclinic, $P\bar{1}$, a = 7.081, b = 6.765, c = 6.465 Å, α = 102.25, β = 91.20, γ = 76.71°,
Z = 2.   R = 0.029 for 1567 reflexions.

The structure contains chains of edge-sharing LaO_{10} polyhedra, linked by edge-
and corner-sharing with SO_4 tetrahedra; Na ions are located in cavities in this framework and have 8-coordination.   La-O = 2.53-2.74, S-O = 1.46-1.50, Na-O = 2.43-2.73 Å.

HYDROGEN CERIUM(III) SULPHATE HYDRATE
$(H_3O)Ce(SO_4)_2 \cdot H_2O$

B.M. GATEHOUSE and A. PRING, 1981.   J. Solid State Chem., 38, 116-120.

Monoclinic, $P2_1/n$, a = 9.359, b = 9.926, c = 8.444 Å, β = 96.53°, D_m = 3.12, Z = 4.
Mo radiation, R = 0.047 for 1787 reflexions.

The structure (Fig. 1) contains sulphate tetrahedra linked by 9-coordinate Ce^{3+}
ions (8 O and 1 H_2O), to form a three-dimensional array with sites containing H_3O^+
ions.   S-O = 1.46-1.50, Ce-O = 2.45-2.63, $H_3O^+\ldots O$ = 2.85-2.94 (5 distances), 3.01,
3.20, 3.28 Å.

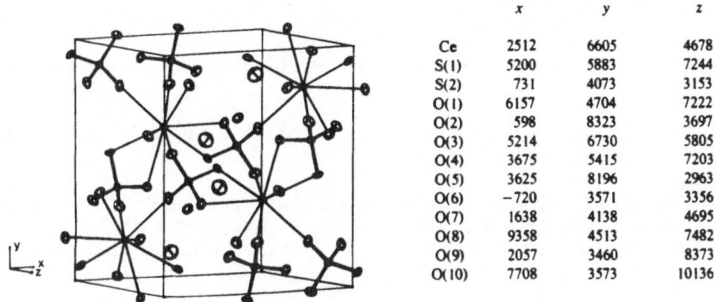

	x	y	z
Ce	2512	6605	4678
S(1)	5200	5883	7244
S(2)	731	4073	3153
O(1)	6157	4704	7222
O(2)	598	8323	3697
O(3)	5214	6730	5805
O(4)	3675	5415	7203
O(5)	3625	8196	2963
O(6)	−720	3571	3356
O(7)	1638	4138	4695
O(8)	9358	4513	7482
O(9)	2057	3460	8373
O(10)	7708	3573	10136

Fig. 1.   Structure of hydrogen cerium(III) sulphate hydrate, and
            atomic positional parameters (x 10^4).

PRASEODYMIUM SULPHATE OCTAHYDRATE
$Pr_2(SO_4)_3 \cdot 8H_2O$

I.S. AHMED FARAG, M.A. EL-KORDY and N.A. AHMED, 1981.   Z. Kristallogr., 155, 165-171.

Monoclinic, C2/c, a = 13.694, b = 6.803, c = 18.061 Å, β = 102°, D_m = 2.83, Z = 4.
Mo radiation, film data, R = 0.12 for 1982 reflexions.   [Previous independent study in
1.]

The structure (Fig. 1) contains 8-coordinate Pr ions, $PrO_4(H_2O)_4$, linked by
bridging sulphate tetrahedra.   S-O = 1.40-1.50, Pr-O = 2.36-2.54(3) Å.

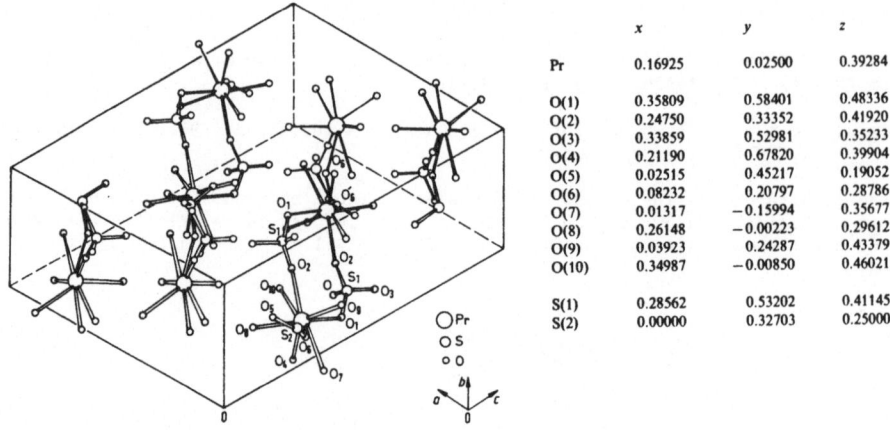

	x	y	z
Pr	0.16925	0.02500	0.39284
O(1)	0.35809	0.58401	0.48336
O(2)	0.24750	0.33352	0.41920
O(3)	0.33859	0.52981	0.35233
O(4)	0.21190	0.67820	0.39904
O(5)	0.02515	0.45217	0.19052
O(6)	0.08232	0.20797	0.28786
O(7)	0.01317	−0.15994	0.35677
O(8)	0.26148	−0.00223	0.29612
O(9)	0.03923	0.24287	0.43379
O(10)	0.34987	−0.00850	0.46021
S(1)	0.28562	0.53202	0.41145
S(2)	0.00000	0.32703	0.25000

Fig. 1.   Structure of praseodymium sulphate octahydrate.

1. Structure Reports, <u>42</u>A, 377.

RUBIDIUM HOLMIUM SULPHATE MONOHYDRATE
$RbHo(SO_4)_2.H_2O$

M.V. PROKOF'EV, 1981. Kristallografija, <u>26</u>, 598-600 [Soviet Physics - Crystallography, <u>26</u>, 337-339].

Monoclinic, $P2_1/b$, a = 10.361, b = 17.775, c = 8.320 Å, γ = 149.90°, Z = 4.   Mo radiation, R = 0.037 for 2031 reflexions.

The structure (Fig. 1) contains a three-dimensional framework of corner-sharing SO_4 tetrahedra and HoO_7 polyhedra; S-O = 1.46-1.49, Ho-O = 2.29-2.40 Å (eighth O at 2.57 Å). Rb has 9-coordination, Rb-O = 2.80-3.32 Å (two further O at 3.47, 3.59 Å).

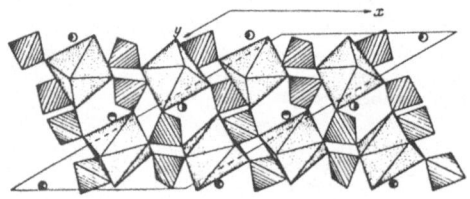

Fig. 1.   Structure of $RbHo(SO_4)_2.H_2O$.

LITHIUM LUTETIUM SULPHATE
$LiLu(SO_4)_2$

ERBIUM SULPHATE
$Er_2(SO_4)_3$

S.P. SIROTINKIN, A.N. POKROVSKIJ and L.M. KOVBA, 1981. Kristallografija, $\underline{26}$, 385-389 [Soviet Physics - Crystallography, $\underline{26}$, 219-221].

$LiLu(SO_4)_2$, orthorhombic, Pbcn, a = 12.575, b = 9.051, c = 9.138 Å, D_m = 3.52, Z = 6. Mo radiation, R = 0.076 for 642 reflexions.

$Er_2(SO_4)_3$, orthorhombic, Pbcn, a = 12.442, b = 9.001, c = 9.837 Å, D_m = 3.64, Z = 4. Mo radiation, R = 0.086 for 426 reflexions.

The structures (Fig. 1) are similar, and contain SO_4 tetrahedra, linked by M (= 3/4 Lu + 1/4 Li) or Er ions which have distorted octahedral coordinations; the double salt contains additional Li in tetrahedral cavities. S-O = 1.34-1.58, M-O = 2.17-2.24, Li-O = 1.77-2.17, Er-O = 2.13-2.50 Å.

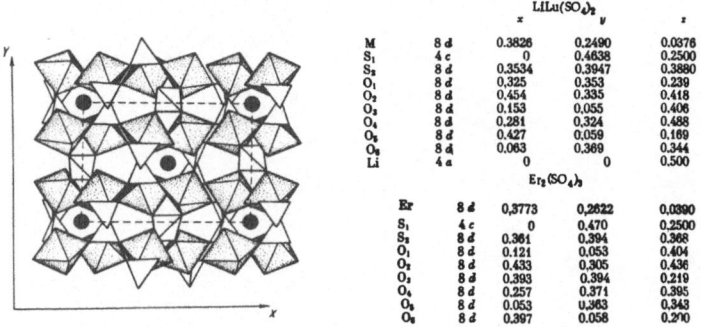

$LiLu(SO_4)_2$

		x	y	z
M	8 d	0.3826	0.2490	0.0376
S_1	4 c	0	0.4638	0.2500
S_2	8 d	0.3534	0.3947	0.3880
O_1	8 d	0.325	0.353	0.239
O_2	8 d	0.454	0.335	0.418
O_3	8 d	0.153	0.055	0.406
O_4	8 d	0.281	0.324	0.488
O_5	8 d	0.427	0.059	0.169
O_6	8 d	0.063	0.369	0.344
Li	4 a	0	0	0.500

$Er_2(SO_4)_3$

		x	y	z
Er	8 d	0.3773	0.2622	0.0390
S_1	4 c	0	0.470	0.2500
S_2	8 d	0.361	0.394	0.368
O_1	8 d	0.121	0.053	0.404
O_2	8 d	0.433	0.305	0.436
O_3	8 d	0.393	0.394	0.219
O_4	8 d	0.257	0.371	0.395
O_5	8 d	0.053	0.363	0.343
O_6	8 d	0.397	0.058	0.270

Fig. 1.   Structure of $LiLu(SO_4)_2$ and atomic positional parameters for $LiLu(SO_4)_2$ and $Er_2(SO_4)_3$.

MAGNESIUM URANYL SULPHATE UNDECAHYDRATE
$MgUO_2(SO_4)_2 \cdot 11H_2O$

V.N. SEREŽKIN, M.A. SOLDATKINA and V.A. EFREMOV, 1981. Ž. Strukt. Khim., $\underline{22}$, No. 3, 174-177 [J. Struct. Chem., $\underline{22}$, 454-457].

Monoclinic, C2/c, a = 11.334, b = 7.715, c = 21.709 Å, β = 102.22°, D_m = 2.37, Z = 4. Mo radiation, R = 0.058 for 1832 reflexions.

The structure (Fig. 1) contains $[UO_2(H_2O)(SO_4)_2^{2-}]_n$ layers, in which U has pentagonal bipyramidal coordination. The layers are linked by hydrogen bonding via $Mg(H_2O)_6$ octahedra and four additional non-coordinated water molecules. U-O = 1.77 (uranyl), 2.34-2.43, S-O = 1.45-1.48, Mg-O = 2.06-2.09(1) Å.

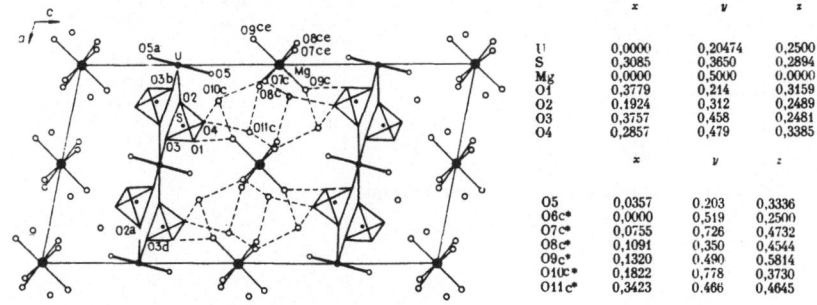

	x	y	z
U	0.0000	0.20474	0.2500
S	0.3085	0.3650	0.2894
Mg	0.0000	0.5000	0.0000
O1	0.3779	0.214	0.3159
O2	0.1924	0.312	0.2489
O3	0.3757	0.458	0.2481
O4	0.2857	0.479	0.3385

	x	y	z
O5	0.0357	0.203	0.3336
O6c*	0.0000	0.519	0.2500
O7c*	0.0755	0.726	0.4732
O8c*	0.1091	0.350	0.4544
O9c*	0.1320	0.490	0.5814
O10c*	0.1822	0.778	0.3730
O11c*	0.3423	0.466	0.4645

Fig. 1.   Structure of $MgUO_2(SO_4)_2 \cdot 11H_2O$.

TELLURIUM(IV) PYROSULPHATE
$Te(S_2O_7)_2$

F.W. EINSTEIN and A.C. WILLIS, 1981. Acta Cryst., B37, 218-220.

Monoclinic, Cc, a = 10.604, b = 8.524, c = 12.052 Å, β = 102.20°, Z = 4. Mo radiation, R = 0.038 for 909 reflexions.

    Te has distorted trigonal-bipyramidal coordination (including an equatorial lone-pair) to two bidentate pyrosulphate groups, one of which is disordered (Fig. 1). Te-O = 1.96 (equatorial), 2.07 Å (axial), O(eq)-Te-O(eq) = 95, O(ax)-Te-O(ax) = 157°. There are four longer Te...O contacts, 2.73-3.02 Å.

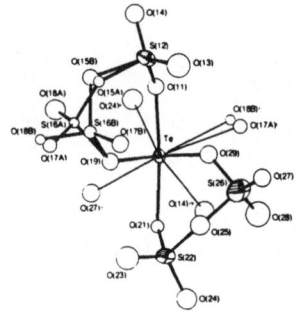

Fig. 1.   Structure of tellurium(IV) pyrosulphate.

AMMONIUM HYDROGEN SELENATE
NH_4HSeO_4

K.S. ALEKSANDROV, A.I. KRUGLIK, S.V. MISJUL' and M.A. SIMONOV, 1980.   Kristallografija, 25, 1142-1147   [Soviet Physics - Crystallography, 25, 654-656].

Monoclinic, B2, a = 19.75, b = 4.61, c = 7.55 Å, γ = 102.6°, Z = 6. Mo radiation, R = 0.037 for 1136 reflexions.

    The structure (Fig. 1) contains SeO_4 tetrahedra linked into chains along b by hydrogen bonds with disordered hydrogen positions; the chains are joined via ammonium ions. Se-O = 1.61-1.71, N...O = 2.86-3.41 Å.

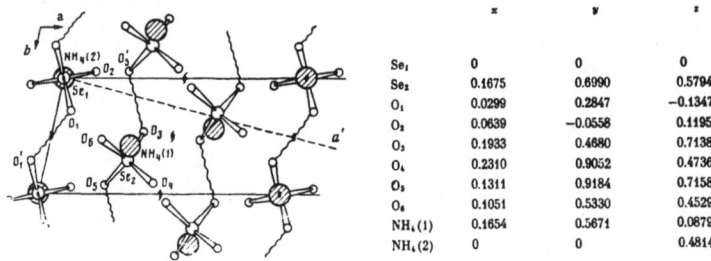

	x	y	z
Se_1	0	0	0
Se_2	0.1675	0.6990	0.5794
O_1	0.0299	0.2847	−0.1347
O_2	0.0639	−0.0558	0.1195
O_3	0.1933	0.4680	0.7138
O_4	0.2310	0.9052	0.4736
O_5	0.1311	0.9184	0.7158
O_6	0.1051	0.5330	0.4529
$NH_4(1)$	0.1654	0.5671	0.0879
$NH_4(2)$	0	0	0.4814

Fig. 1.   Structure of ammonium hydrogen selenate.

SODIUM HYDROGENSELENITE
NaHSeO$_3$

S. CHOMNILPAN, R. LIMINGA, E.J. SONNEVELD and J.W. VISSER, 1981. Acta Cryst., B37, 2217-2220.

Monoclinic, C2/c, a = 21.9799, b = 5.7910, c = 10.2796 Å, β = 105.107°, Z = 16. Mo radiation, R = 0.026 for 1924 reflexions.

The structure (Fig. 1) contains pyramidal SeO$_2$(OH)$^-$ ions, hydrogen bonded to form centrosymmetric dimers; Se-O = 1.652-1.764 Å (probably some H-atom disorder), O-Se-O = 100.3-103.0°. The dimers are linked by Na ions with 5- and 6-coordinations, Na-O = 2.32-2.63 Å. A previous description (1) is incorrect.

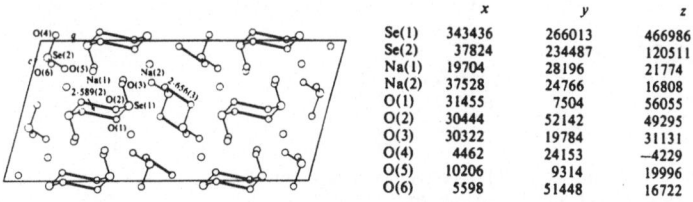

	x	y	z
Se(1)	343436	266013	466986
Se(2)	37824	234487	120511
Na(1)	19704	28196	21774
Na(2)	37528	24766	16808
O(1)	31455	7504	56055
O(2)	30444	52142	49295
O(3)	30322	19784	31131
O(4)	4462	24153	−4229
O(5)	10206	9314	19996
O(6)	5598	51448	16722

Fig. 1.   Structure of NaHSeO$_3$ and atomic positional parameters
           (x 10^6 for Se, x 10^5 for others).

1. Structure Reports, 28, 221.

AMMONIUM COPPER SELENATE HEXAHYDRATE
(NH$_4$)$_2$Cu(SeO$_4$)$_2$.6H$_2$O

A. MONGE and E. GUTIERREZ-PUEBLA, 1981. Acta Cryst., B37, 427-429.

Monoclinic, P2$_1$/n, a = 6.424, b = 12.547, c = 9.351 Å, β = 105.62°, Z = 2. Cu radiation, R = 0.050 for 1269 reflexions.

[Tutton's Salt structure (1).] Cu has distorted octahedral coordination, Cu-OH$_2$ = 1.99, 2.03, 2.24 Å (each x 2). Se-O = 1.61-1.64 Å.

1. Strukturbericht, 2, 93; Structure Reports, 41A, 354.

GOLD(III) SELENITE DISELENITE
Au$_2$(SeO$_3$)$_2$(Se$_2$O$_5$)

P.G. JONES, E. SCHWARZMANN, G.M. SHELDRICK and H. TIMPE, 1981. Z. Naturforsch., 36B, 1050-1051.

Monoclinic, C2/c, a = 20.344, b = 4.130, c = 13.254 Å, β = 115.88°, Z = 4. Mo radiation, R = 0.035 for 1197 reflexions.

The structure (Fig. 1) contains two Au atoms on centres of symmetry, both with square-planar coordination and linked in three-dimensions by selenite and diselenite ions; Se atoms have trigonal pyramidal coordinations. Au-O = 1.97-2.01(1), Se-O = 1.82 (diselenite bridge), 1.63-1.74 Å, O-Se-O = 91-105, Se-O-Se = 115°.

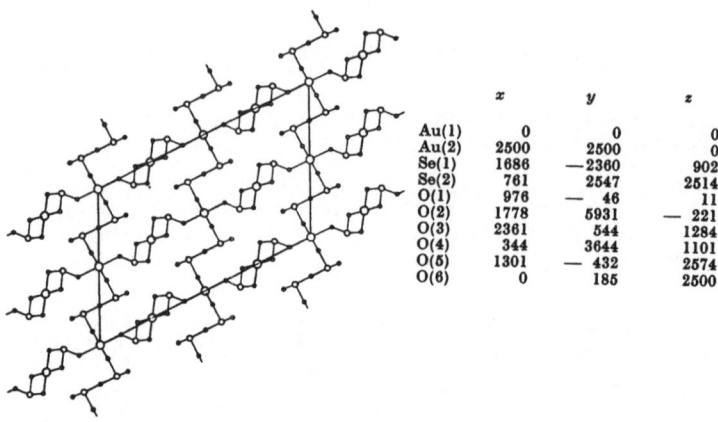

	x	y	z
Au(1)	0	0	0
Au(2)	2500	2500	0
Se(1)	1686	—2360	902
Se(2)	761	2547	2514
O(1)	976	— 46	11
O(2)	1778	5931	— 221
O(3)	2361	544	1284
O(4)	344	3644	1101
O(5)	1301	— 432	2574
O(6)	0	185	2500

Fig. 1.   Structure of gold(III) selenite diselenite, and
atomic positional parameters (x 10^4).

CADMIUM SELENATE MONOHYDRATE
$CdSeO_4.H_2O$

C. STÅLHANDSKE, 1981. Acta Cryst., B$\underline{37}$, 2055-2057.

Monoclinic, $P2_1/c$, a = 7.679, b = 7.723, c = 8.207 Å, β = 120.96°, Z = 4. Mo
radiation, R = 0.023 for 1281 reflexions. Isostructural with $CdSO_4.H_2O$ ($\underline{1}$) and
$HgSeO_4.H_2O$ ($\underline{2}$).

   The structure (Fig. 1) contains SeO_4 tetrahedra, linked by $CdO_4(H_2O)_2$ octahedra.
Se-O = 1.622-1.650, Cd-O = 2.246-2.421, O-H...O = 2.713, 2.791(3) Å.

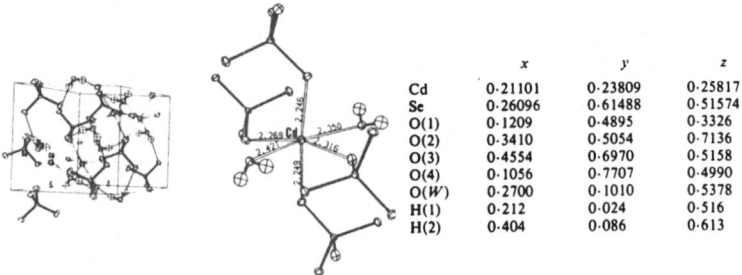

	x	y	z
Cd	0·21101	0·23809	0·25817
Se	0·26096	0·61488	0·51574
O(1)	0·1209	0·4895	0·3326
O(2)	0·3410	0·5054	0·7136
O(3)	0·4554	0·6970	0·5158
O(4)	0·1056	0·7707	0·4990
O(W)	0·2700	0·1010	0·5378
H(1)	0·212	0·024	0·516
H(2)	0·404	0·086	0·613

Fig. 1.   Structure of cadmium selenate monohydrate.

$\underline{1}$. Structure Reports, $\underline{35}$A, 372.
$\underline{2}$. Ibid., $\underline{44}$A, 286.

URANYL SELENATE TETRAHYDRATE
$UO_2SeO_4.4H_2O$

V.N. SEREŽKIN, M.A. SOLDATKINA and V.A. EFREMOV, 1981. Ž. Strukt. Khim., 22, No. 3, 171-174 [J. Struct. Chem., 22, 451-454].

Monoclinic, C2/c, a = 14.653, b = 10.799, c = 12.664 Å, β = 119.95°, D_m = 3.67, Z = 8. Mo radiation. R = 0.046 for 1709 reflexions.

The structure (Fig. 1) contains infinite $[UO_2(H_2O)_2SeO_4]$ ribbons, in which U has pentagonal bipyramidal coordination. The ribbons are linked by hydrogen bonding via the additional water molecules. U-O = 1.75 (uranyl), 2.38-2.46, Se-O = 1.61-1.64(1) Å.

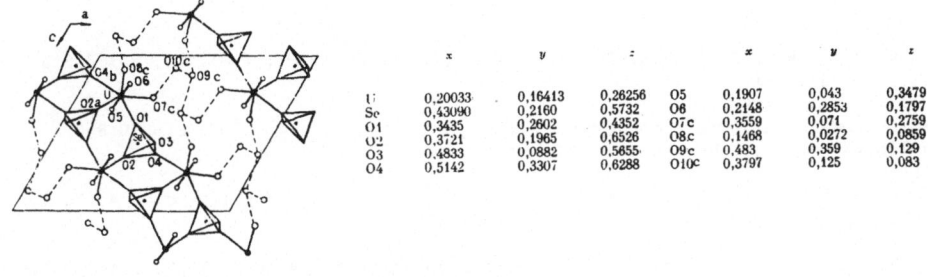

	x	y	z		x	y	z
U	0,20033	0,16413	0,26256	O5	0,1907	0,043	0,3479
Se	0,43090	0,2160	0,5732	O6	0,2148	0,2853	0,1797
O1	0,3435	0,2602	0,4352	O7c	0,3559	0,071	0,2759
O2	0,3721	0,1965	0,6526	O8c	0,1468	0,0272	0,0859
O3	0,4833	0,0882	0,5655	O9c	0,483	0,359	0,129
O4	0,5142	0,3307	0,6288	O10c	0,3797	0,125	0,083

Fig. 1.   Structure of uranyl selenate tetrahydrate.

AMMONIUM TETRATELLURATE
$(NH_4)_2Te_4O_9$

L. BENMILOUD, M. MAURIN, J. MORET and E. PHILIPPOT, 1981. Rev. Chim. Minér., 18, 190-198.

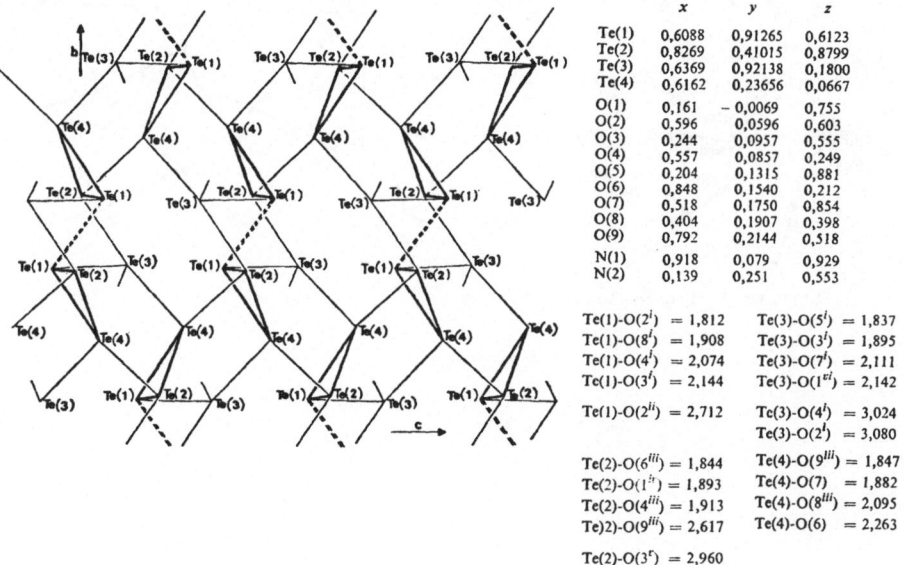

	x	y	z
Te(1)	0,6088	0,91265	0,6123
Te(2)	0,8269	0,41015	0,8799
Te(3)	0,6369	0,92138	0,1800
Te(4)	0,6162	0,23656	0,0667
O(1)	0,161	– 0,0069	0,755
O(2)	0,596	0,0596	0,603
O(3)	0,244	0,0957	0,555
O(4)	0,557	0,0857	0,249
O(5)	0,204	0,1315	0,881
O(6)	0,848	0,1540	0,212
O(7)	0,518	0,1750	0,854
O(8)	0,404	0,1907	0,398
O(9)	0,792	0,2144	0,518
N(1)	0,918	0,079	0,929
N(2)	0,139	0,251	0,553

Te(1)-O(2^i) = 1,812   Te(3)-O(5^i) = 1,837
Te(1)-O(8^i) = 1,908   Te(3)-O(3^i) = 1,895
Te(1)-O(4^i) = 2,074   Te(3)-O(7^i) = 2,111
Te(1)-O(3^i) = 2,144   Te(3)-O(1^{vi}) = 2,142

Te(1)-O(2^{ii}) = 2,712   Te(3)-O(4^i) = 3,024
                              Te(3)-O(2^i) = 3,080

Te(2)-O(6^{iii}) = 1,844   Te(4)-O(9^{iii}) = 1,847
Te(2)-O(1^{iv}) = 1,893   Te(4)-O(7) = 1,882
Te(2)-O(4^{iii}) = 1,913   Te(4)-O(8^{iii}) = 2,095
Te)2)-O(9^{iii}) = 2,617   Te(4)-O(6) = 2,263

Te(2)-O(3^r) = 2,960

Fig. 1.   Structure of $(NH_4)_2Te_4O_9$.

Monoclinic, $P2_1/c$, a = 7.980, b = 18.450, c = 7.926 Å, β = 117.30°, Z = 4. Mo
radiation, R = 0.055 for 1378 reflexions.

The structure contains infinite $[Te_4O_9{}^{2-}]_n$ sheets (Fig. 1), linked by ammonium
cations. Te atoms have TeO_4E coordination (Te(3) and Te(4), trigonal bipyramid with
lone pair, E, equatorial), 3 + 1 + E (Te(2)), and 4 + 1 + E (Te(1)).

LITHIUM DITELLURATE(IV)
$Li_2Te_2O_5$

D. CACHAU-HERREILLAT, A. NORBERT, M. MAURIN and E. PHILIPPOT, 1981. J. Solid State
Chem., 37, 352-361.

α-Form (low-temperature), monoclinic, $P2_1/n$, a = 10.355, b = 4.702, c = 10.860 Å,
β = 110.13°, Z = 4. Mo radiation, R = 0.026 for 832 reflections.

β-Form (high-temperature), orthorhombic, Pnaa, a = 5.194, b = 8.170, c = 24.165 Å,
Z = 8. Mo radiation, R = 0.039 for 569 reflections.

Both structures (Figs. 1 and 2) contain anionic sheets, with Te_6O_6 rings in the
α-form and $Te_{10}O_{12}$ rings in the β-form; Te atoms have 3 + 1 + lone-pair coordinations,
Te-O = 1.84-1.99, 2.25-2.42 Å. The sheets are linked by tetrahedrally coordinated Li
ions, Li-O = 1.93-2.22 Å.

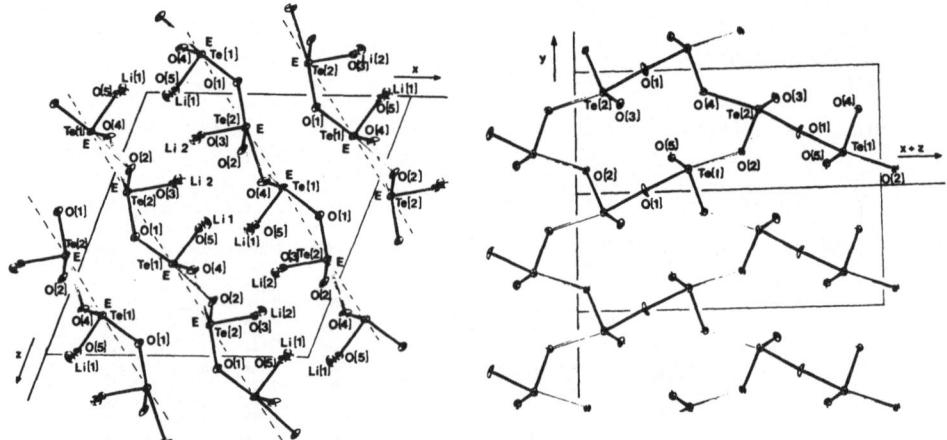

Fig. 1.   Structure of $\alpha\text{-}Li_2Te_2O_5$.

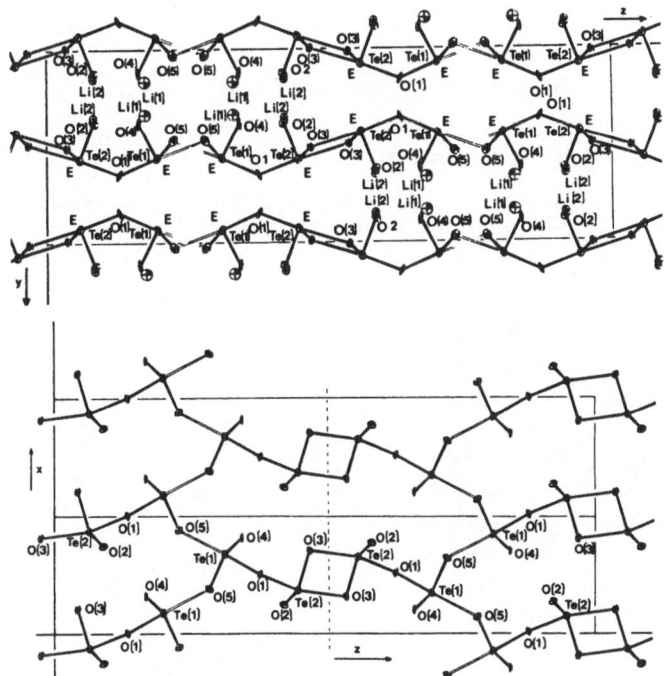

Fig. 1.   Structure of β-Li₂Te₂O₅.

SODIUM DITELLURATE(IV) DIHYDRATE
Na₂Te₂O₅·2H₂O

F. DANIEL, J. MORET, M. MAURIN and E. PHILIPPOT, 1981. Acta Cryst., B**37**, 1278-1281.

Triclinic, P$\bar{1}$, a = 11.632, b = 5.613, c = 6.193 Å, α = 102.51, β = 93.14, γ = 92.23°, Z = 2. Mo radiation, R = 0.015 for 1533 reflexions.

The structure (Fig. 1) contains $(Te_2O_5)_n$ chains along [110], linked by Na ions which have 5- and 6-coordinations. Te atoms have (3+1)- and 4-coordinations, Te-O = 1.84-2.15, 2.55 Å. Na-O = 2.32-2.57, O-H...O = 2.77-2.89 Å.

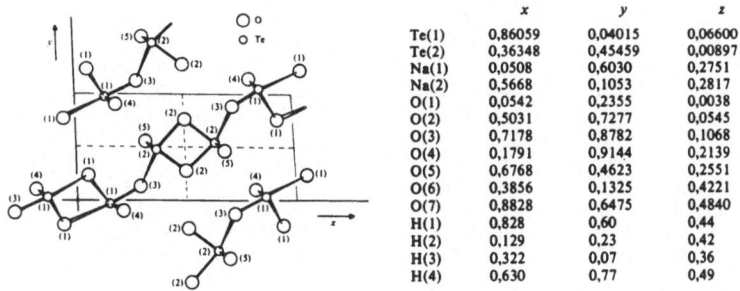

	x	y	z
Te(1)	0,86059	0,04015	0,06600
Te(2)	0,36348	0,45459	0,00897
Na(1)	0,0508	0,6030	0,2751
Na(2)	0,5668	0,1053	0,2817
O(1)	0,0542	0,2355	0,0038
O(2)	0,5031	0,7277	0,0545
O(3)	0,7178	0,8782	0,1068
O(4)	0,1791	0,9144	0,2139
O(5)	0,6768	0,4623	0,2551
O(6)	0,3856	0,1325	0,4221
O(7)	0,8828	0,6475	0,4840
H(1)	0,828	0,60	0,44
H(2)	0,129	0,23	0,42
H(3)	0,322	0,07	0,36
H(4)	0,630	0,77	0,49

Fig. 1.   Tellurate chains and atomic positional parameters in Na₂Te₂O₅·2H₂O.

CALCIUM ORTHOTELLURATE
Ca_3TeO_6

D. HOTTENTOT and B.O. LOOPSTRA, 1981. Acta Cryst., B37, 220-222.

Monoclinic, $P2_1/n$, a = 5.5730, b = 5.7964, c = 8.0113 Å, β = 90.24°, Z = 2. Mo radiation, R = 0.041 for 583 reflexions.

   Isostructural with cryolite (1). Te and Ca(1) have nearly regular octahedral coordinations (Fig. 1), Te-O = 1.92(1), Ca-O = 2.27-2.33 Å, and Ca(2) has 8 oxygen neighbours (12 in cryolite), Ca-O = 2.34-2.95 Å.

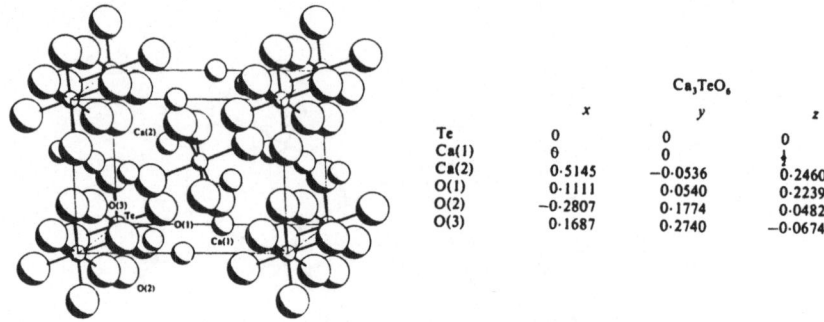

	Ca_3TeO_6		
	x	y	z
Te	0	0	0
Ca(1)	0	0	½
Ca(2)	0·5145	−0·0536	0·2460
O(1)	0·1111	0·0540	0·2239
O(2)	−0·2807	0·1774	0·0482
O(3)	0·1687	0·2740	−0·0674

Fig. 1.   Structure of Ca_3TeO_6.

1. Strukturbericht, 6, 29, 120; Structure Reports, 41A, 154.

BARIUM DITELLURATE(VI)   BARIUM DITUNGSTATE(VI)
$Ba_3Te_2O_9$ $Ba_3W_2O_9$

A.J. JACOBSON, J.C. SCANLON, K.R. POEPPELMEIER, J.M. LONGO and D.E. COX, 1981. Mater. Res. Bull., 16, 359-367.

$Ba_3Te_2O_9$, hexagonal, $P6_3/mmc$, a = 5.8603, c = 14.3037 Å, Z = 2. X-ray and neutron powder data.

$Ba_3W_2O_9$, rhombohedral, $R\bar{3}c$, a = 10.1404, c = 13.9731 Å, Z = 6. Neutron powder data.

Atomic positions

$Ba_3Te_2O_9$			x	y	z
Ba(1)	in	4(f)	1/3	2/3	0.0630
Ba(2)		2(d)	1/3	2/3	3/4
Te		4(e)	0	0	0.1493
O(1)		12(k)	-0.1639	2x	0.0906
O(2)		6(h)	0.1426	2x	1/4
$Ba_3W_2O_9$					
Ba	in	18(e)	0.6697	0	1/4
W		12(c)	0	0	0.1452
O(1)		36(f)	0.1731	0.5044	0.2552
O(2)		18(e)	0.1462	0	1/4

   The $Ba_3Te_2O_9$ structure is related to that of $Cs_3Fe_2F_9$ (1), and $Ba_3W_2O_9$ is isostructural with $Cs_3Tl_2Cl_9$ (2, as previously reported in 3). Both structures contain $M_2O_9^{6-}$ anions which consist of face-sharing octahedra; Te-O = 1.86 (terminal), 2.04

(bridging), W-O = 1.82 (terminal), 2.08 Å (bridging). Ba ions have twelve-coordinations.

1. Structure Reports, 37A, 187.
2. Strukturbericht, 3, 144, 505.
3. Structure Reports, 45A, 259; 46A, 274.

SODIUM TRIMETAPHOSPHATE TELLURATE HEXAHYDRATE
$Te(OH)_6.2Na_3P_3O_9.6H_2O$

POTASSIUM TRIMETAPHOSPHATE TELLURATE DIHYDRATE
$Te(OH)_6.K_3P_3O_9.2H_2O$

AMMONIUM SULPHATE TELLURATE
$Te(OH)_6.(NH_4)_2SO_4$

I.   N. BOUDJADA, M.T. AVERBUCH-POUCHOT and A. DURIF, 1981. Acta Cryst., B37, 645-647.
II. Idem, 1981. Ibid., B37, 647-649.
III. R. ZILBER, A. DURIF and M.T. AVERBUCH-POUCHOT, 1981. Ibid., B37, 650-652.

Sodium salt. Hexagonal, $P6_3/m$, a = 11.67, c = 12.12 Å, Z = 2. Ag radiation, R = 0.04 for 741 reflexions.

Potassium salt. Monoclinic, $P2_1/c$, a = 15.61, b = 7.456, c = 14.84 Å, β = 108.01°, Z = 4. Ag radiation, R = 0.03 for 4271 reflexions.

Ammonium salt. Monoclinic, Cc, a = 13.741, b = 6.631, c = 11.405 Å, β = 106.75°, Z = 4. Ag radiation, R = 0.024 for 1035 reflexions.

   The sodium and potassium salts (Fig. 1) contain cyclic $P_3O_9^{3-}$ anions and TeO_6 octahedra, linked by Na^+ (distorted octahedral coordination) or K^+ ions (7-coordination). The ammonium salt (Fig. 2) contains SO_4^{2-} tetrahedra and TeO_6 octahedra, linked by NH_4^+ ions (7 and 8 oxygen neighbours). Te-O = 1.87-1.94 Å.

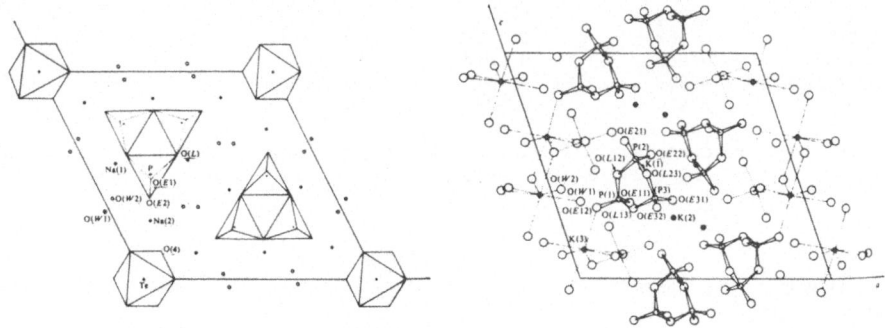

Fig. 1.   Structures of $Te(OH)_6.2Na_3P_3O_9.6H_2O$ (left) and
           $Te(OH)_6.K_3P_3O_9.2H_2O$ (right).

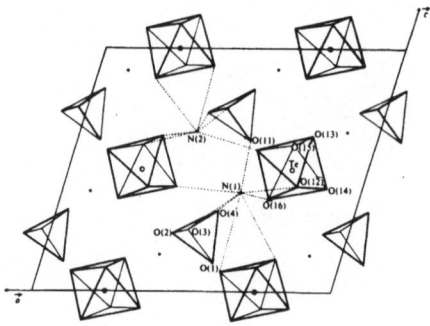

Fig. 2.   Structure of $Te(OH)_6 \cdot (NH_4)_2SO_4$.

THALLIUM(I) PHOSPHATE TELLURATE
$Te(OH)_6 \cdot 2TlH_2PO_4$

M.T. AVERBUCH-POUCHOT and A. DURIF, 1981. Mater. Res. Bull., __16__, 71-76.

Monoclinic, $P2_1/n$, a = 6.285, b = 14.74, c = 7.844 Å, β = 113.38°, Z = 2. Ag
radiation, R = 0.037 for 2014 reflexions.

Atomic positions

	x	y	z
Tl	0.31173	0.08193	0.60412
Te	0	0	0
P	0.7245	0.1831	0.3905
O(1)	0.770	0.0027	0.1017
O(2)	0.382	0.3896	0.3648
O(3)	0.202	0.0693	0.2078
O(4)	0.588	0.1522	0.1930
O(5)	0.889	0.2614	0.3830
O(6)	0.849	0.1077	0.5174
O(7)	0.559	0.2248	0.4775

The structure contains TeO_6 octahedra and PO_4 tetrahedra linked by 7-coordinate
Tl^+ ions. Te-O = 1.91-1.93, P-O = 1.49-1.58, Tl-O = 2.74-3.39 Å.

SILVER PHOSPHATE TELLURATE
$Te(OH)_6 \cdot 2Ag_2HPO_4$

A. DURIF and M.T. AVERBUCH-POUCHOT, 1981. Z. anorg. Chem., __472__, 129-132.

Monoclinic, $P2_1/n$, a = 5.950, b = 20.52, c = 5.829 Å, β = 119.89°, Z = 2. Ag
radiation, R = 0.048 for 2534 reflexions.

The structure (Fig. 1) contains TeO_6 octahedra, PO_4 tetrahedra, and five-coordinate
Ag ions; Te-O = 1.908, 1.911, 1.932, P-O = 1.526, 1.526, 1.535, 1.607, Ag-O = 2.360-
2.701 Å.

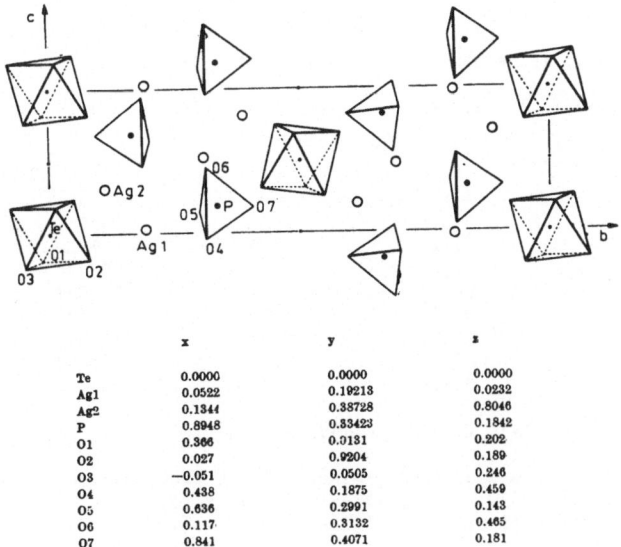

	x	y	z
Te	0.0000	0.0000	0.0000
Ag1	0.0522	0.19213	0.0232
Ag2	0.1344	0.38728	0.8046
P	0.8948	0.83423	0.1842
O1	0.366	0.0131	0.202
O2	0.027	0.9204	0.189
O3	−0.051	0.0505	0.246
O4	0.438	0.1875	0.459
O5	0.636	0.2991	0.143
O6	0.117	0.3132	0.465
O7	0.841	0.4071	0.181

Fig. 1.   Structure of $Te(OH)_6 \cdot 2Ag_2HPO_4$.

NICKEL DITELLURATE(IV)
$NiTe_2O_5$

C. PLATTE and M. TRÖMEL, 1981. Acta Cryst., B37, 1276-1278.

Orthorhombic, Pnma, a = 8.868, b = 12.126, c = 8.452 Å, D_m = 5.71, Z = 8.   Mo
radiation, R = 0.040 for 1230 reflexions.

The structure contains Te_3O_7 chains (Fig. 1) of TeO_3 trigonal pyramids and TeO_5
square pyramids, with additional isolated trigonal-pyramidal TeO_3 groups and octahedrally-coordinated Ni ions. Te-O = 1.86-2.25, Ni-O = 2.04-2.16 Å.

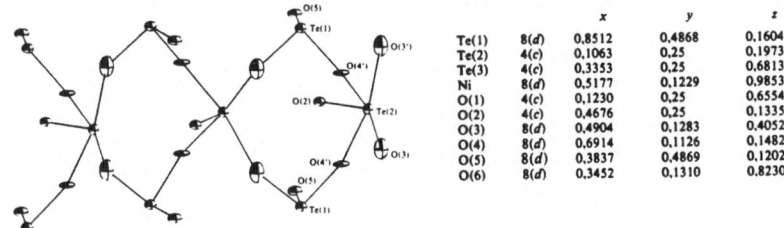

		x	y	z
Te(1)	8(d)	0.8512	0.4868	0.1604
Te(2)	4(c)	0.1063	0.25	0.1973
Te(3)	4(c)	0.3353	0.25	0.6813
Ni	8(d)	0.5177	0.1229	0.9853
O(1)	4(c)	0.1230	0.25	0.6554
O(2)	4(c)	0.4676	0.25	0.1335
O(3)	8(d)	0.4904	0.1283	0.4052
O(4)	8(d)	0.6914	0.1126	0.1482
O(5)	8(d)	0.3837	0.4869	0.1202
O(6)	8(d)	0.3452	0.1310	0.8230

Fig. 1. Te_3O_7 chains and atomic positional parameters in $NiTe_2O_5$.

CLIFFORDITE
UTe_3O_9

F. BRANDSTÄTTER, 1981. Tschermaks Miner. Petr. Mitt., 29, 1-8.

Cubic, Pa3, a = 11.335 Å, Z = 8. Mo radiation, R = 0.053 for 501 reflexions.

Atomic positions

			x	y	z
U(1)*	in	4(a)	0	0	0
U(2)		4(b)	1/2	1/2	1/2
Te		24(d)	0.2626	0.2859	0.0428
O(1)*		8(c)	0.091	0.091	0.091
O(2)		8(c)	0.591	0.591	0.591
O(3)		24(d)	0.178	0.416	0.093
O(4)		24(d)	0.342	0.113	0.409
O(5)		8(c)	0.226	0.226	0.226

* Occupancies = 0.82 for U(1), 0.85 for O(1).

The structure is as previously determined (1). U atoms have hexagonal bipyramidal coordinations, U-O = 1.78, 1.79 (apical), 2.43, 2.47 Å (equatorial). Te has one-sided 5-coordination, Te-O = 1.85-2.22, 2.44 Å, with a sixth O at 2.99 Å.

1. Structure Reports, 37A, 317.

LEAD URANYL TELLURATE
$Pb_2UO_2(TeO_3)_3$

F. BRANDSTÄTTER, 1981. Z. Kristallogr., 155, 193-200.

Monoclinic, $P2_1/n$, a = 11.605, b = 13.389, c = 6.981 Å, β = 91.23°, Z = 4. Mo radiation, R = 0.059 for 2566 reflexions.

The structure (Fig. 1) contains $[UO_2(TeO_3)_3^{4-}]_n$ sheets parallel to (010), consisting of $(UO_2)O_5$ pentagonal bipyramids and TeO_3 trigonal pyramids; U-O = 1.78, 1.83 (uranyl), 2.24-2.43(2), Te-O = 1.83-1.92, with 3 or 4 longer Te...O = 2.43-3.17 Å. The sheets are linked by Pb^{2+} ions with irregular 8- and 7-coordinations, with stereochemically-active lone-pair electrons, Pb-O = 2.37-3.21 Å.

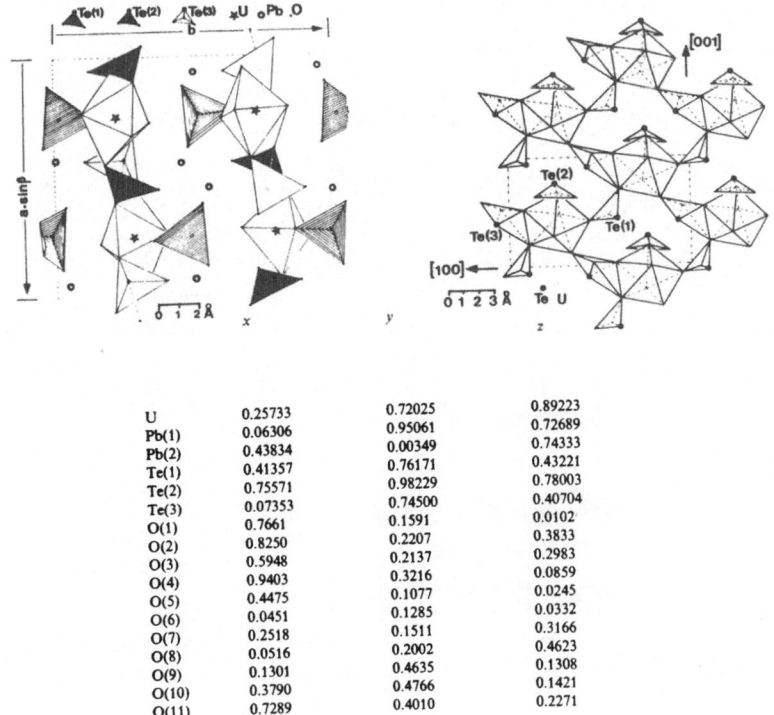

U	0.25733	0.72025	0.89223
Pb(1)	0.06306	0.95061	0.72689
Pb(2)	0.43834	0.00349	0.74333
Te(1)	0.41357	0.76171	0.43221
Te(2)	0.75571	0.98229	0.78003
Te(3)	0.07353	0.74500	0.40704
O(1)	0.7661	0.1591	0.0102
O(2)	0.8250	0.2207	0.3833
O(3)	0.5948	0.2137	0.2983
O(4)	0.9403	0.3216	0.0859
O(5)	0.4475	0.1077	0.0245
O(6)	0.0451	0.1285	0.0332
O(7)	0.2518	0.1511	0.3166
O(8)	0.0516	0.2002	0.4623
O(9)	0.1301	0.4635	0.1308
O(10)	0.3790	0.4766	0.1421
O(11)	0.7289	0.4010	0.2271

Fig. 1.   Structure of $Pb_2UO_2(TeO_3)_3$.

POTASSIUM CHLORATE
$KClO_3$

J. DANIELSEN, A. HAZELL and F.K. LARSEN, 1981. Acta Cryst., B$\underline{37}$, 913-915.

Monoclinic, $P2_1/m$, a = 4.630, 4.657, b = 5.568, 5.591, c = 7.047, 7.099 Å, β = 110.21, 109.65°, at 77, 298K, Z = 2. Mo radiation, R = 0.033, 0.023 for 517, 491 reflexions.

Atomic positions

	77K			298K		
	x	y	z	x	y	z
K	0.3555	1/4	0.7086	0.3537	1/4	0.7091
Cl	0.1210	1/4	0.1745	0.1216	1/4	0.1756
O(1)	0.4009	1/4	0.1165	0.3964	1/4	0.1152
O(2)	0.1456	0.4639	0.3057	0.1465	0.4614	0.3050

Structure as previously described ($\underline{1}$).

$\underline{1}$. Strukturbericht, $\underline{2}$, 66, 408; Structure Reports, $\underline{22}$, 484; $\underline{44A}$, 291.

LANTHANON PERCHLORATE HEXAHYDRATES
$Ln(ClO_4)_3 \cdot 6H_2O$ (Ln = La, Tb, Er)

THALLIUM(III) PERCHLORATE HEXAHYDRATE
$Tl(ClO_4)_3 \cdot 6H_2O$

J. GLASER and G. JOHANSSON, 1981. Acta Chem. Scand., A35, 639-644.

Cubic, Fm3m, a = 12.173, 11.926, 11.900, 11.482 Å, Z = 4. Mo radiation, R = 0.074,
0.063, 0.054, 0.032 for 87, 98, 83, 96 reflexions. Ln or Tl in 4(a): 0,0,0; O (water)
in 24(e): 0,0,z: z = 0.189, 0.186, 0.181, 0.189; Cl(1) in 4(b): 1/2,1/2,1/2; 0.5 O(1)
in 32(f): x,x,x, x = 0.435, 0.431, 0.433, 0.428; Cl(2) in 8(c): 1/4,1/4,1/4; 0.4
O(2A) in 32(f): x,x,x, x = 0.18, 0.18, 0.18, 0.17; 0.4 O(2B) in 48(g): 1/4,1/4,z, z =
0.145, 0.131, 0.146, 0.123.

The structures (Fig. 1) contain $Ln(OH_2)_6^{3+}$ cations and disordered ClO_4^- anions.

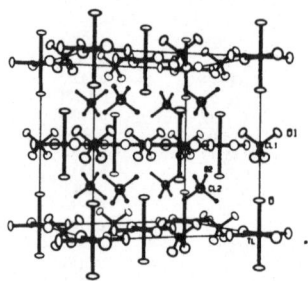

Fig. 1.   Structure of $Tl(ClO_4)_3 \cdot 6H_2O$; only one orientation of each
             perchlorate group is shown.

POTASSIUM MERCURY(II) BROMATE NITRATE
$K_2Hg(BrO_3)_2(NO_3)_2$

MERCURY(II) BROMATE DIHYDRATE
$Hg(BrO_3)_2 \cdot 2H_2O$

K. AURIVILLIUS and C. STÅLHANDSKE, 1981. Acta Chem. Scand., A35, 537-544.

$K_2Hg(BrO_3)_2(NO_3)_2$, orthorhombic, Pnnm, a = 14.170, b = 7.254, c = 5.656 Å, Z = 2.
Mo radiation, R = 0.026 for 887 reflexions.

$Hg(BrO_3)_2 \cdot 2H_2O$, orthorhombic, $P2_12_12_1$, a = 6.2194, b = 9.3139, c = 12.529 Å, Z = 4.
Ag radiation, R = 0.055 for 2083 reflexions.

Hg has 8-coordination in both compounds (Fig. 1), with bridging bromate groups
giving rise to bands in the first compound and to a three-dimensional structure in
the second. K ions are in cavities, and have 10-coordination, K-O = 2.76-3.11 Å.

K$_2$Hg(BrO$_3$)$_2$(NO$_3$)$_2$. Hg in position 2(*a*); Br, K, O(3), O(4) and N in 4(*g*); O(1) and O(2) in 8(*h*).

	x	*y*	*z*
Hg	0	0	0
Br	0.38400	0.73193	0
K	0.2807	0.2438	0
O(1)	0.1248	0.0882	0.2728
O(2)	0.4326	0.2049	0.3156
O(3)	0.2824	0.8385	0
O(4)	0.1285	0.4812	0
N	0.0888	0.6265	0

Hg(BrO$_3$)$_2$·2H$_2$O. All atoms in position 4(*a*).

	x	*y*	*z*
Hg	0.06910	0.07098	0.2172
Br(1)	0.2126	0.2198	0.6118
Br(2)	0.3327	0.3408	0.1154
O(1)	0.3145	0.2425	0.4922
O(2)	0.1519	0.0460	0.6145
O(3)	0.4259	0.2190	0.6945
O(4)	0.2557	0.4932	0.1685
O(5)	0.4009	0.2316	0.2127
O(6)	0.1048	0.2659	0.0793
OW(1)	0.2096	0.5191	0.4022
OW(2)	0.2288	−0.0190	0.3721

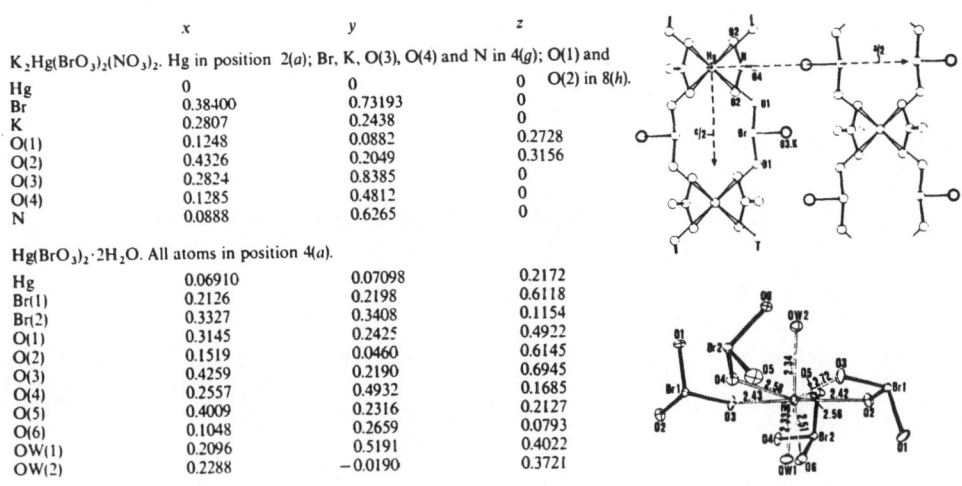

Fig. 1.　Structures of K$_2$Hg(BrO$_3$)$_2$(NO$_3$)$_2$ (top, Hg-O = 2.43, 2.57 Å), and Hg(BrO$_3$)$_2$.2H$_2$O (bottom).

COBALT(II) IODATE
Co(IO$_3$)$_2$

C. SVENSSON, S.C. ABRAHAMS and J.L. BERNSTEIN, 1981. J. Solid State Chem., **36**, 195-204.

Trigonal, P3, a = 10.9597, c = 5.0774 Å, at 298K, Z = 4. Mo radiation, R = 0.050 for 1825 reflexions.

The structure is related to those of α-LiIO$_3$ (1), Fe(IO$_3$)$_3$ (2), and α-Cu(IO$_3$)$_2$ (3) (Fig. 1); it contains IO$_3$ trigonal pyramids linked by three octahedrally-coordinated Co ions, two of which have partial occupancy. I-O = 1.79-1.82 (three longer I...O contacts at 2.66-2.90), Co-O = 2.09-2.23 Å, O-I-O = 97-100°.

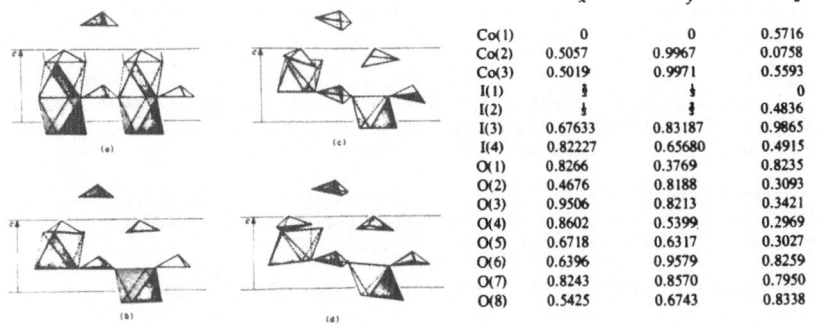

	x	*y*	*z*
Co(1)	0	0	0.5716
Co(2)	0.5057	0.9967	0.0758
Co(3)	0.5019	0.9971	0.5593
I(1)	⅓	⅔	0
I(2)	⅓	⅔	0.4836
I(3)	0.67633	0.83187	0.9865
I(4)	0.82227	0.65680	0.4915
O(1)	0.8266	0.3769	0.8235
O(2)	0.4676	0.8188	0.3093
O(3)	0.9506	0.8213	0.3421
O(4)	0.8602	0.5399	0.2969
O(5)	0.6718	0.6317	0.3027
O(6)	0.6396	0.9579	0.8259
O(7)	0.8243	0.8570	0.7950
O(8)	0.5425	0.6743	0.8338

Fig. 1.　Structures of (a) α-LiIO$_3$, (b) Co(IO$_3$)$_2$, (c) Fe(IO$_3$)$_3$, (d) α-Cu(IO$_3$)$_2$, and atomic positional parameters for Co(IO$_3$)$_2$; Co(2) and Co(3) have 2/3 and 1/3 occupancies, respectively.

1. Strukturbericht, 2, 49, 332; Structure Reports, 31A, 214; 39A, 327; 42A, 425.
2. Structure Reports, 42A, 391.
3. Ibid., 41A, 365.

COPPER(II) PARAPERIODATE DIHYDRATE
$Cu_2HIO_6 \cdot 2H_2O$

V. ADELSKÖLD, P.-E. WERNER, M. SUNDBERG and R. UGGLA, 1981. Acta Chem. Scand., A35, 789-794.

Monoclinic, $P2_1$, a = 10.398, b = 5.114, c = 6.442 Å, β = 114.97°, D_m = 4.0, Z = 2. Powder data.

The structure (Fig. 1) contains two chains along b of corner-sharing square-planar CuO_4 groups; the chains are linked by IO_6 octahedra, and the water molecules complete square-pyramidal 5-coordinations at Cu. I-O = 1.85-2.70, Cu-O = 1.72-2.15, $Cu-OH_2$ = 2.20, 2.66(8) Å.

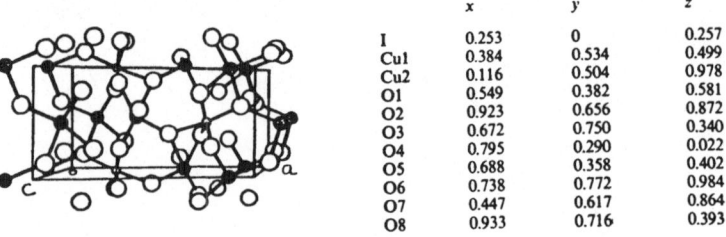

	x	y	z
I	0.253	0	0.257
Cu1	0.384	0.534	0.499
Cu2	0.116	0.504	0.978
O1	0.549	0.382	0.581
O2	0.923	0.656	0.872
O3	0.672	0.750	0.340
O4	0.795	0.290	0.022
O5	0.688	0.358	0.402
O6	0.738	0.772	0.984
O7	0.447	0.617	0.864
O8	0.933	0.716	0.393

Fig. 1.   Structure of $Cu_2HIO_6 \cdot 2H_2O$.

QUARTZ (α and β)
SiO_2

A.F. WRIGHT and M.S. LEHMANN, 1981. J. Solid State Chem., 36, 371-380.

α-Quartz (25°C), trigonal, $P3_221$, a = 4.9134, c = 5.4052 Å, Z = 3. Neutron radiation, R = 0.019 for 204 reflexions. Si in 3(a): 0.4701,0,0; O in 6(c): 0.4136,0.2676,0.1191 [usual non-standard setting, 1].

β-Quartz (590°C), hexagonal, $P6_222$, a = 4.9977, c = 5.4601 Å, Z = 3. Neutron radiation, powder and single-crystal data for synthetic and natural specimens, R = 0.038, 0.035 for 128 reflexions. 3 Si in 6(g): 0.486,0,0; 6 O in 12(k): 0.4166,0.2381, 0.1414 [average of synthetic and natural specimens].

The α-quartz structure is as previously described (1). That of β-quartz is similar to a previous description (2), but with disorder of O around the 6(j) position, with a corresponding small disorder in the Si position.

1. Strukturbericht, 1, 166; 3, 21; Structure Reports, 27, 674; 28, 119; 30A, 420; 42A, 393; 43A, 262.
2. Strukturbericht, 1, 166; Structure Reports, 27, 674.

TRIDYMITE
SiO$_2$

K. KIHARA, 1981. Z. Kristallogr., 157, 93.

Orthorhombic, Cc2m, a = 8.75, b = 5.052, c = 8.27 Å, Z = 8. Data of 1 reinterpreted
in terms of orthorhombic symmetry. Si (1/6,1/2,-0.0624), O(1) (-0.164,0.423,1/4),
O(2) (0,0.592,0), O(3) (0.219,0.256,0.040).

1. Structure Reports, 46A, 405.

COESITE
SiO$_2$

L. LEVIEN and C.T. PREWITT, 1981. Amer. Min., 66, 324-333.

Monoclinic, [C2/c], a = 7.136-6.990, b = 12.369-12.233, c = 7.174-7.111 Å, β = 120.34-
120.74°, at 1 atm - 52 kbar pressure, [Z = 16]. Mo radiation, R = 0.017-0.055 for
260-742 reflexions, at six pressures.

The structure is as previously described (1). Both silicate tetrahedra compress
(mean Si-O = 1.609-1.603(2) Å) but do not distort with increasing pressure. An
increasing temperature factor of the central oxygen atom in the 180° Si-O-Si angle
suggests that this angle becomes unstable at high pressure.

1. Structure Reports, 23, 340; 34A, 363; 43A, 297.

LITHIUM METASILICATE   LITHIUM METAGERMANATE
Li$_2$SiO$_3$ Li$_2$GeO$_3$

H. VÖLLENKLE, 1981. Z. Kristallogr., 154, 77-81.

Orthorhombic, Cmc2$_1$, a = 9.396, 9.634, b = 5.396, 5.481, c = 4.661, 4.843 Å, Z = 4.
Mo radiation, R = 0.022, 0.016 for 415, 447 reflexions.

Atomic positions

			x	y	z
Li$_2$SiO$_3$	Li	8b	0,1738	0,3449	0,0060
	Si	4a	0	0,1709	0,5000
	O(1)	4a	0	0,1129	0,8542
	O(2)	8b	0,1443	0,3087	0,4193
Li$_2$GeO$_3$	Li	8b	0,1758	0,3440	0,0153
	Ge	4a	0	0,1782	0,5000
	O(1)	4a	0	0,1315	0,8703
	O(2)	8b	0,1522	0,3173	0,4086

Structures as previously described (1, 2), containing tetrahedral metasilicate
and metagermanate chains, linked by tetrahedrally-coordinated Li ions. Si-O = 1.677
(bridging), 1.591 (terminal), Ge-O = 1.811 (bridging), 1.711 Å (terminal), Si-O-Si =
124.7, Ge-O-Ge = 118.4°, Li-O = 1.92-1.96 (3 distances), 2.17 Å.

1. Structure Reports, 20, 404; 21, 431; 33A, 452; 43A, 297.
2. Ibid., 33A, 503.

SODIUM ORTHOSILICATE
Na$_4$SiO$_4$

M.G. BARKER and P.G. GADD, 1981. J. Chem. Res., (S), 274; (M), 3446-3466.

Triclinic, P$\bar{1}$, a = 5.576, b = 6.191, c = 8.507 Å, α = 103.13, β = 95.50, γ = 123.72°, Z = 2. Mo radiation, R = 0.054 for 1267 reflexions.

Isostructural with K$_4$SnO$_4$ (1), the structure containing SiO$_4$ tetrahedra connected by 4- and 5-coordinate Na ions. Si-O = 1.630-1.651(2), Na-O = 2.252-2.481 Å.

1. Structure Reports, 41A, 227.

SODIUM HYDROGENSILICATE DIHYDRATE (HIGH-TEMPERATURE FORM)
Na$_3$HSiO$_4$.2H$_2$O

R.L. SCHMID, L. SZOLNAI, J. FELSCHE and G. HUTTNER, 1981. Acta Cryst., B37, 789-792.

Orthorhombic, Pbca, a = 10.380, b = 10.053, c = 11.414 Å, Z = 8. Mo radiation, R = 0.032 for 664 reflexions.

The structure (Fig. 1) contains SiO$_3$(OH)$^{3-}$ tetrahedra, linked into chains along b by O-H...O hydrogen bonds, 2.639 Å (the corresponding bond in the low-temperature form is weaker, 2.917 Å (1)). The chains are linked by Na ions (6- and 7-coordinations) and by hydrogen bonds from the water molecules.

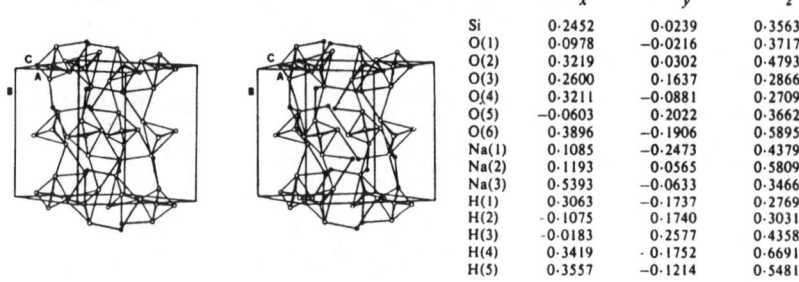

	x	y	z
Si	0·2452	0·0239	0·3563
O(1)	0·0978	−0·0216	0·3717
O(2)	0·3219	0·0302	0·4793
O(3)	0·2600	0·1637	0·2866
O(4)	0·3211	−0·0881	0·2709
O(5)	−0·0603	0·2022	0·3662
O(6)	0·3896	−0·1906	0·5895
Na(1)	0·1085	−0·2473	0·4379
Na(2)	0·1193	0·0565	0·5809
Na(3)	0·5393	−0·0633	0·3466
H(1)	0·3063	−0·1737	0·2769
H(2)	-0·1075	0·1740	0·3031
H(3)	-0·0183	0·2577	0·4358
H(4)	0·3419	-0·1752	0·6691
H(5)	0·3557	−0·1214	0·5481

Fig. 1.   Stereoview of sodium hydrogensilicate dihydrate (high-temperature form) and atomic positional parameters.

1. Structure Reports, 45A, 361.

MAGNESIUM SILICATE
β-Mg$_2$SiO$_4$

H. HORIUCHI and H. SAWAMOTO, 1981. Amer. Min., 66, 568-575.

Orthorhombic, Imma, a = 5.698, b = 11.438, c = 8.257 Å, Z = 8. Mo radiation, R = 0.028 for 438 reflexions.

Atomic positions

			x	y	z
Mg(1)	in	4(a)	0	0	0
Mg(2)		4(e)	0	1/4	0.9701
Mg(3)		8(g)	1/4	0.1276	1/4
Si		8(h)	0	0.1198	0.6168
O(1)		4(e)	0	1/4	0.2166
O(2)		4(e)	0	1/4	0.7164
O(3)		8(h)	0	0.9900	0.2558
O(4)		16(j)	0.2615	0.1225	0.9925

The structure (Fig. 1) is as previously described (1). Mg ions have octahedral coordinations, Mg-O = 2.017-2.129 Å, and Si has tetrahedral coordination, Si-O = 1.631-1.702(3) Å.

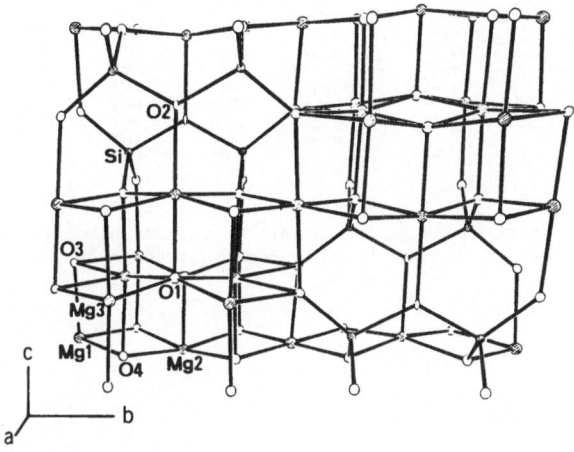

Fig. 1.   Structure of β-Mg$_2$SiO$_4$.

<u>1</u>. Structure Reports, <u>40A</u>, 311..

SODIUM MAGNESIUM SILICATE
Na$_2$MgSiO$_4$

W.H. BAUR, T. OHTA and R.D. SHANNON, 1981. Acta Cryst., B<u>37</u>, 1483-1491.

Monoclinic, Pn, a = 7.015, b = 10.968, c = 5.260 Å, β = 89.97°, Z = 4 (substructure with b = 5.484 Å). Mo radiation, R = 0.029 for 3175 reflexions (0.025 for 2729 substructure reflexions with k = 2n, 0.281 for 446 reflexions with k = 2n+1; twinned crystal).

Isostructural with Na$_2$ZnSiO$_4$ (1), with the superstructure resulting from ordering of deficiencies in the Na sites. All atoms are tetrahedrally coordinated, the structure (Fig. 1) being derived from that of wurtzite. Si-O = 1.61-1.65, Mg-O = 1.91-1.98, Na-O = 2.24-2.44 Å. The ionic conductivity probably results from diffusion of Na along [$\bar{1}$01] via unoccupied tetrahedral or octahedral sites.

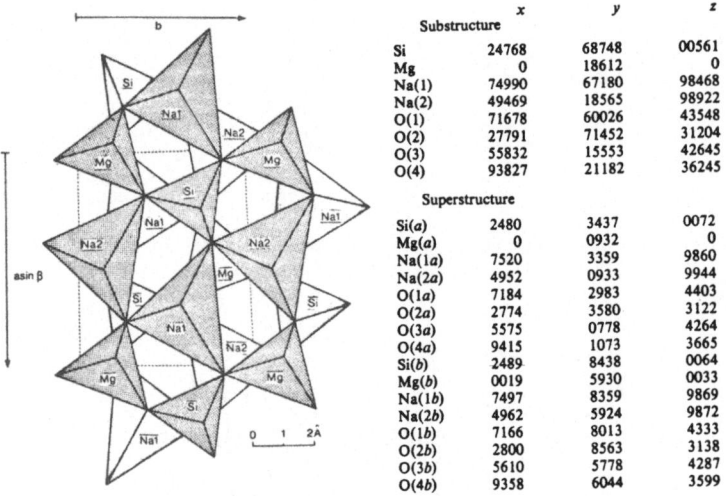

	x	y	z
Substructure			
Si	24768	68748	00561
Mg	0	18612	0
Na(1)	74990	67180	98468
Na(2)	49469	18565	98922
O(1)	71678	60026	43548
O(2)	27791	71452	31204
O(3)	55832	15553	42645
O(4)	93827	21182	36245
Superstructure			
Si(*a*)	2480	3437	0072
Mg(*a*)	0	0932	0
Na(1*a*)	7520	3359	9860
Na(2*a*)	4952	0933	9944
O(1*a*)	7184	2983	4403
O(2*a*)	2774	3580	3122
O(3*a*)	5575	0778	4264
O(4*a*)	9415	1073	3665
Si(*b*)	2489	8438	0064
Mg(*b*)	0019	5930	0033
Na(1*b*)	7497	8359	9869
Na(2*b*)	4962	5924	9872
O(1*b*)	7166	8013	4333
O(2*b*)	2800	8563	3138
O(3*b*)	5610	5778	4287
O(4*b*)	9358	6044	3599

Fig. 1.   Substructure of Na_2MgSiO_4 and atomic positional parameters
in the substructure (x 10^5) and superstructure (x 10^4).

1. Structure Reports, 31A, 218; 42A, 397.

CALCIUM SILICATE PHOSPHATE
$6CaSiO_4 \cdot Ca_3(PO_4)_2$

I.   H. SAALFELD and K.H. KLASKA, 1981. Z. Kristallogr., 154, 323-324.
II.   Idem, 1981. Ibid., 155, 65-73.

Orthorhombic, $Pnm2_1$, a = 9.40, b = 21.71, c = 6.83 Å, Z = 2; there is a sub-cell, Pcmn,
a' = a, b' = b/4, c' = c. Cu radiation, R = 0.16 for 650 weak superlattice reflexions
(the subcell reflexions are overlapped due to twinning).

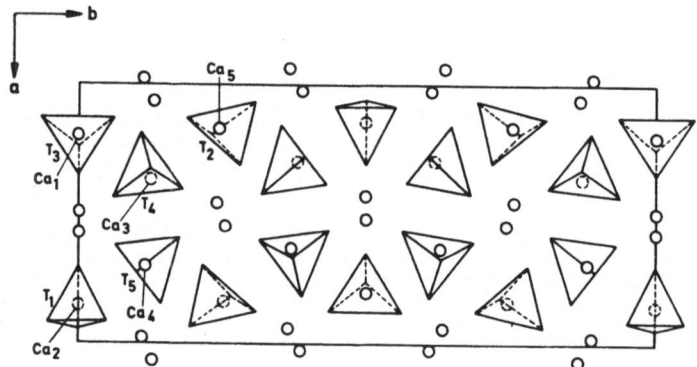

Fig. 1.   Structure of $6CaSiO_4 \cdot Ca_3(PO_4)_2$.

The structure (Fig. 1) is a derivative of that of α'-Ca_2SiO_4, with Si/P and Ca disorder. The superlattice reflexions are satellites, suggesting a modulated structure.

LEAD SILICATE (M' FORM)
Pb_2SiO_4 $2PbO.SiO_2$

L.S. DENT GLASSER, R.A. HOWIE and R.M. SMART, 1981. Acta Cryst., B**37**, 303-306.

Monoclinic, A2, a = 19.43, b = 7.64, c = 12.24 Å, β = 99.33°, D_m = 7.62, Z = 16. Mo radiation, R = 0.086 for 571 reflexions with k even (streaking of k odd reflexions indicates disorder, and limits the accuracy of the analysis).

The structure (Fig. 1) contains $Si_4O_{12}^{8-}$ anions with rings of four SiO_4 tetrahedra, and isolated oxide ions, each coordinated tetrahedrally to four Pb^{2+} ions; the formula is therefore $Pb_8O_4(Si_4O_{12})$. Pb ions have 4- and 5-coordinations, with stereochemically-active lone electron-pairs. Si-O = 1.64-1.69(6) Å, Si-O-Si = 155-166°, Pb-O = 2.09-2.69 Å.

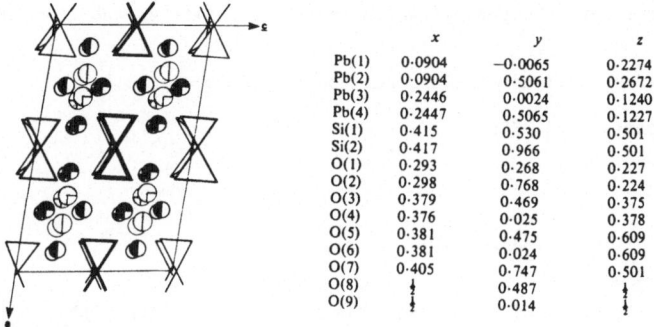

	x	y	z
Pb(1)	0·0904	−0·0065	0·2274
Pb(2)	0·0904	0·5061	0·2672
Pb(3)	0·2446	0·0024	0·1240
Pb(4)	0·2447	0·5065	0·1227
Si(1)	0·415	0·530	0·501
Si(2)	0·417	0·966	0·501
O(1)	0·293	0·268	0·227
O(2)	0·298	0·768	0·224
O(3)	0·379	0·469	0·375
O(4)	0·376	0·025	0·378
O(5)	0·381	0·475	0·609
O(6)	0·381	0·024	0·609
O(7)	0·405	0·747	0·501
O(8)	¼	0·487	¼
O(9)	¼	0·014	¼

Fig. 1.   Structure of lead silicate (M' form).

SODIUM ZIRCONIUM SILICATE
$Na_4Zr_2(SiO_4)_3$

I.   D. TRAN QUI, J.J. CAPPONI, J.C. JOUBERT and R.D. SHANNON, 1981. J. Solid State Chem., **39**, 219-229.
II. R.G. SIZOVA, V.A. BLINOV, A.A. VORONKOV, V.V. ILJUKHIN and N.V. BELOV, 1981. Kristallografija, **26**, 293-300 [Soviet Physics - Crystallography, **26**, 165-169].

I.
Rhombohedral, R$\bar{3}$c, a = 9.1863, c = 22.181 Å, Z = 6. Ag radiation, R = 0.019, 0.026, 0.026 for 554, 498, 476 reflexions at 20, 300, 620°C. Previous study in **1**.

Atomic positions (20°C)

	x	y	z
Zr	0	0	0.1468
Na(1)	0	0	0
Na(2)	-0.3625	0	1/4
O(1)	0.1853	0.1666	0.0849
O(2)	0.1846	-0.0171	0.1912
Si	0.2968	0	1/4

II.
Rhombohedral, R3c, a = 9.189, c = 22.203 Å, Z = 6 (1). Mo radiation, R = 0.028 for
500 reflexions.

Atomic positions

			x	y	z
Zr(1)	in	6(a)	0	0	0.1436
Zr(2)		6(a)	0	0	0.3502
Si		18(b)	0.2975	0.2969	0.7468
Na(1)		6(a)	0	0	0
Na(2)		18(b)	0.365	0.359	0.253
O(1)		18(b)	0.186	0.164	0.083
O(2)		18(b)	0.167	0.184	0.413
O(3)		18(b)	0.182	0.204	0.688
O(4)		18(b)	0.201	0.187	0.806

The structure in I contains a framework of corner-sharing SiO_4 tetrahedra and ZrO_6
octahedra; mean Si-O = 1.624, mean Zr-O = 2.081 Å. Na(1) has distorted octahedral
coordination, Na-O = 2.489, and Na(2) has irregular 8-coordination, Na-O = 2.532-2.662
(6 distances), 3.049 Å (2 distances). The ionic conductivity probably results from
mobility of Na ions through openings which increase in size as the temperature is
raised.

The description in II is similar, but non-centrosymmetric.

1. Structure Reports, 38A, 385.

SODIUM MANGANESE SILICATE (SANEROITE)
$Na_{2.29}Mn_{10}(Si,V)_{12}O_{34}(OH)_4$

I.   R. BASSO and A. DELLA GIUSTA, 1980. Neues Jb. Miner., Abh., 138, 333-342.
II. G. LUCCHETTI, A.M. PENCO and R. RINALDI, 1981. Neues Jb. Miner., Mh., 161-168.

Triclinic, P$\bar{1}$, a = 9.741, b = 9.974, c = 9.108 Å, α = 92.70, β = 117.11, γ = 105.30°,
Z = 1. Mo radiation, R = 0.034 for 4361 reflexions.

The structure (Fig. 1) contains single silicate chains with a repeat of six-
tetrahedra (one partially occupied by V), which form a tetrahedral layer parallel to
(1$\bar{1}$1); these layers alternate with layers of Mn octahedra with channels containing
two partially-occupied Na sites.

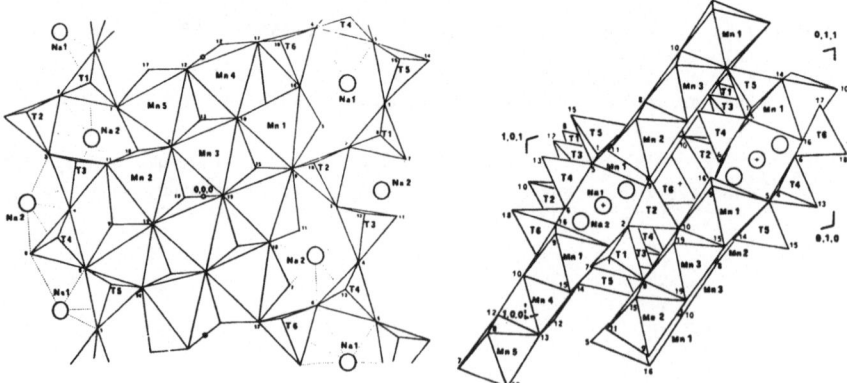

Fig. 1.   Structure of saneroite.

NICKEL ALUMINOSILICATE (PHASES IV and V)
$Ni_3Al_2SiO_8$

I. K. HORIOKA, K.-I. TAKAHASHI, N. MORIMOTO, H. HORIUCHI, M. AKAOGI and S.-I. AKIMOTO, 1981. Acta Cryst., B37, 635-638.
II. K. HORIOKA, M. NISHIGUCHI, N. MORIMOTO, H. HORIUCHI, M. AKAOGI and S.-I. AKIMOTO, 1981. Ibid., B37, 638-640.

Phase IV (quenched from 6.5 x 10^3 MPa and 1553K)
Orthorhombic, Imma, a = 5.665, b = 28.646, c = 8.091 Å, Z = 10. Mo radiation, R = 0.074 for 1115 reflexions.

Phase V (quenched from 8.6 x 10^3 MPa and 1533K)
Orthorhombic, Pmma, a = 5.665, b = 8.590, c = 8.097 Å, Z = 3. Mo radiation, R = 0.059 for 1358 reflexions.

Both structures (Fig. 1) contain slightly-distorted cubic-close-packing of oxygen with spinel-type and modified-spinel-type bands. Isolated TO_4 tetrahedra and T_2O_7 pairs are found in both structures, with single and double columns of MO_6 octahedra. There is partial cation ordering in octahedral and tetrahedral sites.

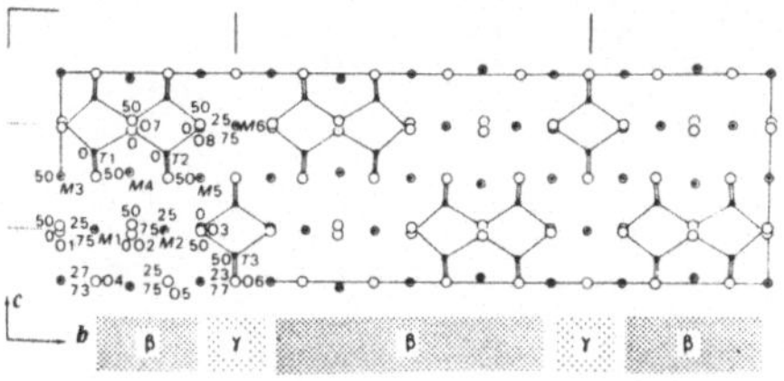

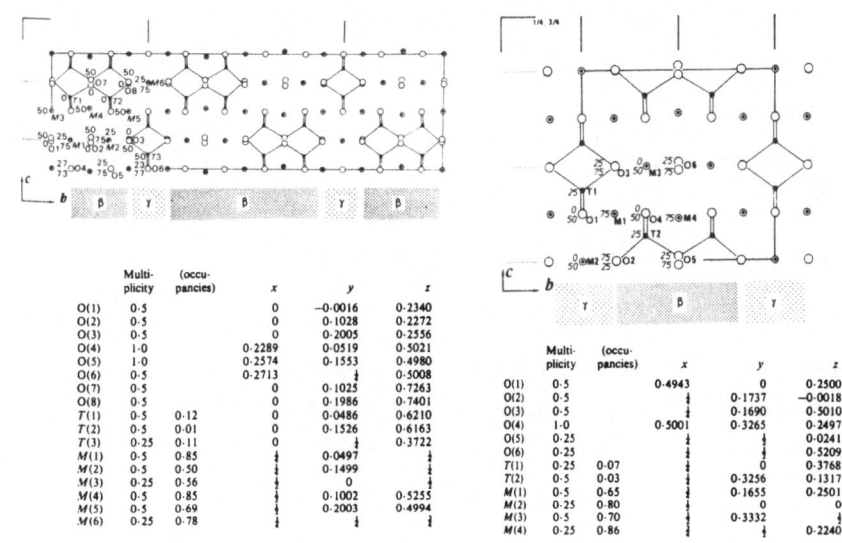

	Multiplicity	(occupancies)	x	y	z
O(1)	0·5		0	−0·0016	0·2340
O(2)	0·5		0	0·1028	0·2272
O(3)	0·5		0	0·2005	0·2556
O(4)	1·0		0·2289	0·0519	0·5021
O(5)	1·0		0·2574	0·1553	0·4980
O(6)	0·5		0·2713	¼	0·5008
O(7)	0·5		0	0·1025	0·7263
O(8)	0·5		0	0·1986	0·7401
T(1)	0·5	0·12	0	0·0486	0·6210
T(2)	0·5	0·01	0	0·1526	0·6163
T(3)	0·25	0·11	0	¼	0·3722
M(1)	0·5	0·85	¼	0·0497	¼
M(2)	0·5	0·50	¼	0·1499	¼
M(3)	0·25	0·56	½	0	½
M(4)	0·5	0·85	½	0·1002	0·5255
M(5)	0·5	0·69	½	0·2003	0·4994
M(6)	0·25	0·78	¼	¼	¾

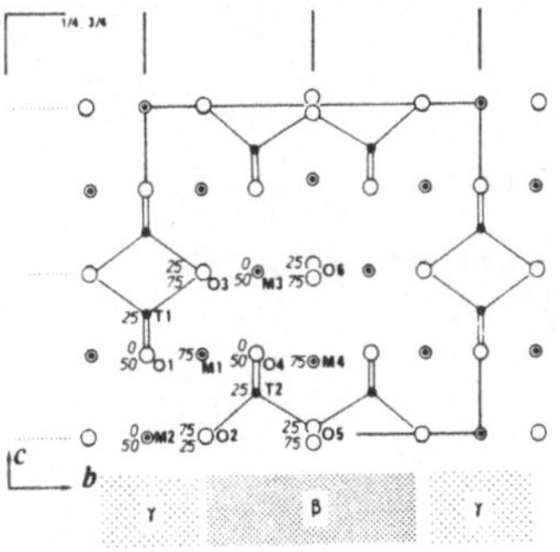

	Multiplicity	(occupancies)	x	y	z
O(1)	0·5		0·4943	0	0·2500
O(2)	0·5		¼	0·1737	−0·0018
O(3)	0·5		¼	0·1690	0·5010
O(4)	1·0		0·5001	0·3265	0·2497
O(5)	0·25		¼	½	0·0241
O(6)	0·25		¼	½	0·5209
T(1)	0·25	0·07	¼	0	0·3768
T(2)	0·5	0·03	¼	0·3256	0·1317
M(1)	0·5	0·65	¼	0·1655	0·2501
M(2)	0·25	0·80	¼	0	0
M(3)	0·5	0·70	¼	0·3332	½
M(4)	0·25	0·86	¼	½	0·2240

Fig. 1. Structures of $Ni_3Al_2SiO_8$, phase IV (left) and phase V (right); spinel-type (γ) and modified-spinel-type (β) bands are indicated.

POTASSIUM COPPER SILICATE
$K_2CuSi_4O_{10}$

K. KAWAMURA and J.T. IIYAMA, 1981. Bull. Minéral., 104, 387-395.

Monoclinic, $P2_1/m$, a = 11.285, b = 8.244, c = 11.065 Å, β = 110.94°, D_m = 2.86, Z = 4. Mo radiation, R = 0.059 for 2706 reflexions.

The structure (Fig. 1) contains tetrahedral $(Si_{16}O_{40}^{16-})_n$ chains along $\underline{b}$; Cu has square-planar coordination, and K ions have 8- and 9-coordinations. Si-O = $\overline{1}$.57-1.65, Cu-O = 1.91-1.95, K-O = 2.24-3.36 Å. The Na_2 ($\underline{1}$) and NaK (litidionite, $\underline{2}$) compounds are triclinic.

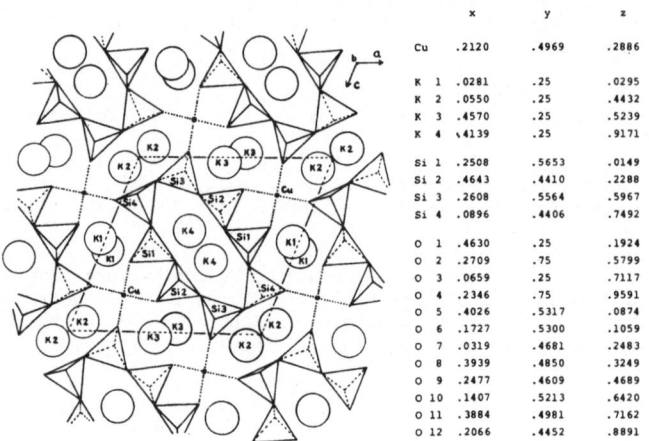

	x	y	z
Cu	.2120	.4969	.2886
K 1	.0281	.25	.0295
K 2	.0550	.25	.4432
K 3	.4570	.25	.5239
K 4	.4139	.25	.9171
Si 1	.2508	.5653	.0149
Si 2	.4643	.4410	.2288
Si 3	.2608	.5564	.5967
Si 4	.0896	.4406	.7492
O 1	.4630	.25	.1924
O 2	.2709	.75	.5799
O 3	.0659	.25	.7117
O 4	.2346	.75	.9591
O 5	.4026	.5317	.0874
O 6	.1727	.5300	.1059
O 7	.0319	.4681	.2483
O 8	.3939	.4850	.3249
O 9	.2477	.4609	.4689
O 10	.1407	.5213	.6420
O 11	.3884	.4981	.7162
O 12	.2066	.4452	.8891

Fig. 1.   Structure of $K_2CuSi_4O_{10}$.

$\underline{1}$. Structure Reports, $\underline{43A}$, 301.
$\underline{2}$. Ibid., $\underline{41A}$, 386.

ÅKERMANITE
$Ca_2MgSi_2O_7$

M. KIMATA and N. II, 1981. Neues Jb. Miner., Mh., 1-10.

Tetragonal, $P\overline{4}2_1m$, a = 7.835, c = 5.010 Å, Z = 2. Mo radiation, R = 0.032 for 409 reflexions.

Atomic positions

		x	y	z
Ca	in	4(e)	0.3318	1/2-x	0.5067
Mg		2(a)	0	0	0
Si		4(e)	0.1397	1/2-x	0.9352
O(1)		2(c)	1/2	0	0.1789
O(2)		4(e)	0.1410	1/2-x	0.2536
O(3)		8(f)	0.0805	0.1867	0.7877

Isostructural with melilite, hardystonite, and gehlinite ($\underline{1}$). Mg-O = 1.915 (tetrahedral), Si-O = 1.595-1.649, Ca-O = 2.424-2.712 Å (8-coordination), Si-O-Si = 139.4°.

$\underline{1}$. Strukturbericht, $\underline{2}$, 146, 541, 542, 543; Structure Reports, $\underline{17}$, 574; $\underline{32A}$, 439; $\underline{35A}$, 455; $\underline{37A}$, 334; $\underline{41A}$, 433.

AMESITE-2H$_2$

$(Mg_2Al)(SiAl)O_5(OH)_4$

C.S. ANDERSON and S.W. BAILEY, 1981. Amer. Min., 66, 185-195.

Triclinic, C1, a = 5.307, b = 9.195, c = 14.068 Å, α = 90.09, β = 90.25, γ = 89.96°, Z = 4. Mo radiation, R = 0.059 for 1713 reflexions.

The material is from the Saranovskoye chromite deposit, North Ural Mountains. Cation ordering reduces the symmetry of the structure from P6$_3$ (1) to C1, the ordering pattern being different from that in a specimen from Antartica (2). Mean T-O = 1.63, 1.74, 1.63, 1.74 Å (indicating substantial but incomplete ordering); mean M-O = 2.06, 2.00, 2.06 Å in one layer, 2.09, 1.94, 2.09 Å in the second layer (the smaller octahedra being interpreted as Al-rich).

1. Structure Reports, 20, 434.
2. Ibid., 46A, 382.

AMPHIBOLES

$(Na,Ca)_3Mg_3Al_2(Si,Al)_8O_{22}(OH)_2$

L. UNGARETTI, D.C. SMITH and G. ROSSI, 1981. Bull. Minéral., 104, 400-412.

Monoclinic, [C2/m], a ∿ 9.7, b ∿ 17.8, c ∿ 5.3 Å, β = 104°, [Z = 2]. Mo radiation, R ∿ 0.02 for ∿1350 reflexions, for 21 amphibole specimens from the Nybö eclogite pod, Norway; the Na$_3$Si$_7$Al compound is named nyböite. Atomic positional parameters are not given, but cation site occupancies are listed.

Amphibole structures (1).

1. Structure Reports, 38A, 361.

ANDALUSITE (MANGANIAN)
$(Al,Mn,Fe)_2SiO_5$ (<0.50 Mn)

KANONAITE (ALUMINOUS)
$(Al,Mn,Fe)_2SiO_5$ (>0.50 Mn)

I. ABS-WURMBACH, K. LANGER, F. SEIFERT and E. TILLMANNS, 1981. Z. Kristallogr., 155, 81-113.

Orthorhombic, Pnnm, a = 7.810-7.961, b = 7.915-8.053, c = 5.570-5.616 Å, for Al:Mn:Fe = 1.920:0.023:0.057 - 1.302:0.680:0.018, Z = 4. Mo radiation, R = 0.018-0.041 for 535-722 reflexions for four natural specimens.

Andalusite structure (1), with Mn^{3+} and Fe^{3+} exclusively in the octahedral Al(1) site, and not in the trigonal-bipyramidal Al(2) site.

1. Strukturbericht, 2, 110, 512; 6, 22; Structure Reports, 26, 502; 27, 710.

AXINITE
$(Fe,Mg,Mn)Ca_2Al_2BSi_4O_{15}(OH)$

J.S. SWINNEA, H. STEINFINK, L.E. RENDON-DIAZMIRON and S. ENCISO de la VEGA, 1981. Amer. Min., 66, 428-431.

Triclinic, $P\bar{1}$, a = 7.1437, b = 9.1898, c = 8.9529 Å, α = 91.86, β = 98.19, γ = 77.36°, Z = 2. Mo radiation, R = 0.019 for 3124 reflexions, for mineral from Mexico.

Structure as previously described (1).

1. Structure Reports, 41A, 376.

BUSTAMITE (MANGANESE-RICH)
$(Mn,Ca)SiO_3$

T. YAMANAKA and Y. TAKÉUCHI, 1981. Z. Kristallogr., 157, 131-145.

Triclinic, $A\bar{1}$, a = 7.605, b = 7.102, c = 13.568 Å, α = 89.95, β = 94.39, γ = 103.53°, Z = 12. Mo radiation, R = 0.041 for 1129 reflexions. Material obtained by heating rhodonite.

Atomic positions

	x	y	z
M1	0.2024	0.4196	0.3741
M2	0.2045	0.9277	0.3739
M3	0.5000	0.2500	0.2500
M4	0.5000	0.7500	0.2500
Si1	0.1809	0.4066	0.6302
Si2	0.1675	0.9353	0.6349
Si3	0.3956	0.7246	0.5186
O1	0.4053	0.2437	0.4239
O2	0.4357	0.7287	0.4002
O3	0.3070	0.4505	0.7266
O4	0.3038	0.9505	0.7328
O5	0.0403	0.6384	0.3659
O6	0.0305	0.1338	0.3541
O7	0.2663	0.5068	0.5361
O8	0.2591	0.8705	0.5379
O9	0.1919	0.1735	0.6092

M1 = 0.16 Ca + 0.84 Mn   M3 = Mn
M2 = 0.26 Ca + 0.74 Mn   M4 = 0.53 Ca + 0.47 Mn

Bustamite structure (1).

1. Structure Reports, 27, 701; 28, 246.

CANCRINITE
$Na_7Ca(AlSiO_4)_6(CO_3)_{1.4} \cdot 2H_2O$

Ju.I. SMOLIN, Ju.F. ŠEPELEV, I.K. BUTIKOVA and I.B. KOBJAKOV, 1981. Kristallografija, 26, 63-66 [Soviet Physics - Crystallography, 26, 33-35].

Hexagonal, $P6_3$, a = 12.635, c = 5.115 Å, D_m = 2.43, Z = 1. Mo radiation, R = 0.04 for 1000 reflexions, for a synthetic sample.

The structure (Fig. 1) is essentially as previously described (1). It contains rings of six Si, Al tetrahedra, carbonate groups, and eight-coordinate Na ions (which must have some replacement by Ca).

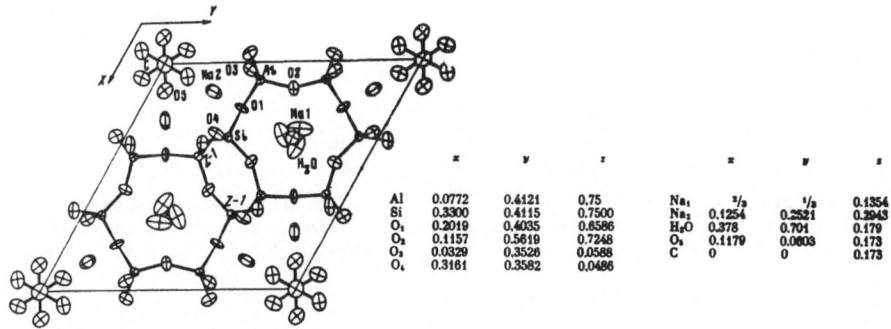

	x	y	z			x	y	z
Al	0.0772	0.4121	0.75		Na₁	¹/₃	¹/₃	0.1354
Si	0.3300	0.4115	0.7500		Na₂	0.1254	0.2521	0.2943
O₁	0.2019	0.4035	0.6586		H₂O	0.378	0.701	0.179
O₂	0.1157	0.5619	0.7248		O₅	0.1179	0.0803	0.173
O₃	0.0329	0.3528	0.0588		C	0	0	0.173
O₄	0.3161	0.3582	0.0486					

Fig. 1.   Structure of cancrinite.

1. Strukturbericht, 3, 150, 524; Structure Reports, 19, 480; 30A, 428; 35A, 458.

CARPHOLITE (MAGNESIUM)
$(Mg,Fe)Al_2Si_2O_6(OH)_4$

K. VISWANATHAN, 1981. Amer. Min., 66, 1080-1085.

Orthorhombic, Ccca, a = 13.714, b = 20.079, c = 5.105 Å, Z = 8. Mo radiation, R = 0.044 for 1163 reflexions.

Atomic positions

	x	y	z
Mg,Fe	0	0.8754	3/4
Al(1)	0.1909	3/4	3/4
Al(2)	0	0.9614	1/4
Si	0.1935	0.8798	0.4172
OH(1)	0.0997	0.8099	0.8794
OH(2)	0.0684	0.9657	0.9276
O(1)	0.2060	0.7997	0.4313
O(2)	0.0803	0.8993	0.4099
O(3)	0.2485	0.9129	0.6707

Isostructural with other carpholites (1). Si-O = 1.602-1.639(3), Al-O = 1.852-1.963, Mg,Fe-O = 2.008-2.233 Å.

1. Structure Reports, 20, 406; 41A, 379; 45A, 369.

CKALOVITE (ZINC)
$Na_2ZnSi_2O_6$

M.A. SIMONOV, E.L. BELOKONEVA and N.V. BELOV, 1980. Kristallografija, 25, 1282-1284 [Soviet Physics - Crystallography, 25, 731-732].

Orthorhombic, Fdd2, a = 21.54, b = 7.139, c = 7.413 Å, D_m = 3.1, Z = 8. Mo radiation, R = 0.040 for 4173 reflexions, for a synthetic specimen.

Atomic positions

	x	y	z
Zn	0	0	0
Si	0.08330	0.28672	0.7731
Na	0.07715	0.2694	0.2273
O₁	0.07787	0.4509	0.6270
O₂	0.12035	0.1004	0.6911
O₃	0.01850	0.2180	0.8544

The structure is as previously described (1), containing a β-cristobalite-like framework of SiO_4 and ZnO_4 tetrahedra, with 5-coordinate Na in gaps in the framework. Si-O = 1.598-1.660, Zn-O = 1.936, 1.955, Na-O = 2.267-2.554 Å (sixth O at 3.061 Å).

1. Structure Reports, 41A, 432.

CLINOPYROXENES (MANGANOAN DIOPSIDE and KANOITE)
$(Ca,Mn,Mg)_2Si_2O_6$

W.A. GORDON, D.R. PEACOR, P.E. BROWN, E.J. ESSENE and L.F. ALLARD, 1981. Amer. Min., 66, 127-141.

Manganoan diopside, $Ca_{0.68}Mn_{0.44}Mg_{0.88}Si_2O_6$, monoclinic, C2/c, a = 9.76, b = 8.93, c = 5.27 Å, β = 106.44°, Z = 4. Mo radiation, R = 0.054 for 374 reflexions.

Kanoite, $Ca_{0.12}Mn_{1.02}Mg_{0.86}Si_2O_6$, monoclinic, $P2_1/c$, a = 9.78, b = 8.93, c = 5.32 Å, β = 108.60°, Z = 4. Mo radiation, R = 0.057 for 982 reflexions.

Atomic positions

	x	y	z			
				Mn-Diopside		
M1	0	0.0929	3/4			
M2	0	0.7072	3/4	M1		Mn 0.06
Si	0.2114	0.4075	0.7635			Mg 0.94
O1	0.3829	0.4121	0.8557			
O2	0.1358	0.2510	0.6715	M2		Ca 0.87
O3	0.1489	0.4812	0.9994			Mn 0.13
M1	0.2505	0.6544	0.2357			
M2	0.2534	0.0235	0.2332	Kanoite		
SiA	0.0413	0.3411	0.2710			
SiB	0.5466	0.8388	0.2408	M1		Mn 0.10
O1A	0.8674	0.3383	0.1652			Mg 0.90
O1B	0.3726	0.8382	0.1367			
O2A	0.1181	0.5009	0.3255	M2		Mn 0.86
O2B	0.6239	0.9916	0.3589			Mg 0.14
O3A	0.1034	0.2590	0.5707			
O3B	0.6038	0.7161	0.4901			

The material contains two co-existing C2/c and $P2_1/c$ phases, with structures as previously described [e.g. 1, 2].

1. Strukturbericht, 2, 130.
2. Structure Reports, 24, 465.

DIOPSIDE
$CaMgSi_2O_6$

L. LEVIEN and C.T. PREWITT, 1981. Amer. Min., 66, 315-323.

Monoclinic, C2/c, a = 9.746-9.612, b = 8.920-8.765, c = 5.252-5.179 Å, β = 105.86-105.32°, at 1 atm - 53.0 kbar pressure, Z = 4. Mo radiation, R = 0.016-0.037 for 424-795 reflexions, at five pressures.

The structure is as previously described (1). Increasing pressure causes a 1% reduction in volume of the SiO_4 tetrahedron, and 5% reductions in the M(1) coordination octahedron and M(2) eight-coordinate polyhedron.

1. Strukturbericht, 2, 130, 528; Structure Reports, 39A, 355; 41A, 394; 42A, 403.

DIXENITE
$CuMn_{14}Fe(OH)_6(AsO_3)_5(SiO_4)_2(AsO_4)$

T. ARAKI and P.B. MOORE, 1981. Amer. Min., 66, 1263-1273.

Rhombohedral, R3, a = 8.233, c = 37.499 Å, Z = 3. Mo radiation, R = 0.064 for 2507 reflexions.

Atomic positions

Atom	Population	x	y	z
M(1)	$1.0Mn^{2+}$	0	0	0
M(2)	$0.90(2)Mn^{2+} + 0.10(2)Mg^{2+}$	1/3	2/3	0.00622
M(3)	$1.0Fe^{2+}$	0	0	0.25749
M(4)	$1.0Mn^{2+}$	0.0408	0.2617	0.06782
M(5)	$1.0Mn^{2+}$	0.4158	0.3359	0.12987
M(6)	$1.0Mn^{2+}$	0.1089	0.3976	0.19230
M(7)	$1.0Mn^{2+}$	0.4226	0.3154	0.26133
Cu(1)	$0.651(9)Cu^{1+} + 0.349Cu$	1/3	2/3	0.51292
Cu(2)	$0.192(9)Cu^{1+} + 0.808Cu$	2/3	1/3	0.0030(
T(1)	$0.86(1)Si^{4+} + 0.14As^{5+}$	2/3	1/3	0.18792
T(2)	$0.60(1)Si^{4+} + 0.40As^{5+}$	0	0	0.14620
T(3)	$0.24(1)Si^{4+} + 0.76As^{5+}$	1/3	2/3	0.11357
As(1)	$1.0As^{5+}$	2/3	1/3	0.06992
As(2)	$1.0As^{5+}$	1/3	2/3	0.25062
As(3)	$1.0As^{5+}$	0.08854	0.37369	0.31589
O(1)		0	0	0.1019
O(2)		1/3	2/3	0.1584
O(3)		2/3	1/3	0.2311
O(4)		0.0899	0.4444	0.0212
O(5)		0.4698	0.1659	0.0949
O(6)		0.2817	0.4545	0.0988
O(7)		0.1655	0.2081	0.1617
O(8)		0.5194	0.1277	0.1707
O(9)		0.3690	0.5010	0.2277
O(10)		0.1468	0.2272	0.2905
O(11)		0.4608	0.1185	0.2956
OH(1)		0.2140	0.1791	0.0375
OH(2)		0.2352	0.0814	0.2271

The structure is related to that of hematolite (1), and contains five layers along c. One layer contains a disordered tetrahedral cluster, $Cu(I)As(III)_4$.

1. Structure Reports, 44A, 265.

FORSTERITE
Mg_2SiO_4

G.A. LAGER, F.K. ROSS, F.J. ROTELLA and J.D. JORGENSEN, 1981. J. Appl. Cryst., 14, 137-139.

Orthorhombic, Pbnm, a = 4.7534, b = 10.1989, c = 5.9813 Å, Z = 4.   Neutron time-of-flight powder data.   Results are in agreement with previous studies (1).

1. Structure Reports, 33A, 468; 39A, 356; 42A, 405.

FORSTERITE
α-Mg$_2$SiO$_4$

FAYALITE
α-Fe$_2$SiO$_4$

TEPHROITE
α-Mn$_2$SiO$_4$

K. FUJINO, S. SASAKI, Y. TAKÉUCHI and R. SADANAGA, 1981.   Acta Cryst., B37, 513-518.

Orthorhombic, Pbnm, a = 4.753, 4.820, 4.902, b = 10.190, 10.479, 10.596, c = 5.978, 6.087, 6.257 Å, Z = 4.   Mo radiation, R = 0.021, 0.026, 0.031 for 2168, 2042, 2088 reflexions.

Atomic positions

	x	y	z
Forsterite			
M(1)	0·0	0·0	0·0
M(2)	0·99169	0·27739	0·25
Si	0·42645	0·09403	0·25
O(1)	0·76594	0·09156	0·25
O(2)	0·22164	0·44705	0·25
O(3)	0·27751	0·16310	0·03304
Fayalite			
M(1)	0·0	0·0	0·0
M(2)	0·98598	0·28026	0·25
Si	0·43122	0·09765	0·25
O(1)	0·76814	0·09217	0·25
O(2)	0·20895	0·45365	0·25
O(3)	0·28897	0·16563	0·03643
Tephroite			
M(1)	0·0	0·0	0·0
M(2)	0·98792	0·28041	0·25
Si	0·42755	0·09643	0·25
O(1)	0·75776	0·09363	0·25
O(2)	0·21088	0·45369	0·25
O(3)	0·28706	0·16384	0·04140

     Olivine structures (1), as previously described (2).   Atomic net charges and electron distributions are determined.

1. Strukturbericht, 1, 352.
2. Structure Reports, 28, 265; 33A, 468; 39A, 356; 41A, 382; 42A, 405.

HASTINGSITE (MAGNESIAN)
(Na,K)Ca$_2$(Mg,Fe)$_5$(Si,Al)$_8$O$_{22}$(OH,O)$_2$

E.M. WALITZI and F. WALTER, 1981.   Z. Kristallogr., 156, 197-208.

Monoclinic, C2/m, a = 9.880, b = 18.012, c = 5.324 Å, β = 105.26°, D$_m$ = 3.225, Z = 2.   Mo radiation, R = 0.047 for 1191 reflexions.

Atomic positions

	x	y	z	
O(1)	0,1044	0,0883	0,218	
O(2)	0,1188	0,1728	0,733	
O(3)(OH)	0,1061	0	0,713	
O(4)	0,3661	0,2505	0,788	
O(5)	0,3498	0,1409	0,114	
O(6)	0,3454	0,1169	0,611	
O(7)	0,3369	0	0,280	
T(1)	0,2802	0,0856	0,3037	$0,600\,Si^{4+} + 0,400\,Al^{3+}$
T(2)	0,2912	0,1728	0,8137	$0,885\,Si^{4+} + 0,115\,Al^{3+}$
M(1)	0	0,0864	0,50	$0,290\,Fe^{2+} + 0,710\,Mg^{2+}$
M(2)	0	0,1772	0	$0,188\,Fe^{2+} + 1,144\,Mg^{2+}$
M(3)	0	0	0	$0,528\,Fe^{3+} + 0,210\,Ti^{4+}$
M(4)	0	0,2792	0,50	$0,950\,Ca^{2+} + 0,050\,Na^{1+}$
A(2)	0	0,489	0	$0,170\,Na^{1+} + 0,075\,K^{1+}$
A(m)	0,044	0,50	0,092	$0,145\,Na^{1+} + 0,100\,K^{1+}$

Amphibole structure (e.g. 1), with positional disorder of the A-site.

1. Structure Reports, 39A, 342.

HEMIMORPHITE

$Zn_4Si_2O_7(OH)_2 \cdot H_2O$

B.J. COOPER, G.V. GIBBS and F.K. ROSS, 1981. Z. Kristallogr., 156, 305-321.

Orthorhombic, Imm2, a = 8.337, 8.268, 8.206, b = 10.724, 10.784, 10.815, c = 5.116, 5.113, 5.089 Å, at 22, 300, 600°C, Z = 2. Mo radiation, R = 0.046, 0.074, 0.061 for 532, 255, 432 reflexions.

The structure is as previously determined (1). Contraction of the structure above 300°C is due to dehydration.

1. Strukturbericht, 2, 125, 524; Structure Reports, 24, 475; 32A, 435; 44A, 306; 45A, 407.

HEULANDITE (CALCIUM-AMMONIUM, DEHYDRATED)

$(NH_4)_{1.2}K_{0.4}Na_{1.1}Ca_{2.8}Si_{27.7}Al_{8.3}O_{72} \cdot xH_2O$

W.J. MORTIER and J.R. PEARCE, 1981. Amer. Min., 66, 309-314.

Monoclinic, C2/m, a = 17.158, b = 17.433, c = 7.388 Å, β = 113.41°, Z = 1. Mo radiation, R = 0.086 for 2046 reflexions.

Atomic positions

		Population	x	y	z
T(1)	8j	8	0.3200	0.1632	0.2343
T(2)	8j	8	0.0668	0.2137	0.2154
T(3)	8j	8	0.2162	0.3145	0.2018
T(4)	8j	8	0.2020	0.0895	0.4121
T(5)	4h	4	0.0	0.2956	0.5
O(1)	8j	8	0.4190	0.1504	0.3841
O(2)	8j	8	0.2906	0.2513	0.2421
O(3)	8j	8	0.3068	0.1409	0.0091
O(4)	8j	8	0.0170	0.2408	0.3462
O(5)	8j	8	0.2627	0.1018	0.2909
O(6)	8j	8	0.1117	0.1337	0.3046
O(7)	8j	8	0.2520	0.3862	0.3558
O(8)	8j	8	0.1345	0.2768	0.2172
O(9)	4i	4	0.1765	0.0	0.4095
O(10)	4g	4	0.0	0.1948	0.0
Ca(1)	4i	2.41	0.2591	0.5	0.0820
Ca(2)	4i	0.76	0.0326	0.0	0.2901

The structure exhibits distortion of the fundamental polyhedral units of natural heulandite (1) in the direction observed for the fully collapsed heulandite-B phase (2). The exchangeable cations are located in the framework eight-rings and are responsible for the distortions, which result in a decrease of the critical channel apertures.

1. Structure Reports, 33A, 484; 38A, 368.
2. Ibid., 39A, 343.

HEULANDITE (SILVER-EXCHANGED)
$Ag_{7.3}Al_{7.2}Si_{28.8}O_{72}.18H_2O$

N. BRESCIANI-PAHOR, M. CALLIGARIS, G. NARDIN and L. RANDACCIO, 1981. J. Chem. Soc., Dalton, 2288-2291.

Monoclinic, C2/m, a = 17.736, b = 18.104, c = 7.447 Å, β = 116.2°, D_m = 2.53, Z = 1. Mo radiation, R = 0.068 and 0.072 for 1602 and 1724 reflexions, for two samples.

The zeolite framework is similar to that previously described (1), with Ag in the CS(1) and CS(2) cation sites occupied by Ca and Na in natural heulandite, and in a low occupancy CS(3) site (Fig. 1).

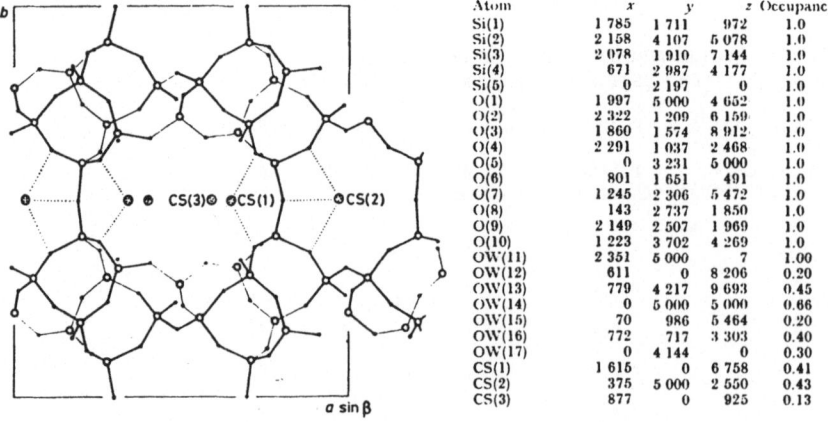

Atom	x	y	z	Occupancy
Si(1)	1 785	1 711	972	1.0
Si(2)	2 158	4 107	5 078	1.0
Si(3)	2 078	1 910	7 144	1.0
Si(4)	671	2 987	4 177	1.0
Si(5)	0	2 197	0	1.0
O(1)	1 997	5 000	4 652	1.0
O(2)	2 322	1 209	6 159	1.0
O(3)	1 860	1 574	8 912	1.0
O(4)	2 291	1 037	2 468	1.0
O(5)	0	3 231	5 000	1.0
O(6)	801	1 651	491	1.0
O(7)	1 245	2 306	5 472	1.0
O(8)	143	2 737	1 850	1.0
O(9)	2 149	2 507	1 969	1.0
O(10)	1 223	3 702	4 269	1.0
OW(11)	2 351	5 000	7	1.00
OW(12)	611	0	8 206	0.20
OW(13)	779	4 217	9 693	0.45
OW(14)	0	5 000	5 000	0.66
OW(15)	70	986	5 464	0.20
OW(16)	772	717	3 303	0.40
OW(17)	0	4 144	0	0.30
CS(1)	1 615	0	6 758	0.41
CS(2)	375	5 000	2 550	0.43
CS(3)	877	0	925	0.13

Fig. 1.   Structure of silver-exchanged heulandite, and atomic positional parameters (x 10^4).

1. Structure Reports, 46A, 387.

HYDROGARNET (TITANIUM)
$Ca_3(Al,Fe,Ti)_2(SiO_4)_{2.85}(OH)_{0.6}$ (idealized)

R. BASSO, A. DELLA GIUSTA and L. ZEFIRO, 1981. Neues Jb. Miner., Mh., 230-236.

Cubic, Ia3d, a = 11.970 Å, Z = 8. Mo radiation, R = 0.024 for 469 reflexions. Ca in the dodecahedral site, (Al,Fe,Ti) in the octahedral site, Si in the tetrahedral site; O at (0.0386,0.0471,0.6534).

Garnet structure, with no Ti in the tetrahedral site. H atom not located.

JAGOITE
$Pb_{18}Fe_4(Si_4T_6)(Pb_4Si_{16}T_4)O_{78}(O,OH)_4Cl_6$ (T = Si/Fe)

M. MELLINI and S. MERLINO, 1981. Amer. Min., 66, 852-858.

Hexagonal, $P\bar{6}2c$, a = 8.528, c = 33.33 Å, Z = 1. Mo radiation, R = 0.057 for 768 reflexions.

The structure (Fig. 1) contains double and single tetrahedral layers, connected by a sheet of Fe and Pb cations; other Pb ions and Cl$^-$ ions are located inside the double layers. The single layer includes an FeO_6 octahedron; the double layer includes a PbO_3E tetrahedron (E = lone-pair) (Fig. 2). Pb(1)-O = 2.29 (3 distances), other Pb-O = 2.39-3.30 (7 and 8 distances), Fe-O = 1.98-2.05, Si/Fe = 1.70-1.77, Si-O = 1.54-1.64 Å.

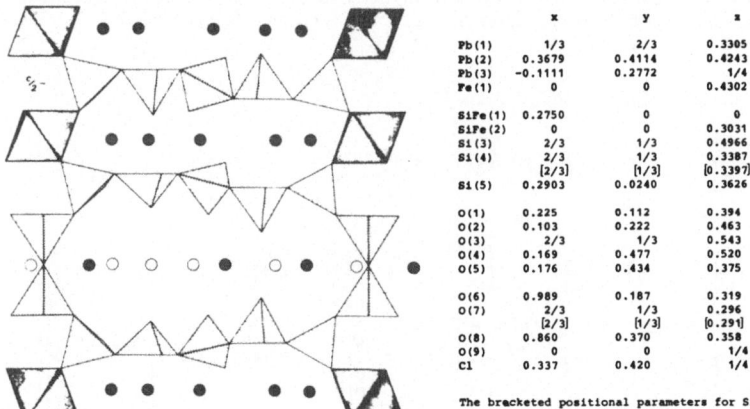

	x	y	z
Pb(1)	1/3	2/3	0.3305
Pb(2)	0.3679	0.4114	0.4243
Pb(3)	-0.1111	0.2772	1/4
Fe(1)	0	0	0.4302
SiFe(1)	0.2750	0	0
SiFe(2)	0	0	0.3031
Si(3)	2/3	1/3	0.4966
Si(4)	2/3	1/3	0.3387
	[2/3]	[1/3]	[0.3397]
Si(5)	0.2903	0.0240	0.3626
O(1)	0.225	0.112	0.394
O(2)	0.103	0.222	0.463
O(3)	2/3	1/3	0.543
O(4)	0.169	0.477	0.520
O(5)	0.176	0.434	0.375
O(6)	0.989	0.187	0.319
O(7)	2/3	1/3	0.296
	[2/3]	[1/3]	[0.291]
O(8)	0.860	0.370	0.358
O(9)	0	0	1/4
Cl	0.337	0.420	1/4

The bracketed positional parameters for Si(4) and O(7) atoms were obtained from ΔF syntheses.

Fig. 1.   Structure of jagoite, viewed along a; open circles are Cl, filled circles Pb.

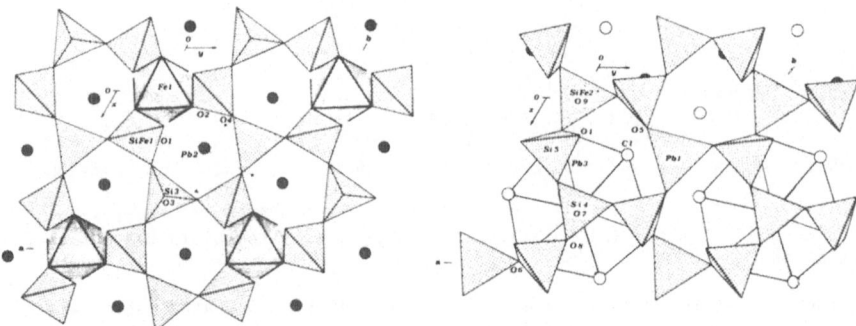

Fig. 2.   The single sheet (left) and one part of the double sheet (right)
            in jagoite, viewed along c.

JASMUNDITE
$Ca_{11}(SiO_4)_4O_2S$

L.S. DENT GLASSER and C.K. LEE, 1981. Acta Cryst., B37, 803-806.

Tetragonal, $I\bar{4}m2$, a = 10.461, c = 8.813 Å, D_m = 3.03, Z = 2. Mo radiation, R = 0.07
for 637 reflexions.

      The structure (Fig. 1) contains an SiO_4 tetrahedron, a sulphide anion, two oxide
anions, and four independent Ca cations. Three Ca have irregular octahedral coordinations, with four contacts to silicate oxygens and two to oxide or sulphide ions;
the fourth Ca has eight-coordination to silicate oxygen atoms. Si-O = 1.635, Ca-O =
2.18-2.72, Ca-S = 3.08, 3.27 Å.

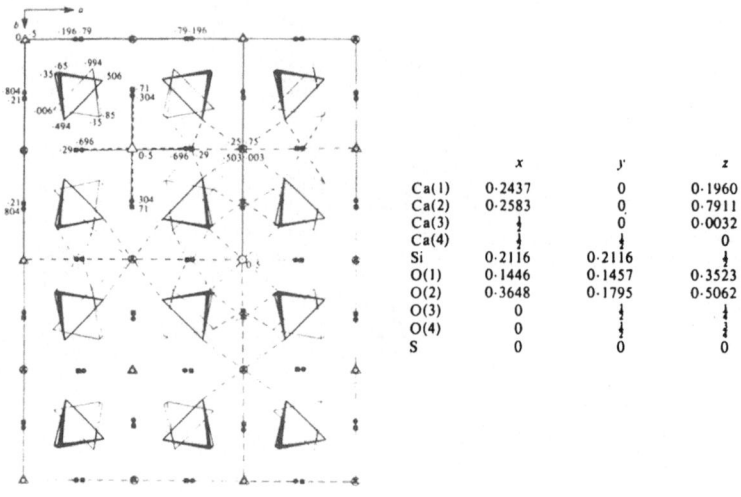

	x	y	z
Ca(1)	0·2437	0	0·1960
Ca(2)	0·2583	0	0·7911
Ca(3)	¼	0	0·0032
Ca(4)	¼	¼	0
Si	0·2116	0·2116	¼
O(1)	0·1446	0·1457	0·3523
O(2)	0·3648	0·1795	0·5062
O(3)	0	¼	¾
O(4)	0	¼	¼
S	0	0	0

Fig. 1.   Structure of jasmundite.

KANONAITE
$MnAlO(SiO_4)$

Z. WEISS, S.W. BAILEY and M. RIEDER, 1981. Amer. Min., 66, 561-567.

Orthorhombic, Pnnm, a = 7.959, b = 8.047, c = 5.616 Å, Z = 4. Mo radiation, R = 0.031 for 472 reflexions.

Atomic positions

			x	y	z
M(1)	in	4(e)	0	0	0.2429
M(2)		4(g)	-0.1252	0.3630	0
Si		4(g)	0.2494	0.2549	0
O(1)		4(g)	0.0743	-0.1369	0
O(2)		4(g)	0.4243	0.3626	0
O(3)		4(g)	0.1042	0.3989	0
O(4)		8(h)	0.2430	0.1413	0.2383

M(1) = 0.74Mn + 0.26Al
M(2) = 0.12Mn + 0.88Al

Isostructural with andalusite (1), with considerable Jahn-Teller distortion of the $M(1)O_6$ octahedron, M(1)-O = 1.850, 1.916, 2.242(3) Å (each x2); M(2)-O = 1.806-1.921 (5-coordination), Si-O = 1.622-1.638 Å (tetrahedral).

1. Strukturbericht, 2, 110, 512; 6, 22; Structure Reports, 26, 502; 27, 710; 45A, 367.

KOASHVITE
$Na_6(Ca,Mn)(Ti,Fe)[Si_6O_{18}]$

N.M. ČERNICOVA, Z.V. PUDOVKINA, A.A. VORONKOV and Ju.A. PJATENKO, 1980. Mineral. Ž., 2, 40-44.

Orthorhombic, Pmnb, a = 10.179, b = 20.899, c = 7.335 Å, Z = 4. R = 0.108.

The structure contains a three-dimensional skeleton of Ti,Fe octahedra and Si_6O_{18} rings, similar to lovozerite (1).

1. Structure Reports, 24, 498.

LÅVENITE
$(Na,Ca)_2(Ca,Mn,Fe,Ti)(Zr,Nb)(Si_2O_7)OF$

M. MELLINI, 1981. Tschermaks Min. Petr. Mitt., 28, 99-112.

Monoclinic, $P2_1/a$, a = 10.83, b = 9.98, c = 7.174 Å, β = 108.1°, Z = 4. Mo radiation, R = 0.032 for 1743 reflexions.

The structure is as previously described (1, except for x for Fe,Mn); it contains walls of large cation polyhedra, linked directly and via Si_2O_7 groups (Fig. 1).

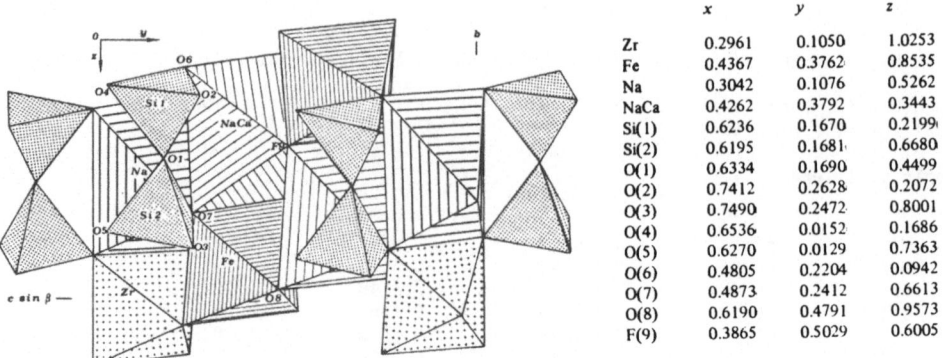

	x	*y*	*z*
Zr	0.2961	0.1050	1.0253
Fe	0.4367	0.3762	0.8535
Na	0.3042	0.1076	0.5262
NaCa	0.4262	0.3792	0.3443
Si(1)	0.6236	0.1670	0.2199
Si(2)	0.6195	0.1681	0.6680
O(1)	0.6334	0.1690	0.4499
O(2)	0.7412	0.2628	0.2072
O(3)	0.7490	0.2472	0.8001
O(4)	0.6536	0.0152	0.1686
O(5)	0.6270	0.0129	0.7363
O(6)	0.4805	0.2204	0.0942
O(7)	0.4873	0.2412	0.6613
O(8)	0.6190	0.4791	0.9573
F(9)	0.3865	0.5029	0.6005

Fig. 1.   Structure of lavenite.

1.   Structure Reports, 24, 499.

LEPIDOLITE
$K(Al,Li)_2(Li,Mn,Mg)(Si,Al)_4O_{10}(OH,F)_2$

S. GUGGENHEIM, 1981. Amer. Min., 66, 1221-1232.

1M
Monoclinic, C2/m or C2, a = 5.20-5.242, b = 9.01-9.055, c = 10.09-10.149 Å, β = 99.3-
100.77°, for three samples, Z = 2. Mo radiation, R = 0.054-0.100 for C2/m, 0.043-
0.062 for C2, for 400-1164 reflexions.

$2M_2$
Monoclinic, C2/c, a = 9.023, b = 5.197, c = 20.171 Å, β = 99.48°, Z = 4. Mo radiation,
R = 0.048 for 2764 reflexions.

Atomic positions

	x	*y*	*z*

Radkovice Lepidolite-1M

	x	*y*	*z*
K	0.0	0.5	0.0
M(1)	0.0	0.0	0.5
M(2)	0.0	0.3289	0.5
T	0.08100	0.16860	0.23203
O(1)	0.0218	0.0	0.1750
O(2)	0.3252	0.2319	0.1680
O(3)	0.1418	0.1768	0.3945
F	0.1076	0.5	0.4017

Radkovice Lepidolite-$2M_2$

	x	*y*	*z*
K	0.5	0.4097	0.25
M(1)	0.25	0.25	0.0
M(2)	0.58561	0.2437	0.00005
T(1)	0.79426	0.4078	0.13397
T(2)	0.12556	0.4136	0.13394
O(1)	0.7676	0.3937	0.05266
O(2)	0.0905	0.4261	0.05283
O(3),F	0.4468	0.4291	0.04923
O(4)	0.7058	0.1774	0.16634
O(5)	0.2380	0.1785	0.16219
O(6)	0.9719	0.3737	0.16611

The structures are essentially as previously described ($\underline{1}$, $\underline{2}$). 1M and $2M_2$ crystals from Radkovice, Czechoslovakia, are ordered in C2/m and C2/c, respectively, M(1) = $Li_{0.91}(Mn,Mg)_{0.09}$ for 1M and $Li_{1.0}$ for $2M_2$. Lepidolite-1M from Japan is similar topologically to zinnwaldite in C2.

$\underline{1}$. Structure Reports, $\underline{42A}$, 407.
$\underline{2}$. Ibid., $\underline{39A}$, 346.

LIDDICOATITE
$Ca(Li,Al)_3Al_6(BO_3)_3Si_6O_{18}(O,OH,F)_4$

B. NUBER and K. SCHMETZER, 1981. Neues Jb. Miner., Mh., 215-219.

[Rhombohedral, $R3m$], a = 15.875, c = 7.126 Å, Z = 3. Mo radiation, R = 0.033 for 1608 reflexions.

Atomic positions

			x	y	z
Ca	in	3(a)	0	0	0.8548
Al(1)		9(b)	0.0619	-0.0619	0.4582
Al(2)		18(c)	0.2600	0.2969	0.4806
Si		18(c)*	0.1900	0.1920	0.0922
B		9(b)*	-0.1090	0.1090	0.6376
O(1)		3(a)	0	0	0.3036
O(2)		9(b)	-0.0601	0.0601	0.6108
O(3)		9(b)	0.1345	-0.1345	0.5831
O(4)		9(b)	-0.0926	0.0926	0.0183
O(5)		9(b)	0.0924	-0.0924	-0.0030
O(6)		18(c)	0.1862	0.1962	0.3158
O(7)		18(c)	0.2855	0.2859	0.0112
O(8)		18(c)	0.2703	0.2097	0.6509

* interchanged in original paper

The mineral from Madagascar is a tourmaline, with a structure similar to that of elbaite ($\underline{1}$).

$\underline{1}$. Structure Reports, $\underline{39A}$, 340.

LIEBENBERGITE
$(Ni,Mg)_2SiO_4$

D.L. BISH, 1981. Amer. Min., $\underline{66}$, 770-776.

Orthorhombic, [Pbnm], a = 4.731, 4.737, b = 10.180, 10.164, c = 5.941, 5.932 Å, for natural ($Ni_{1.52}Co_{0.05}Fe_{0.09}Mg_{0.34}$) and synthetic ($Ni_{1.16}Mg_{0.84}$) specimens, Z = 4. Mo radiation, R = 0.029, 0.047 for 860, 1079 reflexions.

Atomic positions

			Natural			Synthetic		
			x	y	z	x	y	z
M(1)	in	4(a)	0	0	0	0	0	0
M(2)		4(c)	0.9903	0.2747	1/4	0.9909	0.2748	1/4
Si		4(c)	0.4265	0.0938	1/4	0.4263	0.0937	1/4
O(1)		4(c)	0.7671	0.0925	1/4	0.7676	0.0931	1/4
O(2)		4(c)	0.2190	0.4455	1/4	0.2199	0.4456	1/4
O(3)		8(d)	0.2755	0.1625	0.0310	0.2752	0.1625	0.0311

M(1) =	Ni	0.83Ni + 0.17Mg
M(2) =	0.66Ni + 0.34Mg	0.33Ni + 0.67Mg

Olivine structure (1).

1. Strukturbericht, 1, 352; Structure Reports, 41A, 392.

MEDAITE
$Mn_6[VSi_5O_{18}(OH)]$

C.M. GRAMACCIOLI, G. LIBORIO and T. PILATI, 1981. Acta Cryst., B37, 1972-1978.

Monoclinic, $P2_1/n$, a = 6.712, b = 28.948, c = 7.578 Å, β = 95.40°, D_m = 3.70, Z = 4.
Mo radiation, R = 0.059 for 3350 reflexions.

    The structure (Fig. 1) contains discrete linear vanadatopentasilicate anions (with some substitution of As for V), linked by Mn^{2+} ions with octahedral (Mn-O = 2.056-2.341 Å) and (4+2+1)-coordinations (Mn-O = 2.095-2.202; 2.360-2.613; 2.976 and 3.002 Å).

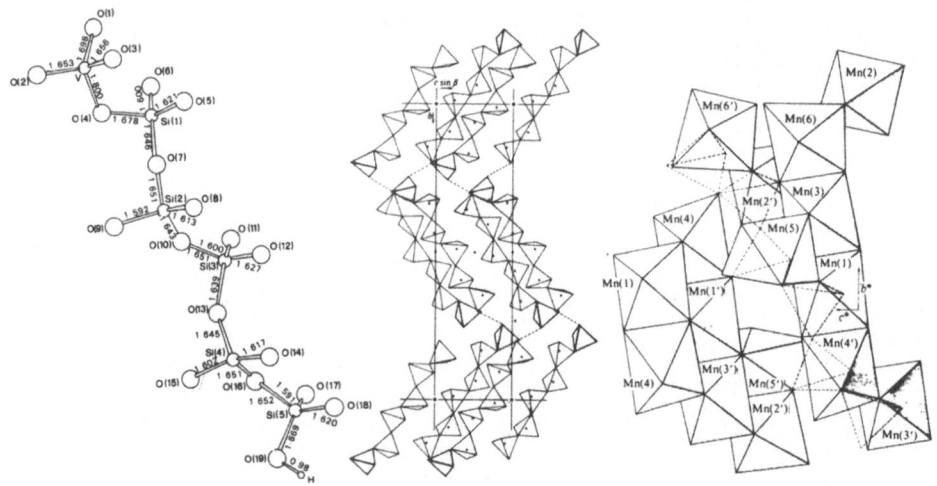

Fig. 1.   Structure of medaite.

MICA (TRIOCTAHEDRAL)
$KMg_3FeSi_3O_{10}F_2$ (idealized)

R.M. HAZEN, L.W. FINGER and D. VELDE, 1981. Amer. Min., <u>66</u>, 586-591.

Monoclinic, C2/m, a = 5.329, b = 9.230, c = 10.219 Å, β = 99.98°, Z = 2. Mo radiation, R = 0.030 for 625 reflexions.

Atomic positions

	x	y	z
T	0.5749	0.1667	0.2255
M1	0	1/2	1/2
M2	0	0.8337	1/2
K	0	0	0
O1	0.8198	0.2355	0.1672
O2	0.5264	0	0.1669
O3	0.6298	0.1670	0.3901
F	0.1336	0	0.3992

Structure as previously described for trioctahedral micas (<u>1</u>).

1. Structure Reports, <u>19</u>, 469; <u>39A</u>, 337.

MICROCLINE
$(K,Na)AlSi_3O_8$

A. BLASI, C. DE POL BLASI and P.F. ZANAZZI, 1981. Neues Jb. Miner., Abh., <u>142</u>, 71-90.

Triclinic, C$\overline{1}$, a = 8.583, b = 12.975, c = 7.208 Å, α = 90.10, β = 116.02, γ = 89.78°, Z = 4. Mo radiation, R = 0.047 for 1207 reflexions. Material from a microperthite.

Feldspar structure (<u>1</u>), with some Al/Si ordering; mean T-O = 1.619, 1.619, 1.660, 1.677(4) Å.

1. Structure Reports, <u>19</u>, 474; <u>27</u>, 678; <u>29</u>, 407; <u>41A</u>, 434.

MULLITE (3:2)
$Al(Al,Si)_2O_{4.86}$

H. SAALFELD and W. GUSE, 1981. Neues. Jb. Miner., Mh., 145-150.

Orthorhombic, Pbam, a = 7.553, b = 7.686, c = 2.8864 Å, Z = 2. Mo radiation, R = 0.027 for 660 reflexions.

Atomic positions

		x	y	z
1	Al(1)	0	0	0
0.137	Al*	0.2379	0.2943	1/2
0.863	Si/Al	0.3515	0.1589	1/2
1	O(1)	0.3735	0.2799	0
1	O(2)	0.1421	0.0779	1/2
0.590	O(3)	0	1/2	1/2
0.137	O*	0.0493	0.4482	1/2

Isostructural with a 2:1 mullite (1) and a 3:2 germanium-mullite (2), apart from differences in occupancies. The Al*-tetrahedral site is more distorted than in 1.

1. Structure Reports, 28, 233; 34A, 366.
2. Ibid., 42A, 446.

NATROAPOPHYLLITE

$NaCa_4Si_8O_{20}F \cdot 8H_2O$

I.   H. MATSUEDA, Y. MIURA and J. RUCKLIDGE, 1981. Amer. Min., 66, 410-415.
II. Y. MIURA, T. KATO, J. RUCKLIDGE and H. MATSUEDA, 1981. Ibid., 66, 416-423.

Orthorhombic, Pnnm, a = 8.875, b = 8.881, c = 15.79 Å, D_m = 2.50, Z = 2. Mo radiation, R = 0.056 for 563 (diffuse) reflexions.

The structure (Fig. 1) is very similar to that of (tetragonal) apophyllite (1). SiO_4 tetrahedra are linked into layers, with rings of four and eight tetrahedra. The layers are linked by NaO_8 and $Ca(O,F)_7$ polyhedra. Si-O = 1.58-1.63, Na-O = 2.76, 2.85, Ca-O = 2.21-2.49, Ca-F = 2.41 Å.

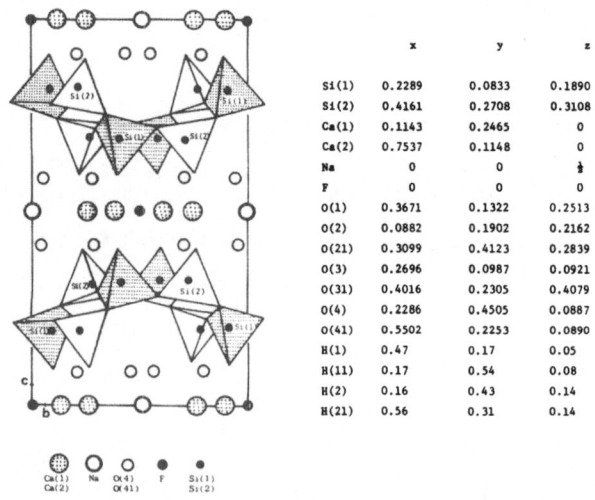

	x	y	z
Si(1)	0.2289	0.0833	0.1890
Si(2)	0.4161	0.2708	0.3108
Ca(1)	0.1143	0.2465	0
Ca(2)	0.7537	0.1148	0
Na	0	0	½
F	0	0	0
O(1)	0.3671	0.1322	0.2513
O(2)	0.0882	0.1902	0.2162
O(21)	0.3099	0.4123	0.2839
O(3)	0.2696	0.0987	0.0921
O(31)	0.4016	0.2305	0.4079
O(4)	0.2286	0.4505	0.0887
O(41)	0.5502	0.2253	0.0890
H(1)	0.47	0.17	0.05
H(11)	0.17	0.54	0.08
H(2)	0.16	0.43	0.14
H(21)	0.56	0.31	0.14

Fig. 1.   Structure of natroapophyllite.

1. Strukturbericht, 2, 145, 545; Structure Reports, 37A, 329; 42A, 399; 44A, 303.

NATROLITE

$Na_2Al_2Si_3O_{10} \cdot 2H_2O$

I.   A. ALBERTI and G. VEZZALINI, 1981. Acta Cryst., B37, 781-788.
II. F. PECHAR, 1981. Ibid., B37, 1909-1911.

Orthorhombic, Fdd2, a = 18.354, 18.325, b = 18.587, 18.653, c = 6.608, 6.601 Å, in I, II, D_m = 2.15, Z = 8. Mo, Cu radiations, R = 0.054, 0.056 for 463, 400 (films) reflexions.

The structure (Fig. 1) is as previously described (1); the sample in I shows partial Si/Al disorder (Al content of Si(1), Si(2), Al sites being 12, 12, 84%, respectively).

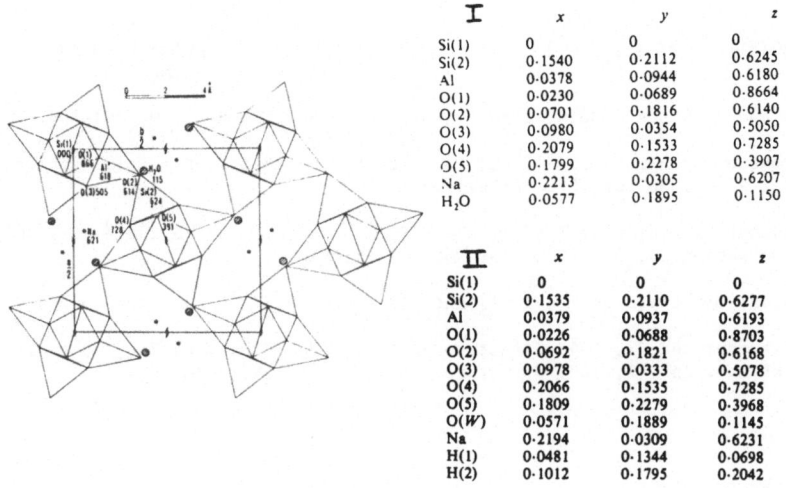

I	x	y	z
Si(1)	0	0	0
Si(2)	0·1540	0·2112	0·6245
Al	0·0378	0·0944	0·6180
O(1)	0·0230	0·0689	0·8664
O(2)	0·0701	0·1816	0·6140
O(3)	0·0980	0·0354	0·5050
O(4)	0·2079	0·1533	0·7285
O(5)	0·1799	0·2278	0·3907
Na	0·2213	0·0305	0·6207
H_2O	0·0577	0·1895	0·1150

II	x	y	z
Si(1)	0	0	0
Si(2)	0·1535	0·2110	0·6277
Al	0·0379	0·0937	0·6193
O(1)	0·0226	0·0688	0·8703
O(2)	0·0692	0·1821	0·6168
O(3)	0·0978	0·0333	0·5078
O(4)	0·2066	0·1535	0·7285
O(5)	0·1809	0·2279	0·3968
O(W)	0·0571	0·1889	0·1145
Na	0·2194	0·0309	0·6231
H(1)	0·0481	0·1344	0·0698
H(2)	0·1012	0·1795	0·2042

Fig. 1.   Structure of natrolite.

1. Strukturbericht, 3, 168, 529; Structure Reports, 24, 477; 28, 244; 29, 409; 39A, 357.

OLIVINE
$(Mg,Fe)_2SiO_4$

G. NOVER and G. WILL, 1981. Z. Kristallogr., 155, 27-45.

Orthorhombic, [Pmcn], a = 6.003-6.005, b = 4.767-4.769, c = 10.239-10.277 Å, for seven natural crystals, Z = 4. Mo radiation, R = 0.023-0.039 for 975-980 reflexions; five of the crystals had been previously subjected to oxygen partial pressures of 10^{-16} to 10^{-21} bar.

Olivine structure (1); low oxygen pressure increases the amount of Fe in the M(1) site.

1. Strukturbericht, 1, 352; Structure Reports, 39A, 348.

ORTHOERICSSONITE
$(Ba,Sr)FeMn_3Si_2O_7(OH)$

S. MATSUBARA, 1980. Mineral. J., 10, 107-121.

Orthorhombic, Pnmm, a = 20.230, b = 6.979, c = 5.392 Å, Z = 4. R = 0.054 for 1626 reflexions.

The structure contains composite sheets of Si_2O_7 groups, $Fe(III)O_5$ square pyramids, and $Mn(II)O_6$ octahedra; the sheets are linked by (Ba,Sr) ions.

ORTHOPYROXENE   CLINOPYROXENE
$MgSiO_3$ $CaMgSi_2O_6$

R. BASSO and A. DELLA GIUSTA, 1980. Neues Jb. Miner., Abh., 139, 254-264.

Orthopyroxene, orthorhombic, [Pbca], a = 18.251, b = 8.814, c = 5.199 Å, Z = 16. Mo radiation, R = 0.030 for 1852 reflexions.

Clinopyroxene, monoclinic, [C2/c], a = 9.717, b = 8.888, c = 5.258 Å, β = 106.22°, Z = 4. Mo radiation, R = 0.034 for 966 reflexions.

Typical orthopyroxene (e.g. 1) and clinopyroxene (e.g. 2) structures.

1. Strukturbericht, 2, 134; Structure Reports, 30A, 425; 40A, 288; 42A, 411.
2. Strukturbericht, 2, 130; Structure Reports, 39A, 356; 41A, 394.

PLAGIOCLASE
$(Na,Ca)(Al,Si)_4O_8$

W. HORST, T. TAGAI, M. KOREKAWA and H. JAGODZINSKI, 1981. Z. Kristallogr., 157, 233-250.

The superstructure of a plagioclase An_{52} from Labrador is described as a periodic antiphase domain structure, consisting of two centrosymmetric structure elements [see also 1]. Al/Si and Ca/Na distributions are determined.

1. Structure Reports, 38A, 372; 39A, 345; 42A, 412; 46A, 388, 391.

PSEUDOWOLLASTONITE
α-$CaSiO_3$

T. YAMANAKA and H. MORI, 1981. Acta Cryst., B37, 1010-1017.

Triclinic, C$\bar{1}$, a = 6.853, b = 11.895, c = 19.674 Å, α = 90.12, β = 90.55, γ = 90.00°, Z = 24. Mo radiation, R = 0.040 for 1935 reflexions.

The structure is a four-layer polytype (six-layer and disordered stacking polytypes also being found). It contains four Ca-octahedra layers with Si_3O_9 rings of three tetrahedra between the layers (Fig. 1).

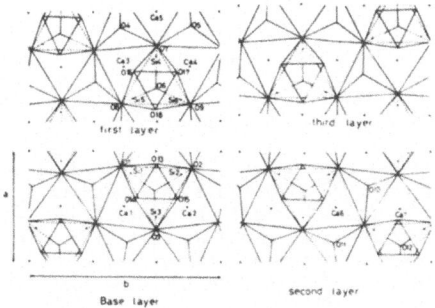

Fig. 1.   Layers in pseudowollastonite.

PYROPHYLLITE-1Tc
AlSi$_2$O$_5$(OH)

J.H. LEE and S. GUGGENHEIM, 1981. Amer. Min., <u>66</u>, 350-357.

Triclinic, C$\bar{1}$, a = 5.160, b = 8.966, c = 9.347 Å, α = 91.18, β = 100.46, γ = 89.64°,
Z = 4. Mo radiation, R = 0.060 for 1574 reflexions.

The structure (Fig. 1) is as previously described (<u>1</u>). Mean Al-O = 1.912, Si-O
= 1.618 Å.

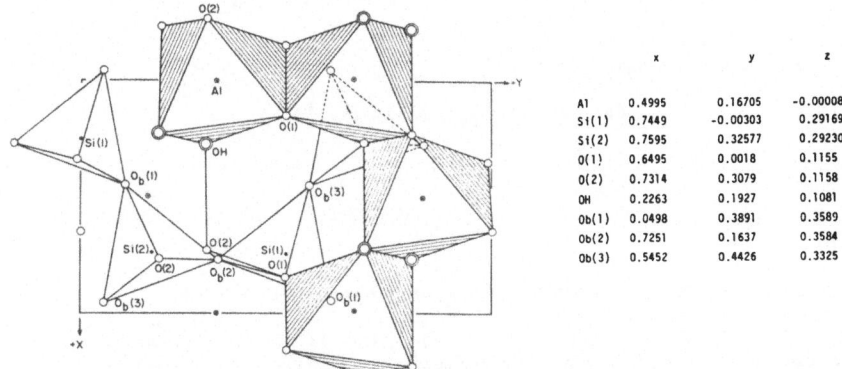

	x	y	z
Al	0.4995	0.16705	-0.00008
Si(1)	0.7449	-0.00303	0.29169
Si(2)	0.7595	0.32577	0.29230
O(1)	0.6495	0.0018	0.1155
O(2)	0.7314	0.3079	0.1158
OH	0.2263	0.1927	0.1081
Ob(1)	0.0498	0.3891	0.3589
Ob(2)	0.7251	0.1637	0.3584
Ob(3)	0.5452	0.4426	0.3325

Fig. 1.   Structure of pyrophyllite-1Tc.

<u>1</u>.   Structure Reports, <u>38</u>A, 373.

PYROXENES

M. CAMERON and J.J. PAPIKE, 1981. Amer. Min., <u>66</u>, 1-50.

Review of pyroxene structures.

SANTACLARAITE

$CaMn_4[Si_5O_{14}(OH)]OH.H_2O$

Y. OHASHI and L.W. FINGER, 1981. Amer. Min., 66, 154-168.

Triclinic, I$\bar{1}$, a = 10.273, b = 11.910, c = 12.001 Å, α = 105.77, β = 110.64, γ = 87.13°, Z = 4. Mo radiation, R = 0.036 for 3307 reflexions.

The structure (Fig. 1) contains alternating tetrahedral and octahedral layers, similar to but differently arranged from those in rhodonite (1). The tetrahedral layer is made up of infinite single chains of silicate tetrahedra, with a repeat period of five tetrahedra. The octahedral layer contains rows of ten octahedra, with adjacent rows displaced to form bands two or three octahedra wide; four octahedral sites contain Mn, with Ca ordered in the fifth site. Si-O = 1.58-1.67, Mn-O = 2.06-2.35, Ca-O = 2.30-2.46 (seventh O at 2.60) Å.

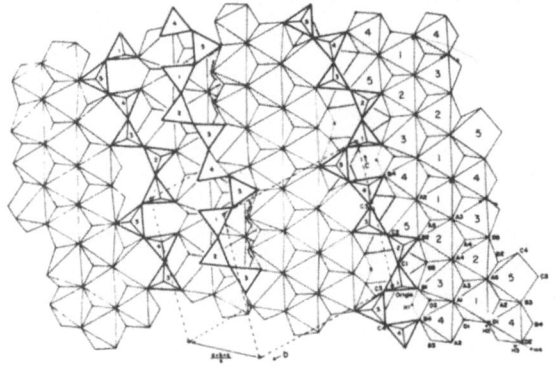

Fig. 1.   Structure of santaclaraite.

1.   Structure Reports, 22, 506; 23, 476; 28, 258; 42A, 422.

TALC

$Mg_3Si_4O_{10}(OH)_2$

B. PERDIKATSIS and H. BURZLAFF, 1981. Z. Kristallogr., 156, 177-186.

Triclinic, P$\bar{1}$, a = 5.291, b = 9.460, c = 5.290 Å, α = 98.68, β = 119.90, γ = 85.27°, Z = 2. Mo radiation, R = 0.05 for 2300 reflexions. Pseudo-monoclinic C$\bar{1}$ cell, a' = -c, b' = 2a + c, c' = -b.

Atomic positions

	C$\bar{1}$			P$\bar{1}$		
	x	y	z	x	y	z
Si(1)	0,24527	0,50259	0,29093	0,00518	0,70927	0,25732
Si(2)	0,24590	0,83587	0,29108	0,6714	0,70892	0,58997
Mg(1)	0,00000	0,0000	0,0000	0,00000	0,00000	0,00000
Mg(2)	0,50012	0,83332	0,99994	0,66664	0,00006	0,33320
O(1)	0,1991	0,8344	0,1176	0,6688	0,8824	0,6453
O(2)	0,6970	0,6674	0,1126	0,3348	0,8874	0,9704
O(3)	0,1980	0,5012	0,1176	0,0024	0,8824	0,3032
O(4)	0,0199	0,9287	0,3481	0,8574	0,6519	0,9088
O(5)	0,5202	0,9109	0,3494	0,8218	0,6506	0,3907
O(6)	0,2429	0,6699	0,3484	0,3398	0,6516	0,4270
H	0,719	0,669	0,203	0,338	0,797	0,950

The structure is as previously described (1). Two oppositely-oriented silicate layers provide octahedral coordination for Mg (Fig. 1). Si-O = 1.621-1.625, Mg-O = 2.052-2.081(2) Å.

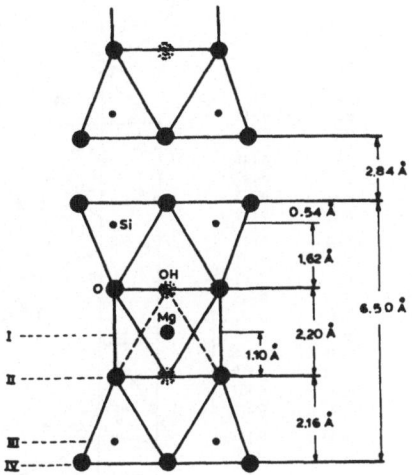

Fig. 1.   Structure of talc.

<u>1</u>. Structure Reports, <u>39</u>A, 351.

TOBERMORITE

$Ca_{2.25}[Si_3O_{7.5}(OH)_{1.5}] \cdot H_2O$

I.   S.A. HAMID, 1981. Z. Kristallogr., <u>154</u>, 189-198.
II.   E. EBERHARD and S.A. HAMID, 1981. Ibid., <u>154</u>, 268-269.
III. Idem, 1981. Ibid., <u>154</u>, 269-271.

Orthorhombic sub-cell, Imm2, a' = a/2 = 5.586, b' = b/2 = 3.696, c = 22.779 Å, D_m = 2.43, Z = 2 (a possible ordered structure is monoclinic, $P2_1$, a_m = 6.69, b_m = 7.39, c_m = 22.779 Å, γ = 123.49°). R_w = 0.085 for 513 sharp reflexions. Previous study in <u>1</u>.

The average structure contains disordered, superimposed $Si_3(O/OH)_9$ chains, from which a possible ordered monoclinic structure can be derived (Fig. 1); the chains are linked by 7-coordinate Ca ions.

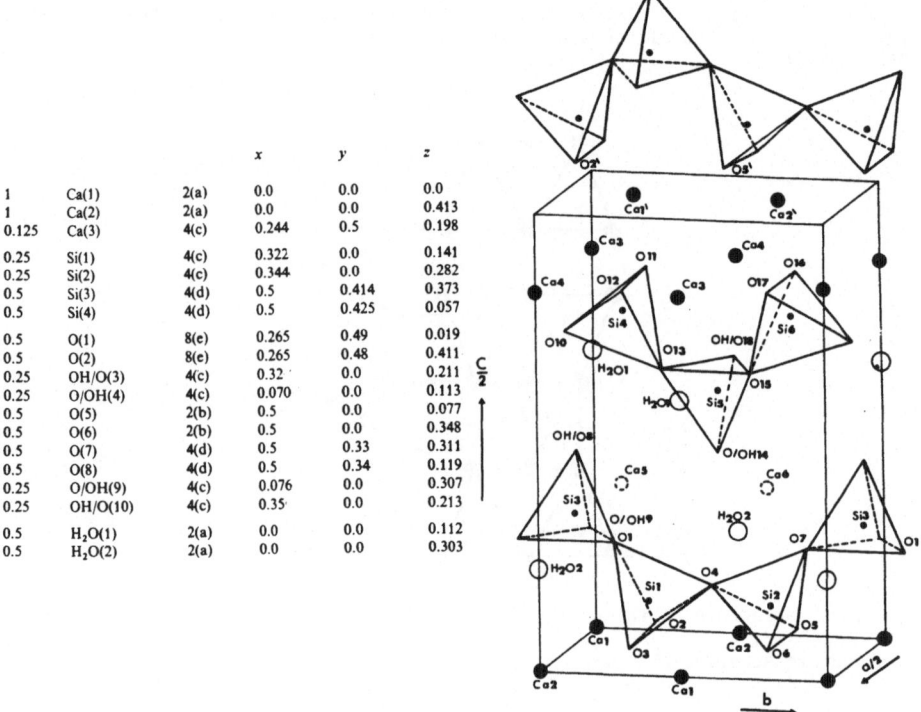

			x	y	z
1	Ca(1)	2(a)	0.0	0.0	0.0
1	Ca(2)	2(a)	0.0	0.0	0.413
0.125	Ca(3)	4(c)	0.244	0.5	0.198
0.25	Si(1)	4(c)	0.322	0.0	0.141
0.25	Si(2)	4(c)	0.344	0.0	0.282
0.5	Si(3)	4(d)	0.414	0.373	
0.5	Si(4)	4(d)	0.5	0.425	0.057
0.5	O(1)	8(e)	0.265	0.49	0.019
0.5	O(2)	8(e)	0.265	0.48	0.411
0.25	OH/O(3)	4(c)	0.32	0.0	0.211
0.25	O/OH(4)	4(c)	0.070	0.0	0.113
0.5	O(5)	2(b)	0.5	0.0	0.077
0.5	O(6)	2(b)	0.5	0.0	0.348
0.5	O(7)	4(d)	0.5	0.33	0.311
0.5	O(8)	4(d)	0.5	0.34	0.119
0.25	O/OH(9)	4(c)	0.076	0.0	0.307
0.25	OH/O(10)	4(c)	0.35	0.0	0.213
0.5	$H_2O(1)$	2(a)	0.0	0.0	0.112
0.5	$H_2O(2)$	2(a)	0.0	0.0	0.303

Fig. 1.   Atomic positional parameters in the Imm2 sub-cell, and derived structure of tobermorite (a/2 = 5.58, b = 7.39, c/2 = 11.389 Å; dotted Ca ions are disordered).

1. Structure Reports, 20, 412.

URANOPHANE
$Ca(H_3O)_2(UO_2)_2(SiO_4)_2 \cdot 2H_2O$

BOLTWOODITE
$K(H_3O)(UO_2)(SiO_4)$

WEEKSITE
$K_2(UO_2)_2(Si_2O_5)_3 \cdot 4H_2O$

F.V. STOHL and D.K. SMITH, 1981. Amer. Min., 66, 610-625.

Uranophane, monoclinic, $P2_1$, a = 15.858, b = 6.985, c = 6.641 Å, β = 97.55°, Z = 2. Mo radiation, R = 0.081 for 525 reflexions.

Boltwoodite, monoclinic, $P2_1$, a = 7.073, b = 7.064, c = 6.638 Å, β = 105.75°, Z = 2. Mo radiation, R = 0.109 for 223 reflexions (twinned crystal).

Weeksite, orthorhombic, Amm2 pseudo-cell, a = 7.106, b = 17.90, c = 7.087 Å, Z = 2
(true cell is face-centred, with doubled cell parameters). Mo radiation, R = 0.15
for 270 pseudo-cell reflexions; structure not fully established.

Atomic positions

uranophane.

	x	y	z
U(1)	0.2557	0.7822	0.1344
U(2)	-0.2557	-0.7822	-0.1344
Si(1)	0.284	0.281	0.339
Si(2)	-0.284	-0.281	-0.339
O(1)	0.371	0.795	0.138
O(2)	-0.371	-0.795	-0.138
O(3)	0.143	0.757	0.123
O(4)	-0.143	-0.757	-0.123
O(5)	0.271	0.453	0.187
O(6)	-0.271	-0.453	-0.187
O(7)	0.258	0.095	0.181
O(8)	-0.258	-0.095	-0.181
O(9)	0.229	0.292	0.523
O(10)	-0.229	-0.292	-0.523
O(11)	0.381	0.272	0.432
O(12)	-0.381	-0.272	-0.432
$H_2O(1)$	0.069	0.345	0.359
$H_2O(2)$	0.991	0.010	0.196
$H_2O(3)$	0.936	0.610	0.558
$H_2O(4)$	0.993	0.507	0.945
Ca	0.019	0.672	0.280

boltwoodite.

	x	y	z
U	0.0252	0.25	0.1385
K	0.541	0.548	0.153
Si	0.933	0.25	0.637
O(1)	0.290	0.25	0.147
O(2)	0.227	0.75	0.885
O(3)	0.022	-0.079	0.194
O(3B)	-0.022	0.079	-0.194
O(5)	0.060	0.25	0.484
O(6)	0.294	0.75	0.490
H_2O	0.400	0.413	0.689

weeksite partial structure analysis

	x	y	z
U	0.0	0.198	0.0
Si	0.0	0.130	0.523
O(1)	0.25	0.307	0.546
O(2)	0.0	0.180	0.693
O(2B)	0.0	0.180	0.307
O(3)	0.194	0.071	0.487
O(4)	0.0	0.429	0.473

The uranophane structure is as previously described (1). The structures (Fig. 1)
contain sheets of SiO₄ tetrahedra and UO₇ pentagonal bipyramids, linked by the other
cations.

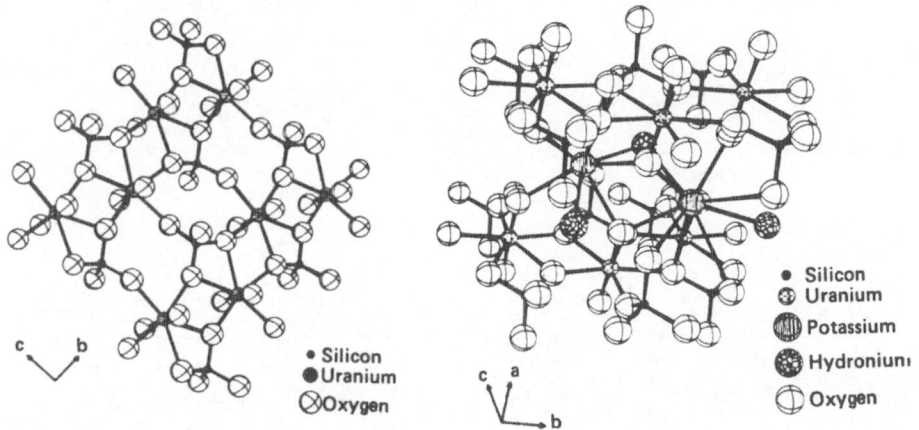

Fig. 1.   Structures of uranophane (left) and boltwoodite (right).

1. Structure Reports, 21, 338.

WONESITE
$(Na,K)(Mg,Fe,Al)_6(Al,Si)_8O_{20}(OH,F)_4$

F.S. SPEAR, R.M. HAZEN and D. RUMBLE, 1981. Amer. Min., <u>66</u>, 100-105.

Monoclinic, C2/m, a = 5.312, b = 9.163, c = 9.825 Å, β = 103.18°, Z = 2. Mo radiation, data for 10 00ℓ reflexions. The material is a 1Md mica with stacking disorder.

ZEOLITES

ZEOLITES A (DEHYDRATED)
$Na_{12}Al_{12}Si_{12}O_{48}$

L.A. BURSILL, E.A. LODGE, J.M. THOMAS and A.K. CHEETHAM, 1981. J. Phys. Chem., <u>85</u>, 2409-2421.

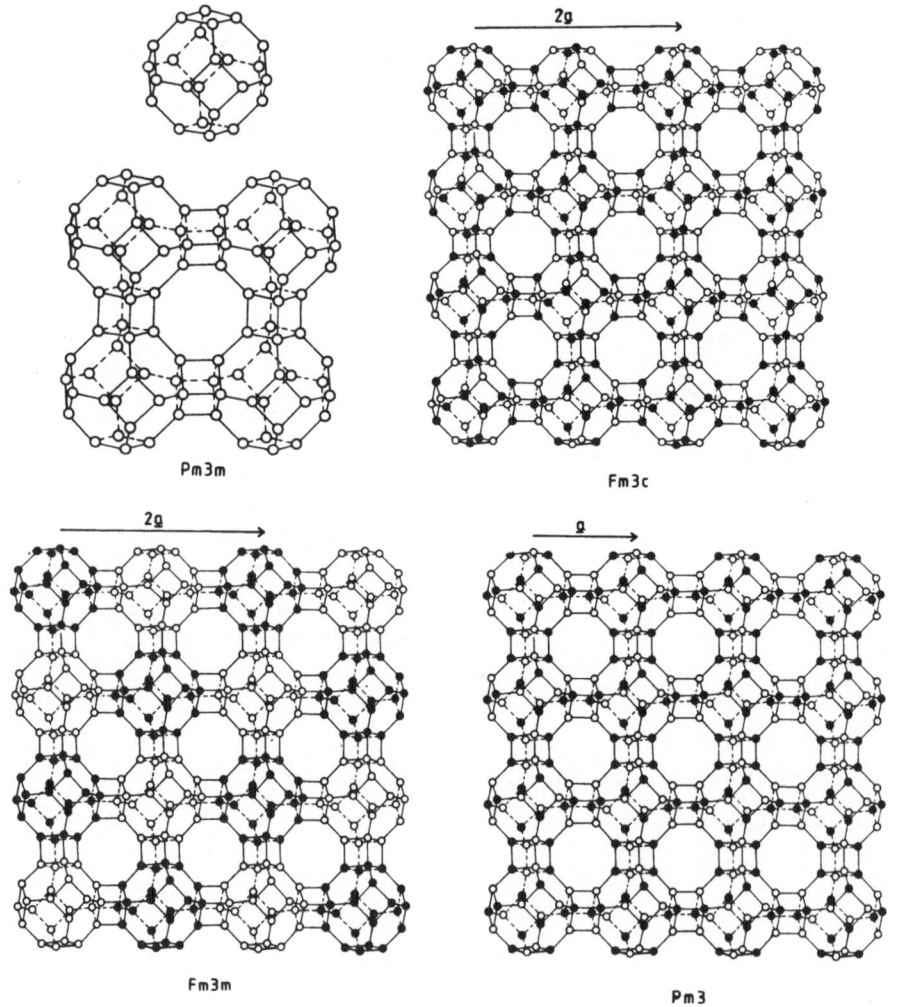

Fig. 1.   Si,Al framework in A-type zeolites.

The principal features of the framework structure of Na-A zeolite have been described previously in terms of a Si,Al cuboctahedron (Fig. 1) in a cubic unit cell, a = 12 Å, Pm3m, and also in a doubled 24 Å cell, space group Fm3c, or Fm3m, or Fm3, or in a 12 Å, Pm3 cell.

The present paper describes neutron diffraction, electron microscope, and NMR results which indicate that the true symmetry is rhombohedral, R$\overline{3}$, a = 17.401 Å, α = 59.53°, at 5K, for Si:Al = 1.00. A 3:1 ordering scheme is indicated (each Si bridged to 3 Al and 1 Si, and vice versa) and a unique model is derived (Fig. 2). Further X-ray and neutron studies are required to establish the detailed structure.

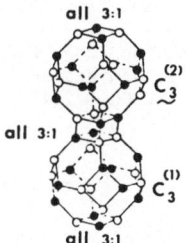

Fig. 2.   The asymmetric unit of the rhombohedral structure for A-type zeolites.

ZEOLITE A (AMMONIUM-EXCHANGED, DEHYDRATED)
$(NH_4)_{12}Al_{12}Si_{12}O_{48}$

L.B. McCUSKER and K. SEFF, 1981. J. Amer. Chem. Soc., 103, 3441-3446.

Cubic, Fm3c, a = 24.568 Å, Z = 8. Mo radiation, R = 0.059 for 265 reflexions in a Pm3m pseudo-cell; Cu radiation, R = 0.053 for 467 reflexions in Fm3c (an additional 30 reflexions violate the c-glide condition).

The aluminosilicate framework is as previously described (e.g. 1), with ammonium ions distributed over four sites.

1. Structure Reports, 37A, 346; 44A, 318; 45A, 378.

ZEOLITE A (SILVER-EXCHANGED, DEHYDRATED)
$Ag_{12}Al_{12}Si_{12}O_{48} \cdot xH_2O$ (3 samples)

$Ag_{2.4}K_{9.6}Al_{12}Si_{12}O_{48} \cdot xH_2O$

L.R. GELLENS, W.J. MORTIER, R.A. SCHOONHEYDT and J.B. UYTTERHOEVEN, 1981. J. Phys. Chem., 85, 2783-2788.

The description is in terms of the 12 Å, Pm3m cell, and cations positions are interpreted in terms of Ag_3^{2+} linear clusters, Ag^0, and Ag^+ (compare 1).

1. Structure Reports, 43A, 327; 44A, 318, 350.

ZEOLITE A (CADMIUM-EXCHANGED, HYDRATED and PARTIALLY-DEHYDRATED)
$Cd_6Al_{12}Si_{12}O_{48} \cdot xH_2O$ (x = 31 for hydrated sample)

ZEOLITE A (COPPER-EXCHANGED, DEHYDRATED)
$Cu_8Al_{12}Si_{12}O_{48}(OH)_y \cdot xH_2O$ (4 samples)

ZEOLITE A (ZINC-EXCHANGED, HYDRATED and PARTIALLY-DEHYDRATED)
$Zn_6Al_{12}Si_{12}O_{48} \cdot xH_2O$ (x = 29 for hydrated sample)

I.   L.B. McCUSKER and K. SEFF, 1981. J. Phys. Chem., 85, 166-174.
II. H.S. LEE and K. SEFF, 1981. Ibid., 85, 397-405.
III. L.B. McCUSKER and K. SEFF, 1981. Ibid., 85, 405-410.

All structures are described in terms of the 12 Å, Pm3m cell, and the cation
positions are located. In the more-fully dehydrated Cu compounds, the data are
interpreted in terms of a Cu_3^+ cluster (Cu-Cu = 2.49 Å) and Cu atoms.

ZEOLITE 4A
$Li_{12}Al_{12}Si_{12}O_{48} \cdot 9 \cdot 8LiNO_3 \cdot 9 \cdot 3H_2O$

$Na_{12}Al_{12}Si_{12}O_{48} \cdot 10NaNO_3 \cdot 6 \cdot 6H_2O$

$Ag_{12}Al_{12}Si_{12}O_{48} \cdot 9 \cdot 5AgNO_3 \cdot 5 \cdot 9H_2O$

N. PETRANOVIĆ, U. MIOČ, M. ŠUŠIĆ, R. DIMITRIJEVIĆ and I. KRSTANOVIĆ, 1981. J. Chem.
Soc., Faraday I, 77, 379-389.

Cubic, Pm3m, a = 12.075, 12.295, 12.340 Å (weak reflexions indicate a 24 Å supercell),
Z = 1. Mo radiation, R = 0.099, 0.095 for 132, 191 reflexions for Li, Na compounds;
refinement unsatisfactory for Ag compound.

    Framework structure is as previously described (1).

1. Structure Reports, 24, 480; 37A, 346; 41, 398.

ZEOLITE E (SODIUM, TETRAMETHYLAMMONIUM)
$Na_{8.1}(Me_4N)_{2.2}Al_{9.4}Si_{26.6}O_{72}(OH)_{0.9} \cdot 25H_2O$

W.M. MEIER and M. GRONER, 1981. J. Solid State Chem., 37, 204-218.

Hexagonal, P6_3/mmc, a = 13.28, 13.00, 12.86, c = 15.21, 15.47, 15.51 Å, at 20, 220,
350°C, Z = 1. Powder data for synthetic material.

    The structure is a new zeolite type of the chabazite family, related to that of
erionite (1); the parallel 6-rings of the silicate framework have sequences AABAAC
in erionite and ABBACC in the present compound.

1. Structure Reports, 23, 494; 34A, 375.

ZEOLITE X (NICKEL CERIUM, DEHYDRATED)
$Ni_{21}Ce_{6.5}Na_{20}H_4Al_{88}Si_{104}O_{384}$

J. JEANJEAN, S. DJEMEL, M.F. GUILLEUX and D. DELAFOSSE, 1981. J. Phys. Chem., $\underline{85}$, 4145-4147.

Structure as previously described ($\underline{1}$). Study of cation positions during dehydration, and reduction and reoxidation.

$\underline{1}$. Structure Reports, $\underline{40A}$, 313.

ZEOLITE Y (Na and NaH)
$(Na,H)_{55}Al_{55}Si_{137}O_{384}$ (4 samples)

I. Z. JIRÁK, S. VRATISLAV and V. BOSÁČEK, 1980. J. Phys. Chem. Solids, $\underline{41}$, 1089-1095.
II. V. BOSÁČEK, S. BERAN and Z. JIRÁK, 1981. J. Phys. Chem., $\underline{85}$, 3856-3859.

Cubic, Fd3m, neutron powder study of cation distributions.

ZEOLITE Y
$Fe_{13}Na_{24}H_6Al_{56}Si_{136}O_{384} \cdot 250H_2O$

J.R. PEARCE, W.J. MORTIER, J.B. UYTTERHOEVEN and J.H. LUNSFORD, 1981. J. Chem. Soc., Faraday I, $\underline{77}$, 937-946.

Cubic, Fd3m, $\underline{a}$ not given, four samples, Z = 1. Powder data. [Structure as in $\underline{1}$.] Fe positions are determined.

$\underline{1}$. Structure Reports, $\underline{32A}$, 484.

ZEOLITE Y (PALLADIUM-EXCHANGED)
$[Pd(NH_3)_4,Pd,Na,H]_{42}Al_{56}Si_{136}O_{384} \cdot xH_2O$ (3 samples)

G. BERGERET, P. GALLEZOT and B. IMELIK, 1981. J. Phys. Chem., $\underline{85}$, 411-416.

Cubic, Fd3m, powder data. Faujasite-type structures; Pd locations are described.

ZEOLITE ZSM-5
$Na_xAl_{1.1}Si_{94.9}O_{192}$

D.H. OLSON, G.T. KOKOTAILO, S.L. LAWTON and W.M. MEIER, 1981. J. Phys. Chem., $\underline{85}$, 2238-2243.

Orthorhombic, Pnma, a = 20.07, b = 19.92, c = 13.42 Å, Z = 1. Cu radiation, R = 0.119 for 1026 reflexions. Preliminary study in $\underline{1}$.

The three-dimensional framework (Fig. 1) contains straight channels parallel to $\underline{b}$ having rings of 10 tetrahedra, free diameter 5.4 x 5.6 Å, and sinusoidal channels along $\underline{a}$ with 10-ring openings of 5.1 x 5.4 Å. Two non-framework peaks were found, but could not be definitely assigned.

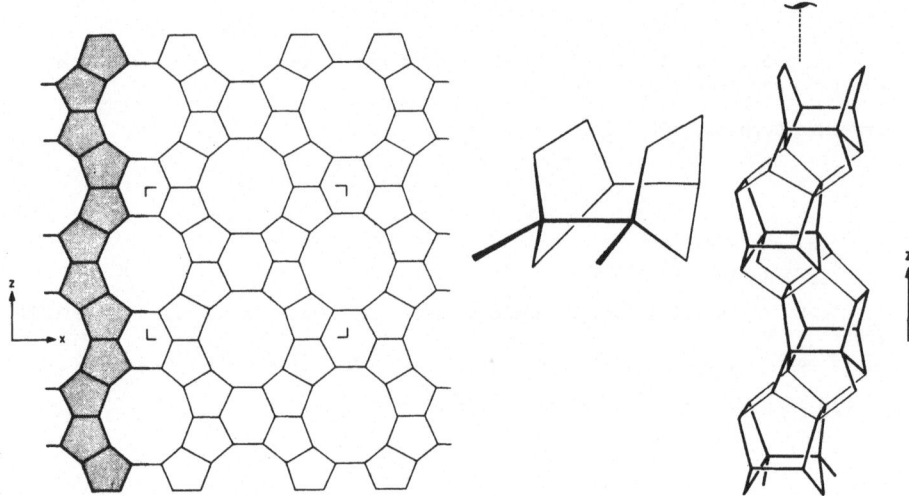

Fig. 1.   Structure of zeolite ZSM-5, and the building units of the framework.

1. Structure Reports, 44A, 331.

TABLE I

Some structural information has also been given for the following materials (listed with abbreviated 1981 reference).

Compound	Structure	Reference
$Ca_4YFe_5O_{13}$	Structure proposed based on unit-cell twinning of the $Ca_2Fe_2O_5$ structure. Electron microscopy and diffraction study	Acta Cryst., A37, 723
Magnesium fluoride, MgF_2	Study of charge distribution [structure in Structure Reports, 45A, 143]	Ibid., A37, 826
Anhydrite, $CaSO_4$	Charge density is studied	Ibid., B37, 525
$(Cr,Fe)_2O_3$-$(Ti,Zr)O_2$	α-PbO_2-related intergrowth structures	Ibid., B37, 793
Mullite, Al_2SiO_5	Electron microscope study of twinning and super-structure in Al-rich material	Amer. Min., 66, 142
$(H_3O)_2Mg_6Si_8O_{22}(OH)_2$ (high-pressure)	Trioctahedral 2:1 phyllosilicate	Ibid., 66, 576

TABLE I   371

Compound	Structure	Reference
Asbecasite, $Ca_3(Ti,Sn)As_6Si_2Be_2O_{20}$	Refinement of the structure [Structure Reports, 34A, 381] indicates possible disorder of O(4) or a lower symmetry space group, giving a bent Be-O-Si grouping	Amer. Min., 66, 819
Graphite inclusion compounds	Graphite, with insertion of XF_6^- ions (X = P, As, Sb)	Carbon, 19, 193
Graphite - vanadyl fluoride, $C_{40}VOF_3$	c = 18.07 Å	Ibid., 19, 249
$Ba_2(CrTa)O_6$	Hexagonal 8-layer perovskite, with cation ordering - Cr^{3+} in face-shared octahedra, Ta^{5+} in corner-shared octahedra	Chem. Lett., 1185 (1980)
$FeCr_3O_6$	Spinel	C.R. Acad. Sci. Paris, 293, 437
$CsScBr_3$ $CsScI_3$ $RbScCl_3$ $RbScBr_3$	$CsNiCl_3$-type	Inorg. Chem., 20, 2627
$LaTiO_3$	Orthorhombic perovskite	Izv. Akad. Nauk SSSR, Neorg. Mater., 16, 2069
$(La,Sr)TiO_3$	Perovskite	J. Amer. Ceram. Soc., 64, C75
Lead(IV) oxide, β-PbO_2	P4/mnm, a = 4.961, c = 3.385 Å, Z = 2. Powder data. Rutile structure, x(O) = 0.305	J. Appl. Cryst., 14, 141
Fe_2O_3 Cr_2O_3 V_2O_3	High-pressure studies. α-Corundum structure, $R\bar{3}c$, a ∼ 5, c ∼ 14 Å, z(M) = 0.35, x(O) = 0.31.	J. Appl. Phys., 51, 5362 (1980)
$Ln_2Mo_3O_9$	Scheelite-structure, with cation and anion vacancies	J. Chem. Soc., Dalton, 668 (1981)
UOCl UOBr UOI	PbFCl-type	J. Inorg. Nucl. Chem., 43, 1841
$LnNbF_7$, Ln = Ce-Lu,Y	$LnZrF_7$	J. Less-Common Metals, 79, 39

Compound	Structure	Reference
Ba_2NaReO_6	Perovskite, Fm3m, a = 8.278 Å. Ba in 8(c); Na in 4(b); Re in 4(a); O in 24(e): x = 0.224	J. Less-Common Metals, <u>79</u>, 165
$La_3Ni_2O_7$ $La_4Ni_3O_{10}$	$Sr_3Ti_2O_7$ $Sr_4Ti_3O_{10}$	Ibid., <u>79</u>, 215
Lithium sulphate (high-temperature), Li_2SO_4	Cubic, Fm3m, a = 7.07 Å. S at 0,0,0, with rotational disorder of the SO_4 group; Li at ±(1/4,1/4,1/4), possibly with statistical displacement from these sites. Neutron powder data at 908K. Previous study in Structure Reports, <u>21</u>, 364	J. Phys. C: Solid State Phys., <u>13</u>, 6441 (1980)
$CoFe_2O_4$	Spinel, with 80% of Co on octahedral sites, x(O) = 0.3808. Study with synchrotron radiation	J. Phys. Chem. Solids, <u>41</u>, 1097 (1980)
TiO_2 SnO_2 GeO_2 RuO_2 MnF_2	High-pressure studies. Rutile structures, <u>a</u> compresses about twice as much as <u>c</u>, x(O or F) changes little with pressure	Ibid., <u>42</u>, 143
$Cd_2Ge_{1-x}Si_xO_4$, x = 0 -0.4	Olivine	J. Solid State Chem., <u>36</u>, 241
$Bi_3Ti_2O_8F$ $PbBi_3Ti_3O_{11}F$	Structures proposed	Ibid., <u>36</u>, 349
$Cu(V,Mo)_2O_6$	Triclinically-distorted brannerite	Ibid., <u>38</u>, 97
$Ca_3Ln(PO_4)_3$, Ln = La-Gd	Eulytite	Ibid., <u>38</u>, 128
$(Mg,M)_2SnO_4$, M = Zn, Co, Ni	Inverse spinels, u(O) = 0.38. Zn in tetrahedral sites, Co in tetrahedral and octahedral sites, Ni in octahedral sites. Powder data	Ibid., <u>38</u>, 173
Iron tellurate, $Fe_2Te_4O_{11}$	Structure as previously described [Structure Reports, <u>38A</u>, 348]. Magnetic structure at 4.2K is described	Ibid., <u>39</u>, 39

TABLE I

373

Compound	Structure	Reference
$Na_2Te_2O_7$	Weberite-type	J. Solid State Chem., $\underline{39}$, 94
$\lambda-MnO_2$	Spinel-type (as for $LiMn_2O_4$, but with the Li removed from the tetrahedral sites)	Ibid., $\underline{39}$, 142
$(Fe,V)NbO_4$ $(Fe,Cr)NbO_4$	Wolframite-type for Fe-rich, rutile-type for V/Cr-rich phases	Ibid., $\underline{39}$, 294, 395
$(Li,M)Co_2O_4$, $M = Co, Zn$	Spinels	Ibid., $\underline{40}$, 117
$La_{1-x}Th_xNbO_{4+x/2}$	Distorted oxygen-excess scheelite	Ibid., $\underline{40}$, 318
Ammonium tetrachlorozincate, $(NH_4)_2ZnCl_4$	Study of superstructures	Krist. Tech., $\underline{15}$, 667 (1980)
Terbium gallate, $Tb_3Ga_5O_{12}$	Garnet, Ia3d, a = 12.340 Å, $\{Tb\}_3[Ga]_3(Ga)_2O_{12}$, O parameters are (0.0285, 0.0546,0.6493)	Kristallografija, $\underline{25}$, 1155 (1980) [Soviet Physics - Crystallography, $\underline{25}$, 661]
La_2WO_6	Intergrowths of various hexagonal polytypes	Ibid., $\underline{26}$, 604 [Ibid., $\underline{26}$, 341]
$Bi_{18}Mg_8O_{36}$ $Bi_{18}Ni_8O_{36}$	$\gamma-Bi_2O_3$ structure assumed, with the parameters of $Bi_{12}GeO_{20}$ [Structure Reports, $\underline{32A}$, 292] with random occupation of the 24(f) Bi sites	Mater. Res. Bull., $\underline{16}$, 169
Li_5NI_2 Li_7N_2I	F$\bar{4}$3m and Fd3m structures [H. SATTLEGGER and H. HAHN, 1964. Naturwissenschaften, $\underline{51}$, 534; Structure Reports, $\underline{29}$, 440], with additional anions in 4(d) sites	Ibid., $\underline{16}$, 581 and 587
Thomsonite, $NaCa_2Al_5Si_5O_{20} \cdot 6H_2O$	Structure proposed, but not fully determined	Miner. Mag., $\underline{44}$, 231
$Fe_{1.4}(Mg,Zn)_{1.3}Ti_{0.3}O_4$	Spinel, x(O) = 0.383-0.387	Phys. Status Solidi, A, $\underline{59}$, 581 (1980)
$Mg(Al,Fe)_2O_4$	Spinel, x(O) = 0.381-0.386	Ibid., A, $\underline{59}$, 817 (1980)
$Nd_2Zr_2O_7$	Pyrochlore, with vacant 8(b) site	Ibid., B, $\underline{101}$, 765 (1980)
$Nd_xZr_{1-x}O_{2-x/2}$	Pyrochlore and defect fluorite phases	

Compound	Structure	Reference
$Li_2(N,Cl)$	Antifluorite	Solid State Comm., <u>36</u>, 703 (1980)
$Na_{3-3x}Al_x\square_{2x}PO_4$	At low Al concentrations the structure is that of Na_3PO_4-II; additional Al stabilizes the high-temperature f.c.c. phase of Na_3PO_4	Solid State Ionics, <u>1</u>, 395 (1980)
Ba_3UO_6	Tetragonal perovskite, with ordered cation vacancies	Z. anorg. Chem., <u>473</u>, 171
Ba_2InRuO_6 $Ba_3InRu_2O_9$	Hexagonal $BaTiO_3$ structure with ordered In/Ru distribution	Ibid., <u>473</u>, 178
$Ca_2[Fe(CN)_6]$ $Cd_2[Fe(CN)_6]$	$K_2Pt(SCN)_6$ type [Strukturbericht, <u>1</u>, 433; Structure Reports, <u>32A</u>, 236]	Ibid., <u>474</u>, 96
$Zn_2[Fe(CN)_6].2H_2O$ $Zn_2[Fe(CN)_6].2NH_3$ $Sn_2[Fe(CN)_6]$ $Pb_2[Fe(CN)_6]$	$Cu_2[Pt(CN)_6].2NH_3$- or $M(I)_2M(IV)F_6$-type	
$Ba_{12}Ba_{2.7}M_{7.3}O_{33}$, M = Nb, Ta	Rhombohedral 36-layer perovskites	Ibid., <u>476</u>, 109
$Ba_4MRe_2O_{12}$ and related materials, M = Co, Ni	Rhombohedral 12-layer perovskites	Ibid., <u>476</u>, 115
$Ba_3SmRu_2O_9$ $Ba_3SmIrRuO_9$	6-layer hexagonal $BaTiO_3$ polytypes ($P6_3/mmc$)	Ibid., <u>477</u>, 161
Ba_2InIrO_6	Trigonal polytype of hexagonal $BaTiO_3$ structure ($P\bar{3}m1$)	
$Ba_3(W,Nb)_2O_9$ $Ba_3Nb_2O_8$	Rhombohedral 12-layer perovskites, $R\bar{3}m$, stacking sequence $(hhc)_3$	Ibid., <u>478</u>, 198
Ba_2YOsO_6 Ba_2LaOsO_6	Cubic perovskite Rhombohedral perovskite	Ibid., <u>478</u>, 223
$(Ca,Ln)F_{2.33}$ Ca_2LnF_7	Anion-excess fluorite structure Tetragonal superstructure	Ibid., <u>479</u>, 165
$Ba_3LnPtRuO_9$	Hexagonal $BaTiO_3$-type, with ordered cation distribution	Ibid., <u>479</u>, 171

TABLE I

Compound	Structure	Reference

$Ba_3M(II)M(V)_2O_9$,
  M(II) = Mg, Ca, Sr, Ba
  M(V) = Nb, Ta
	Hexagonal and orthorhombic
	3-layer and 6-layer
	perovskites
		Z. anorg. Chem., **479**,
		177

$Sr_9Ln_2W_4O_{24}$	Distorted perovskites with cation vacancies	Ibid., **479**, 184
$(NH_4)_2[Mo_3S(S_2)_6] \cdot nH_2O$	Structure as previously described [Structure Reports, **45A**, 280; **46A**, 297]. Water molecule located	Ibid., **479**, 191
$Ba_4ScReW\square O_{12}$	Rhombohedral 12L perovskite, with cation vacancies; $R\bar{3}m$, stacking sequence (hhcc)$_3$	Ibid., **480**, 171
$Ba_3LnRu_2O_9$	Hexagonal $BaTiO_3$ type, sequence 6L, (hcc)$_2$, with LnO_6 single octahedra and Ru_2O_9 face-sharing octahedra	Ibid., **481**, 143

$Ba_2LnM(V)O_6$
$Sr_2LnM(V)O_6$
  Ln = lanthanons, In, Ga
  M = Sb, Nb, Ta
	Distorted perovskites	Ibid., **482**, 143

$Ba_2Ln_2MgW_2O_{12}$ $Ba_6(Y,Ln)_2W_3O_{18}$ $Sr_9Ln_2W_4O_{24}$	Perovskite stacking polytypes	Ibid., **483**, 126
$Ba_6(Re,Sb)_3O_{15}$ $Ba_6(W,Sb)_3O_{15}$	Distorted 5-layer perovskites, with cation vacancies	Ibid., **483**, 161
Barium rhenium niobates and tantalates	Distorted perovskites with cation vacancies	Ibid., **483**, 165
Iron(II) perchlorate hexahydrate, $Fe(ClO_4)_2 \cdot 6H_2O$	$Pmn2_1$ structure, related to that of $P6_3mc$ $LiClO_4 \cdot 3H_2O$, with disorder of Fe positions	Z. Kristallogr., **155**, 129
$Li_2Al_2Si_3O_{10}$	Stuffed quartz structure. Diffuse scattering is interpreted	Ibid., **157**, 27
Tetracyanoplatinates	Summary of the known structures	Ibid., **157**, 47
Uranium dioxide, UO_2	Neutron powder data. Fluorite-type [Struktur- bericht, **1**, 150]	Ibid., **157**, 101
α-Corundum, Al_2O_3	Neutron powder data. Structure as in Struktur- bericht, **1**, 240. x(Al) = 0.3523, x($\bar{0}$) = 0.5563	

Compound	Structure	Reference
$(Zn,Fe)(Fe,V)_2O_4$	Spinel, a = 8.40-8.43 Å, x(O) = 0.377-0.386	Z. Naturforsch., **36B**, 1228
$Rb_3Ln(VO_4)_2$, Ln = Gd-Lu, Y, Sc	Glaserite	Ž. Neorg. Khim., **26**, 264
$(Sr,Ln)_2VO_4$	K_2NiF_4	Ibid., **26**, 881

ELECTRON DIFFRACTION

The following compounds have been studied by electron diffraction of the vapours (listed with abbreviated 1981 reference). Bond lengths are in Å, angles in degrees.

Sulphur dioxide, SO_2	S-O O-S-O	1.434 119.5	J. Chem. Phys., **75**, 5323
Sulphur hexafluoride, SF_6	S-F	1.562	Ibid., **75**, 5326
Difluoro(isoselenocyanato)-phosphine, F_2P-NCSe	P-F P-N N=C C=Se P-N-C F-P-N F-P-F	1.53 1.65 1.21 1.68 149 99 98	J. Chem. Soc., Dalton, 187 (1981)
Difluorophosphino(disilyl)-amine, F_2P-N$(SiH_3)_2$	P-F P-N Si-N Si-H Planar at N	1.585 1.680 1.755 1.49	Ibid., 425 (1981)
Bis(difluorophosphino)-silylamine, $(F_2P)_2NSiH_3$	P-F P-N Si-N Si-H Planar at N	1.570 1.691 1.767 1.46	
Tetraborane(10), B_4H_{10}	Structure as previously described [see this volume, p. 378]; H bridges are considered to be asymmetric, $B(1)-H_b$ = 1.32, $B(2)-H_b$ = 1.48 Å. Electron diffraction and microwave data		Ibid., 472 (1981)
Bis(difluorophosphino)-germylamine, $(F_2P)_2NGeH_3$	Plane P_2NGe skeleton Ge-N N-P P-F	 1.889 1.698 1.592	Ibid., 1047 (1981)
Carbonyl bromide, $OCBr_2$	C-Br C-O Br-C-Br	1.923 1.178 112.3	J. Molec. Struct., **71**, 195

Difluorophosphine selenide, PF_2HSe	P-Se P-F P-H Se-P-F F-P-F Se-P-H	2.026 1.557 1.422 116.8 98.1 118.6	J. Molec. Struct., <u>71</u>, 217
Phosphorus pentafluoride, PF_5	Trigonal bipyramid P-F(ax) P-F(eq)	 1.576 1.530	Ibid., <u>72</u>, 153
Difluorophosphine sulphide, PF_2HS	P-S P-F S-P-F F-P-F	1.876 1.551 115.9 98.3	Ibid., <u>73</u>, 111
Chlorodifluorophosphine sulphide, $PClF_2S$	P-S P-F P-Cl S-P-F F-P-F S-P-Cl	1.864 1.535 1.985 116.2 100.5 118.0	
Bromodifluorophosphine sulphide, $PBrF_2S$	P-S P-F P-Br S-P-F F-P-F S-P-Br	1.881 1.543 2.155 118.2 98.3 118.2	
Tungsten oxide tetrabromide, $WOBr_4$	C_{4V} symmetry W-Br W-O Br-W-Br Br-W-O	 2.455 1.695 87.4 103.0	Ibid., <u>73</u>, 249
Sulphuryl chloride SO_2Cl_2	S-Cl S-O Cl-S-Cl O-S-O Cl-S-O	2.012 1.418 100.3 123.5 108.0	Ibid., <u>73</u>, 253
Sulphur chloride pentafluoride, SF_5Cl	S-Cl S-F (eq and ax) Cl-S-F(eq) Electron diffraction and microwave data	2.055 1.571 90.4	Ibid., <u>75</u>, 271
Octachlorotrisilane, Si_3Cl_8, $Cl_3Si-SiCl_2SiCl_3$	Si-Si Si-Cl Si-Si-Si Si-Si-Cl	2.33 2.03 119 108, 109	Ibid., <u>77</u>, 315
Rubidium nitrate, $RbNO_3$	C_{2V} symmetry Rb-2 O N-O O-Rb-O	 2.65 1.25 47	Ž. Strukt. Khim., <u>22</u>, No. 2, 196 [J. Struct. Chem., <u>22</u>, 310]

Thallium(I) nitrate, Tl-2 O   2.46   Z. Strukt. Khim., 22,
 TlNO$_3$ N-O   1.25   No. 3, 166 [J. Struct.
                       O-Tl-O   51   Chem., 22, 446]

Caesium nitrate,   Cs-2 O   2.80   Ibid., 22, No. 3, 168
 CsNO$_3$ N-O   1.25   [Ibid., 22, 448]
                       O-Cs-O   45

Sodium metaphosphate, C$_{2v}$ symmetry   Ibid., 22, No. 4, 158
 NaPO$_3$ Na-2 O   2.20
                       P-O   1.48

MICROWAVE SPECTRA

Tetraborane(10),   B(1)-B(2)   1.854   Inorg. Chem., 20, 533
 B$_4$H$_{10}$ B(1)-B(3)   1.718
                       B(2)...B(4) 2.806
                       B-H(μ)   1.43
                       Previous studies in
                       Strukturbericht, 6, 78;
                       Structure Reports, 17, 317;
                       21, 197 [see also this
                       volume, p. 376]

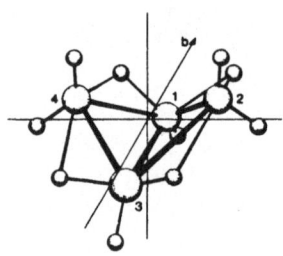

Sulphuric acid,   C$_2$ symmetry   J. Amer. Chem. Soc.,
 H$_2$SO$_4$, O$_2$S(OH)$_2$ S-OH   1.57   103, 2561
                       S-O   1.42
                       O-H   0.97
                       HO-S-OH   101.3
                       O-S-O   123.3
                       S-O-H   108.5

FSO radical   S-F   1.602   J. Chem. Phys., 74,
                       S-O   1.452   1568
                       F-S-O   108.3

Borinic acid,   B-O   1.352   Ibid., 74, 5430
 BH$_2$OH   O-H   0.967
                       B-O-H   112.0
                       H-B-O   117, 122
                       B-H (assumed) 1.20

CO$_2$.HF   Linear O=C=O...H-F   Ibid., 74, 6544
                       O-H   ∼1.9

N$_2$O.HF	Non-linear, N-N-O.		J. Chem. Phys., $\underline{74}$ 6550
	$\overset{\cdot}{\underset{F}{H}}$		

Germyl bromide, GeH$_3$Br	Ge-Br Ge-H H-Ge-Br	2.297 1.527 106.3	Ibid., $\underline{75}$, 2147
HCO$^+$	C-O C-H	1.107 1.093	Ibid., $\underline{75}$, 4256
N$_2$H$^+$	N-N N-H	1.095 1.032	Ibid., $\underline{75}$, 4261
Thionyl chloride, SOCl$_2$	S-Cl S-O Cl-S-Cl Cl-S-O	2.072 1.435 97.2 108.0	J. Molec. Struct., $\underline{73}$, 41
Germyl halides, GeH$_3$X, X = F, Cl, Br, I	Ge-F Ge-Cl Ge-Br Ge-I Ge-H	1.734 2.149 2.299 2.509 1.52	Ibid., $\underline{74}$, 265

PAPERS REFERRED TO LATER YEARS

Many preliminary notes have not been reported, since fuller accounts will appear at a later date. The compounds studied, and abbreviated 1981 references, are listed below.

P$_4$O$_7$	Angew. Chem., $\underline{93}$, 120
(NH$_4$)$_4$P$_4$S$_8$·2H$_2$O	Ibid., $\underline{93}$, 121
α-AsTeI β-AsTeI As$_4$Te$_5$I$_2$ α-AsSeI β-AsSeI AsSI	Ibid., $\underline{93}$, 218
[Cu$_4$(CN)$_6$Cu(NH$_3$)$_2$]n [Cu$_3$(CN)$_5$Cu(NH$_3$)$_2$(OH$_2$)]n	Chem. Comm., 1116 (1980)
Fe$_4$(CO)$_{12}$(CS)S	Ibid., 812 (1980)
K$_5$Nb(CN)$_8$	Ibid., 1167 (1980)
(NH$_4$)$_3$HThUMo$_{12}$O$_{42}$·15H$_2$O	Ibid., 93 (1981)
Os$_5$(CO)$_{19}$	Ibid., 273 (1981)
Zeolite A	Ibid., 276 (1981)

$Na_7Al_3O_8$ Chem. Comm., 379 (1981)
Na_5AlO_4

$Ru_5C(CO)_{15}$ Ibid., 415 (1981)

Tetrasulphur dinitride, S_4N_2 Ibid., 584 (1981) [See also this
                                             volume, p. 125]

Ag_6O_2 Ibid., 664 (1981)

$La_5V_3S_6O_7$ C.R. Acad. Sci. Paris, __292__, 957

$K_4Zr_2Si_6O_{18}\cdot2H_2O$ Dokl. Akad. Nauk SSSR, __256__, 860

Magnesium molybdouranylate   Ibid., __256__, 888
Zinc molybdouranylate

Cement phase Y   Ibid., __256__, 1387

Iron cryolite, Na_3FeF_6 Ibid., __257__, 105

Lithium borate, $Li_3B_5O_8(OH)_2$ Ibid., __257__, 111

β-$CsNdP_4O_{12}$ Ibid., __257__, 357

$K_2ZrSi_3O_9\cdot H_2O$ Ibid., __257__, 608

$Li_3Cr_4PO_{16}\cdot3H_2O$ Ibid., __257__, 619

$Na_5Zr_2Si_6O_{18}Cl\cdot H_2O$ Ibid., __257__, 622

$ReOF_4$ Ibid., __257__, 625
$OsOF_4$

Leucosphenite   Ibid., __257__, 1128

Clinopyroxene   Ibid., __258__, 99

Metaborates   Ibid., __258__, 103

$K_8Yb_3(S_6O_{16})_2(OH)$ Ibid., __258__, 1111

$Cs_3V_3P_{12}O_{36}$ Ibid., __259__, 103

$KFe_4(PO_4)_3$ Ibid., __259__, 591

Potassium neodymium tetraphosphate   Ibid., __259__, 1102

Cerium tetrafluoride (gas), CeF_4 Ibid., __259__, 1399

Ammonium tetrachlorozincate (ferroelectric), Ibid., __260__, 620
    $(NH_4)_2ZnCl_4$

Catapleite, $Na_2ZrSi_3O_9\cdot2H_2O$ Ibid., __260__, 623

Hilairite, $Na_2ZrSi_3O_9\cdot3H_2O$ Ibid., __260__, 1118

$Na_3YSi_2O_7$ Ibid., __260__, 1128

$Ca_2Ga_2GeO_7$	Dokl. Akad. Nauk SSSR, $\underline{260}$, 1363
$GdAl_3(BO_3)_4$	Ibid., $\underline{261}$, 361
Nastrophite, $Na(Sr,Ba)PO_4 \cdot 9H_2O$	Ibid., $\underline{261}$, 619
$Na_5YSi_4O_{12}$	Ibid., $\underline{261}$, 623
$Na_5LnSi_4O_{12}$	Ibid., $\underline{261}$, 874
$I_3Cl_2{}^+SbCl_6{}^-$	Inorg. Nucl. Chem. Letters, $\underline{17}$, 193 [this volume, p. 153]
Ammonia-carboxyborane, $H_3N.BH_2CO_2H$	J. Amer. Chem. Soc., $\underline{102}$, 6343 (1980)
Potassium niobium sulphate hydrate, $K_4(H_5O_2)[Nb_3O_2(SO_4)_6(H_2O)_3] \cdot 5H_2O$	Ibid., $\underline{102}$, 7990 (1980)
V_9O_{17}	J. Solid State Chem., $\underline{36}$, 133
Barium zinc decametaphosphate, $Ba_2Zn_3P_{10}O_{30}$	Ibid., $\underline{40}$, 248
Neodymium decavanadate hydrate, $Nd_2V_{10}O_{28} \cdot 28H_2O$	Kristallografija, $\underline{26}$, 47 [Soviet Physics - Crystallography, $\underline{26}$, 24]
Zinc oxide sulphate, $Zn_3O(SO_4)_2$	Naturwissenschaften, $\underline{68}$, 39
$LnAlGeO_5$	Ibid., $\underline{68}$, 92
Li_2MnBr_4	Ibid., $\underline{68}$, 328
Li_2CdBr_4	
Braunite, $Mn_7O_8SiO_4$	Z. Kristallogr., $\underline{154}$, 240
$K_2Zn(CN)_4$	Ibid., $\underline{154}$, 243
Newberyite, $MgHPO_4 \cdot 3H_2O$	Ibid., $\underline{154}$, 249
$3CuO.B_2O_3$	Ibid., $\underline{154}$, 251
$K_2Li_{14}[Pb_3O_{14}]$	Ibid., $\underline{154}$, 259
Xonotlite, $Ca_6[Si_6O_{17}](OH)_2$	Ibid., $\underline{154}$, 271
$[Pd(NH_3)_4]I_8$	Ibid., $\underline{154}$, 274
$(H_9O_4)_2SnCl_6 \cdot 2H_2O$	Ibid., $\underline{154}$, 285
β-NH_4LiSO_4	Ibid., $\underline{154}$, 286
$Al_2Ge_2O_7$	Ibid., $\underline{154}$, 292
$RbAlGeO_4$	Ibid., $\underline{154}$, 294
$Cl_2Sn[Co(CO)_4]_2$	Ibid., $\underline{154}$, 296
$ClSn[Co(CO)_4]_3$	

$K_8Ga_4S_{10} \cdot 14H_2O$ Z. Kristallogr., $\underline{154}$, 297
$Na_8In_4S_{10} \cdot 14H_2O$
$Na_2Ge(OH)_2S_2 \cdot 3H_2O$
$Na_3Ge(OH)S_3 \cdot 8H_2O$
$Cs_6Ge_8S_{19} \cdot 12H_2O$

$HBF_4 \cdot H_2O$ Ibid., $\underline{154}$, 306
$HBF_4 \cdot 2H_2O$

Clinoenstatite   Ibid., $\underline{154}$, 313

Nickel ditellurate(IV), $NiTe_2O_5$ Ibid., $\underline{154}$, 316 [This volume, p. 329]

$NdAlGe_2O_7$ Ibid., $\underline{154}$, 326

Petalite, $LiAlSi_4O_{10}$ Ibid., $\underline{154}$, 334

Ammonium thiosulphate, $(NH_4)_2S_2O_3$ Ibid., $\underline{154}$, 337
Sodium thiosulphate, $Na_2S_2O_3$

$(W,Mo)_{14}O_{41}$ Ibid., $\underline{154}$, 340

$YAlGeO_5$ Ibid., $\underline{154}$, 341

InSeBr   Ibid., $\underline{154}$, 343

$Zn_3O(SO_4)_2$ Ibid., $\underline{156}$, 9

$Pb_{2.8}Nb_2O_{7.8}$ Ibid., $\underline{156}$, 13

PdB_2O_4 Ibid., $\underline{156}$, 28

Tunisite, $NaCa_2Al_4(OH)_8(CO_3)_4Cl$ Ibid., $\underline{156}$, 31 [This volume, p. 268]

Barium titanate, $BaTiO_3$ Ibid., $\underline{156}$, 31

Sodium thiosulphate, $Na_2S_2O_3$ Ibid., $\underline{156}$, 34
Magnesium thiosulphate hexahydrate,
  $MgS_2O_3 \cdot 6H_2O$

$Cs[I(ICN)_2]$ Ibid., $\underline{156}$, 39
$[Fe(NCBr)_6][FeBr_4]_2 \cdot 2BrCN$

Fayalite, α-Fe_2SiO_4 Ibid., $\underline{156}$, 41 [This volume, p. 348]

$Rb_4V_{16}O_{42}$ Ibid., $\underline{156}$, 43

$Sr(NO_3)_2$ Ibid., $\underline{156}$, 54
$Ba(NO_3)_2$

Rb_2NaHoF_6 Ibid., $\underline{156}$, 58

Caesium hydroxide, CsOH   Ibid., $\underline{156}$, 59 [This volume, p. 247]
Caesium hydroxide monohydrate, $CsOH \cdot H_2O$

P_4O_6	Z. Kristallogr., 156, 60
P_4O_7	
$Rb_2Ga_2Ge_3O_{10}$	Ibid., 156, 64
$(Nd,La)_2AlGe_2O_8(OH)$	Ibid., 156, 65
Prosperite, $Ca_2Zn_4(AsO_4)_4 \cdot H_2O$	Ibid., 156, 70
Gaitite, $Ca_2Zn(AsO_4)_2 \cdot 2H_2O$	
Anhydrite, $CaSO_4$	Ibid., 156, 71
$Bi[Co(CO)_4]_3$	Ibid., 156, 74
$\alpha\text{-}SeCl_4$	Ibid., 156, 75
$MoCl_4$	Ibid., 156, 84
Cu_2O	Ibid., 156, 85
$Si_2S_2Cl_4$	Ibid., 156, 90
$Si_2S_2Br_4$	
$Si_4S_6Br_4$	
SiS_2	
$SiSe_2$	
Hydroxoplatinic acid, H_8PtO_6	Ibid., 156, 107 [This volume, p. 247]
$Ca[Ag(CN)_2]_2 \cdot 2H_2O$	Ibid., 156, 113
$Sr[Ag(CN)_2]_2 \cdot 2H_2O$	
Ice-IV	Ibid., 156, 116 [This volume, p. 192]

ADDITIONAL PAPERS

The following reports were prepared too late for inclusion in the main text.

NEPTUNIUM(IV) BOROHYDRIDE
$Np(BH_4)_4$

R.H. BANKS, N.M. EDELSTEIN, B. SPENCER, D.H. TEMPLETON and A. ZALKIN, 1980. J. Amer. Chem. Soc., 102, 620-623.

Tetragonal, $P4_2/nmc$, a = 8.559, c = 6.017 Å, Z = 2. Mo radiation, R = 0.114 for 352 reflexions.

Atomic positions

			x	y	z
Np	in	2(a)	0	0	0
B		8(g)	0	0.235	0.233
H(1)		8(g)	0	0.33	0.33
H(2)		8(g)	0	0.26	0.05
H(3)		16(h)	0.10	0.15	0.24

The structure contains monomeric molecules. Np is coordinated tetrahedrally by four BH_4 groups, Np-12H = 2.1-2.3, Np...4B = 2.46, B-H = 1.0-1.2 Å.

TRICARBONYLDICHLORORUTHENIUM DIMER - ANTIMONY TRICHLORIDE
$[Ru(CO)_3Cl_2]_2SbCl_3$

P. TEULON and J. ROZIERE, 1981. J. Organometal. Chem., 214, 391-397.

Monoclinic, $P2_1/c$, a = 9.289, b = 11.986, c = 16.932 Å, β = 100.31°, D_m = 2.60, Z = 4. Mo radiation, R = 0.057 for 1969 reflexions.

The structure contains two crystallographically independent centrosymmetric $Cl(CO)_3RuCl_2Ru(CO)_3Cl$ dimers with a double Ru-Cl-Ru bridge, and the Cl atoms all cis; each dimer interacts with $SbCl_3$ trigonal pyramids via Ru-Cl...Sb contacts.

HYDROXONIUM TRICHLOROTRICARBONYLRUTHENATE(II) - ANTIMONY(III) CHLORIDE
$[H_5O_2][Ru(CO)_3Cl_3].SbCl_3$

P. TEULON and J. ROZIERE, 1981. Z. anorg. Chem., 483, 219-224.

Triclinic, $P\bar{1}$, a = 7.129, b = 10.129, c = 10.997 Å, α = 75.40, β = 97.17, γ = 120.94°, D_m = 2.50, Z = 2. Mo radiation, R = 0.030 for 3268 reflexions.

The structure (Fig. 1) contains fac-octahedral $Ru(CO_3)Cl_3^-$ anions and trigonal pyramidal $SbCl_3$ molecules, linked by longer Sb...Cl contacts; the $H_5O_2^+$ ion has a very short O-H...O hydrogen bond, 2.37(1) Å, and forms bifurcated O-H...Cl bonds. Ru-Cl = 2.406-2.424, Ru-C = 1.905-1.909, Sb-Cl = 2.364-2.387, Sb...Cl = 3.186-3.419 Å.

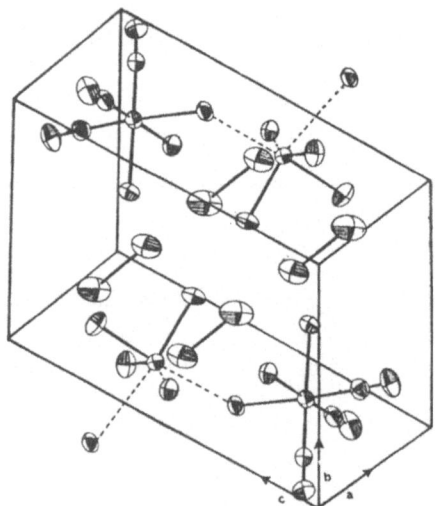

Fig. 1.   Structure of $[H_5O_2][Ru(CO)_3Cl_3].SbCl_3$.

CARBIDOBISSULPHIDODODECACARBONYLHEXACOBALT
$Co_6C(CO)_{12}S_2$

G. BOR, U.K. DIETLER, P.L. STANGHELLINI, G. GERVASIO, R. ROSSETTI, G. SBRIGNADELLO and G.A. BATTISTON, 1981. J. Organometal. Chem., 213, 277-292.

Monoclinic, Cc, a = 16.250, b = 9.413, c = 16.036 Å, β = 116.77°, D_m = 2.33, Z = 4. Mo radiation, R = 0.034 for 1974 reflexions.

The structure contains a Co_6 trigonal prism, with a C atom at its centre and the triangular faces each capped by a S atom; each Co has two terminal CO groups. Co-Co = 2.432 (triangular edges), 2.669 (rectangular edges), Co-C = 1.94, Co-S = 2.192 Å.

TRISULPHUR DINITROGEN DIOXIDE
$S_3N_2O_2$

G. MacLEAN, J. PASSMORE, P.S. WHITE, A. BANISTER and J.A. DURRANT, 1981. Canad. J. Chem., 59, 187-190.

Monoclinic, C2/c, a = 16.389, b = 4.572, c = 6.790 Å, β = 96.64°, Z = 4. Mo radiation, R = 0.057 for 550 reflexions.

The structure (Fig. 1) is essentially as previously described (1).

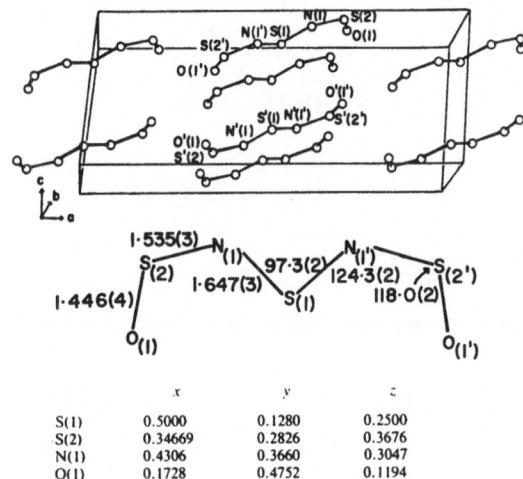

	x	y	z
S(1)	0.5000	0.1280	0.2500
S(2)	0.34669	0.2826	0.3676
N(1)	0.4306	0.3660	0.3047
O(1)	0.1728	0.4752	0.1194

Fig. 1.   Structure of $S_3N_2O_2$.

1. Structure Reports, 26, 345; 46A, 145.

BORON TRIFLUORIDE
$\gamma\text{-}BF_3$

HYDROXOTRIFLUOROBORIC ACID
BF_3OH_2

D. MOOTZ and M. STEFFEN, 1981. Z. anorg. Chem., 483, 171-180.

BF_3, monoclinic, $P2_1/c$, a = 4.779, b = 14.00, c = 7.430 Å, β = 107.60°, at -131°C,
Z = 8 (2 molecules per asymmetric unit). Mo radiation, R = 0.096 for 650 reflexions.

BF_3OH_2, monoclinic, $P2_1/n$, a = 7.641, b = 7.957, c = 7.864 Å, β = 94.80°, at -35°C,
Z = 4. Mo radiation, R = 0.097 for 609 reflexions.

The BF_3 structure (Fig. 1) contains two independent trigonal planar molecules,
mean B-F = 1.32 Å (corrected for libration), linked by longer B...F interactions.
The BF_3OH_2 molecules are linked by a three-dimensional hydrogen bond system (Fig. 1).
The structure of $BF_3O(CH_3)H$ is also described [see 1].

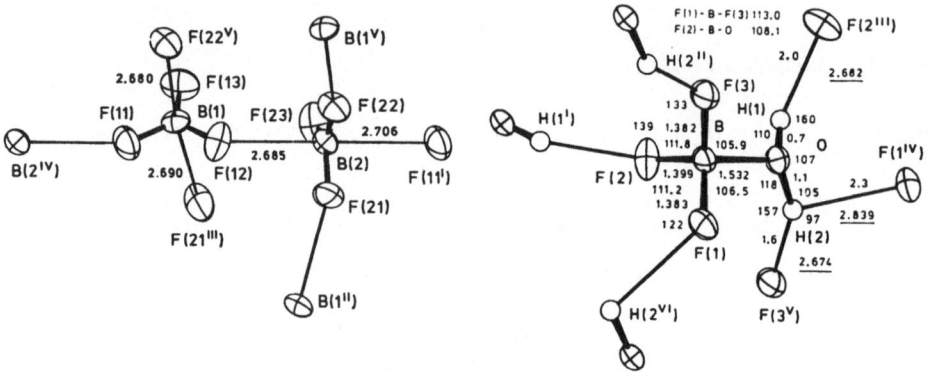

Fig. 1.   Structures of γ-BF_3 (left) and BF_3OH_2 (right).

1. Structure Reports, 48B.

CAESIUM RUBIDIUM TRIFLUOROCOBALTATE(II)
$Cs_{0.5}Rb_{0.5}CoF_3$

J.-M. DANCE, J.-L. SOUBEYROUX, N. KERKOURI and A. TRESSAUD, 1981. C.R. Acad. Sci.
Paris, 293, 279-282.

Hexagonal, $P6_3/mmc$, a = 5.984, c = 14.560 Å, at 80K, Z = 6. Neutron powder data.

Atomic positions

			x	y	z
Cs,Rb(1)	in	2(b)	0	0	1/4
Cs,Rb(2)		4(f)	1/3	2/3	0.0891
Co(1)		2(a)	0	0	0
Co(2)		4(f)	1/3	2/3	0.8551
F(1)		6(h)	0.5144	2x	1/4
F(2)		12(k)*	0.8362	2x	*

*[not 6(h), not z = 1/4; probably z ∿ 0.08, see 1]

Hexagonal 6-layer perovskite (1). Below 62K, antiparallel magnetic moments in
the Co 2(a) and 4(f) sites.

1. Structure Reports, 34A, 203; 35A, 158.

ANTIMONY(III) CHLORIDE - GALLIUM(III) CHLORIDE
(DICHLOROANTIMONY(III) TETRACHLOROGALLATE(III))
GaSbCl$_6$

C. PEYLHARD, P. TEULON and A. POTIER, 1981. Z. anorg. Chem., <u>483</u>, 236-240.

Monoclinic, P2$_1$/c, a = 9.837, b = 7.812, c = 12.504 Å, β = 101.0°, Z = 4. Mo
radiation, R = 0.052 for 1051 reflexions.

The structure (Fig. 1) contains distorted tetrahedral GaCl$_4^-$ anions and bent
SbCl$_2^+$ cations, with interionic Sb...Cl interactions giving infinite chains and
distorted trigonal bipyramidal geometry at Sb, SbCl$_4$E (E = lone electron-pair).

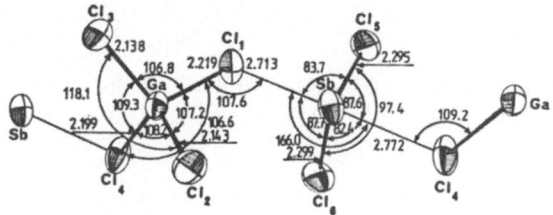

Fig. 1.   Structure of GaSbCl$_6$.

MOLYBDENUM DIOXIDE DICHLORIDE - PHOSPHORUS OXIDE TRICHLORIDE
MoO$_2$Cl$_2$.POCl$_3$

G. BEYENDORFF-GULBA, J. STRÄHLE, A. LIEBELT and K. DEHNICKE, 1981. Z. anorg. Chem.,
<u>483</u>, 26-32.

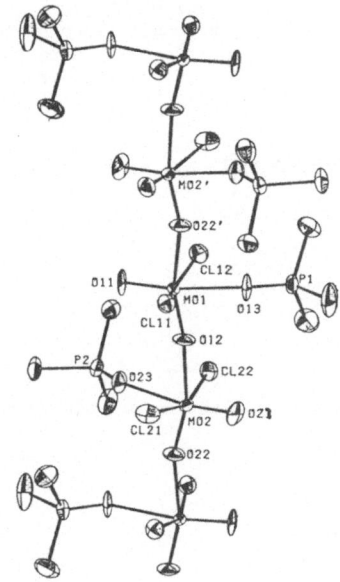

Fig. 1.   Structure of MoO$_2$Cl$_2$.POCl$_3$.

Monoclinic, $P2_1/c$, a = 8.413, b = 15.395, c = 15.595 Å, β = 115.90°, Z = 8. Mo radiation, R = 0.046 for 2497 reflexions.

The structure (Fig. 1) contains chains of Mo atoms linked by asymmetric oxo bridges, Mo-O = 1.72 and 2.18 Å; one terminal O, two Cl, and an $OPCl_3$ molecule complete distorted octahedral coordination at Mo, Mo-O = 1.66, Mo-Cl = 2.30, 2.31, Mo-OP = 2.33, O-P = 1.46, P-Cl = 1.94 Å.

POTASSIUM DIBROMOCYANOMERCURATE(II) MONOHYDRATE
$KHgBr_2CN.H_2O$

K. BRODERSEN and H.M. FROHRING, 1981. Z. anorg. Chem., **483**, 86-94.

Orthorhombic, Pmmm, a = 4.542, b = 17.381, c = 4.651 Å, D_m = 3.94, Z = 2. Ag radiation, R = 0.052 for 392 reflexions.

Atomic positions

			x	y	z
Hg(1)	in	1(a)	0	0	0
0.5 Hg(2)		2(k)	0.4156	1/2	0
0.5 Hg(2')		2(t)	1/2	1/2	0.0865
Br(1)		1(g)	0	1/2	1/2
Br(2)		1(d)	1/2	0	1/2
Br(3)		2(o)	1/2	0.3596	0
K		2(n)	0	0.2956	1/2
O		2(p)	1/2	0.2025	1/2
C		2(m)	0	0.1166	0
N		2(m)	0	0.1820	0

The structure (Fig. 1) contains linear $Hg(CN)_2$ and $HgBr_2$ molecules, K^+ and Br^- ions, and water molecules; longer Hg...Br contacts complete distorted octahedral coordinations at each Hg. The Hg(2) position is disordered. Hg-C = 2.03, Hg-Br = 2.47, Hg...Br = 2.98-3.55 Å. The chloro compound is isostructural.

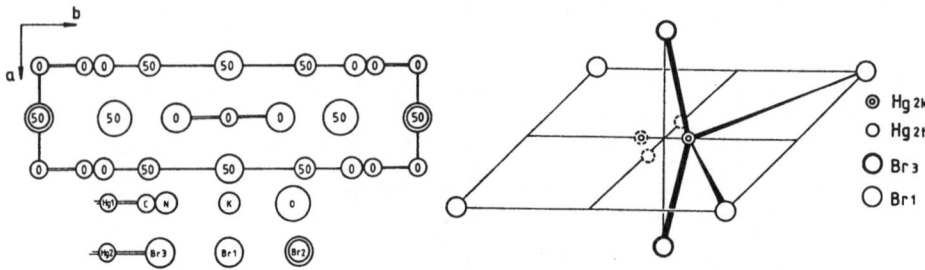

Fig. 1.   Structure of $KHgBr_2CN.H_2O$ and the disordered Hg(2) position.

HYDROGEN PEROXIDE
H_2O_2

J.-M. SAVARIAULT and M.S. LEHMANN, 1980. J. Amer. Chem. Soc., **102**, 1298-1303.

Tetragonal, $P4_12_12$, a = 4.016, 4.004, c = 7.855, 7.832 Å, from X-ray and neutron measurements at 110K, Z = 4. Mo radiation and neutron radiation, R = 0.039, 0.063 for 420, 240 reflexions. O at (0.0762,0.1670,0.2204); H at (-0.0483,0.2834,0.1296), from neutron data.

Structure as previously described (1); electron density distribution is studied.

1. Structure Reports, 15, 171; 30A, 303; 41A, 212.

β''-ALUMINA
$M_2Al_{11}O_{17}$ (M = Na, K)

J.P. BOILOT, G. COLLIN, P. COLOMBAN and R. COMES, 1980. Phys. Rev., B, 22, 5912-5923.

Rhombohedral, $R\bar{3}m$, a = 5.61, 5.63, c = 33.54, 34.01 Å, Z = 3. Mo radiation, R = 0.016, 0.023 for 419, 433 reflexions.

β''-Alumina structure (1), with Na and K distributed in 6(c) and 18(h) sites.

1. Structure Reports, 33A, 277; 34A, 250; 46A, 233.

SILVER β-ALUMINA (LOW-TEMPERATURE)
$AgAl_{11}O_{17}$

J.M. NEWSAM and B.C. TOFIELD, 1981. J. Phys. C: Solid State Phys., 14, 1545-1554.

Hexagonal, $P6_3/mmc$, a = 5.5871, c = 22.5131 Å, at 4.2K, Z = 2. Neutron powder data. Ag in 6(h): x = -0.2237.

β-Alumina structure (1), with only one Ag site.

1. Strukturbericht, 5, 72; Structure Reports, 33A, 275; 46A, 233.

SILVER ALUMINATE (2H)
$AgAlO_2$

G. BRACHTEL and M. JANSEN, 1981. Cryst. Struct. Comm., 10, 173-174.

Hexagonal, $P6_3/mmc$, a = 2.896, c = 12.219 Å, Z = 2. Mo radiation, R = 0.021 for 66 reflexions. Ag in 2(d); Al in 2(a); O in 4(f): z = 0.0781.

Isostructural with $2H-AgFeO_2$ (1), a stacking variant of delafossite, with ABBA oxygen stacking, Al in octahedral sites (AB) and Ag with linear coordination (BB). Ag-O = 2.10, Al-O = 1.93 Å. $\beta-RbScO_2$ (2) has a similar structure.

1. Structure Reports, 38A, 277.
2. Ibid., 43A, 349.

CAESIUM CHROMATE
Cs_2CrO_4

A.J. MORRIS, C.H.L. KENNARD, F.H. MOORE, G. SMITH and H. MONTGOMERY, 1981. Cryst. Struct. Comm., $\underline{10}$, 529-532.

Orthorhombic, Pnma, a = 8.427, b = 6.300, c = 11.200 Å, Z = 4. Neutron radiation, R = 0.049 for 395 reflexions.

Atomic positions

	x	y	z
Cs(1)	0.6689	1/4	0.4099
Cs(2)	-0.0153	1/4	-0.3037
Cr	0.2359	1/4	0.4196
O(1)	0.0410	1/4	0.4124
O(2)	0.2934	1/4	0.5611
O(3)	0.3054	0.0356	0.3532

Structure as previously described ($\underline{1}$), containing CrO_4 tetrahedra linked by 9- and 11-coordinated Cs ions. Cr-O = 1.64$\overline{4}$, 1.649, 1.657(2) Å, O-Cr-O = 109.1-110.0°, Cs-O = 3.0-3.5 Å.

$\underline{1}$.   Strukturbericht, $\underline{6}$, 21, 101.

DIASPORE
α-AlOOH

A. KLUG and L. FARKAS, 1981. Phys. Chem. Miner., $\underline{7}$, 138-140.

Orthorhombic, Pbnm, a = 4.403, b = 9.428, c = 2.846 Å, Z = 4. Powder data for four natural specimens. Atoms in 4(i): x,y,1/4, x = 0.046, 0.713, 0.208, y = -0.144, 0.198, 0.054 for Al, O, OH.

Structure as previously described ($\underline{1}$).

$\underline{1}$.   Strukturbericht, $\underline{2}$, 46, 350; $\underline{3}$, 64, 371; Structure Reports, $\underline{9}$, 150; $\underline{10}$, 98; $\underline{22}$, 276; $\underline{46A}$, 293.

SODIUM PHOSPHATE CHLORIDE HYDRATE
$Na_3PO_4.1/5NaCl.11H_2O$

A. LARBOT, J. DURAND and L. COT, 1981. Cryst. Struct. Comm., $\underline{10}$, 55-57.

Trigonal, P$\overline{3}$c1, a = 11.880, c = 12.680 Å, D_m = 1.60, Z = 4. Cu radiation, R = 0.050 for 756 reflexions.

The structure (Fig. 1) contains $Na(H_2O)_6$ octahedra and PO_4 tetrahedra linked by hydrogen bonding, and Cl^- ions. P-O = 1.52, 1.55, Na-O = 2.37-2.57, O-H...O = 2.68-2.82 Å.

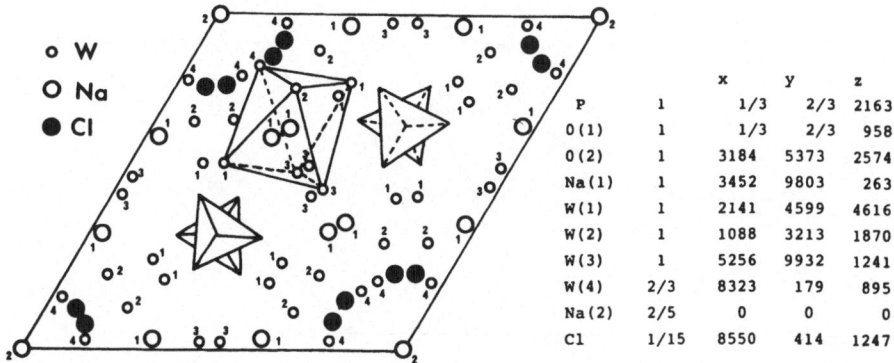

		x	y	z
P	1	1/3	2/3	2163
O(1)	1	1/3	2/3	958
O(2)	1	3184	5373	2574
Na(1)	1	3452	9803	263
W(1)	1	2141	4599	4616
W(2)	1	1088	3213	1870
W(3)	1	5256	9932	1241
W(4)	2/3	8323	179	895
Na(2)	2/5	0	0	0
Cl	1/15	8550	414	1247

o W
O Na
● Cl

Fig. 1.   Structure of $Na_3PO_4.1/5NaCl.11H_2O$, and atomic positional
parameters (decimal fractions x 10^4).

COBALT(II) PHOSPHATE CHLORIDE
$Co_2(PO_4)Cl$

I.   A.G. NORD and T. STEFANIDIS, 1980.   Chem. Scripta, 15, 180-181.
II.   Idem, 1981.   Cryst. Struct. Comm., 10, 1251-1257.

Monoclinic, C2/c, a = 13.514, b = 9.132, c = 9.220 Å, β = 132.90°, D_m = 3.92, Z = 8.
Mo radiation, R = 0.065 for 831 reflexions.

Isostructural with $Fe_2(PO_4)Cl$ (1), the structure containing chains of face-
sharing trans-CoO_4Cl_2 octahedra, and PO_4 tetrahedra.   P-O = 1.48-1.57, Co-O = 1.94-
2.24, Co-Cl = 2.43-2.54 Å.   See Fig. 1.

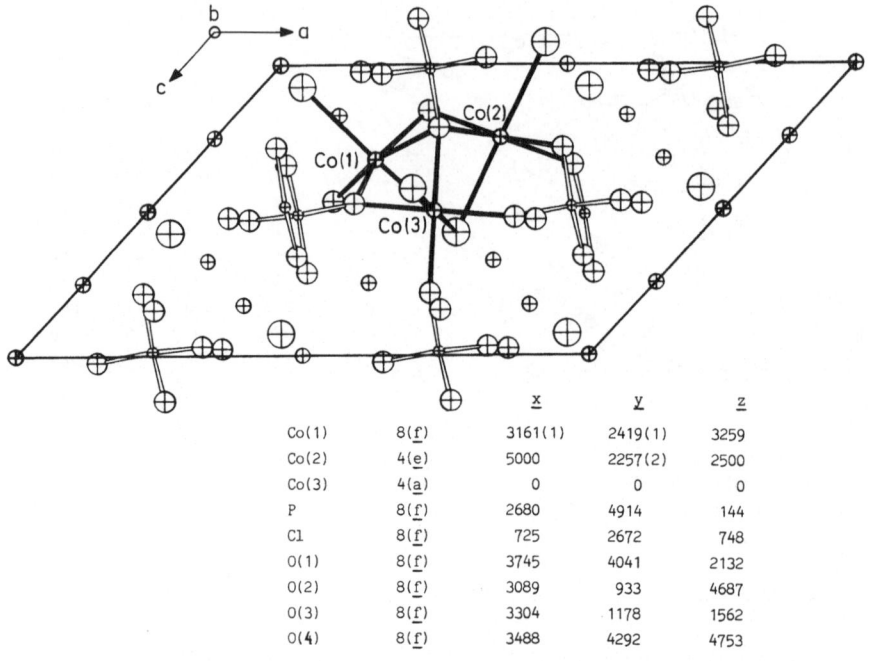

		x	y	z
Co(1)	8(f)	3161(1)	2419(1)	3259
Co(2)	4(e)	5000	2257(2)	2500
Co(3)	4(a)	0	0	0
P	8(f)	2680	4914	144
Cl	8(f)	725	2672	748
O(1)	8(f)	3745	4041	2132
O(2)	8(f)	3089	933	4687
O(3)	8(f)	3304	1178	1562
O(4)	8(f)	3488	4292	4753

Fig. 1.   Structure of $Co_2(PO_4)Cl$, and atomic positional parameters (x 10^4).

1. Structure Reports, 42A, 347.

CAESIUM DIHYDROGEN ARSENATE
CsD_2AsO_4

W.J. HAY and R.J. NELMES, 1981. J. Phys. C: Solid State Phys., 14, 1043-1052.

Paraelectric phase (room temperature and T_C+5K (∿210K))
Tetragonal, I42d, a = 7.985, 7.982, c = 7.893, 7.889 Å, Z = 4. Neutron radiation, R
= 0.034, 0.039 for 265, 163 reflexions. Cs in 4(b): 0,0,1/2; As in 4(a): 0,0,0; O in
16(e): 0.1456,0.0918,0.1232; 0.5 D in 16(e): 0.1381,0.2184,0.1171, at 210K.

Ferroelectric phase (77K)
Orthorhombic, Fdd2, a = 11.516, b = 11.103, c = 7.87 Å, Z = 8. Neutron radiation,
R = 0.052 for 120 reflexions from a multi-domained crystal. Cs in 8(a): 0,0,0.5123;
As in 8(a): 0,0,0; O(1) in 16(b): 0.1194,-0.0281,-0.1254; O(2) in 16(b): 0.0277,
0.1196,0.1229; D in 16(b): -0.0403,0.1797,0.1337.

     The structures are similar to those of K and Rb dihydrogen phosphates (1).

1. This volume, p. 280.

SUBJECT INDEX

This index contains the names of substances printed
at the head of the reports, and some additional general
entries. Greek letter and numerical prefixes, and prefixes
such as cis, trans etc are disregarded in fixing the
alphabetical order.

The entries are in alphabetical order by formula.
Compounds in Table I of the Metals Section are excluded
from this index, and that Table (pp. 94-107), which serves
as its own index, should be consulted for additional
metallic structures.

AUTHOR INDEX

Names beginning with a separated prefix are listed before single-word names beginning with the same letters; accents are omitted.

CORRIGENDA

Change required on	From	To
<u>34A</u>, 296	$Ag_2S.MnS.Sb_2S_3$	$2Ag_2S.MnS.Sb_2S_3$
400	$Ag_2MnS_5Sb_2$	$Ag_4MnS_6Sb_2$
<u>39A</u>, 282 and 386	K. HANDLOVIC	M. HANDLOVIC
<u>41A</u>, 402, Rutile, TiO_2	$z = 0.30493$	$x = 0.30493$
<u>43A</u>, 364	Iron(II) nitrate mono-hydrate, 241	Iron(III)
389	Ray, M.	Ray, S.
<u>45A</u>, 54 and 442	S.Š. ŠILŠTEIN	S.Š. ŠIL'ŠTEIN
265 and 442	A.A.R. SMITH	A.R.R. SMITH or A.R RAE SMITH
383, Cycloheptasulphur	<u>43A</u>, 328	<u>44A</u>, 328
<u>46A</u>, 287	LANTHANUM LITHIUM COBALT(II) OXIDE	COBALT(III)